Autodesk Inventor

James M. Leake
University of Illinois at Urbana-Champaign

Boston Burr Ridge, IL Dubuque, IA Madison, WI New York San Francisco St. Louis
Bangkok Bogotá Caracas Kuala Lumpur Lisbon London Madrid Mexico City
Milan Montreal New Delhi Santiago Seoul Singapore Sydney Taipei Toronto

AUTODESK INVENTOR

Published by McGraw-Hill, a business unit of The McGraw-Hill Companies, Inc., 1221 Avenue of the Americas, New York, NY 10020.

 This book is printed on recycled, acid-free paper containing 10% postconsumer waste.

1 2 3 4 5 6 7 8 9 0 QPD/QPD 0 9 8 7 6 5 4 3

ISBN 0–07–282205–8

Publisher: *Elizabeth A. Jones*
Senior sponsoring editor: *Suzanne Jeans*
Developmental editor: *Kate Scheinman*
Marketing manager: *Sarah Martin*
Project manager: *Joyce Watters*
Production supervisor: *Kara Kudronowicz*
Media project manager: *Sandra M. Schnee*
Senior media technology producer: *Phillip Meek*
Senior designer: *David W. Hash*
Cover designer: *Rokusek Design*
Compositor: *Interactive Composition Corporation*
Typeface: *10.5/12 Times*
Printer: *Quebecor World Dubuque, IA*

Library of Congress Cataloging-in-Publication Data

Leake, James M.
Autodesk inventor / Leake, James M. — 1st ed.
p. cm.
Includes index.
ISBN 0–07–282205–8
1. Autodesk inventor. 2. Engineering graphics. 3. Engineering models — Data processing. I. Title.

T353 .L43 2004
620.0042—dc21 20030059951
CIP

www.mhhe.com

DEDICATION

To put your own life on hold
So that others can realize their dreams
This is surely a mark of enlightenment
A thousand thanks
To Stephanie

I would also like to thank my colleague
Jacob Borgerson
The future looks bright

CONTENTS

PREFACE

This work aims to be a hands-on, tutorial-driven introduction to *Autodesk Inventor*. Although often described as 3D parametric modeling software, it is perhaps more accurate to think of Inventor as product development software. Parametric modelers like Autodesk Inventor focus on creating virtual assemblies of parts (i.e., products) rather than standalone parts; for this reason the book is built around a couple of different product assemblies. The tutorials at the end of each chapter take the user through part and assembly modeling, drawing documentation, and finally, the simulation, analysis, and presentation of these products.

All of the print tutorials are duplicated in video on a CD that comes packaged with the book. This is done in part to accommodate different learning styles. Some users may well prefer the video approach to that of the text-based tutorials. Another reason for this approach is that the level of modeling difficulty encountered in these tutorials is perhaps more advanced than that typically found in an introductory CAD text. The built-in usability of the Inventor software, coupled with the step-by-step tutorials available both in print and video, should allow users to rapidly develop their modeling skills.

In another effort to accommodate different teaching and learning styles, the tutorials at the end of each chapter—while building on chapter content—can be done without prior reading of the chapter. With Inventor it is relatively easy to get new users up and running fast, allowing them to experience first-hand the thrill and excitement of 3D modeling.

Another characteristic of this book is an emphasis on *build strategy*. The steps required to complete each tutorial are summarized at the beginning of the tutorial. Taken together, these summary steps amount to a strategy for building each model. It is important for users new to 3D modeling to learn to think in terms of features, and to plan out a feature-based build strategy before starting to model. This format should also be helpful to users already familiar with 3D modeling. Rather than following the tutorials step-by-step, they can use the build strategy as a roadmap for building the part and assembly models.

To keep pace with future Inventor releases, a website is being maintained in conjunction with this book. Material on new tools and features, as well as other tools not addressed in the text, will be available for download from the website (www.mhhe.com/leake). Autodesk Inventor Release 6 was used in developing the material for this book.

In addition to the 3D parametric modeling material contained in the main body of each chapter, lecture notes in the form of PowerPoint slides are also included on the website. These slides cover a range of topics in engineering graphics, and are based on lecture notes currently used in an introductory engineering graphics course taught at the University of Illinois Urbana-Champaign.

The author and publisher would like to thank the following reviewers for their thoughtful comments on this text: Tom Bledsaw, ITT Educational Services; Joan Davis, Pellissippi State Technical Community College, Dan G. Dimitriu, San Antonio College, James B. Higley, Purdue University Calumet; Minh Quoc Pham, Houston Community Colleges/University of Houston; and James Shahan, Iowa State University.

James M. Leake
Department of General Engineering—UIUC
April 25, 2003

CHAPTER 1

Getting Started

LEARNING OBJECTIVES

- Use the Open dialog box to access Inventor help, start new files, open existing Inventor files, and manage projects
- List the four file formats used in Inventor
- Create a new Inventor project using the Project Editor
- Locate Inventor tools from the panel bar and right-click context menus
- Identify the Browser bar on the Autodesk Inventor interface
- Access and list the seven default reference work features
- List several ways to access the Inventor Help system
- Use the viewing tools to control the viewpoint of the observer
- Use the display tools to change the display of the model

Introduction

Commercial Computer-Aided-Design (CAD) software has come a long way in twenty years. Originally developed to perform 2D manual drafting tasks, CAD software rapidly advanced in the 1990's to include 3D surface and solid modeling. Today mid-range Windows CAD software packages perform parametric assembly modeling, and are rapidly evolving into virtual product development tools. *Parametric solid modeling* employs parametric constraints to define part features, and to create relationships between these features in order to create intelligent part models. These parts can then be combined to form virtual assembly models.

Autodesk Inventor is a prime example of this next-generation CAD software. Inventor is the result of Autodesk's 1) twenty years' experience developing CAD products like AutoCAD and Mechanical Desktop, 2) making the decision to wipe the slate clean and start fresh, and 3) setting out to create a powerful, easy to use parametric solid modeler. Now with Release 6, Inventor also has added the capacity to handle hybrid (or nonmanifold) modeling, where freeform surfaces and parametric solids can be combined seamlessly to create complex parts.

Parametric modelers like Autodesk Inventor allow the user to model parts, and then combine these parts into an assembly. Further combinations of these component parts and sub-assemblies result in a virtual product model. The product examples

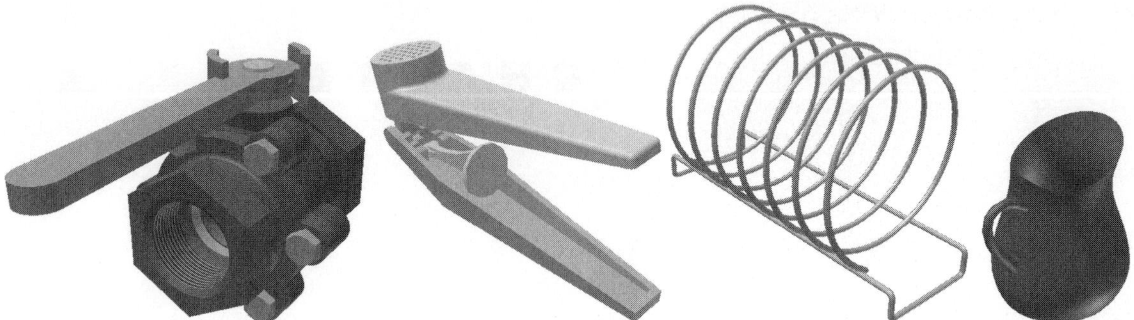

Figure 1.1 - Ball valve, garlic press, toast rack, creamer

shown in Figure 1.1 (ball valve, garlic press, toast rack, creamer) will all be modeled in the course of completing the tutorials that form the basis of this book.

Included among the by-products of the parametric modeling process are part and assembly drawings, exploded view drawings along with a Bill of Materials (BOM), and animation files showing either the assembly of a product, or the motion of a mechanism. Although not covered in this work, these same part and assembly files can then be used for such downstream applications as rapid prototyping, NC tool path generation, as well as finite element and kinematics analysis.

The fact that Inventor is a parametric modeler means that the parts and assemblies are easily edited and modified. This has a tremendous impact on the product development process. Families of parts can be created, products can be iteratively improved upon, multiple prototypes can be created and tested, etc.

Open Dialog Box

When Autodesk Inventor is first launched the Open dialog box appears, as shown in Figure 1.2. In the What To Do column on the left, four options are available: Getting Started, New, Open, and Projects. Several links appear on the Getting Started page. Of particular interest is the "Learn how to build models quickly" link, a gateway to several Inventor tutorials. The New option is used to start a new Inventor file, the Open option to open an existing Inventor file. The Projects option is used to open the Project Editor.

Inventor File Types

If New is selected from the Open dialog box, the right side of the dialog box changes to that seen in Figure 1.3. Three Tabs are now visible: Default on top, then English and Metric. Different icons are visible on the Default tab, each one corresponding to a template file. Inventor uses different file types to create and edit models. Each file type has a unique file format. These are (file extension in parenthesis):

- Part (ipt)
- Assembly (iam)
- Presentation (ipn)
- Drawing (idw)

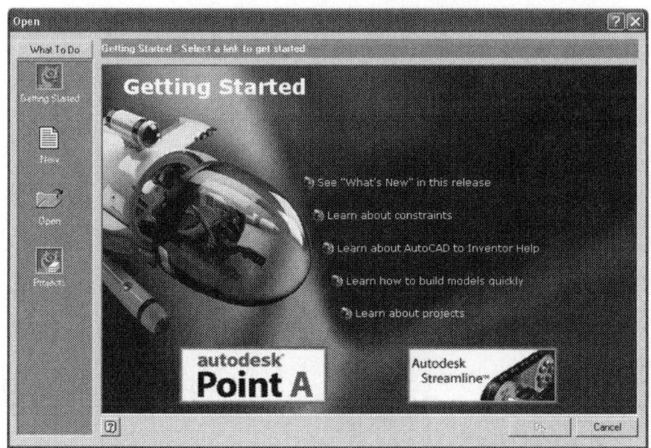

Figure 1.2 - Open dialog box: getting started

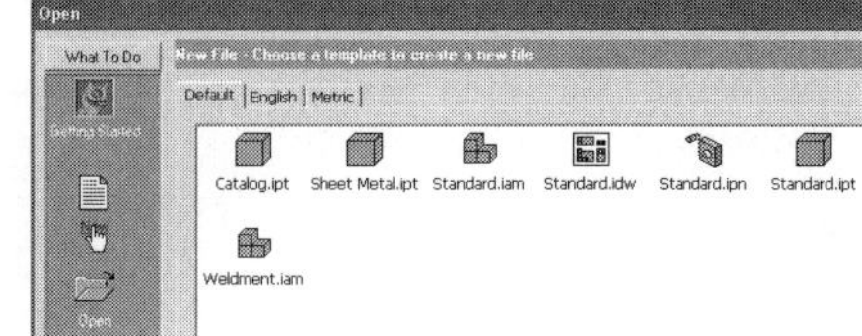

Figure 1.3 - Open dialog box: new

There are two kinds of part files, standard and sheet metal, as well as two kinds of assembly files, standard and weldment.

When Inventor is first installed the user is prompted to specify the default units system, either millimeters or inches. This selection determines the units of the templates available on the Default tab. Select either the Metric or the English tab and other template icons appear. Naturally the units of these templates correspond to that of the tab name. In this book metric predominates as the working unit.

Projects

Since Inventor employs several different file types, with each of these file types needing to share data with one another, file management becomes a significant issue. This is all the more the case in situations where many individuals work collaboratively on the same project. For the purposes of this book though, projects will simply be used to create a project folder (i.e., a workspace). When a given project is active, all file open and save operations default to this project folder. Only one project can be active at a time.

An Inventor project is really an ASCII text file, with an ipj file extension. This means that Inventor projects can be created and modified outside Inventor in a text editor. You will notice these ipj project files in your project folders.

User Interface

The composition of the Autodesk Inventor user interface is shown in Figure 1.4. Although this figure depicts the part file interface, the other Inventor file type interfaces are similarly arranged and configured.

Panel Bar

Regardless of file type, Inventor tools are most conveniently accessed from the Panel Bar. One of the nice things about Inventor is the clean, uncluttered interface. This is mostly due to the modularized "similar but different" work environments. Different tool sets are displayed in the Panel Bar depending upon the type of file (i.e., part, assembly, drawing, presentation) open, and also according to the active work environment (e.g., 2D sketch, 3D sketch, feature). The Panel Bar is normally docked in the upper left of the interface. Shown in Figure 1.5 are the sketch and feature tools as they appear in the (floating, expanded) Panel Bar for a part file.

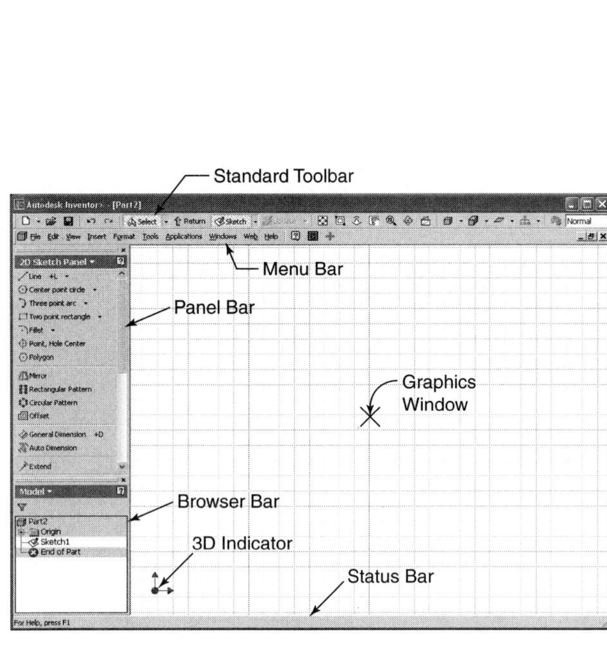

Figure 1.4 - Inventor interface

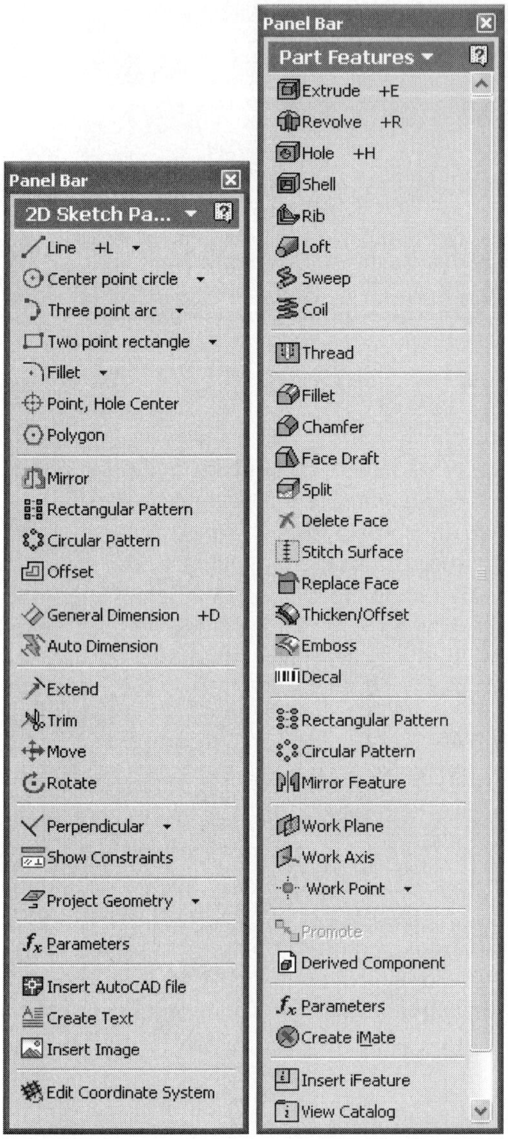

Figure 1.5 - Panel bars: 2D sketch and feature

In addition to the Panel Bar, Inventor tools are also frequently accessed from right-click context menus. Two such context menus are shown in Figure 1.6.

Browser Bar

The Browser Bar, as it is called in Inventor, is a hierarchical tree structure used to display file information. Most parametric modelers employ tree structures similar to the Browser Bar. The Inventor browser changes according to file type (i.e., part, assembly, drawing, presentation). In the case of a part file, the browser would generally be referred to as a feature tree. In Inventor the part file feature tree is called the Part Browser.

TIP: The visibility of both the Panel Bar and the Browser Bar, as well as other toolbars, can all be controlled using View > Toolbar from the menu bar.

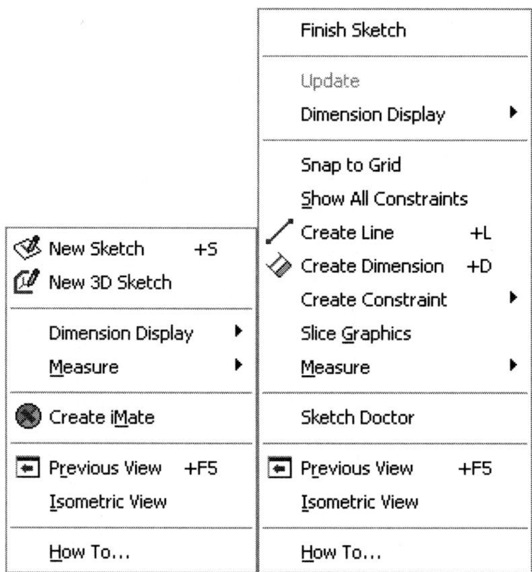

Figure 1.6 - Right-click context menus

Default Reference Work Features

The first element in the Browser Bar (part browser) is the Origin folder. If the Origin folder is expanded by clicking on the + symbol at the left, the folder expands to reveal three work planes (YZ, XZ, and XY), three work axes (X, Y, and Z), and a work point (Center Point). By default, these reference work features are not visible. By moving the mouse over one of the features, however, the feature becomes visible in the graphics window. By right-clicking on any of these work features and selecting Visibility, the feature remains visible in the graphics area. Figure 1.7 shows an expanded Origin folder with visible default work features, and an isometric view. From this it can be seen that the work planes and axes are all mutually perpendicular and that the Center Point is at their intersection.

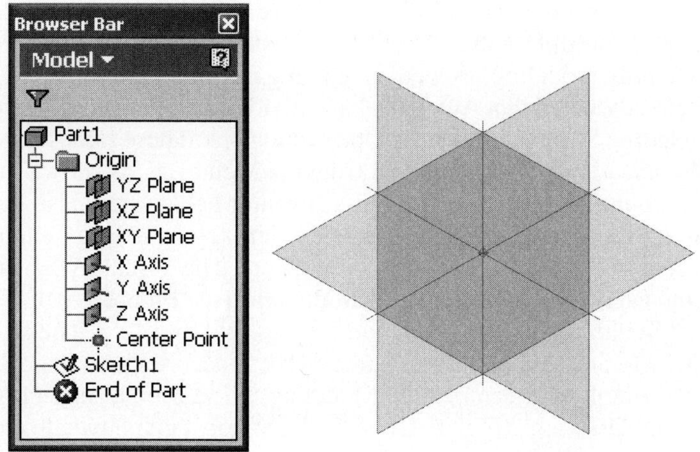

Figure 1.7 - Expanded origin folder with visible work features

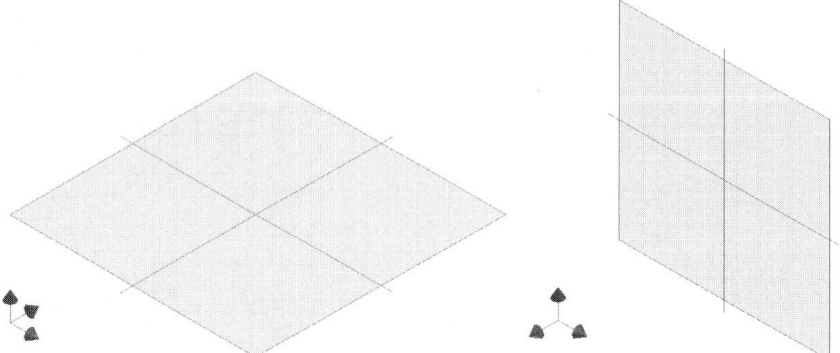

Figure 1.8 - Isometric view of XY Plane; metric template on left, English template on right

These reference work features will be used extensively in this book. The default work planes are useful as sketch planes, or as an aid in creating other (e.g., offset, inclined) work planes. In addition, by constraining sketch geometry to the default work features, the construction of symmetrical parts is considerably simplified. Using the Project Geometry tool, the default work axes and center point can be projected on any sketch plane.

By default, new part files open in 2D sketch mode. This default sketch plane is the XY Plane seen in the Origin folder. In fact though, any of the three reference work planes can be used for the first (or subsequent) sketches. In this book the first part sketch will almost always be drawn on the default XY Plane.

Another point to note is that the XY Plane in an Inventor metric template corresponds to a top, or plan view (see Figure 1.8 on left), whereas in an English template the XY Plane corresponds with a front view (see Figure 1.8 on right).

Autodesk Inventor Help

Just as CAD systems have made significant advances in the past twenty years, so have the software Help systems that support them. Autodesk's built-in Help for Inventor is excellent, so good in fact that it significantly impacts the format of this book. Rather than repeat what is already embedded in the software help, readers are strongly advised to make use of what is already available. This book in turn attempts to cut through the confusing array of multiple options, and provide straightforward hands-on examples of part and assembly modeling, as well as drawing documentation.

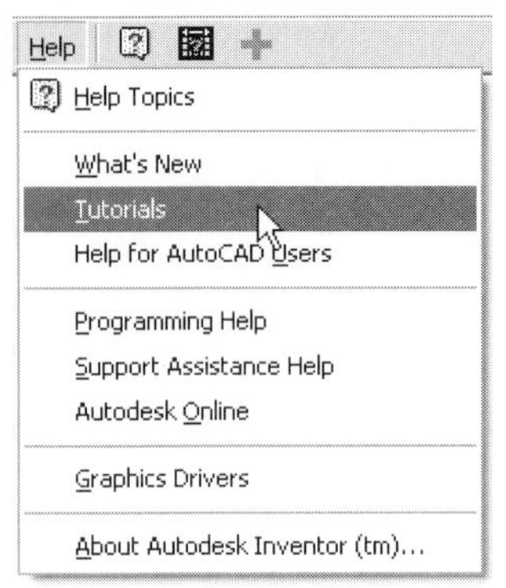

Figure 1.9 - Help from Menu Bar

We have already noted that Autodesk has provided fairly extensive tutorials available from the Getting Started page on the Open dialog box. These same tutorials can also be accessed by selecting Help > Tutorials from the Menu Bar, as shown in Figure 1.9.

In addition to the tutorials, the Autodesk Inventor help system also includes literally hundreds of Show Me animations. These brief step-by-step animations demonstrate how to complete a task or understand a concept. The Visual Syllabus icon on the Menu Bar is the gateway to the Show Me animations (see Figure 1.10).

The Visual Syllabus palette is shown in Figures 1.11 and 1.12. The Syllabus organizes the Show Me animations into separate palettes according to the different work environments available in Inventor: Part Modeling, Sheet Metal, Assembly Modeling, Presentations, and Drawings (Figure 1.11). Each icon on a given palette either directly launches an animation, or opens a window with a menu of animations (Figure 1.12).

The main Inventor Help system can be accessed using the Help Topics icon on the menu bar. The Inventor Help Topics, shown in the Figure 1.13, are similar in

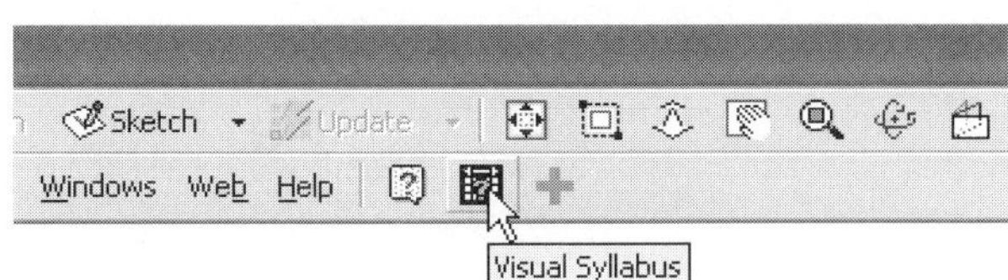

Figure 1.10 - Visual syllabus icon on Inventor Menu Bar

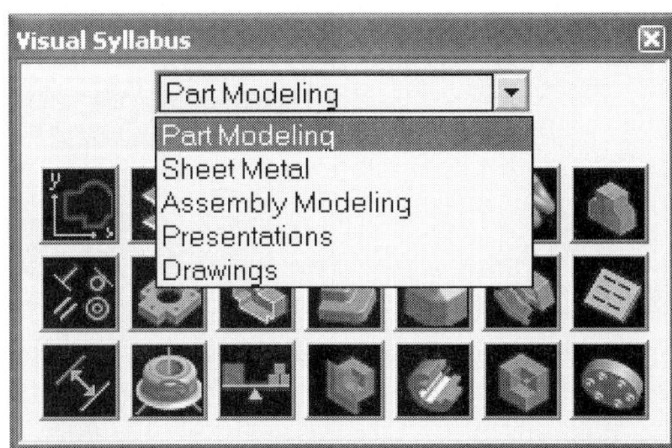

Figure 1.11 - Visual syllabus palette options

Figure 1.12 - Visual Syllabus icons and Show Me animation menus

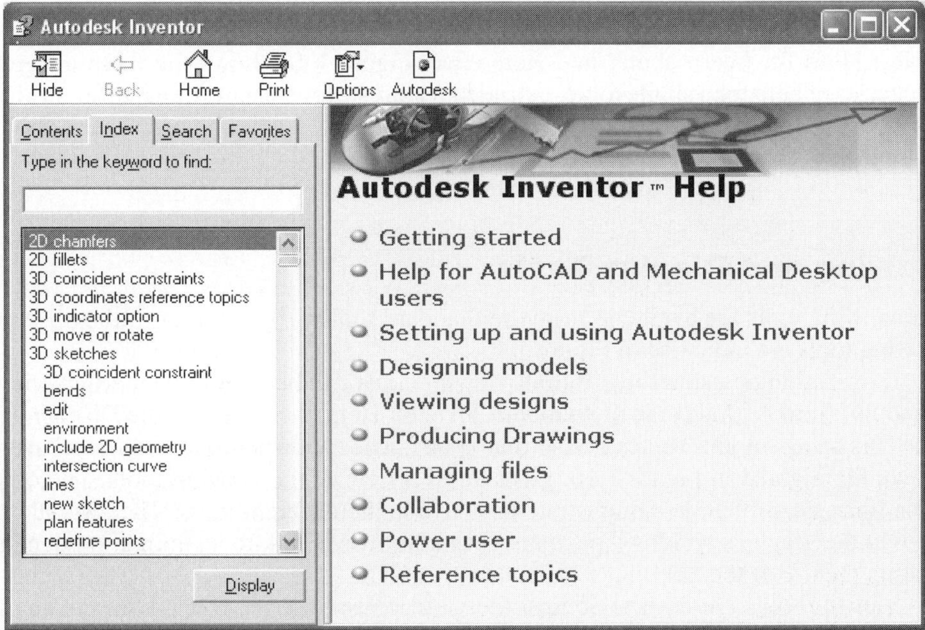

Figure 1.13 - Help topics

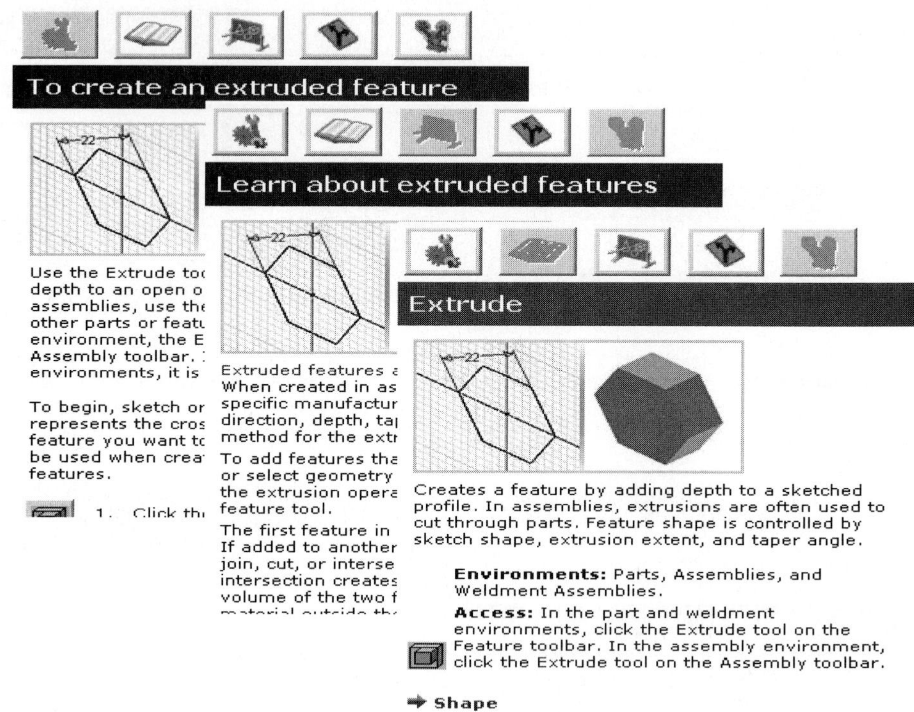

Figure 1.14 - Color coded help page

organization to other Windows software programs. Four tabs are provided in the Help window: Contents, Index, Search, and Favorites. The Contents tab gives access to a Table of Contents format, Index to an Index format, and Search to a search engine format. Favorites can be used to bookmark particularly useful help pages.

The different Help Topics pages are organized into one of three different categories; How To, Learn about, and Reference. Figure 1.14 shows the three different categories of help for a single topic, extruded features. Note that the three different Help Topic categories are color-coded; "How to" pages are a navy color, "Learn about" pages a rust color, and "Reference" pages a teal color.

Viewing and Display Tools

The viewing tools are available from the Standard toolbar. The names of the different viewing tools are indicated in Figure 1.15.

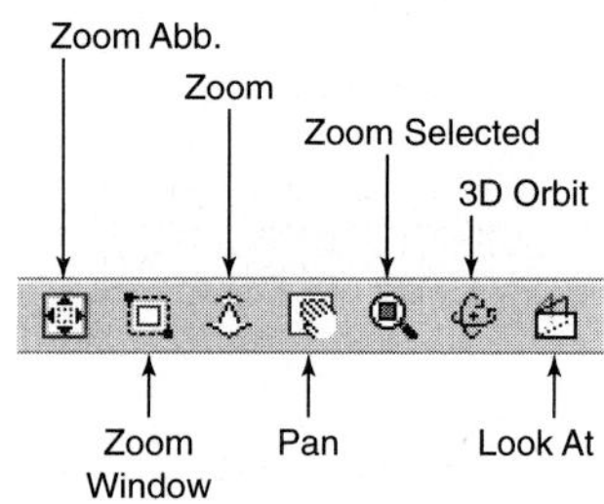

Figure 1.15 - Viewing tools

Information on using the different viewing tools can be found by entering "viewing tools" from the Index tab of Autodesk Inventor Help, then clicking the Display button. This brings up the Topics Found dialog box. The "Viewing tools reference" page, shown highlighted in Figure 1.16, gives an overview of the viewing tools. The other topic links shown below demonstrate how to use the different tools. The "Rotate the view in the graphics window" page also has a Show Me link to an animation demonstrating the use of the 3D Rotate tool.

Viewing tools are used to control the viewpoint of the observer. The size and/or location of a model are not changed by the different viewing commands.

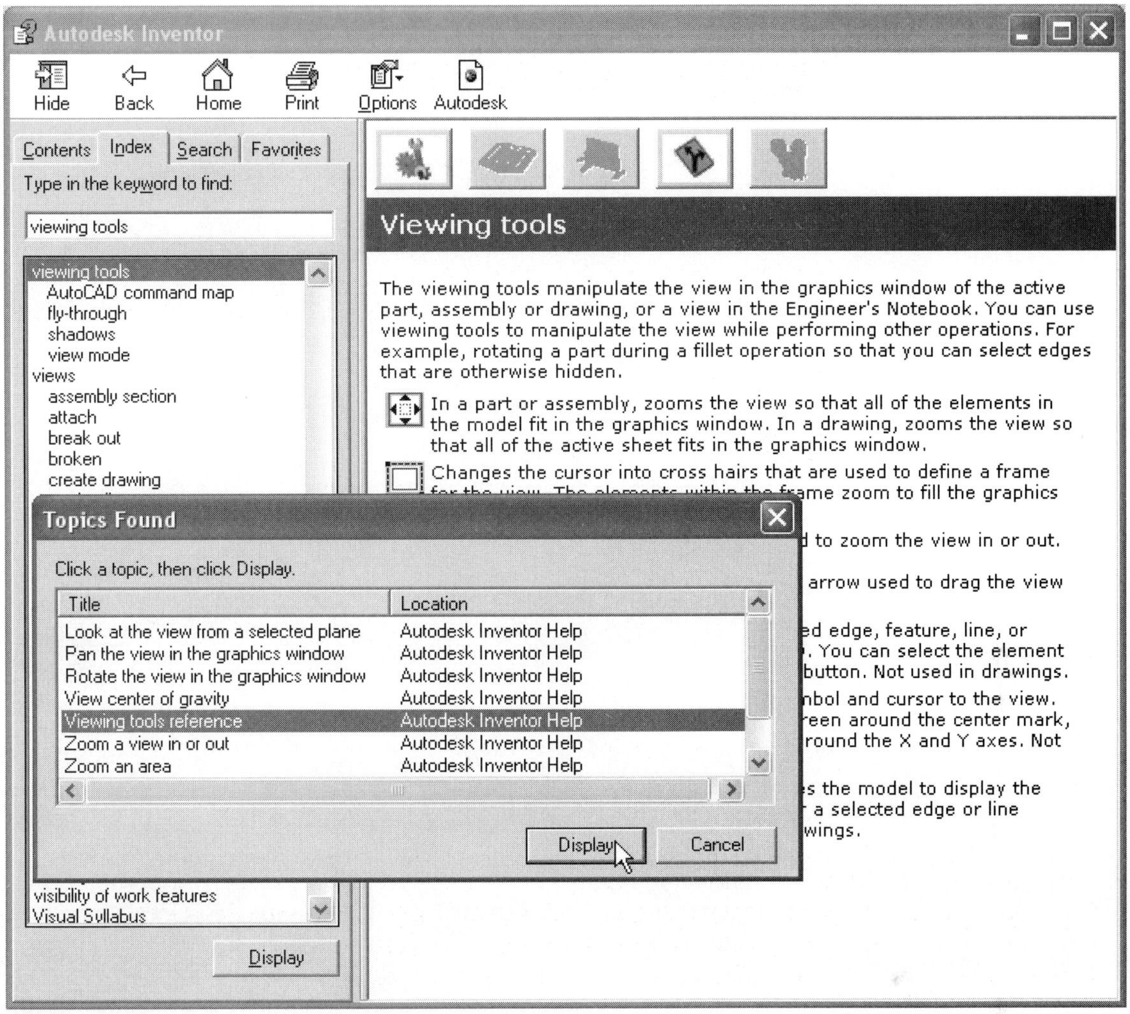

Figure 1.16 - Viewing tools help

Viewing tools can be used in the middle of another operation. Once the view has been changed, the operation resumes.

Right-clicking in the graphics area opens a context menu. Select Isometric View, as shown in Figure 1.17, and an isometric view is composed and centered within the graphics window.

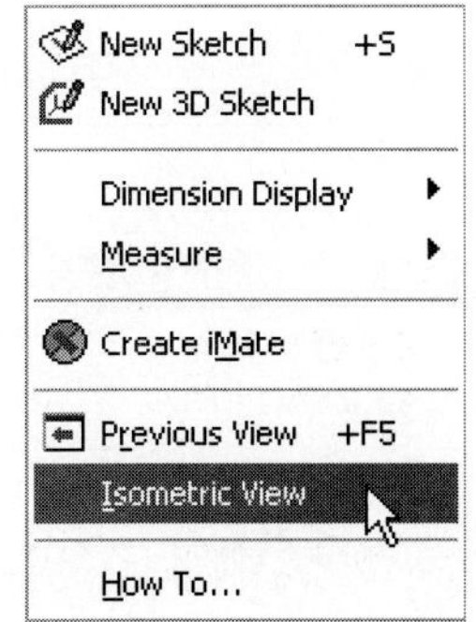

Figure 1.17 - Isometric view access using the context menu

With the 3D Orbit tool in operation, the SPACE bar acts as a toggle between the Free Rotate and Common View modes (Figure 1.18).

Common View can be used to access the eight different *isometric* and six different *principal views*. Highlight one of the Common View arrows so that its color turns red, and then left-click to change to one of these standard views.

While Common View is in operation, it is also possible to redefine the default isometric view. For example, it may be desirable to change the default isometric view on the left in Figure 1.19 to that shown on the right.

To accomplish this, proceed as shown in Figure 1.20. In Figure 1.20A, with Common View in operation, select the highlighted arrow. The view changes to an

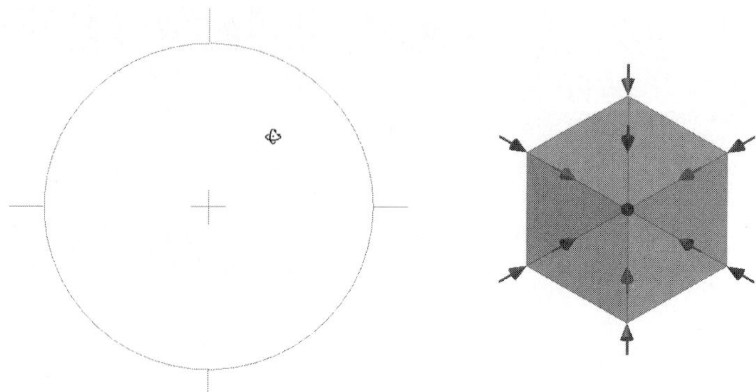

Figure 1.18 - 3D Orbit: free rotate and common view

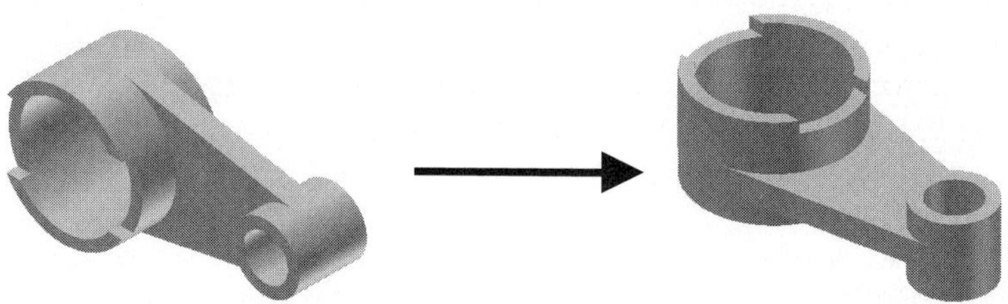

Figure 1.19 - Two different isometric views

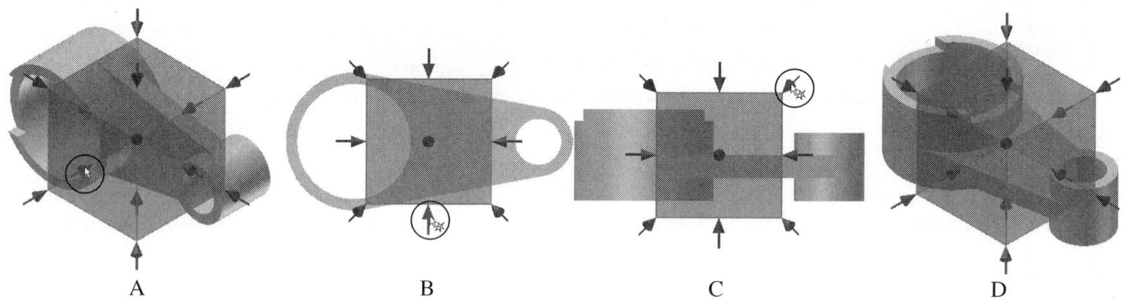

Figure 1.20 - Use of common view tool

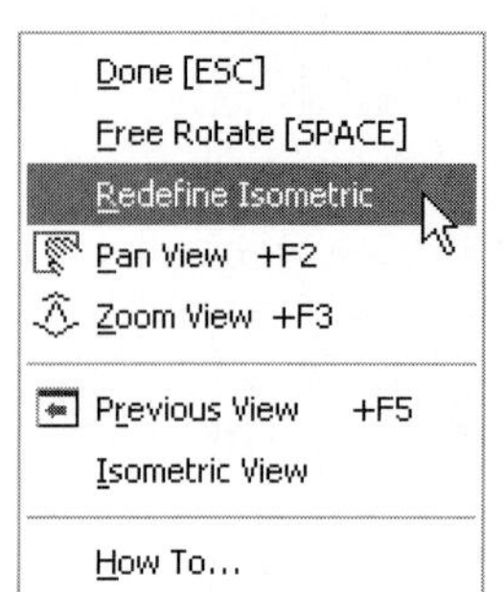

Figure 1.21 - Redefine isometric

orthographic view (Figure 1.20B). Clicking the highlighted arrow shown Figure 1.20B causes the view to change again, to that shown in Figure 1.20C. Now select the isometric view arrow shown in Figure 1.20C. The resulting view is shown in Figure 1.20D.

Finally, to redefine the default isometric view, right-click. The context menu shown in Figure 1.21 appears. Select Redefine Isometric.

The display, camera view, and shadow tools available in Autodesk Inventor are accessed from the Standard toolbar, and are shown in Figure 1.22.

Inventor display options include shaded, hidden edge, and wireframe, as shown in Figure 1.23.

By default, Inventor uses *orthographic projection*. It is also possible to view models using *perspective projection* techniques. Switching between orthographic and perspective projection is also done from the Standard toolbar, as shown in Figure 1.24.

New with Release 6 is the ability to display shadows of part and assembly models. These options are also accessed from the Standard toolbar, as shown in Figure 1.25.

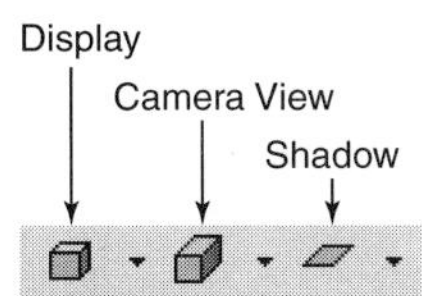

Figure 1.22 - Display tools

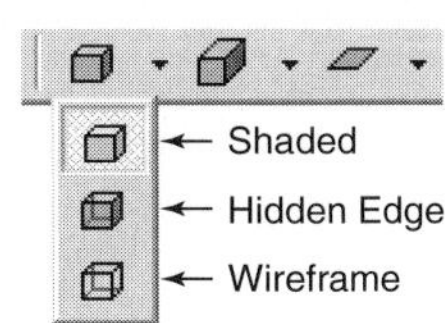

Figure 1.23 - Display options

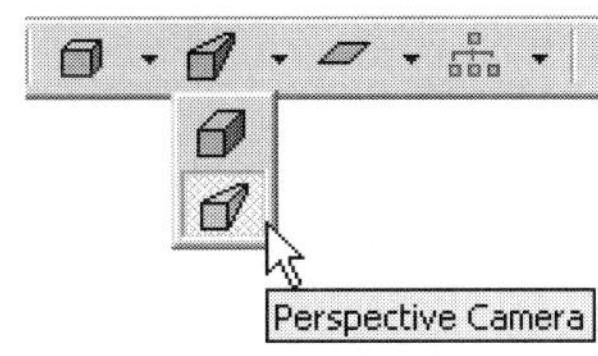

Figure 1.24 - Orthographic and perspective camera modes

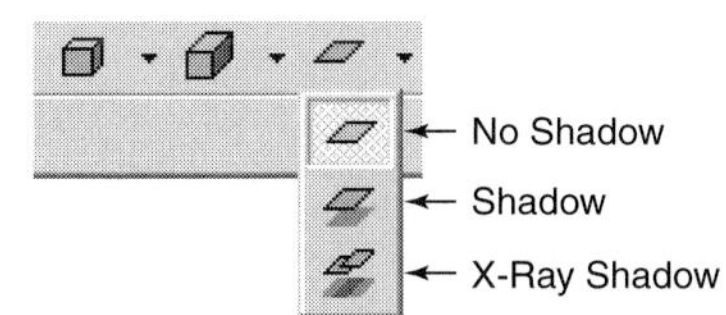

Figure 1.25 - Shadow display

Creating a New Project — TUTORIAL 1

1. With the Project Editor open (File > Projects from the menu bar), click the New button, as shown in Figure 1.26.
2. In the Inventor Project Wizard dialog box (Figure 1.27), accept the default, Personal Workspace for Group Project, then select Next.
3. In the Inventor project wizard dialog box, enter **Ball Valve** in the Name field, as shown in Figure 1.28. Note that this also affects other fields in the dialog box—Ball Valve has been appended to them. If you like, you can also specify a different name and/or location for the project folder, but it is not necessary. Select the Finish button when ready. You are prompted to create a project path. Choose OK.
4. The Ball Valve project name now appears in the upper section of the Project Editor. To make Ball Valve the active project, select it, and then click Apply, as shown in Figure 1.29. Newly created Inventor files will now be saved to the Ball Valve project folder. When opening a file, the contents of the Ball Valve folder will be shown first. To be strictly correct, working files will actually be saved to (or opened from) a subfolder inside the Ball Valve project folder. The name of this folder is Workspace. A workspace subfolder is created inside each new project folder. The first part to be placed in this folder will be an O-ring, created in Tutorial 2, Chapter 2.

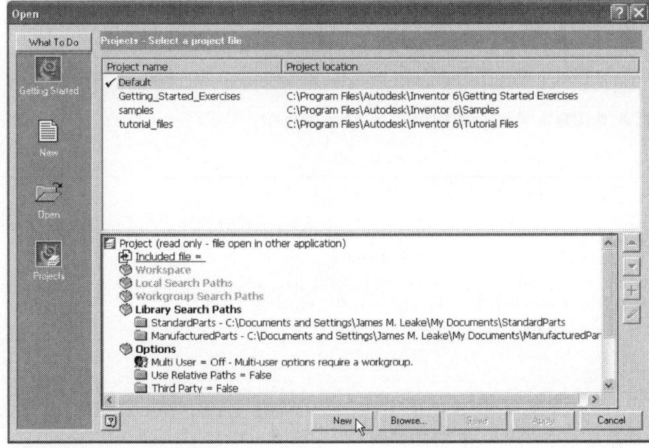

Figure 1.26 - Project editor

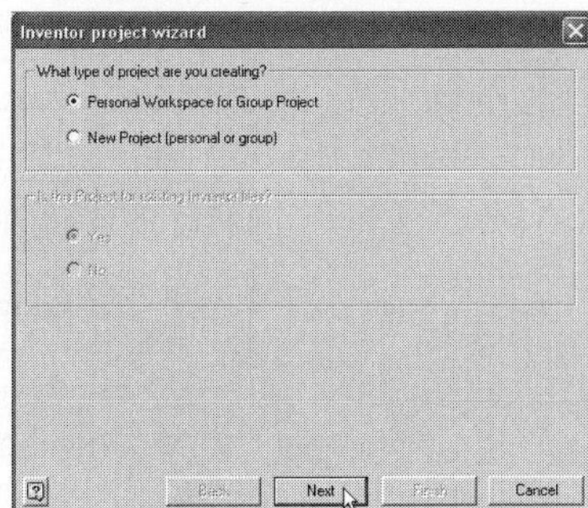

Figure 1.27 - Project type

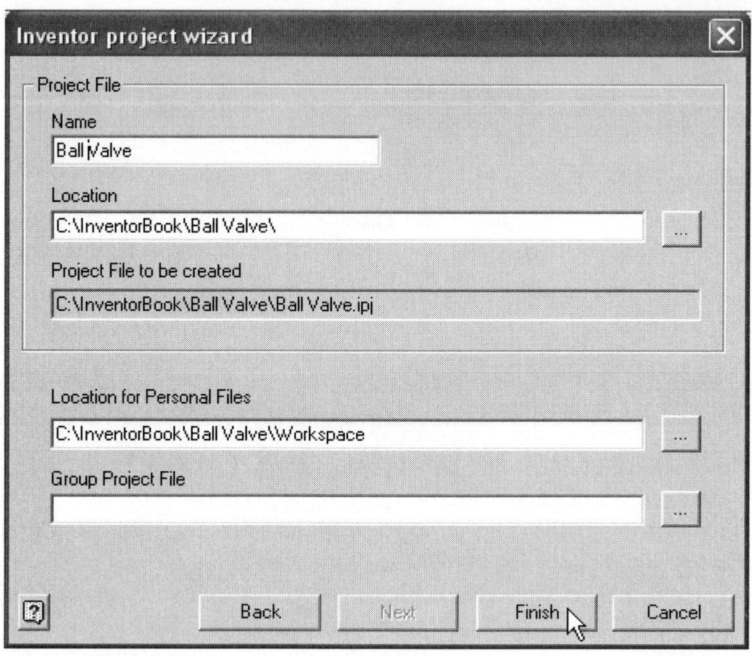

Figure 1.28 - Inventor project wizard

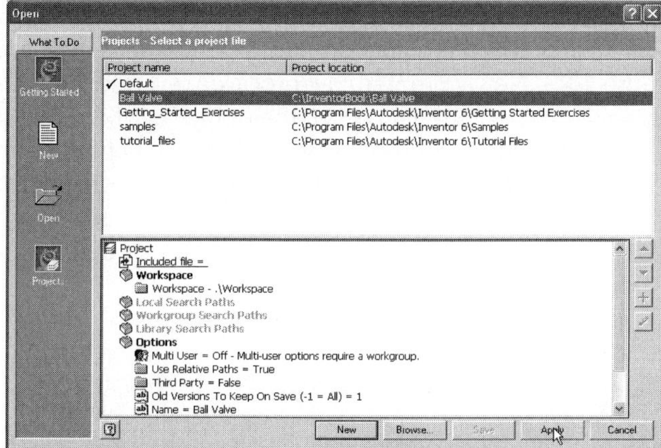

Figure 1.29 - Activate ball valve project

QUESTIONS

1. What are the four options available in the Open dialog box?
2. Which is not a file format for Inventor?
   a. Assembly (iam)
   b. Drawing (idw)
   c. Spreadsheet (isp)
   d. Presentation (ipn)
   e. Part (ipt)

3. Give the names of each of the labeled areas shown below.

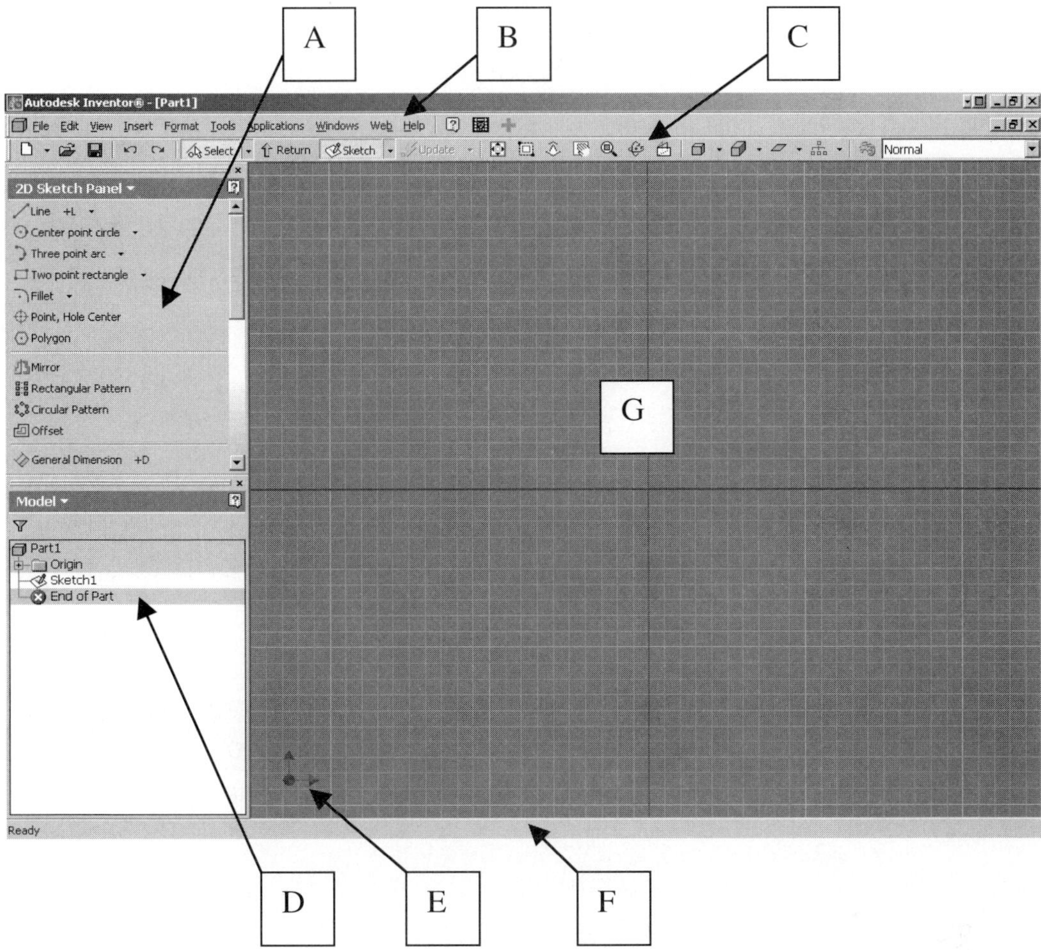

4. List the seven basic reference work features.

PROBLEMS

Use the part file ViewDisplay.ipt included on the CD in the back of the book to investigate Inventor viewing and display tools. Try the following:

1. Viewing tools, including:
   a. Zoom Window
   b. Zoom
   c. Zoom All
   d. Pan
   e. Rotate
      i. Also try Common View. While the 3D Rotate tool is active, the Space bar toggles between the Free Rotate and Common View modes.
      ii. Redefine the default isometric view. Use Common View to select a different isometric view, then right-click, and select Redefine Isometric.

f. Look At
   i. Click on a planar face to align the viewing direction so that it is normal to the planar face.

   Note that to center the display and return to the default isometric view, right-click in the graphics window and choose Isometric.

2. Display tools, including:
   a. Shaded , hidden edge and wireframe display
   b. Orthographic versus perspective camera
   c. No ground shadow , grounded shadow and x-ray ground shadow .
3. From the Menu bar, select Tools > Application Options . . .
   a. Colors tab—change the color scheme. Click Apply to view the new color scheme. Also, in the Background area, select Gradient from the drop down list.
   b. Display tab—under Shaded Display Modes, Active, remove the check from Edge Display, and then click Apply. All of the figures in this book have Edge display turned off.

CHAPTER 2

Sketching and the Base Feature

LEARNING OBJECTIVES

- Name two ways to access sketching tools available in Inventor
- Recognize and name two ways to access geometric constraints available in Inventor
- Give three methods for accessing the General Dimension tool in Inventor
- Describe several good sketching practices
- Describe the algorithm used to create a base feature
- List the different linetypes and use them to help create sketches
- Use the Inventor sketch tools to create closed profile sketches
- Use geometric constraints to help define the shape of a profile sketch
- Control the display of applied geometric sketch constraints
- Use parametric dimensions to define the size of a profile sketch
- Use projected default work features to tie sketch geometry to the reference coordinate system
- Use the Line tool to:
  - Draw a line segment
  - Sketch a closed profile
  - Create a polyline containing line and (tangent) arc segments
- Use the Center point circle tool to draw a circle
- Use the Offset tool to offset existing sketch geometry
- Use the Trim tool to trim existing sketch geometry
- Use the Revolve tool to create a base feature
- Use the Extrude tool to create a base feature

Parametric Solid Modeling

Parts built using a parametric solid modeler like Autodesk Inventor employ parameters to constrain the size and shape of the different features of the part. *Features* are the basic building blocks for creating parts. A feature is a three dimensional (3D) shape.

Model features available in a parametric modeler typically simulate actual design and manufacturing features. For example, features available in Inventor include fillet, counterbore hole, and draft angle. There are essentially two kinds of model features, sketched and placed. Sketched features (e.g., extrude, revolve) require that a 2D sketch be made before the feature can be created. Placed (or built-in) features (e.g., hole, chamfer, fillet, shell, face draft) can be created without sketch geometry. Since features are so important to the part modeling process, parametric modeling is sometimes called *feature based modeling*.

A *parameter* is a named quantity whose value can be changed; for example, d0 = 10. Here d0 is the name of the parameter, 10 is its value. A parametric modeler uses parametric dimensions to control the size and position of features. When the value of the parameter changes, the feature geometry changes as well. Because of this, we say that parametric models are dimensionally driven. In addition, because a parameter has a name as well as a value, we can establish parametric relationships across features, parts, and assemblies. A simple example of this would be d0 = 2*d1. If d1 is 5, then d0 is 10. If d1 is changed, d0 will change by a factor of two.

Constraints are mathematical requirements placed on the geometry of a 3D model. Regarding part models, there are two kinds of constraints, dimensional and geometric. Dimensional constraints (also called parametric dimensions or parametric constraints) place limits on the size or position of a feature. Geometric constraints (e.g., parallelism, tangency, concentricity) place limits on the shape or position of a feature. In addition to part constraints there are also assembly constraints. These constraints determine how different parts are positioned with respect to one another. Examples of assembly constraints include mate and insert. Parametric modeling is sometimes referred to as *constraint based modeling*.

Base Feature Creation

In the part modeling process the first feature created serves as the foundation for all of the subsequent features. This first feature is known as the *base feature*, and is almost always a sketched feature. The sketch for the base feature deserves careful consideration. This 2D *profile* will hopefully capture the general shape of the part, without being too complicated. An algorithm for creating the base feature includes the following steps:

1. Make a 2D sketch of the feature geometry. As the sketch is made some geometric constraints are applied automatically.
2. Manually add and/or delete other geometric constraints.
3. Add parametric dimensions until the sketch is fully constrained.
4. Create the feature.

The first three steps described in this algorithm occur in the 2D sketch environment. This environment can be recognized by the fact that a grid is visible in the graphics area, and that the 2D Sketching Tools menu appears in the panel bar. Step 4 takes place in part mode. The grid disappears, and the Part Features menu appears in the panel bar.

An example of this process using Inventor is illustrated below.

1. Make a rough 2D sketch (Figure 2.1). Some constraints (e.g., parallel, perpendicular, tangent, coincident) are applied automatically.
2. Add geometric constraints (Figure 2.2).
3. Add parametric dimensions (Figure 2.3).
4. Create a feature (extrude) (Figure 2.4).

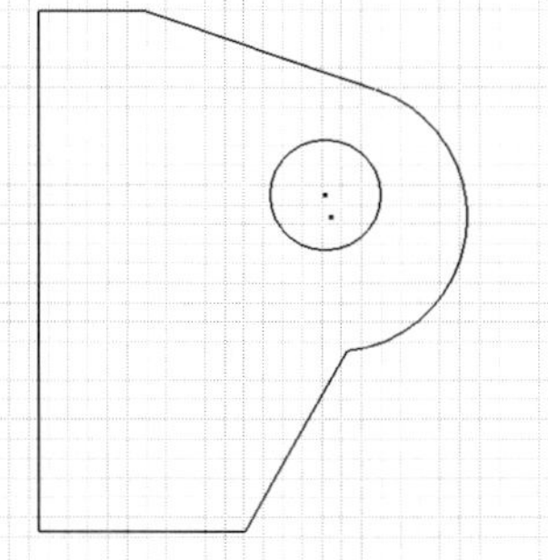

Figure 2.1 - Rough sketch

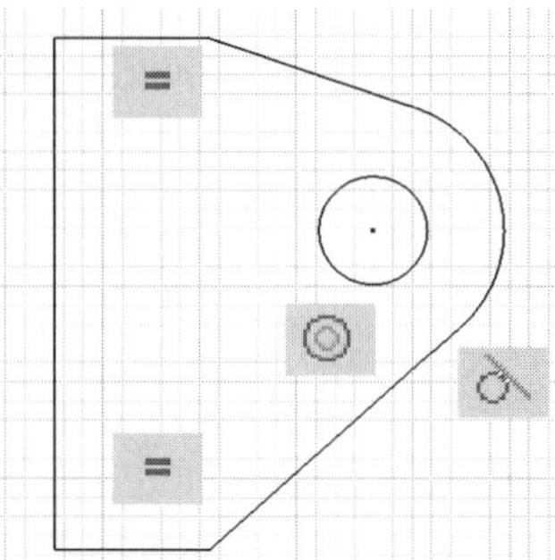

Figure 2.2 - Add geometric constraints

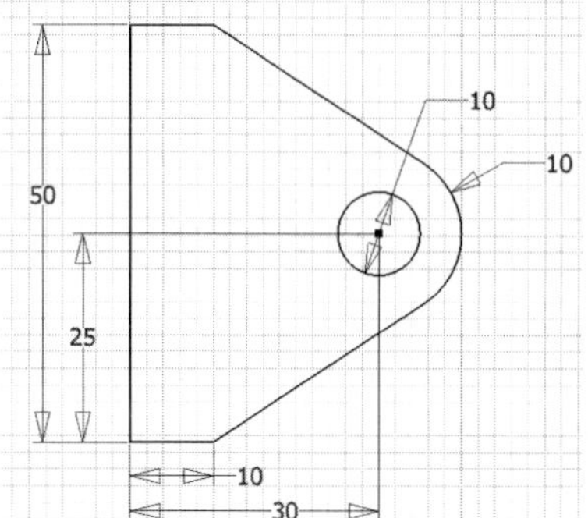

Figure 2.3 - Add parametric dimensions

Sketching Tools

Tools currently available in Inventor for creating sketch geometry in the 2D sketch environment include:

Figure 2.4 - Base feature

2D Sketch Tool	Panel Bar Display
Line, Spline	Line Spline
Circle	Center point circle Tangent circle Ellipse
Arc	Three point arc Tangent arc Center point arc
Rectangle	Two point rectangle Three point rectangle
Point, Hole Center	Point, Hole Center
Polygon	Polygon

Note that the Line tool is really a polyline tool, allowing the user to create both lines and arcs.

Tools currently available in Inventor for modifying sketch geometry include:

2D Sketch Tool	Panel Bar Display
Fillet, Chamfer	Fillet Chamfer
Mirror	Mirror
Rectangular Pattern	Rectangular Pattern
Circular Pattern	Circular Pattern
Offset	Offset
Extend	Extend
Trim	Trim
Move	Move
Rotate	Rotate

Most of these tools will be demonstrated in the course of the tutorials. The Inventor Help provides additional information on any of these sketching tools.

Geometric Constraints

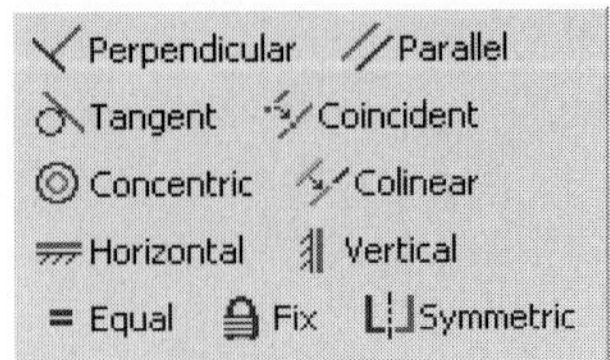

Figure 2.5 - Geometric constraints

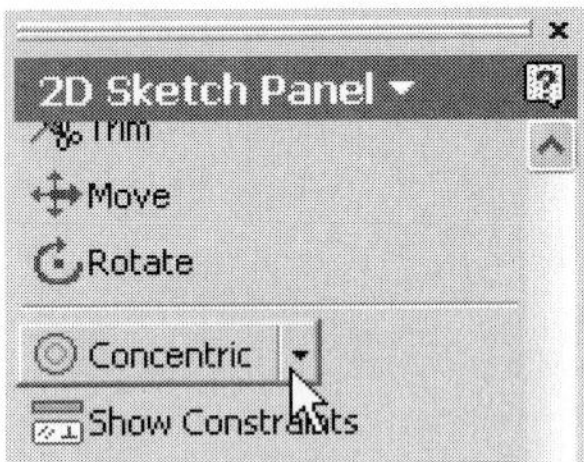

Figure 2.6 - Accessing constraint tools from panel bar

The geometric constraints currently available in Inventor include those shown in Figure 2.5.

These constraint tools are easily accessed from the 2D Sketch panel (i.e., the panel bar in sketch mode), as shown in Figure 2.5, or from the right-click context menu. Only the most recently used constraint tool appears in the panel bar. It is necessary to left-click on the down arrow to access the other constraint tools, as shown in Figure 2.6.

The number of dimensions required to fully constrain a sketch is reduced when constraints are applied. The software applies some constraints automatically while in the course of sketching. This is illustrated in Figure 2.7. Starting at the left, a horizontal constraint is applied to the line segment; moving to the right a perpendicular constraint is applied between the two line segments; then a coincident constraint is applied to the endpoints of the first and third line segments; the rightmost image shows the resulting closed profile. Note that it is possible to override a constraint by holding down the CTRL key while clicking endpoints.

Figure 2.8 shows the effect of manually adding geometric constraints: applying an equal constraint and colinear constraints to the two horizontal lines, and a tangent constraint to the arc and the vertical line.

There are a couple of different ways to display the constraints that have been applied in a sketch. In order to display the constraints applied to all sketch entities, click the right mouse button in graphics area, and then select Show All Constraints. This has been done for the right triangle sketch seen in Figure 2.9. To turn off the displayed constraints, click the right mouse button and select Hide All Constraints.

To display the constraints of one sketch entity at a time, choose Show Constraints from the panel bar, and then select an entity(s).

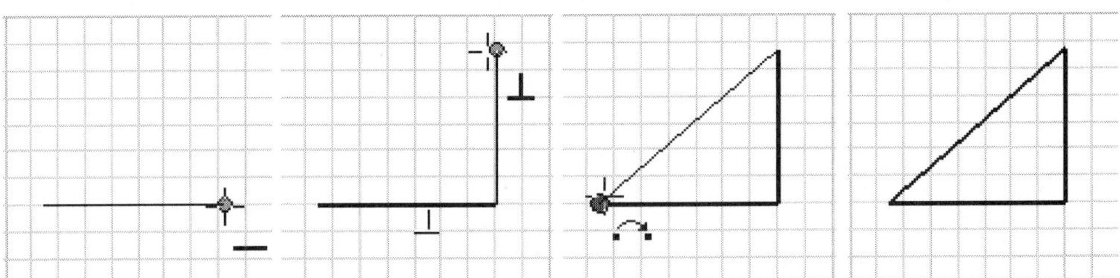

Figure 2.7 - Constraints being applied automatically

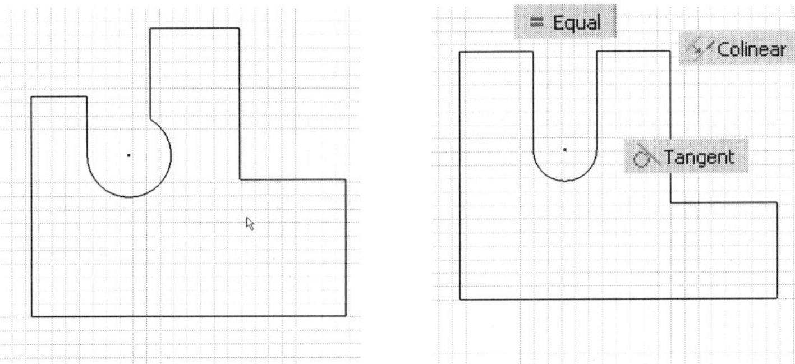

Figure 2.8 - Constraints applied manually

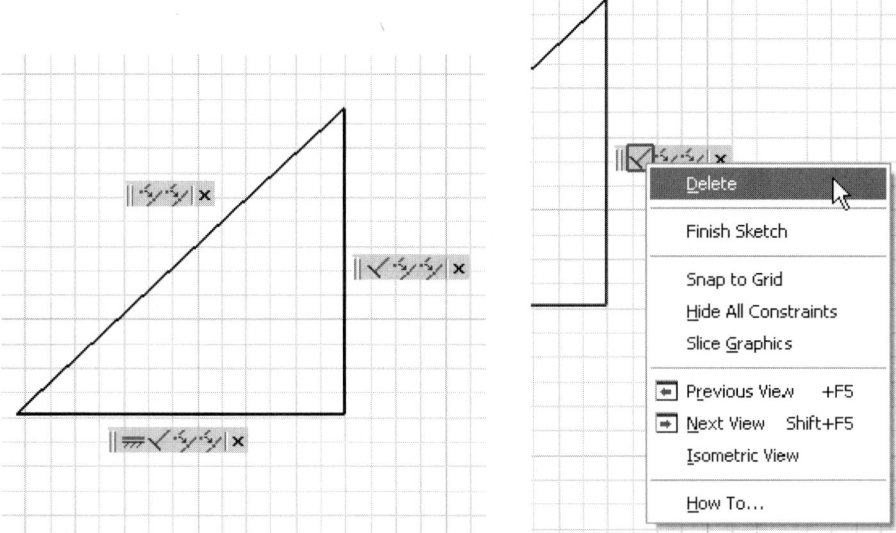

Figure 2.9 - Show all constraints (right-click context menu)

Figure 2.10 - Deleting geometric constraints

A constraint can be deleted by first displaying the constraints, and then selecting the constraint to be removed with the right mouse button, then selecting Delete from the context menu, as illustrated in Figure 2.10.

Parametric Dimensions

Dimensioning is straightforward in Inventor. The General Dimension tool on the panel bar is used to add all dimensions. The General Dimension tool can also be accessed by pressing D from the keyboard, or by right-clicking in the graphics area while in sketch mode, and then selecting Create Dimension.

A linear dimension is added by selecting the length to be dimensioned. This length can either be represented by a line, or by two points. In the case of an existing line, as soon as the mouse is moved, the dimension appears. Click the left mouse button a second time to place the dimension.

In Figure 2.11 on the left, the horizontal leg of the triangle has been dimensioned. The dimension (31.07) is the arbitrary length of the line as it was initially drawn. In the

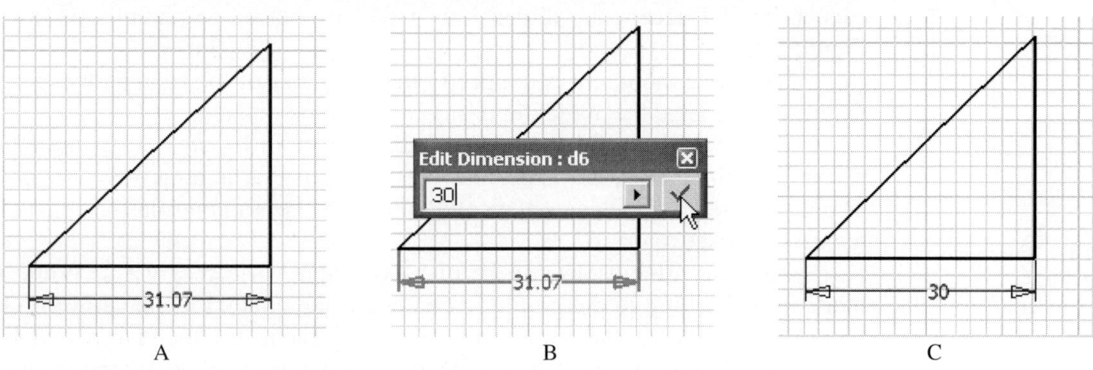

Figure 2.11 - Adding and editing linear dimension

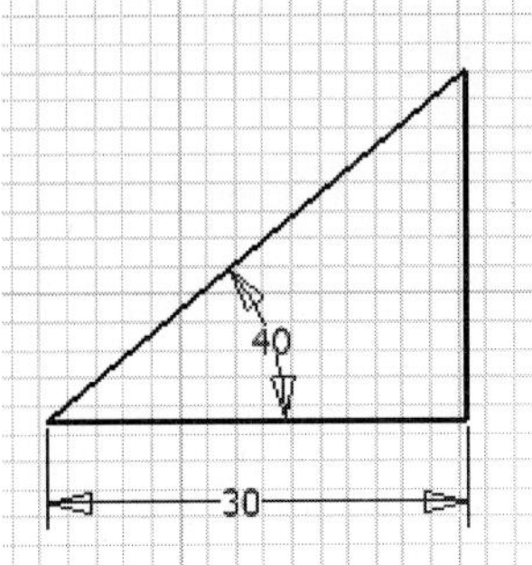

Figure 2.12 - Adding an angle dimension

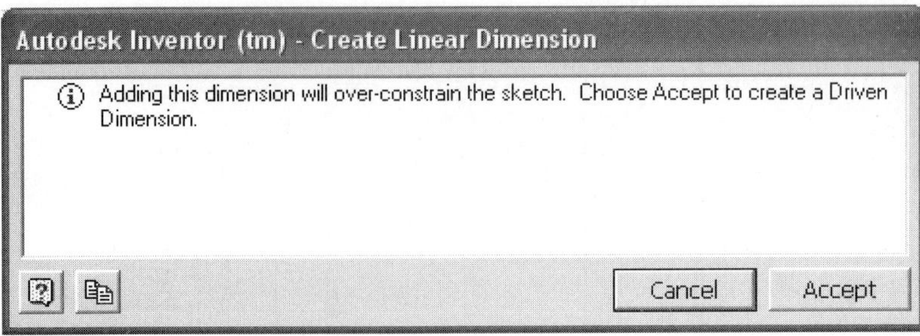

Figure 2.13 - Over-constraint message

middle the dimension is in the process of being modified. Left-click on the dimension to be edited and the Edit Dimension dialog box opens. From the keyboard enter another value, in this case 30. Once the Apply (green check mark) button is selected, the dialog box closes and the sketch updates, as shown on the right.

TIP: If the "Edit dimension when created" option on the Display tab of the Options dialog box (Tools > Application Options . . . on the menu bar) is checked, the Edit Dimension dialog box appears as soon as a new dimension is added.

Angle, radius, diameter, and aligned dimensions are all added using the same General Dimension tool. To add an angular dimension for example, select the two lines forming the angle. This has been done to the right triangle sketch in Figure 2.12.

TIP: If the intended dimension type is not in effect, the right mouse button context menu may provide the desired type (e.g., radius versus diameter).

Having added the angle dimension, our sketch is now *fully constrained.* This means that the geometry of the sketch, both in size and shape, is fully determined. Although in Inventor it is not necessary that a sketch be fully constrained before creating a feature from the sketch, it is, generally speaking, good practice to do so.

If we now try to add a linear dimension to the vertical leg of the triangle, Inventor responds with the message in Figure 2.13.

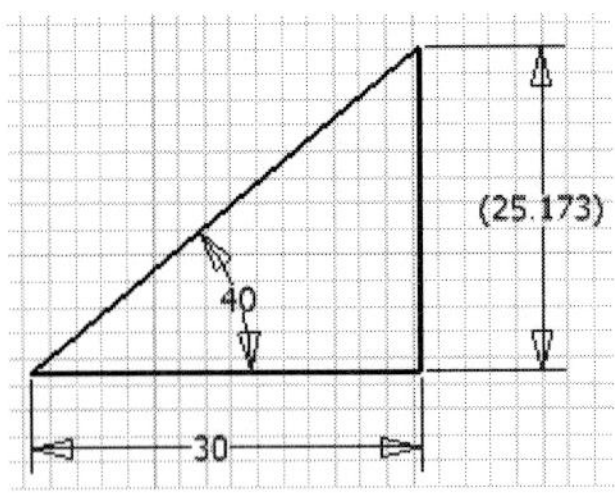

Figure 2.14 - Reference dimension added

The vertical dimension is not required to fully define the geometry. I can either cancel, or, if I accept the dimension, it will be placed in parenthesis, as shown in Figure 2.14.

Of course if the intent is to specify the geometry of the triangle by specifying the lengths of the legs, the angle dimension can also be deleted.

Notice that in the Edit Dimension dialog box in Figure 2.11B, the name of the parametric dimension, d6, appears in the dialog box title area. Inventor provides indexed parameter names automatically, although these names can later be changed if the user chooses to do so.

Parametric dimensions can be displayed in a variety of ways. While in sketch mode right-clicking in the graphics area opens the context menu shown on the left in Figure 2.15. Highlighting Dimension Display expands the menu to reveal several options. The figure shows the triangle sketch dimensions displayed respectively as Value, Name, and Expression.

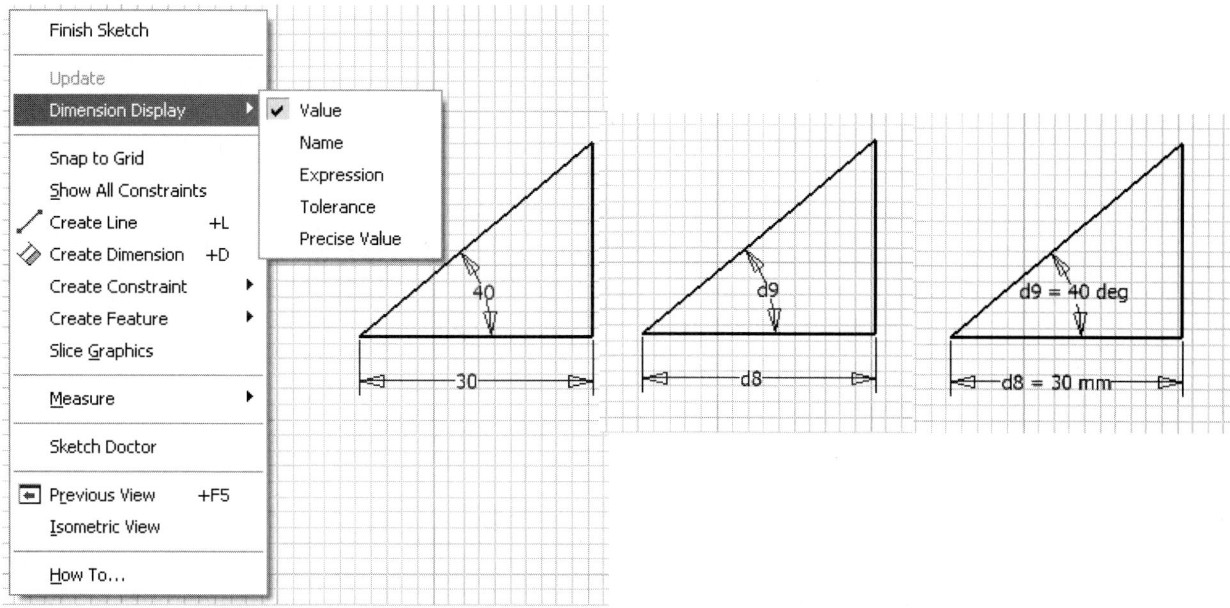

Figure 2.15 - Dimension display

Sketching Tips

1. Look for the shape that best describes the part to be created, and use it for your initial sketch.
2. Keep your sketches simple. In particular, fillets and chamfers can always be added later as placed features. A simple sketch is easier to fully constrain.
3. Only closed profile sketches can be used to create solid features.[1] For this reason it is important that all consecutive line and/or arc segments forming a profile share a common endpoint. In Inventor this means that a coincident constraint must be shared by adjoining line and/or arc segments. Note that in Release 6 it is now possible to use open profile sketches to create surfaces. This will be taken up in Chapter 10.
4. Make your sketches roughly proportional to the final desired shape. Start by drawing the first line or arc at approximately the correct length. This can be accomplished using the status bar, which reports the characteristics of a line (or other entity) as it is being constructed, as shown in Figure 2.16.
5. Sketch geometry should not overlap. Sometimes as dimensions are added a sketch will unexpectedly cross over upon itself. Depending on how the sketch is constrained, it is often possible to click and drag an entity in such a way that the overlap is eliminated. Otherwise it may be necessary to undo the last dimension, and proceed differently.
6. Make use of the default reference axes and center point by projecting them on to the sketch plane. The sketch can then be constrained to the reference geometry. This is a particularly useful approach when parts have considerable symmetry (e.g., the ball valve).

[1]An exception to this is the Rib feature, described in Chapter 5.

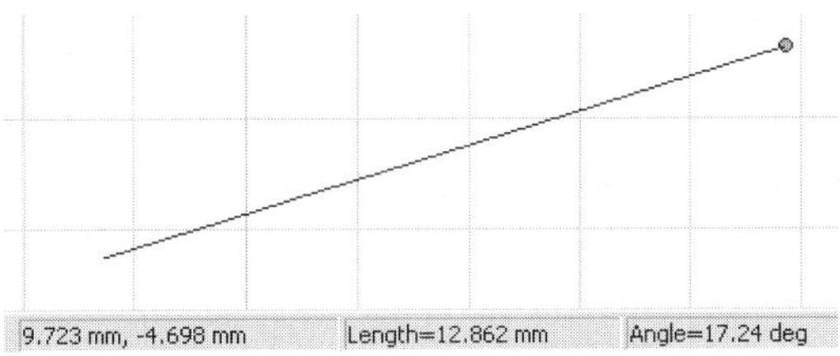

Figure 2.16 - Using the status bar to approximate lengths

Project Geometry

Most of the tutorial sketches found in this book will start by projecting work features available from the Origin folder in the part browser onto the sketch plane. This is accomplished with the Project Geometry tool, available from the 2D Sketch Panel.

By first projecting work axes and center point onto the sketch plane (see Figure 2.17), subsequent sketch geometry can either be made coincident with this work geometry, or dimensioned with respect to it. This in turn makes the positioning of other features and parts easier. It will also considerably simplify the assembly of components in the assembly environment.

Figure 2.17 - Projected default work axes

Linetypes

There are four different sketch linetypes available in Inventor: normal, construction, centerline, and reference. Normal is the default linetype. Features consume sketch geometry that uses the normal linetype. Geometry projected on to the sketch plane using the Project Geometry tool is assigned the reference linetype.

The linetype of a line can be changed using the Style drop-down list on the Standard toolbar. In Figure 2.18, the projected vertical axis is selected, and then its linetype is changed to centerline. Centerlines have certain properties that will be useful in executing the tutorials to follow.

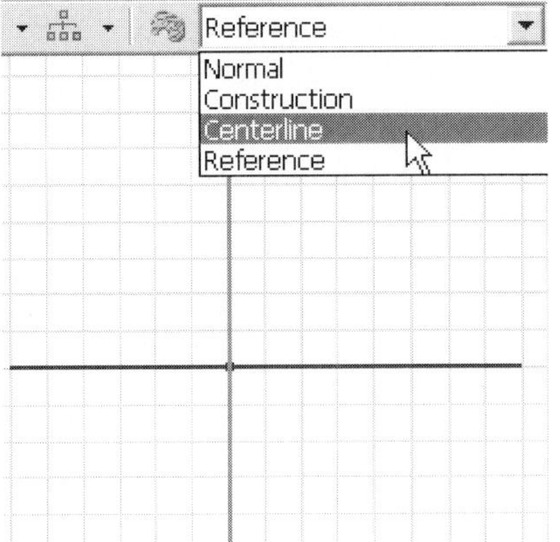

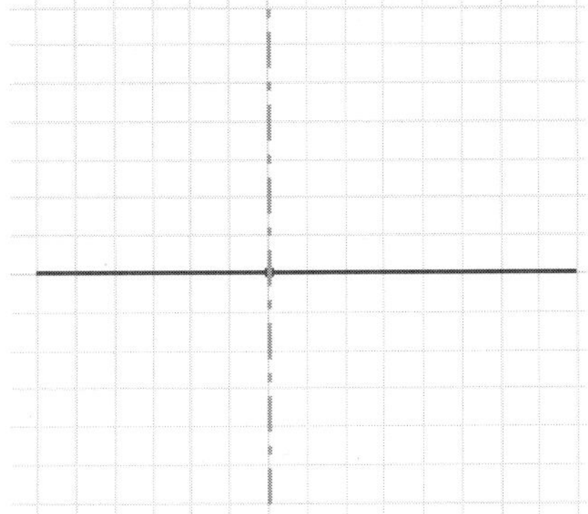

Figure 2.18 - Changing a linetype

Tutorial Notes

1. Most of the tutorials in this book will start with a build strategy, where the modeling approach as well as some key dimensions are provided. Afterwards detailed modeling steps are provided.
2. In a few cases a tutorial started in one chapter is not completed until a subsequent chapter.
3. Video versions of all of the tutorials are included on the accompanying CD.
4. Completed versions of all tutorial parts, assemblies, drawings and presentations are included on the CD.
5. Save early, save often.

O-ring 1 — TUTORIAL 2

NOTE: This is the only tutorial in the book where the base feature sketch is not constrained to the default work geometry. This is done deliberately, in order to illustrate the problems that this approach can cause later in the modeling process. All other base feature sketches will be tied to the default reference geometry.

ANOTHER NOTE: Be sure to complete Tutorial 1 at the end of Chapter 1 before continuing.

Build Strategy

1. Create the sketch shown in Figure 2.19. The sketch consists of a circular profile and a centerline.
2. Use the revolve feature tool to create the part (Figure 2.20).

Detailed Modeling Steps

1. Start a new metric part file. From the Open dialog box, click New, select the Metric tab, highlight the Standard (mm).ipt template icon, and click OK. See Figure 2.21.

   TIP: If upon installation millimeters were specified as the default units, then the file templates on the Default tab can be used directly in order to work in metric units.

2. A new Inventor part file opens, as shown in Figure 2.22. The rectangular grid overlaying the graphics area indicates that we are in sketch mode.

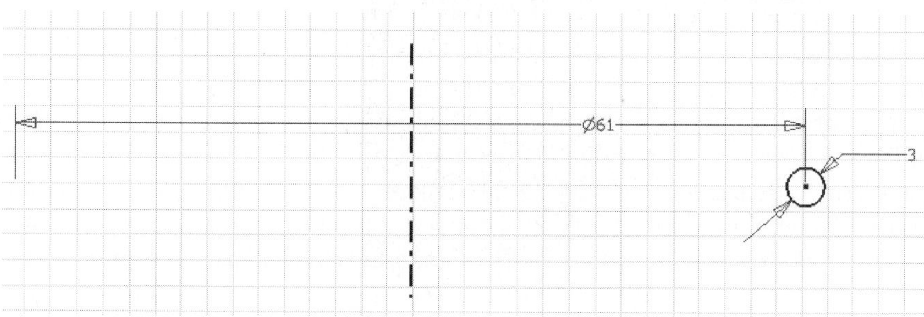

Figure 2.19 - Base feature sketch

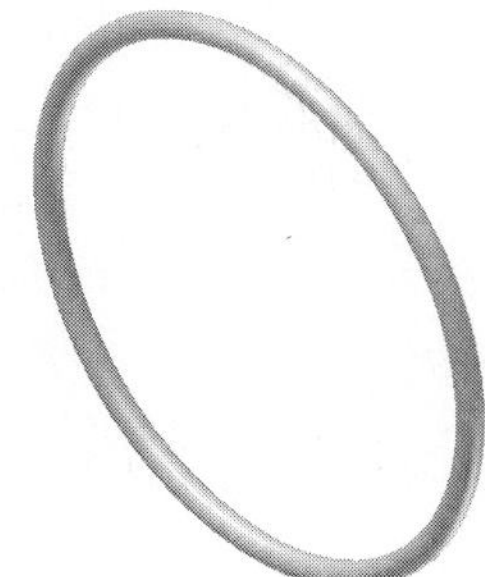

Figure 2.20 - Base feature

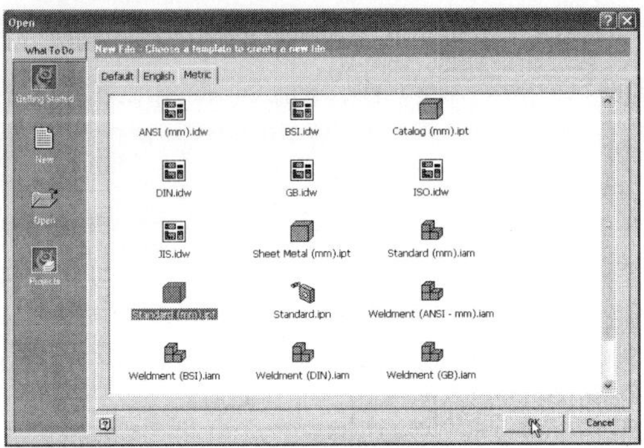

Figure 2.21 - New (metric) part file

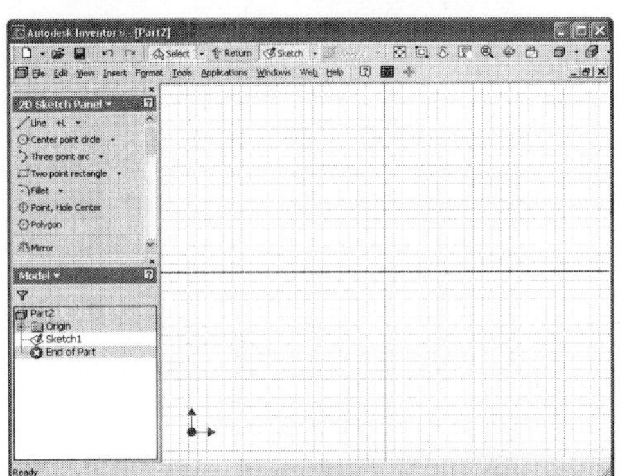

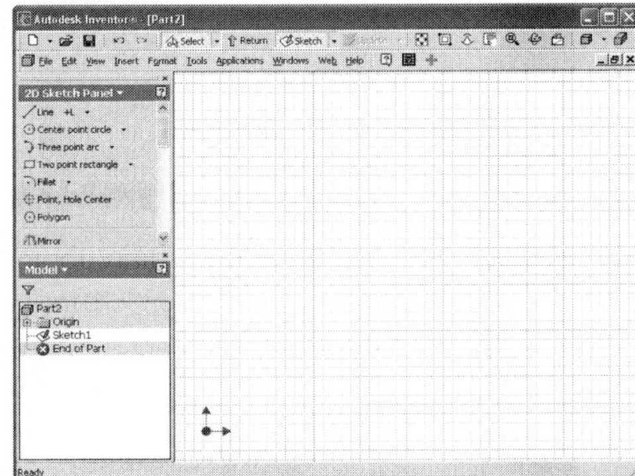

Figure 2.22 - Part file interface; Axes on, Axes off

3. Turn off the coordinate axes (Figure 2.22 on the right). To do so, choose Tools > Application Options . . . from the menu bar. The Options dialog box opens. Select the Sketch tab. In the Display area in the upper right corner, remove the check mark next to Axes, and then click OK.
4. We are now ready to start sketching. Using the panel bar, select the Center point circle tool. In the drawing area click the left mouse button to select the center point of the circle. Move the mouse. A rubber-banding circle appears. Click the left mouse button to create the circle. See Figure 2.23.

   TIP: Note that on the status bar in the lower right corner the circle's center point and radius are displayed. To ensure that the sketch updates predictably when dimensions are added, it is good practice to sketch objects at roughly their final size.

   To terminate the Center point circle command either hit the Escape key, or click the right mouse button, then choose Done.
5. Choose the Line tool from the panel bar. Select a point above and to the left of the circle as a starting point for the line. Move the mouse down. A rubber-band line appears. In order to draw a vertical line segment, wait until the vertical constraint symbol appears before selecting the other endpoint (see Figure 2.24). To terminate the line command, hit the Escape key, or click the right mouse button, then choose Done from the context menu.
6. Change the line to a centerline by selecting the line, then changing the linetype to centerline from the drop-down Style field on the menu bar, as shown in Figure 2.25.

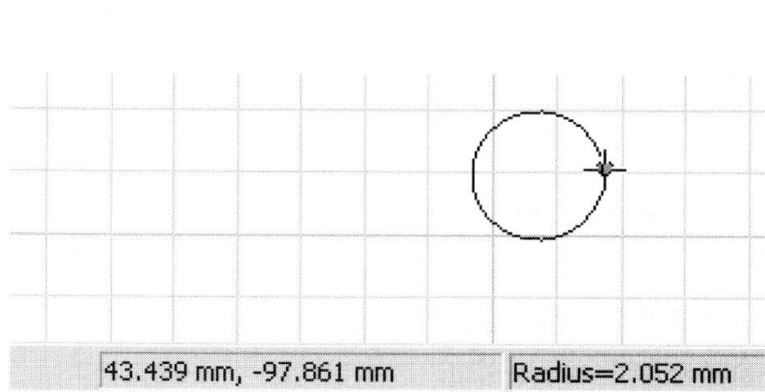

Figure 2.23 - Sketched circle

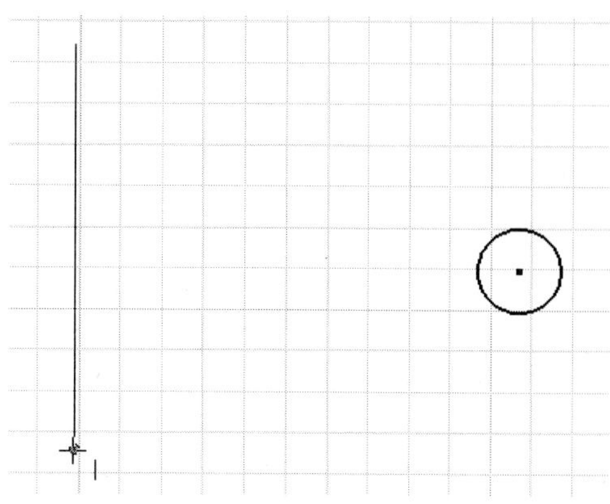

Figure 2.24 - Add a vertical line

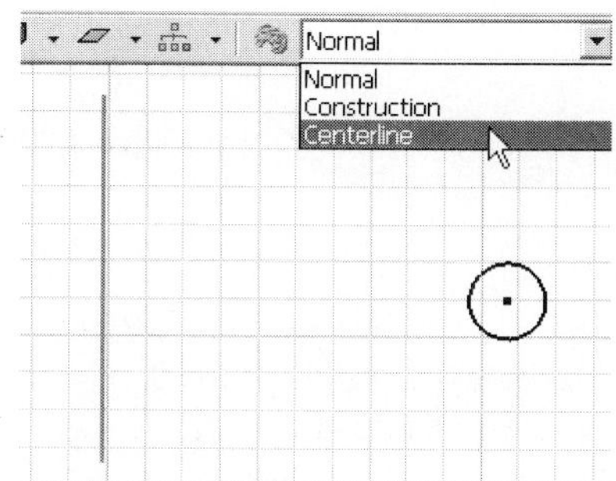

Figure 2.25 - Change linetype to centerline

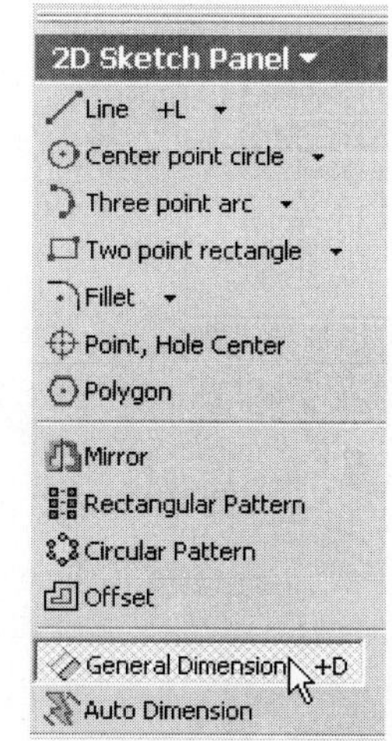

Finish Sketch
Update
Dimension Display
Snap to Grid
Show All Constraints
Create Line +L
Create Dimension +D
Create Constraint
Create Feature
Slice Graphics
Measure
Sketch Doctor
Previous View +F5
Isometric View
How To...

Figure 2.26 - General dimension tool; 2D sketch panel on left, or context menu on right

7. Now add dimensions. Select the General Dimension tool, either from the panel bar or from the context menu, accessed by clicking the right mouse button in the drawing area. See Figure 2.26.
8. Click on the circle, and then move the mouse. A dimension appears. Click the left mouse button to accept and position the dimension.

   TIP: To avoid having to click on placed dimensions in order to edit them, do the following: from the menu bar select Tools > Application Options. . . . The Options dialog box appears. Click on the sketch tab. About half way down on the left, turn on "Edit Dimension When Created" by clicking the adjacent check box. Click OK. Now whenever dimensions are added, the edit Dimension box appears automatically.

9. To edit the dimension, click once on the dimension. The Edit Dimension box appears. Enter 3, and then click the green check box. See Figure 2.27. The diameter of the circle is now 3.

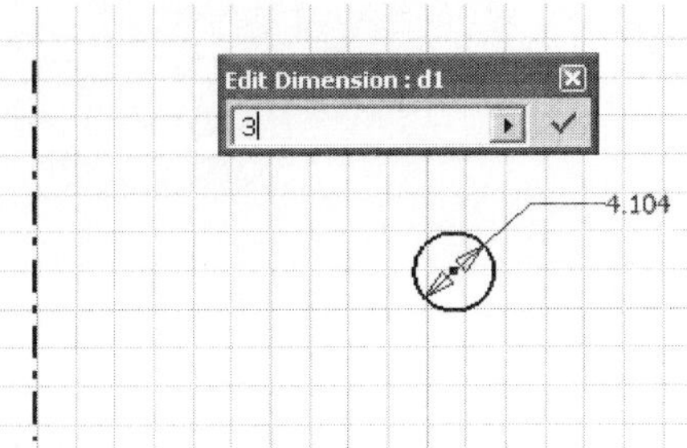

Figure 2.27 - Dimension circle

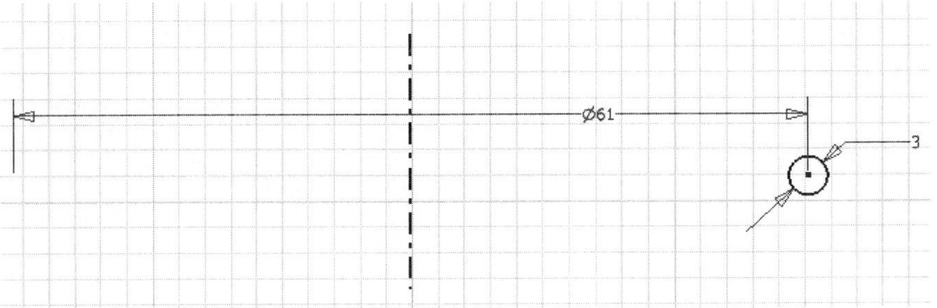

Figure 2.28 - Completed sketch

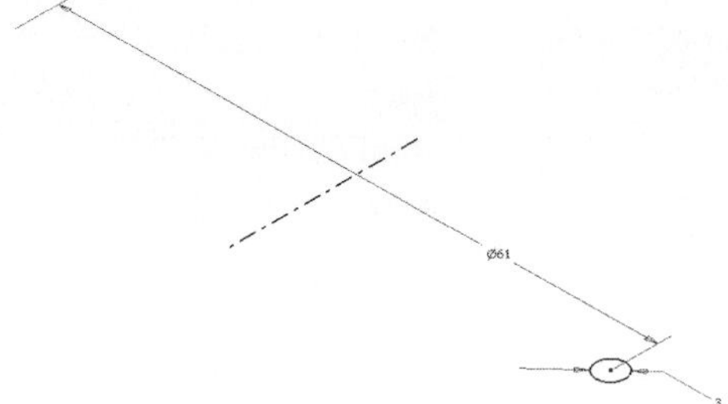

Figure 2.29 - Completed sketch, isometric view

10. Next add the 61 mm diameter dimension. The General Dimension tool should still be active. Start by clicking the centerline, then by clicking on the circle (or the center of the circle). A third click places the dimension. The Edit Dimension box should then appear. Enter 61, then either hit the Enter key or click the Apply (green check mark) button. Your sketch should now be similar to Figure 2.28.
11. To exit General Dimensioning, right-click, then select Done.
12. To exit Sketch mode, right-click, then select Finish Sketch. The grid disappears, indicating that you are no longer in sketch mode. Notice also that the panel bar has also changed; the Sketch tools are no longer visible. In their place are the Part Feature tools.
13. Before creating the revolved feature, change to an isometric view. This can be done by right-clicking, and then selecting Isometric View. Your sketch should now look like Figure 2.29.

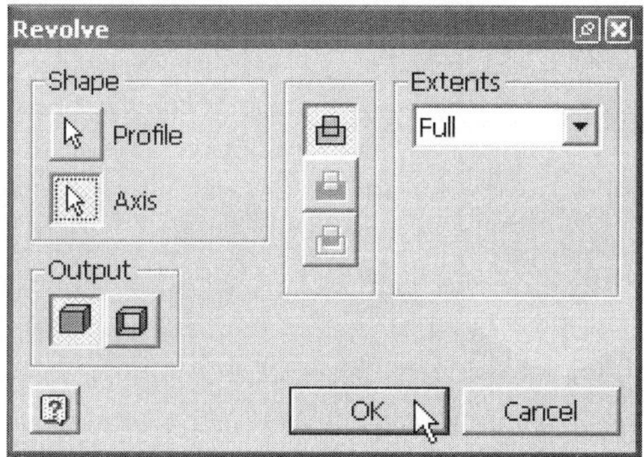

Figure 2.30 - Revolve dialog box

Figure 2.31 - O-ring

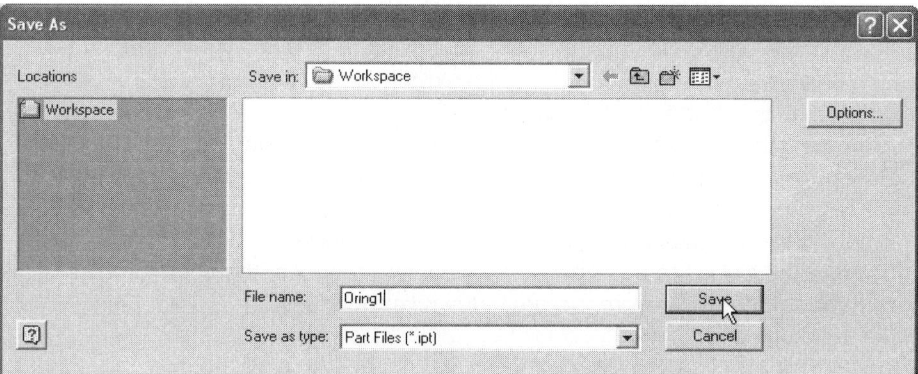

Figure 2.32 - Save in ball valve workspace

14. Select Revolve from the panel bar. The Revolve Dialog Box appears, as shown in Figure 2.30. Accept the defaults by clicking the OK button.
15. Right-click, and then select Isometric View in order to center the image. Your screen should resemble Figure 2.31.
16. Save the file by choosing File > Save from the menu bar. The dialog box shown in Figure 2.32 appears. If you successfully completed Tutorial 1 at the end of Chapter 1, the Save in: field will display the Workspace folder inside the Ball Valve project folder. In the File name field, enter the name **Oring1,** and click the Save button. This completes Tutorial 2.

Seat — TUTORIAL 3

Build Strategy

1. Create the sketch shown in Figure 2.33. The sketch consists of a profile and a centerline.
2. Revolve the profile about the centerline (Figure 2.34).

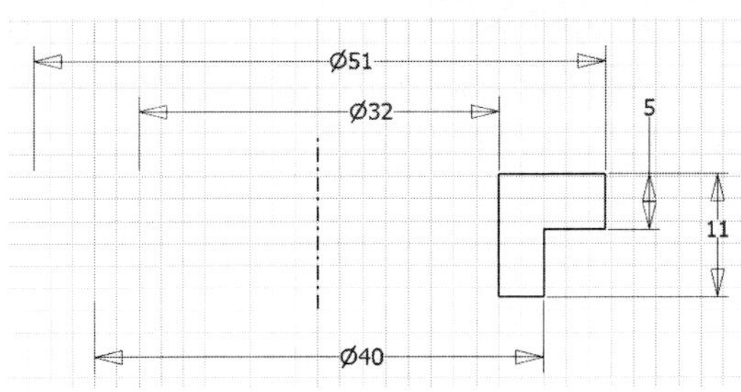

Figure 2.33 - Dimensioned base feature sketch

Figure 2.34 - Base revolved feature

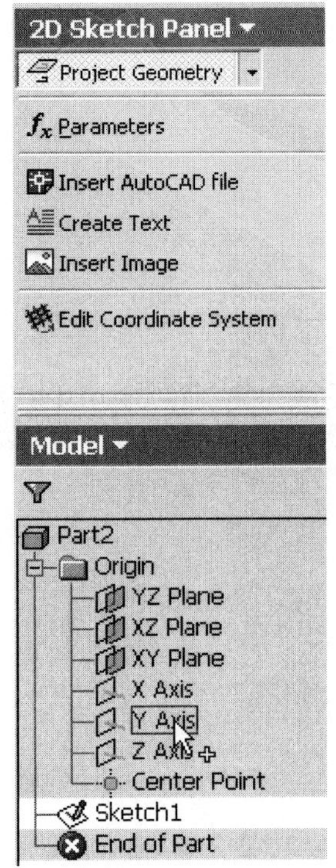

Figure 2.35 - Project geometry, expand origin folder, and select Y Axis

Detailed Modeling Steps

1. Start a new metric part file.
2. For this sketch we will project some construction geometry (available in the Origin folder of the Browser) on to the sketch plane. On the panel bar, scroll down until you find Project Geometry.
3. After selecting Project Geometry, expand the Origin folder in the part browser. Project the Y Axis on to the sketch plane by selecting it (see Figure 2.35) with the left mouse button. The graphics area should now be similar to Figure 2.36. Right-click, then select Done to exit projection mode.
4. Change the Y Axis to a centerline by selecting the axis, then changing its linetype to centerline from the drop-down list on the Standard bar (see Figures 2.18 and 2.25).
5. Use the Line tool to sketch the closed profile shown in Figure 2.37. When sketching the line segments, either parallel or perpendicular constraint symbols should be visible, indicating that all lines are either horizontal or vertical. In Figure 2.37, the sketch is started in the upper left corner and proceeds counterclockwise. Upon reaching the upper right corner (shown on the left), both a yellow circle and a dashed horizontal projection line from the starting point appear, indicating that the Y values of these two points are equivalent. Left-click to select this point. Now move the mouse to the sketch start point. A green circle and the coincident symbol (shown on the right) indicate that by left-clicking now, the sketch start and endpoints will be coincident, resulting in a closed profile.

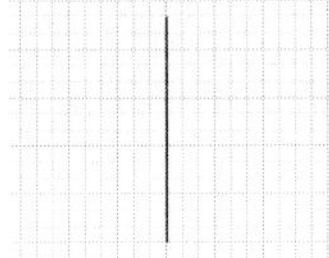

Figure 2.36 - Projected Y Axis

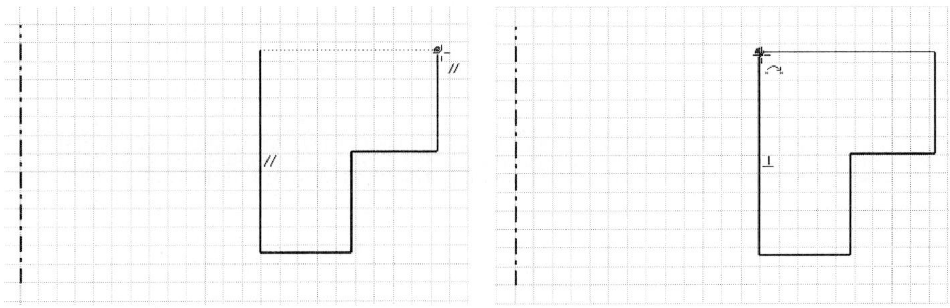

Figure 2.37 - Base feature sketch

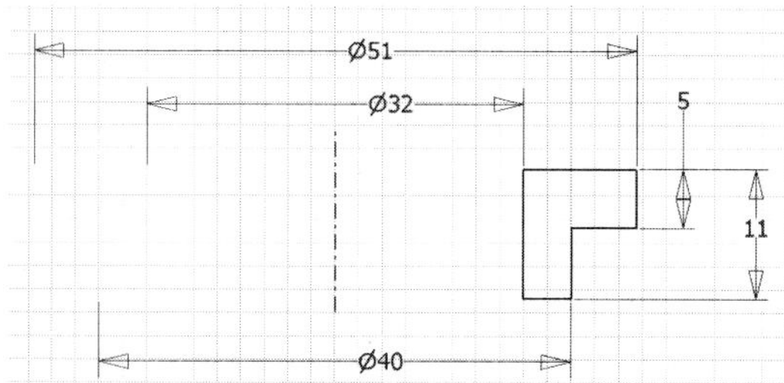

Figure 2.38 - Dimensioned base feature sketch

6. Use the General Dimension tool to add the five dimensions shown in Figure 2.38.
7. Right-click, then select Done.
8. Right-click, then select Finish Sketch to exit sketch mode.
9. Right-click and then select Isometric View.
10. Select the Revolve tool from the panel bar. Accepting the defaults, click OK.
11. To center the feature in the graphics area, right-click, and then select Isometric View once again.
12. Save the file as **Seat** in the Ball Valve project folder. This part will be completed in Chapter 3.

Bumper — Tutorial 4

Build Strategy

1. Sketch and dimension a circle (Figure 2.39).
2. Extrude the circle (Figure 2.40).

Detailed Modeling Steps

1. Start a new metric part file
2. Use the Project Geometry tool (available on the panel bar) to project the X and Y Axes, as well as the Center Point (from the Origin folder on the Browser), on to the sketch plane (Figure 2.41). Right-click and then select Done to exit Project Geometry.

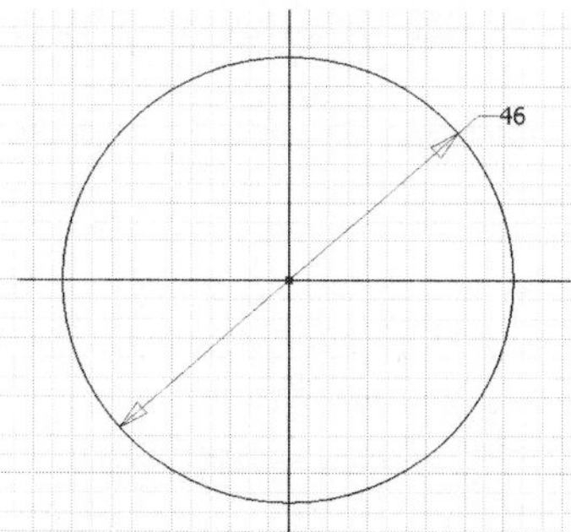

Figure 2.39 - Dimensioned base feature sketch

Figure 2.40 - Extruded base feature

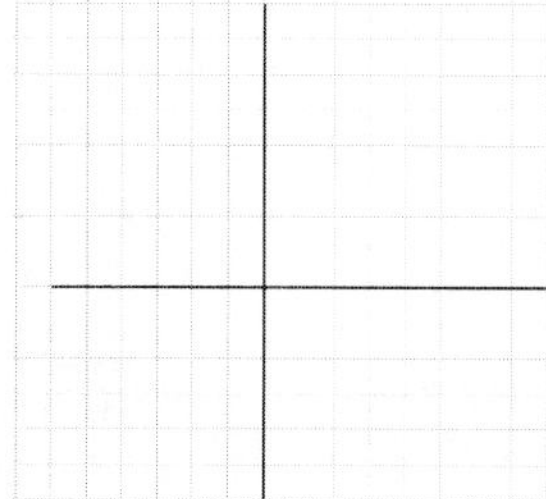

Figure 2.41 - Projected X and Y Axes, Center Point

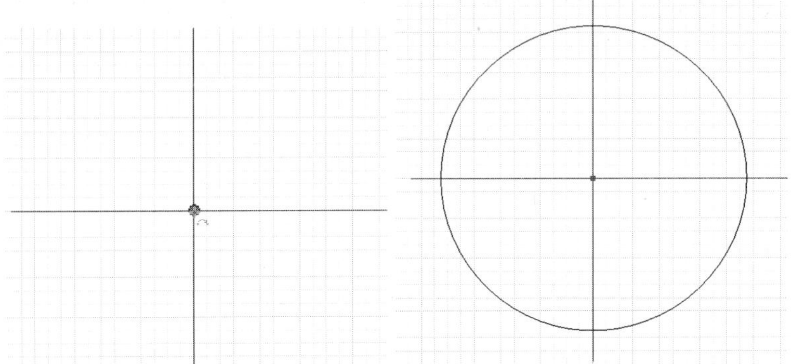

Figure 2.42 - Circle, center at origin

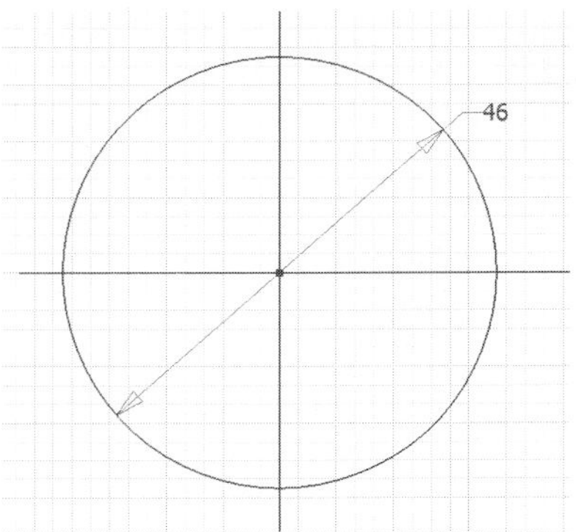

Figure 2.43 - Dimensioned base feature sketch

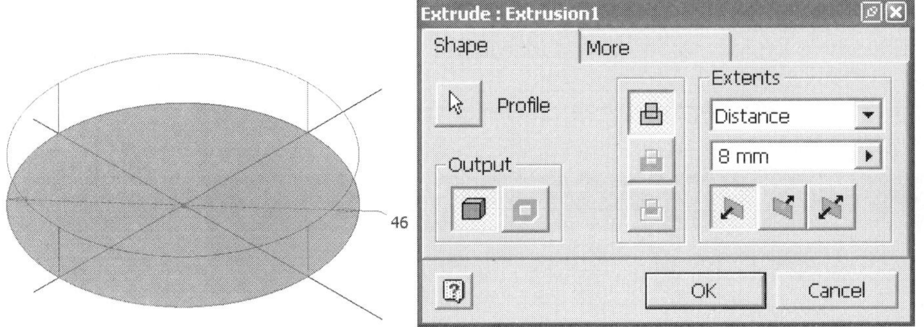

Figure 2.44 - Base feature extrude operation

Figure 2.45 - Base feature

3. Use the Center point circle tool to draw a circle. Left-click to locate the circle's center point at the origin of the projected coordinated system (You will see a green circle and the coincident constraint symbol). Move the mouse, and then left-click again to select a radius. Right-click, and then select Done to exit Center point circle. See Figure 2.42.
4. Use the General Dimension tool to dimension the circle (diameter = 46). Your sketch should now resemble Figure 2.43.
5. Right-click, Finish Sketch.
6. Right-click, Isometric View.
7. From the Features panel bar, select Extrude. Enter an extrusion distance of 8, and then select OK, as shown in Figure 2.44.
8. Right-click, Isometric View. The model should now be similar to Figure 2.45. We will complete the remaining features in the next chapter.
9. Save the part using the name **Bumper.**

Handle — Tutorial 5

Build Strategy

1. Create sketch (Figure 2.46).
2. Extrude the profile (Figure 2.47).

Detailed Modeling Steps

1. Start a new metric part file.
2. Use Project Geometry tool to project the X and Y Axes and Center Point onto sketch plane.
3. Using Center point circle tool, sketch a circle with center at origin of projected coordinated system (Green circle, coincident constraint symbol) and a radius of about 8.
4. Use the Offset tool available on the 2D Sketch Panel to offset 3 circles concentric to the first circle. Offset circles should be larger than the original circle. See Figure 2.48.
5. Use the Line tool to sketch four lines, as shown in Figure 2.49. These lines should either be horizontal or vertical (e.g., horizontal, vertical, parallel, or perpendicular) *and* they should all be coincident to both the largest and the third largest circles. The constraints applied to the four lines for this particular sketch are also displayed.
6. Using the Trim tool on the panel bar, modify the sketch so that it appears like the one shown in Figure 2.50. Note that in using the Trim tool, it is only necessary to trim the bits that are to be removed from the sketch. There is no need to define cutting edges.

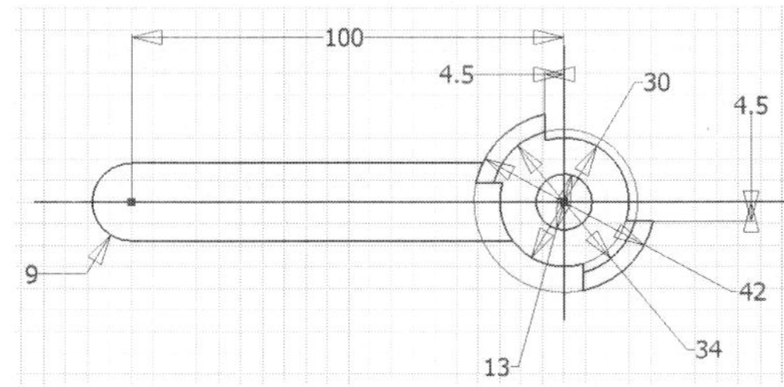

Figure 2.46 - Dimensioned base feature sketch

Figure 2.47 - Base feature

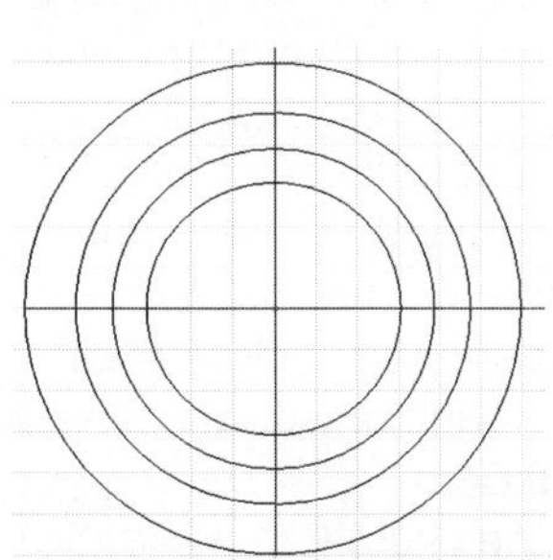

Figure 2.48 - Circle, 3 offsets

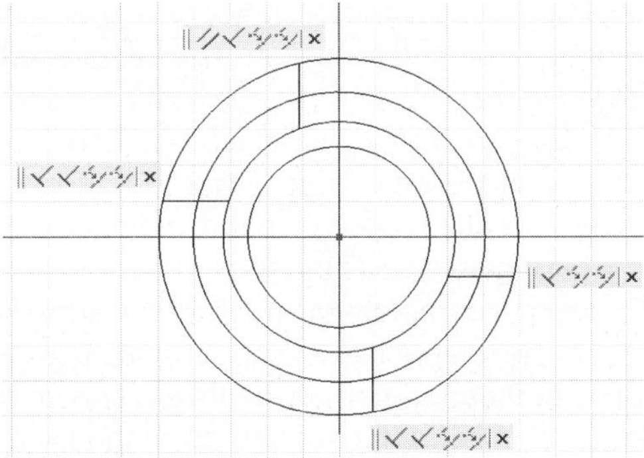

Figure 2.49 - Add coincident lines

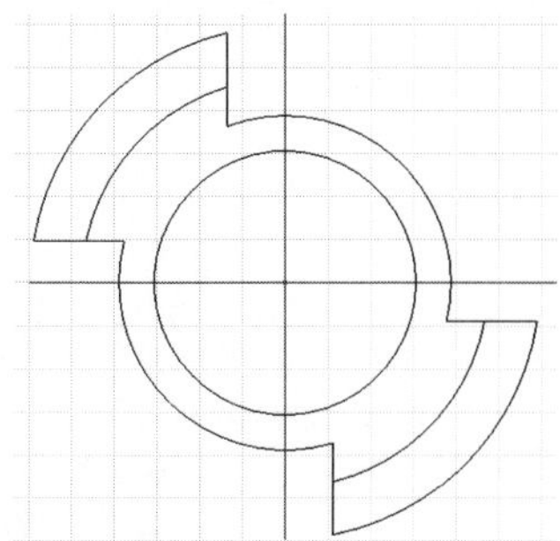

Figure 2.50 - Trim

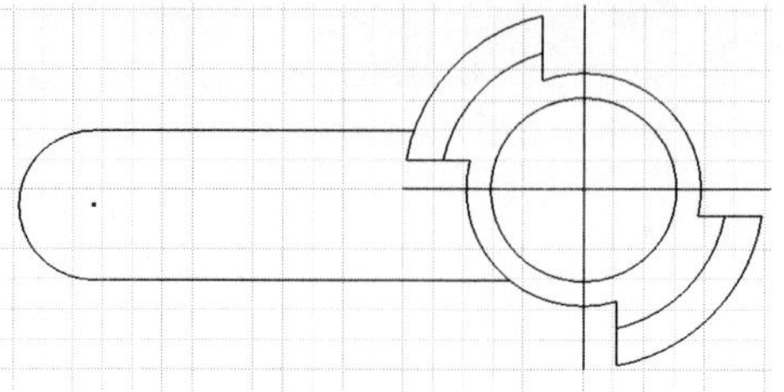

Figure 2.51 - Coincident lines, tangent arc

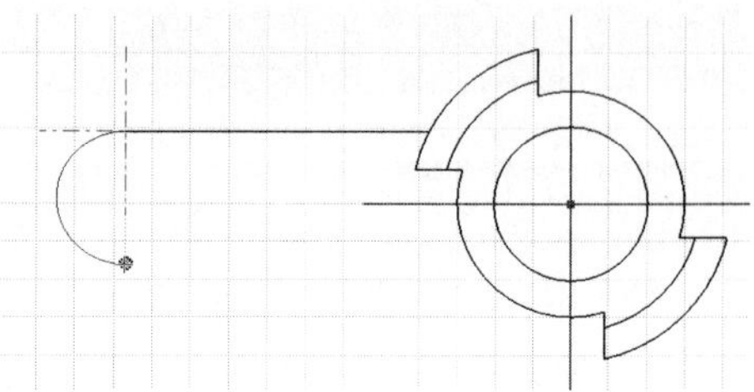

Figure 2.52 - Tangent arc construction

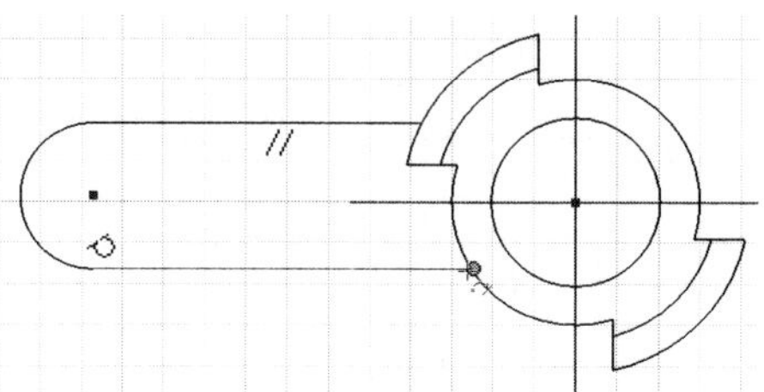

Figure 2.53 - Completed sketch

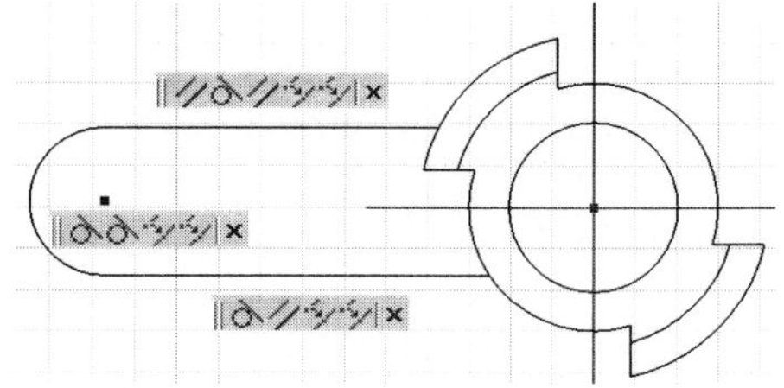

Figure 2.54 - Applied constraints

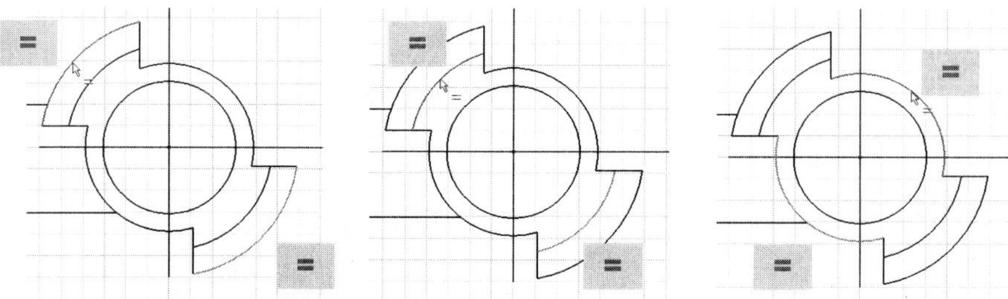

Figure 2.55 - Add equal constraints to 3 arc pairs

7. Save the file as **Handle.** It is recommended that the file be regularly saved every five to ten minutes.
8. Use the Line tool to add the geometry shown in Figure 2.51. Start with the upper horizontal line. The first endpoint must be coincident with the largest diameter circle. Now make a horizontal line by selecting a second endpoint. The length is not important. With the cursor over this endpoint, depress the left mouse button and drag down and to the left. The Line tool changes from line to arc mode, as seen in Figure 2.52. Drag the arc until it is similar to that shown in the figure (i.e., half circular arc visible, vertical projection line from the upper arc endpoint), and then release the left mouse button. This fixes the other endpoint of the arc. With the Line tool still in operation, move the cursor to the right until the constraints shown in Figure 2.53 (Tangent , parallel , coincident) are visible. Click the left button to complete the second horizontal line segment. Right-click, then select Done to end the Line command. The geometric constraints that have been applied to these sketch elements are shown in Figure 2.54.
9. Additional constraints need to be manually added to the sketch. In using the trim tool, three pairs of arcs were created. Equal constraints need to be added to each of these arc pairs. See Figure 2.55. In addition, the four line segments connecting the inner and outer arcs should all be of equal length. See Figure 2.56.

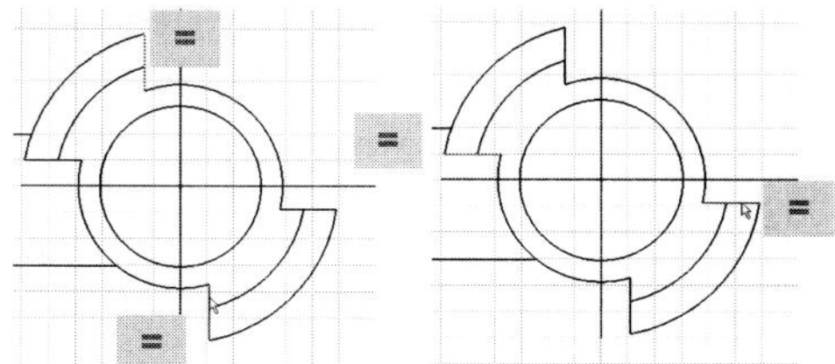

Figure 2.56 - Add equal constraints to 2 line pairs

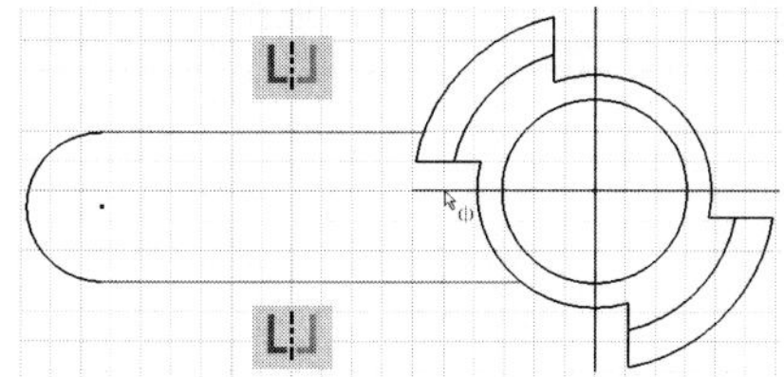

Figure 2.57 - Add symmetric constraint

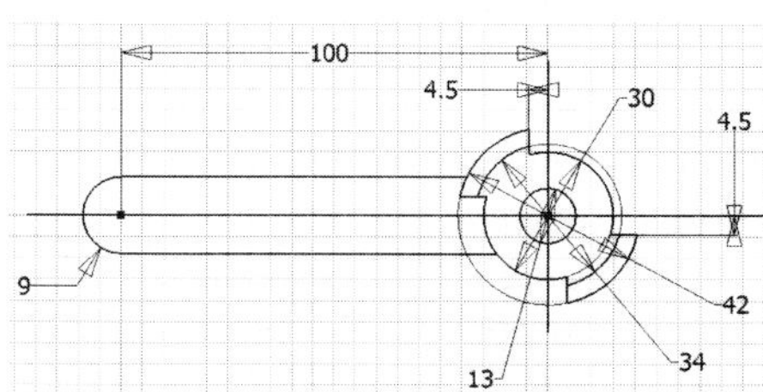

Figure 2.58 - Dimensioned base feature sketch

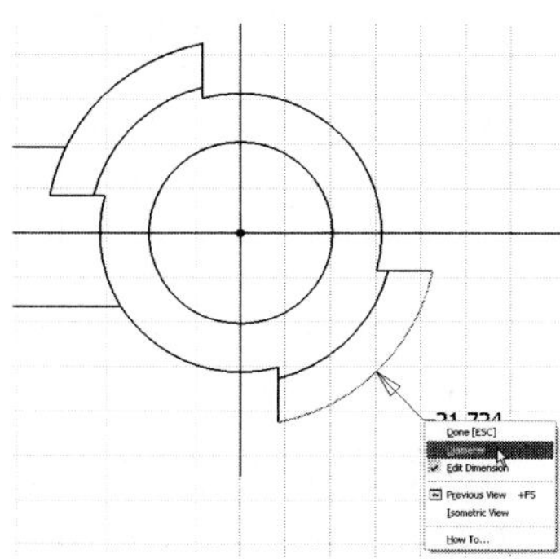

Figure 2.59 - Override dimension type

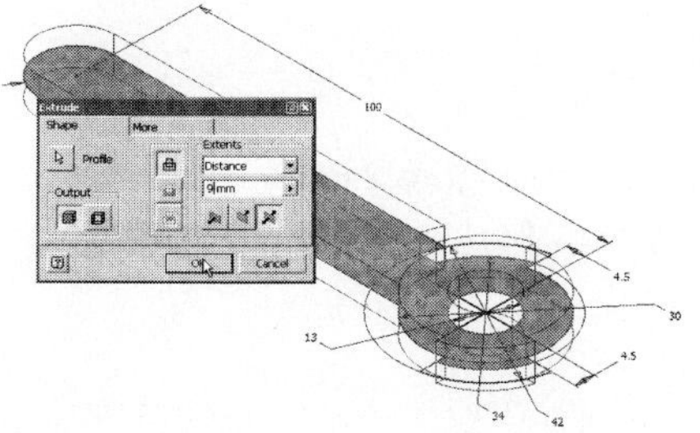

Figure 2.60 - Extruded base feature construction

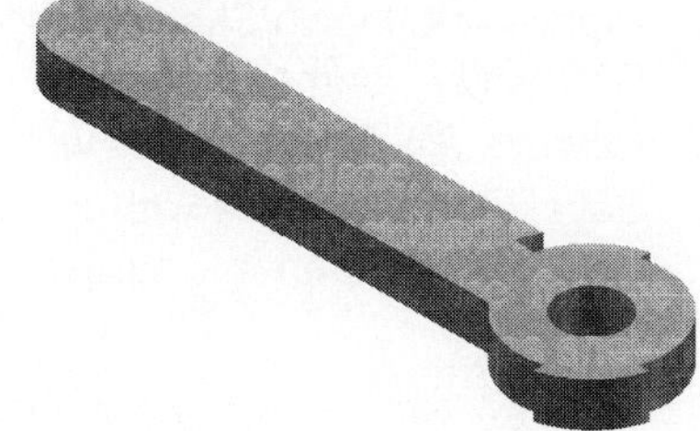

Figure 2.61 - Base feature

10. In addition to the equal constraints, a symmetric constraint should also be added to the pair of horizontal lines shown in Figure 2.57. After selecting the two lines, select the projected X Axis (symmetry axis).
11. Use the General Dimension tool to add the dimensions shown in Figure 2.58.

    TIP: When dimensioning an arc, the radius is specified by default. To change to diameter, right-click and select Diameter while in dimensioning mode, as shown in Figure 2.59.
12. Right-click, Finish Sketch.
13. Right-click, Isometric View.
14. Extrude the two areas shown in Figure 2.60. Because there is more than one area that can be extruded, it may be necessary to click on the Profile button, and then select the areas. The extrusion distance is 9, and the profiles are extruded symmetrically about a mid-plane.
15. After selecting OK, the screen should resemble Figure 2.61.
16. Save the file. The other part features will be completed in Chapter 4.

TUTORIAL 6 Packing Nut

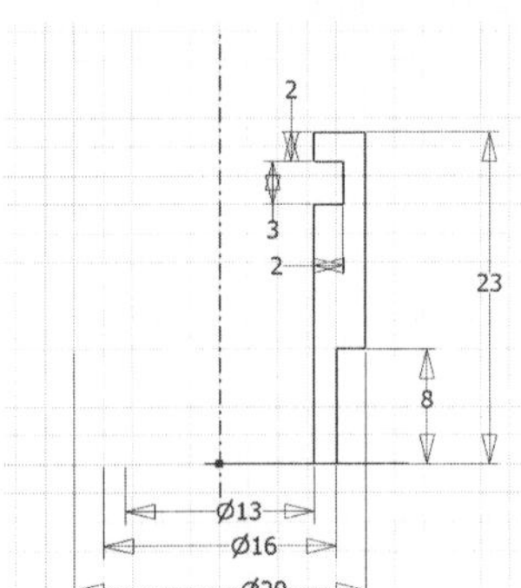

Figure 2.62 - Dimensioned base feature sketch

Figure 2.63 - Revolved base feature

Build Strategy

1. Sketch profile (Figure 2.62).
2. Revolve (Figure 2.63).

Detailed Modeling Steps

1. Start a new metric part file.
2. Project X, Y Axes and Center Point.
3. Use the drop-down list on the standard toolbar to change the projected vertical axis to a centerline (see Figures 2.18 and 2.25).
4. Use Line tool to sketch the closed profile in Figure 2.62. The completed sketch should resemble Figure 2.64.
   HINTS:
   - Use the length information provided on the status bar (at the lower right of the screen) to approximate the line segment lengths as you sketch. Notice that the sketch coordinate system origin is the projected center point. See Figure 2.16 along with the accompanying discussion for more information.
   - The bottom horizontal edge of the profile should be coincident with the projected X Axis.
5. Use General Dimension tool to add dimensions. The completed sketch should resemble Figure 2.65.
6. Right-click, Finish Sketch
7. Right-click, Isometric View
8. Revolve. The graphics window should resemble Figure 2.66.
9. Save the part file with the name **Packing Nut.** Additional features will be added to this part in Chapter 3.

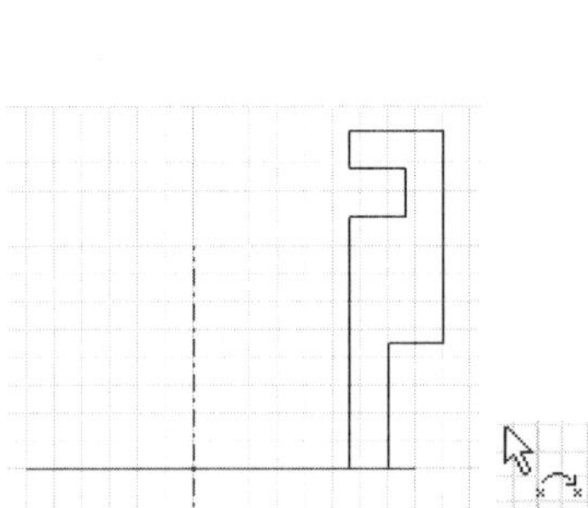

Figure 2.64 - Rough sketch

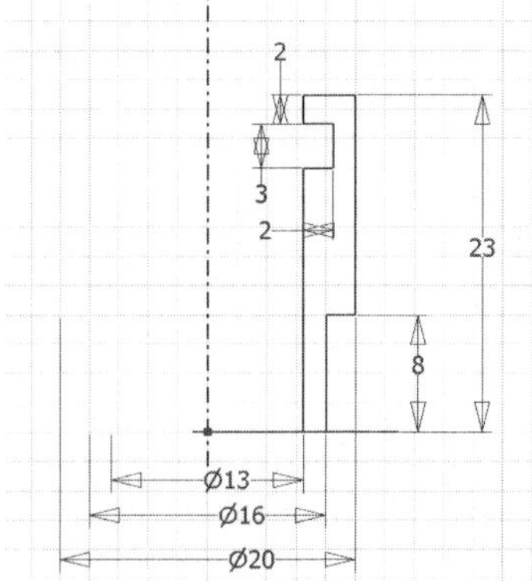

Figure 2.65 - Base sketch

Figure 2.66 - Revolved base feature

Body — TUTORIAL 7

■ Build Strategy

1. Sketch (Figure 2.67).
2. Revolve (Figure 2.68).

■ Detailed Modeling Steps

1. Start a new metric part file.
2. Project X and Y Axes and Center Point.
3. Change the vertical Y Axis to a centerline (see Figure 2.25).
4. Use the Line tool to sketch the closed profile shown in Figure 2.67. The completed sketch should resemble Figure 2.69.
   HINTS:
   - Use the length information provided on status bar to approximate the line segment lengths as you sketch. Notice that the sketch coordinate system origin is the projected center point. See Figure 2.16 and the accompanying discussion for more information.
   - The bottom horizontal edge of the profile should be coincident with the projected X Axis.

   Also note that:
   - The inside arc is tangent to the (upper) vertical line segment
   - The two outside coincident arcs are tangent to one another, as well as to the adjoining vertical line segments.
5. Use the General Dimension tool to constrain the sketch. When finished, the sketch should look like Figure 2.70.
6. Right-click Finish Sketch.
7. Right-click Isometric View.
8. Revolve. When finished the feature should look like Figure 2.71.
9. Save the part file as **Body.** Several additional features will be added in the next chapter.

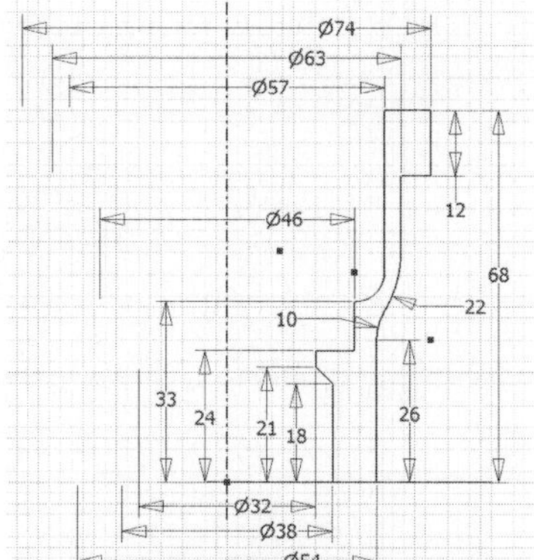

Figure 2.67 - Dimensioned base feature sketch

Figure 2.68 - Base revolved feature

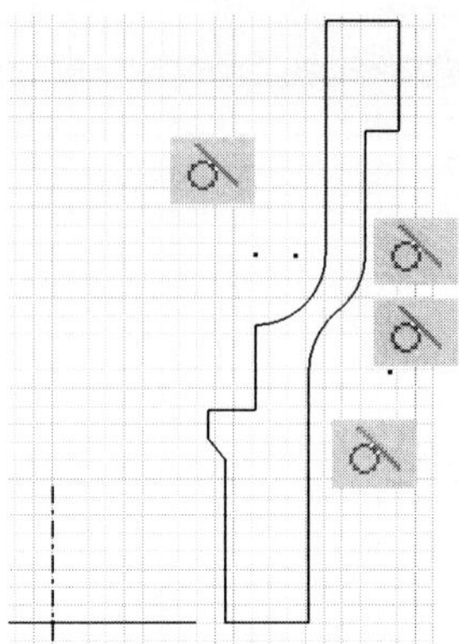

Figure 2.69 - Rough sketch

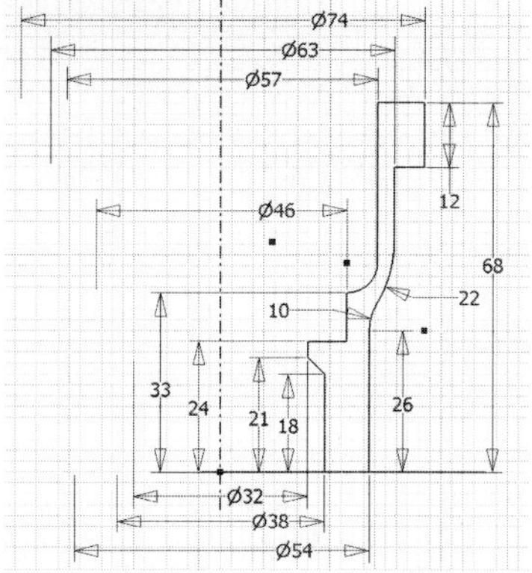

Figure 2.70 - Fully constrained sketch

Figure 2.71 - Base revolved feature

QUESTIONS

1. T F The sketching tools available in Inventor can be accessed via the 2D Sketch panel or by right-clicking the mouse button in the graphics window to view the context menu.
2. T F The geometric constraints available in Inventor can only be accessed from the 2D Sketch panel.
3. Which is a method for accessing the General Dimension tool in Inventor?
   a. Panel bar
   b. D from the keyboard
   c. Context menu
   d. All of the above
4. T F Despite the advantages of projecting work features while in sketch mode, it will make the assembly process more difficult.
5. Which of the following is a linetype available in Inventor:
   a. Phantom
   b. Invisible
   c. Centerline
   d. Baseline
6. T F The Extrude tool needs which of the following inputs to create a feature:
   a. Closed profile
   b. Axis
   c. Path
   d. Distance
   e. a and b
   f. a and c
   g. a and d
7. T F The Revolve tool needs which of the following inputs to create a feature:
   a. Closed profile
   b. Axis
   c. Path
   d. Distance
   e. a and b
   f. a and c
   g. a and d

PROBLEMS

1. Use a metric part template file to create the sketch and revolved base feature shown in the figure below. Note that the center point of the circle should be coincident with the projected X axis. Save the part as **Oring2** in the BallValve Project Workspace.

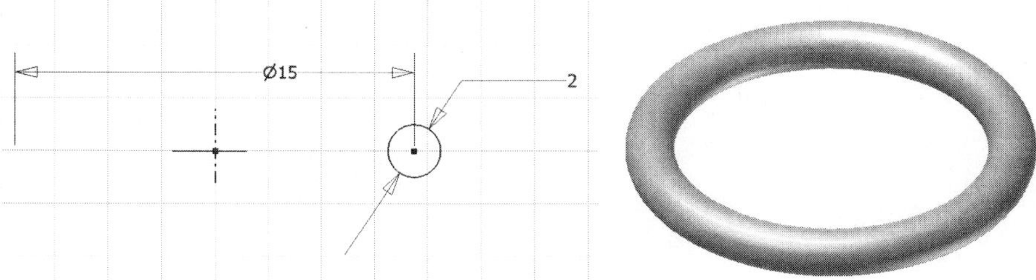

2. Use a metric part template file to create the sketch and revolved base feature shown in the figure below. Note that the center point of the circle should be coincident with the projected X axis. Save the part as **Oring3** in the BallValve Project Workspace.

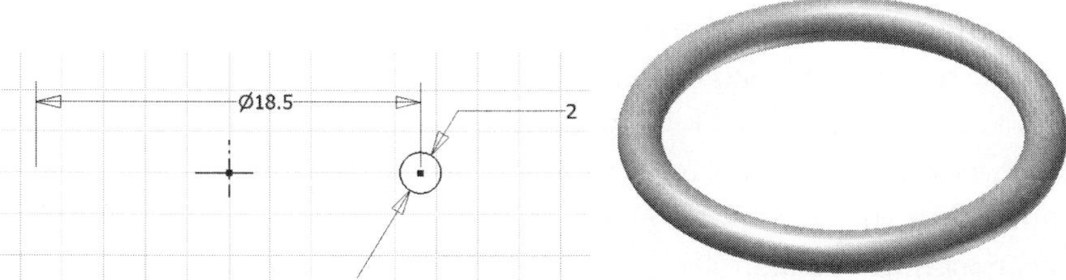

3. Use a metric part template file to create the sketch and revolved base feature shown in the figure below. Save the part as **Body Cap** in the BallValve Project Workspace.

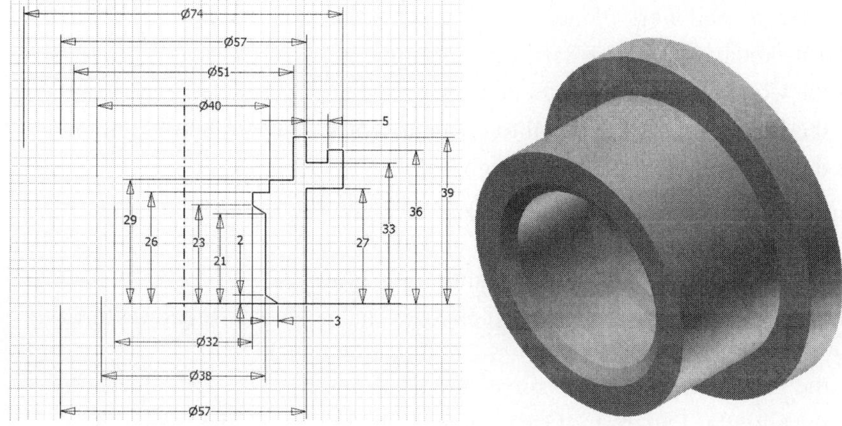

3 CHAPTER

Feature Creation

LEARNING OBJECTIVES

- List common sketched features
- Name the two possible output options for sketched features
- Use Boolean operations to combine sketched features in order to create parts
- Use the termination options for the Extrude and the Revolve tools
- List the common placed features
- List common work features
- Create a sketched feature from a closed profile sketch made on a:
  - Default reference work plane
  - Planar face of an existing feature
  - User-created work plane
- Control sketch and feature visibility from the browser
- Use the Chamfer tool to create a beveled edge
- Use the Point, Hole Center tool to create a center point for a hole feature
- Use the Hole tool to create a through hole
- Use the Polygon tool to create a polygon sketch
- Use the Extrude tool to add features to a part
- Use the Fillet tool to create a constant radius fillet
- Use the Thread tool to add internal and external cosmetic thread features to a part
- Use the Hole tool to create a threaded hole
- Use the Circular Pattern tool to array an existing feature about an axis
- Use the Work Plane tool to create an offset work plane
- Use the Hole tool to create a counterbore hole
- Use the Rectangle tool to create rectangular sketch geometry

Introduction

Parts are typically composed of several features. Once the base feature is constructed, additional features are added. Recall that part geometry is made up of two different kinds of features, sketched and placed. In addition, there is also a third kind of feature

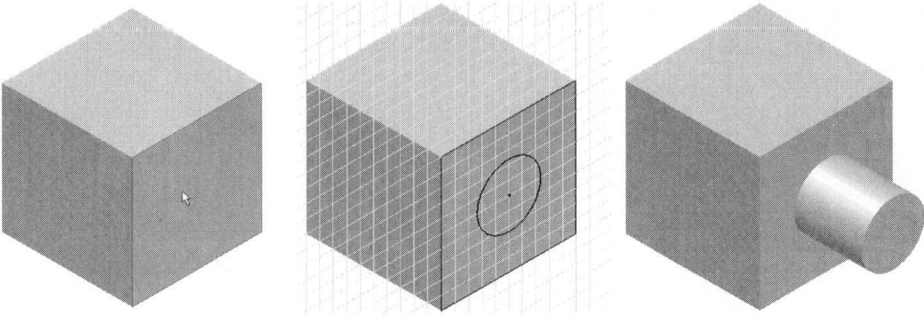

Figure 3.1 - From left to right; base feature, sketch plane on face, second feature addition

called a *work feature.* These work features are used to assist in the construction of other features.[1] They do not add or subtract volume from the part. All of the different kinds of features, sketched, placed, and work, appear in the part browser.

In this chapter basic feature creation tools are discussed. Other more advanced feature creation tools are described in the chapter on advanced part modeling. In addition, a new hybrid solid/surface modeling environment, including several new feature creation tools, has been added in Autodesk Inventor Release 6. These tools will be discussed in Chapter 10 on hybrid modeling.

Sketched Features

We have already used the two most common sketched features, Extrude and Revolve, to create base features. An extruded feature sweeps a profile along a linear path. This path is typically perpendicular to the sketch plane. A revolved feature sweeps the profile about an axis along a circular path. Other sketched features, including Sweep, Loft, Coil, Rib, and Split, will be discussed in Chapter 5 on advanced part modeling.

Once a base feature is created, other features are added sequentially until the part is complete. In the event that a sketched feature is to be added, it is first necessary to select a sketch plane. This sketch plane is commonly a planar face of the existing part geometry. Figure 3.1 depicts this process, where one face of the base feature (i.e., the cube) is used as a sketch plane to create a second feature, a cylinder.

Whenever a sketched feature tool is selected, a dialog box opens. Although the required inputs vary depending upon the type of feature, there are some inputs common to a number of different features. These include output, shape, operation, and extents. Dialog boxes for the Extrude, Revolve, and Sweep features are shown in the Figure 3.2.

Shape

Shape is used to specify the sketch geometry employed in the creation of the feature. At least one profile sketch must be available to create a sketched feature. If the sketch

[1]The default reference work features discussed in Chapter 1 are included with all Inventor files, but the user is also capable of creating his/her own work features.

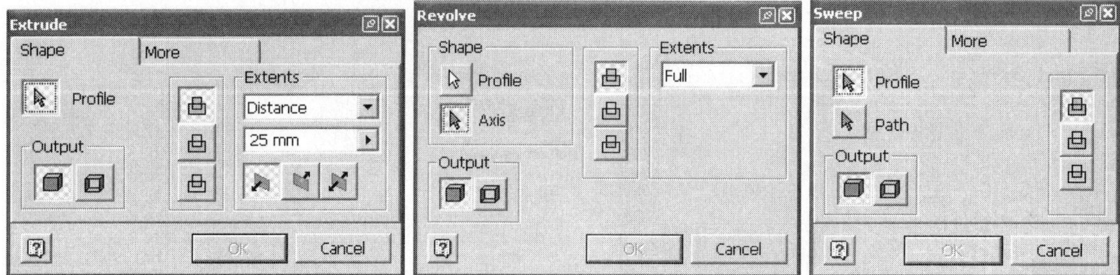

Figure 3.2 - Extrude, revolve, and sweep *dialog* boxes

contains a single profile, it is selected by default. If the sketch contains multiple closed profiles, the user must select the intended profile(s).

An axis of revolution must be specified as part of the shape definition for the Revolve feature. If the axis is a centerline, it is selected by default. Otherwise it must be manually selected.

A path must be included as part of the shape definition for the Sweep feature. If the path sketch contains a single path, it is selected automatically. If the sketch contains multiple paths, the user must select the intended profile(s).

Both the Extrude and Sweep features include a taper option (available from the More tab) as part of the shape definition.

Output

The output option refers to whether the sketched feature is to be created as a solid or as a surface. This option is by and large new to Release 6. Surface features will be covered in Chapter 10. Sketched features created in the tutorials in all other chapters are solid features. Note that while all[2] solid sketched features must be based on a closed profile sketch, open profiles can be used to create surface features.

Operation

Simple features are combined with one another using *Boolean operations.* There are three Boolean operations: join, cut, and intersect.

Figure 3.3 shows the result of the join, cut, and intersect operations when the base feature A is a cube and the new feature B is a sphere. Join (or union) combines the volume of the new feature with the volume of the part. Any volume common to both is only counted once in the updated part. The cut (subtract) operation removes volume. Any new feature volume common to the part will be removed in the updated part. In the intersect operation only the volume common to both features is retained in the resulting part.

Extents (or Termination)

Whereas sketches are two-dimensional, features are three-dimensional shapes. Extents options are used to specify the termination of the third dimension of a feature. The

[2]Apart from the Rib feature, for which an open profile sketch is acceptable.

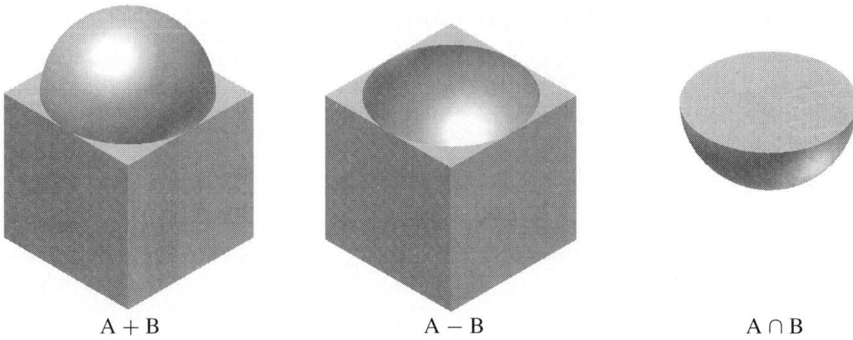

Figure 3.3 - Boolean operations; join, cut, intersect

different extents options for an extruded feature are shown in Figure 3.4. Most of these extents options will be used in the tutorials.

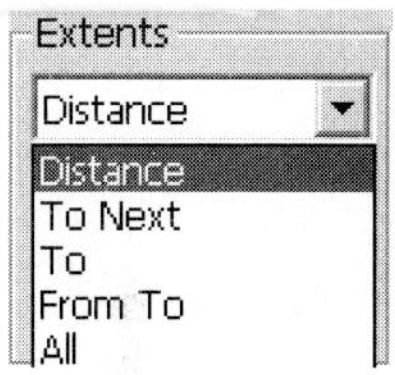

Figure 3.4 - Termination options available for extrude

> TIP: Entering "feature extents" on the Index tab of the Inventor Help links to a page with descriptions of the different terminations.

Some extents options, in particular Distance (Extrude and Sweep) and Angle (Revolve), also require that a direction from the sketch plane be specified. This can be in either direction from the sketch plane, or equally in both directions. Upon selecting one of these direction options, the feature to be created is either previewed or an arrow is displayed indicating the direction.

Placed Features

Placed features, once again, are features that do not require a sketch in order to be created. Placed features include Hole, Shell, Thread, Fillet, Chamfer, and Face Draft. Selecting any of these placed feature tools in Inventor opens a dialog box where the feature parameters are specified. The Hole, Thread, Fillet, and Chamfer feature tools are all used in tutorials appearing at the end of this chapter. The Shell and Face Draft feature tools are used in the tutorials appearing at the end of the chapter on advanced modeling.

In point of fact the Hole tool in Inventor requires that a hole center (made using the Point, Hole Center tool) be first added on a sketch plane before the hole is created. Consequently the Hole feature tool in Inventor is not strictly a placed feature, although it will be treated as such in this book.

Work Features

Work features are abstract geometric tools used to create and position other features. Work features are used in situations where the existing part geometry is insufficient for constructing a required feature. Work features include Work Plane, Work Axis, and Work Points. We have already seen that the Origin folder in the part browser provides several default reference work features.

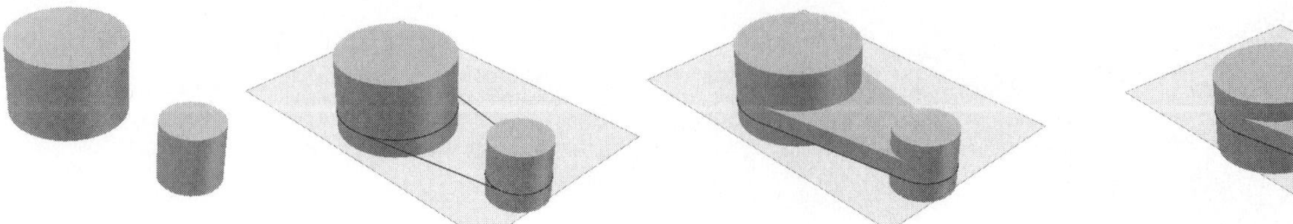

Figure 3.5 - Example where work plane is required to create a feature

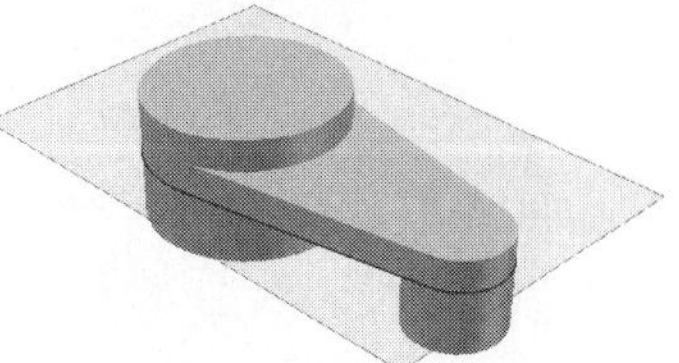

Figure 3.6 - Sketched feature dependent upon work feature

Work Plane

A work plane is an infinite construction plane that can be placed in space at any orientation. Work planes can be designated as sketch planes. This is useful in situations where the existing feature geometry fails to provide a suitable sketch plane for the next feature to be added. An example of this is shown in Figure 3.5. Starting with two cylindrical features, a work plane is created parallel to and above the bottom base of the cylinders. The work plane is then designated as a sketch plane, the sketch is made, and an extruded plate feature is derived from the sketch.

Sketched features are parametrically attached to related work features. This is demonstrated in Figure 3.6, where, by raising the height of the work plane, the height of the sketched plate feature is also raised.

Work planes are established in the graphics area by selecting certain combinations of existing planes, edges, and/or vertices that serve to define the orientation of the work plane. There are several different ways to specify this work plane orientation. Some of these techniques are demonstrated in the tutorials.

TIP: For a complete description of the different ways to define a work plane, use the Autodesk Inventor Help. This topic is covered at the bottom of the Work Plane Reference page. Also try the Visual Syllabus. Several Show Me animations illustrating the different ways to create a work plane are available on the Part Modeling palette. Look for the Work Plane icon.

Work Axis and Work Points

A work axis is an infinite construction line that serves as a centerline, line of symmetry, etc. A work point is a parametric construction point. As of Release 6, there is also a fixed, nonparametric, or grounded work point. Examples depicting the usage of work axes and work points are provided in the tutorials. Work points and grounded work points are also discussed in Chapter 10.

TUTORIAL 3 Seat (continued)

In Chapter 2 the base feature shown in Figure 3.7 had been created and saved. We will now add another feature in order to complete the part.

1. Open the part file **Seat.**
2. Use the 3D Rotate tool to change the view to something like that shown in Figure 3.8.

Figure 3.7 - Base feature for Seat part

Figure 3.8 - View rotated

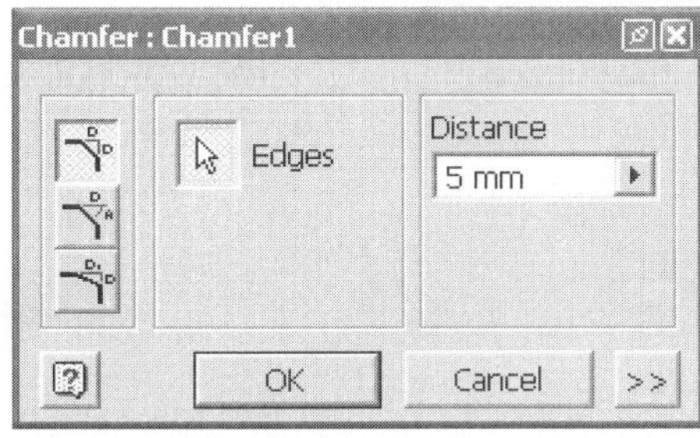

Figure 3.9 - Chamfer feature operation

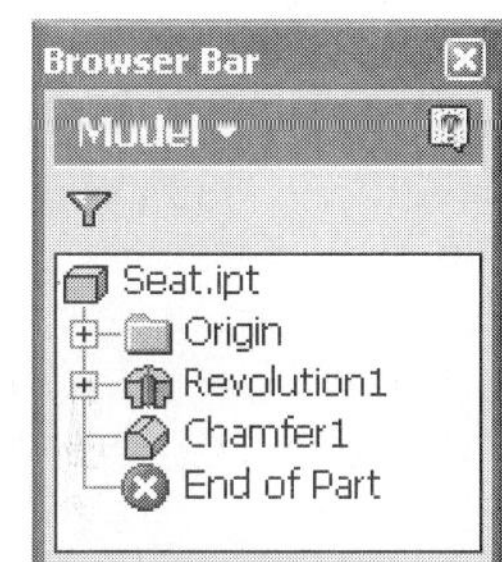

Figure 3.10 - Completed Seat part

3. From the Part Features panel bar, find and select Chamfer. The Chamfer dialog box opens. Enter a chamfer distance of 5, and select the edge highlighted in Figure 3.9, then click OK. Note that the chamfer tool is used to bevel an edge.
4. The part should now resemble Figure 3.10. Save the file. This completes Tutorial 2.

Bumper (continued) — TUTORIAL 4

In Chapter 2 the base feature shown in Figure 3.11 had been created and saved. We will now add two more features in order to complete the part.

1. Open the part file **Bumper.**
2. Right-click, select New Sketch. Select the top face of the cylinder, as shown in Figure 3.12.
3. We are now ready to sketch a second feature on the top face of the cylinder. Since the next feature is a hole, the sketch geometry consists of nothing more than the center point of the hole. Select Point, Hole Center from the 2D Sketch panel. The center point of the original sketched circle is visible, as can be seen in Figure 3.12. Place the hole center so that it is coincident with this center point (green circle, coincident constraint) as shown in Figure 3.13.
4. Right-click, Done.
5. Right-click, Finish Sketch.

Figure 3.11 - Extruded base feature

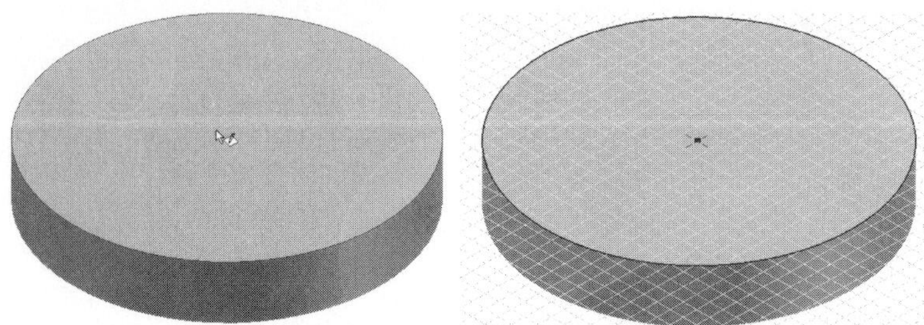

Figure 3.12 - New sketch plane

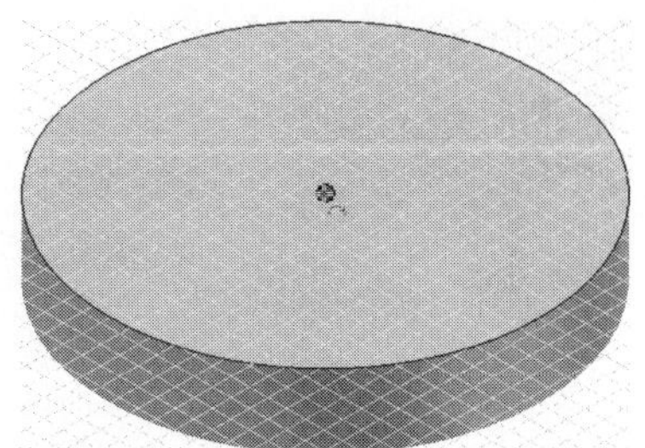

Figure 3.13 - Hole center selection, coincident with base feature sketch center

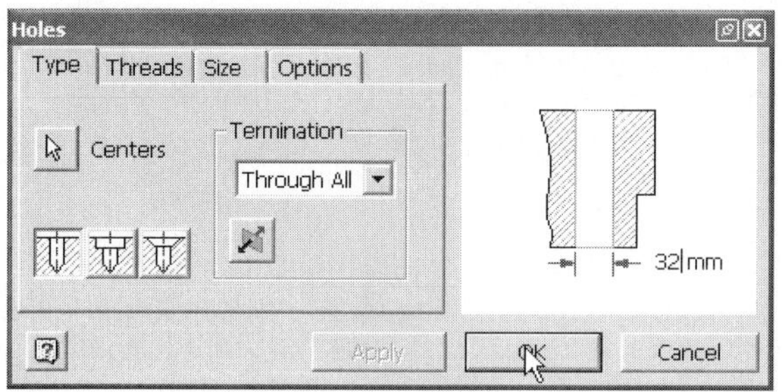

Figure 3.14 - Hole feature dialog box with required inputs

Figure 3.15 - Hole feature added

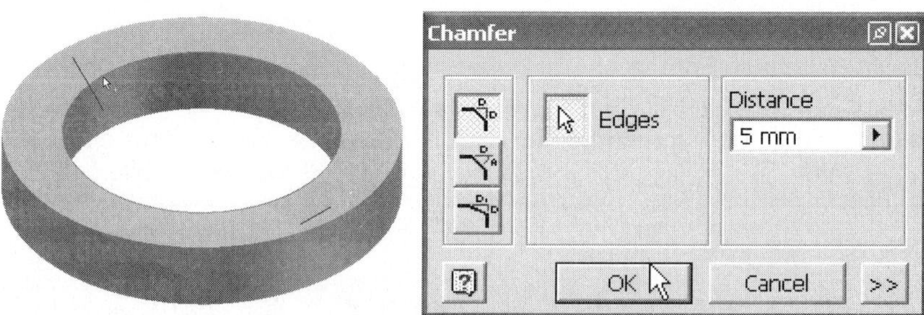

Figure 3.16 - Chamfer operation

6. Select the Hole tool from the Part Features panel bar. The Hole dialog box opens, and a hole is immediately highlighted in the drawing area where the hole center was placed. Change the Termination to Through All, change the diameter to 32 (do this by clicking on the dimension in the schematic on the right side of the dialog box), and then click OK, as shown in Figure 3.14. Your model should be similar to Figure 3.15.
7. Select the Chamfer tool, using a chamfer distance of 5, and the edge highlighted in Figure 3.16. Click OK.
8. Right-click, Isometric View.
9. The model should now be similar to Figure 3.17. Save the part. This completes Tutorial 3.

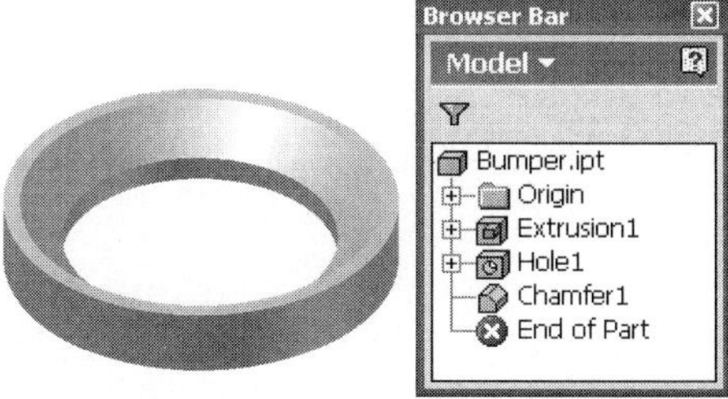

Figure 3.17 - Completed Bumper part

Packing Nut (continued) TUTORIAL 6

In Chapter 2 the packing nut base feature shown in Figure 3.18 had been created and saved. We will now add several additional features in order to complete the part.

Build Strategy

1. Sketch, extrude (Figure 3.19).
2. Fillet (Figure 3.20).
3. Chamfer (Figure 3.21).
4. Chamfer (Figure 3.22).
5. Extrude (Figure 3.23).
6. Thread (Figure 3.24).

Detailed Modeling Steps

1. Open the part file **Packing Nut.**
2. Use the 3D Rotate tool to change the view to something similar to that shown in Figure 3.25.

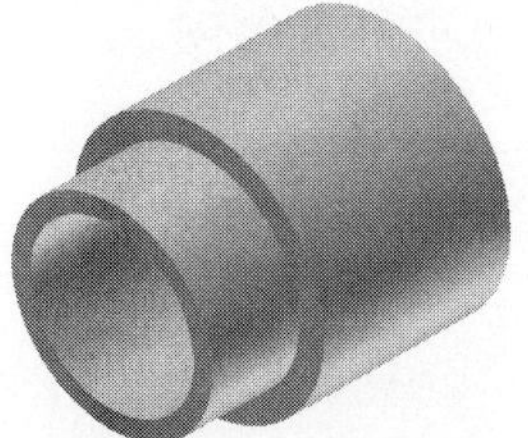

Figure 3.18 - Base feature

Figure 3.19 - Extruded join feature

Figure 3.20 - Fillet edges

Figure 3.21 - Chamfer inside edge

Figure 3.22 - Chamfer outside edge

Figure 3.23 - Extrude (join)

Figure 3.24 - External cosmetic threads

Figure 3.25 - View modified

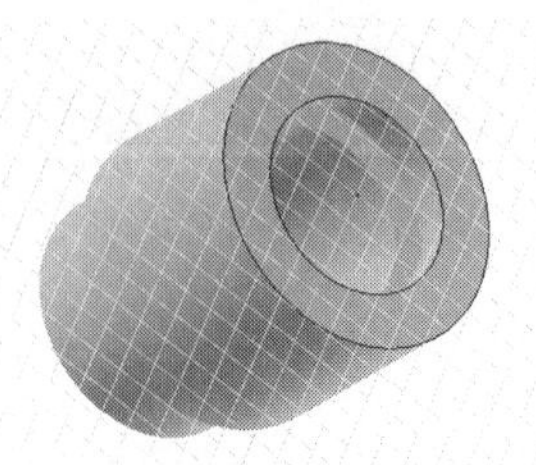

Figure 3.26 - New sketch plane

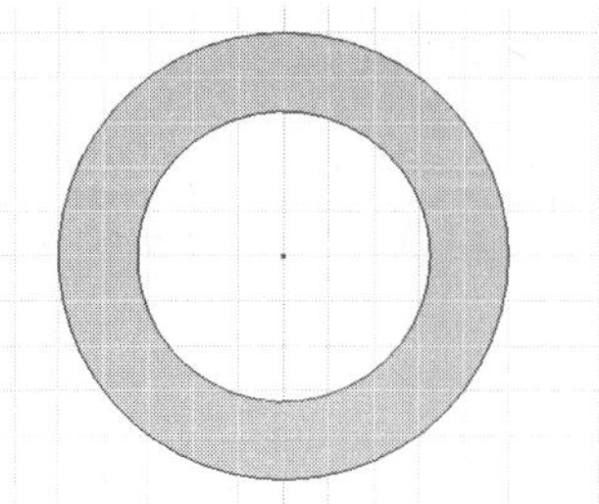

Figure 3.27 - Change view using Look At tool

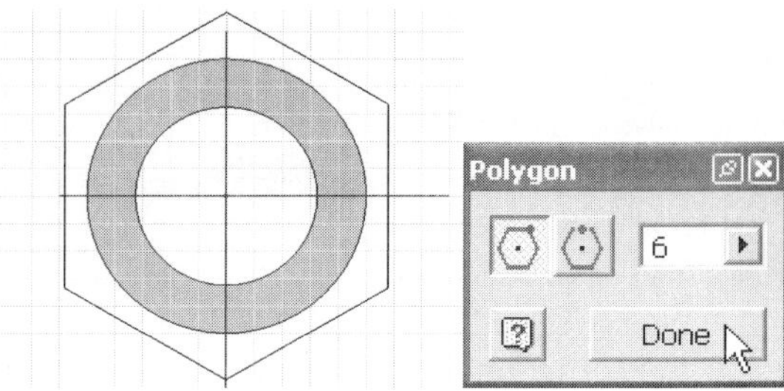

Figure 3.28 - Polygon sketch

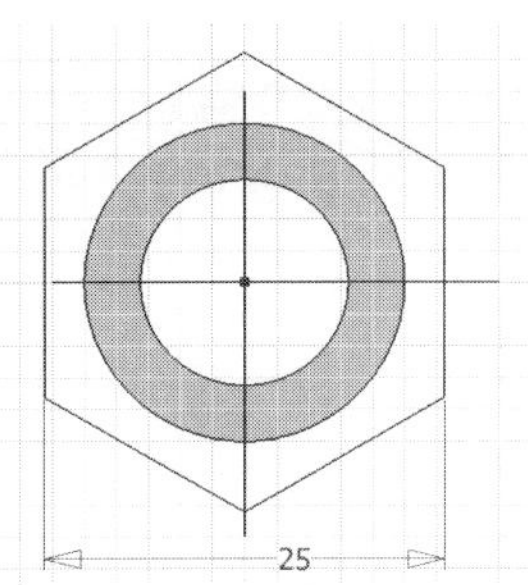

Figure 3.29 - Dimensioned polygon sketch

3. Right-click, New Sketch. Select the top face to be the new sketch plane, as shown in Figure 3.26.
4. Select the Look At tool from the standard toolbar, and then select the top face of the revolved feature. The Look At tool is used to change the view line of sight so that it is perpendicular to the selected planar face. This means that the selected face will be parallel to the screen. The screen should now resemble Figure 3.27.
5. Project the X and Z Axes, and the Center Point.
6. Use the Polygon tool from the panel bar to create the sketch shown in Figure 3.28. Parameters include 6 sides, the center point of the polygon is coincident with the origin of the coordinate system, and a polygon vertex is coincident with the vertical Y Axis. Click Done to create the polygon.
7. Use the General Dimension tool to add the 25 on the flat-to-flat dimension. See Figure 3.29.
8. Right-click, Finish Sketch.
9. Right-click, Isometric View.
10. Extrude the sketch, selecting as a profile the area between the polygon and the outermost diameter of the revolved solid. The extrusion distance is 5, the operation a join. Note also the direction in Figure 3.30. Click OK to create the feature. The resulting solid should be similar to that shown in the figure on the right.
11. Use the Common View mode of the 3D Rotate tool to verify that the hole still passes all the way through the part, as shown in Figure 3.31.
12. Isometric View.
13. Remember to save !

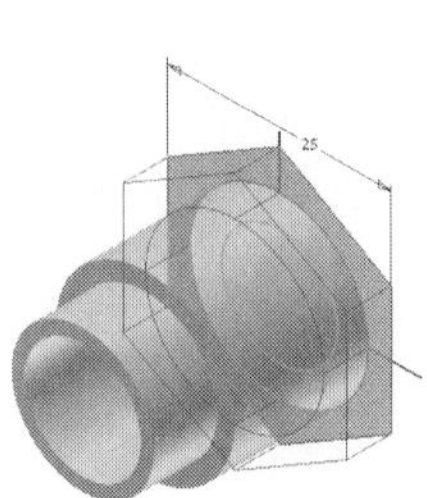

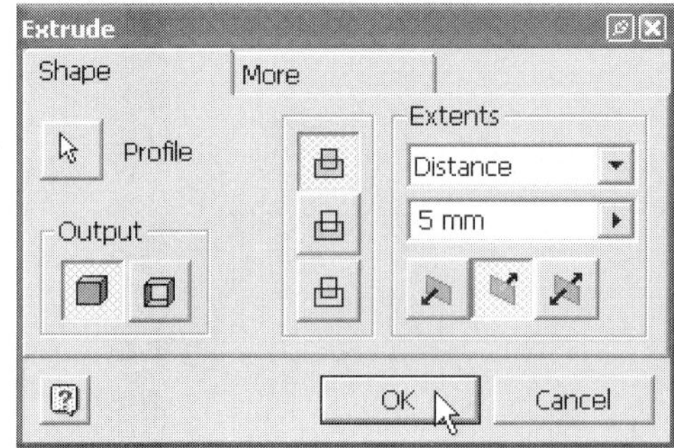

Figure 3.30 - Extruded (join) feature

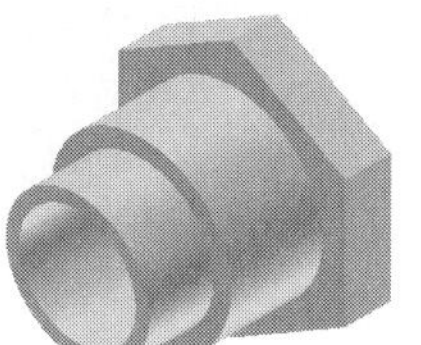

Figure 3.31 - Change view

14. Use the standard toolbar to change to a Wireframe display, as shown in Figure 3.32.
15. Choose the Fillet tool from the Part Features panel. Set the fillet radius to 3, and then select the six polygon edges to be filleted. Click OK when done. See Figure 3.33.
16. Change the display back to Shaded .
17. Use the Chamfer tool to chamfer the bottom **inside** edge of the solid. The (equal) chamfer distance is 0.5 mm. See Figure 3.34.
18. Again use the Chamfer tool, this time applied to the bottom **outside** edge of the solid. The (unequal) chamfer distances are 0.5 and 1 mm. The 0.5 mm distance is in a radial direction; the 1 mm distance is in an axial direction. If necessary, use the Flip icon to reverse the directions. Click OK when finished. The resulting solid should be similar to Figure 3.35.

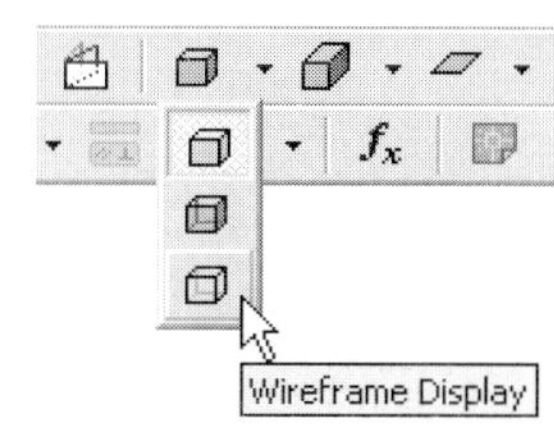

Figure 3.32 - Change to wireframe

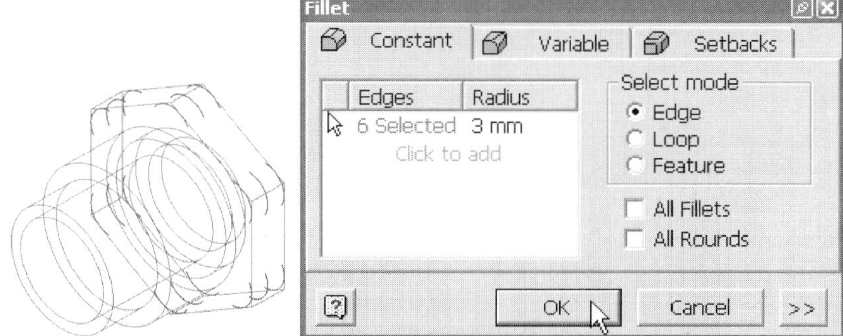

Figure 3.33 - Fillet edges

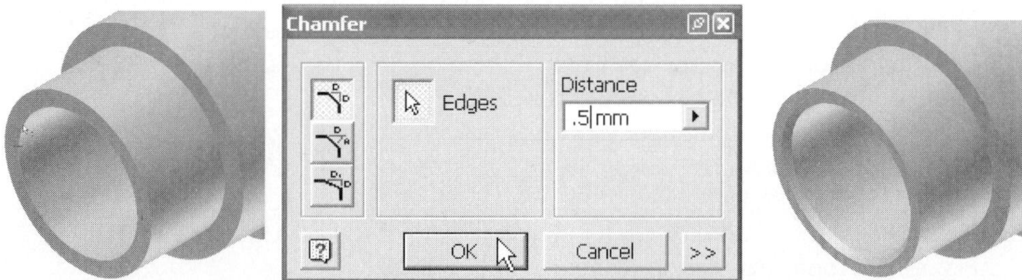

Figure 3.34 - (Equal) chamfer inside edge

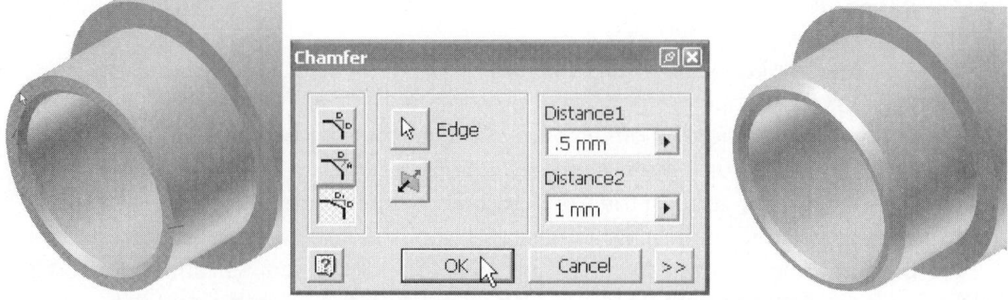

Figure 3.35 - (Unequal) chamfer outside edge

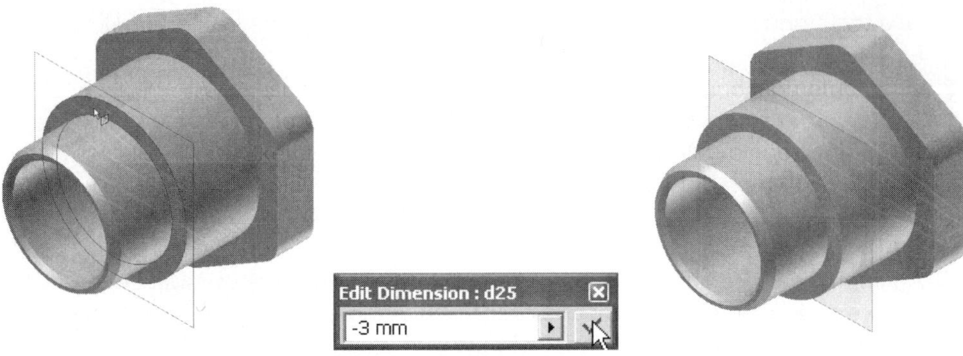

Figure 3.36 - Offset work plane

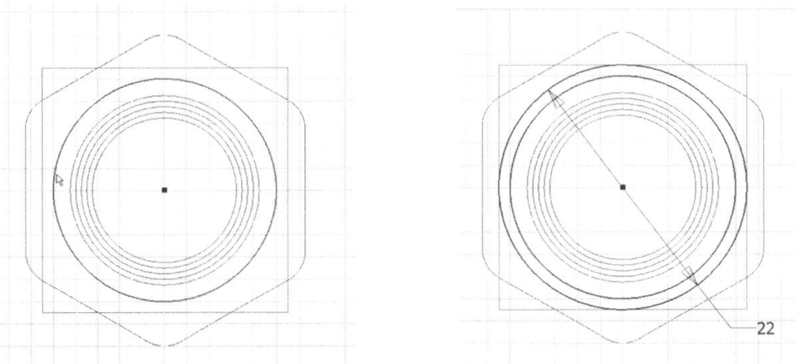

Figure 3.37 - Projected edge

Figure 3.38 - Dimensioned circle

19. Create a new offset work plane, as shown in Figure 3.36. After selecting the Work Plane tool in the part browser, select the annular face shown in Figure 3.35 on the left. A work plane appears. Dragging the mouse causes the Edit Dimension box to appear. Enter **-3**, and click on the Apply (green check mark) button to create the work plane.
20. New sketch. Select the offset work plane.
21. Look At the sketch.
22. Wireframe display.
23. Use the Project Geometry tool to project the circular edge shown in Figure 3.37 onto the sketch plane.
24. Use the Center point circle tool to create a circle coincident with the projected circular edge. Use the General Dimension tool to dimension the circle, as shown in Figure 3.38.
25. Finish Sketch.
26. Isometric View.
27. Change the display back to Shaded .
28. Extrude the annular region, as shown in Figure 3.39.
29. Turn off the visibility of the offset work plane (in the part browser, right-click on the Work Plane feature, then de-select Visibility).
30. Use the Thread tool on the Part Features panel bar to add **external** threads to the newly added feature. If necessary, select the Face button on the Location tab of the Thread dialog box, and then select the cylindrical surface indicated in Figure 3.40. Add the parameters indicated on both the Location and the Specification tabs, and then click OK.

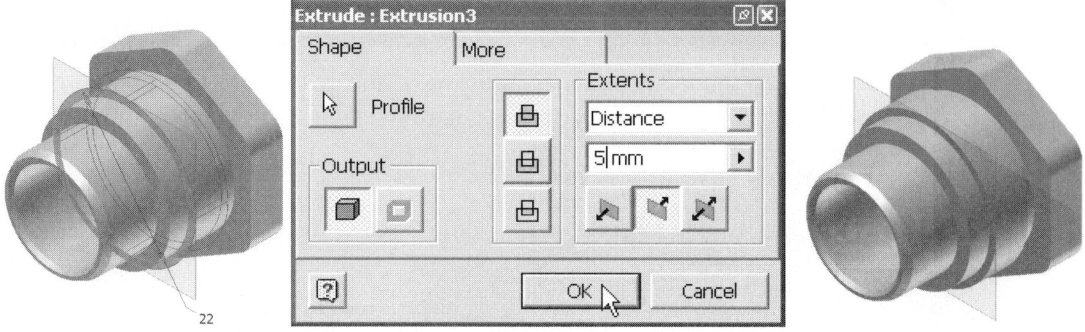

Figure 3.39 - Extruded operation (annular area, join)

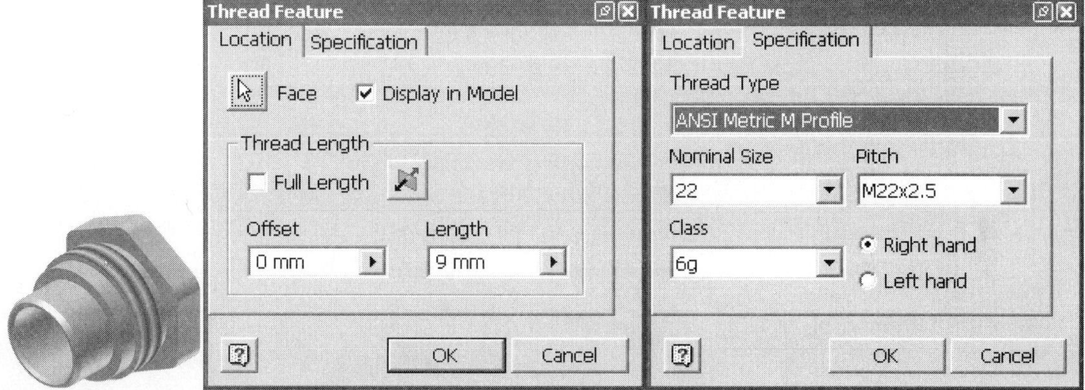

Figure 3.40 - Addition of external cosmetic threads

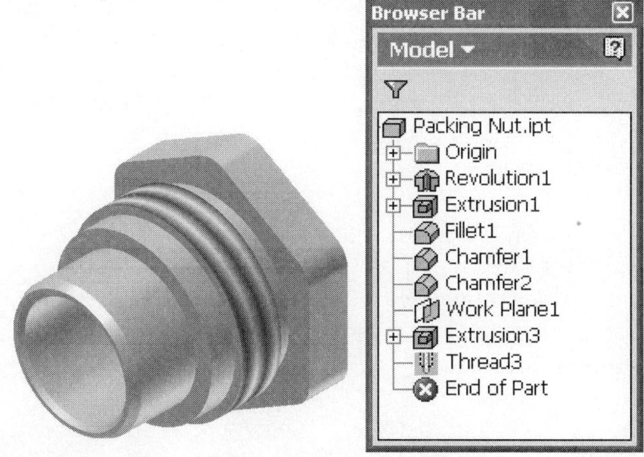

Figure 3.41 - Completed packing nut part

Note that the Thread tool creates a *cosmetic* feature; the actual 3D shape of the threads has not been modeled, but rather a 2D texture has been applied to the thread face to give the appearance of a threaded part. This is a common practice among parametric modelers.

31. The finished part should resemble Figure 3.41. Save the part file. This completes Tutorial 6.

TUTORIAL 7 Body (continued)

In Chapter 2 the body base feature shown in Figure 3.42 had been created and saved. We will now add several additional features in order to complete the part.

Build Strategy

1. Sketch, then extrude (join) (Figure 3.43).
2. Sketch, then extrude (join) (Figure 3.44).
3. Fillet (Figure 3.45).
4. Threaded hole (Figure 3.46).
5. Circular pattern (Figure 3.47).
6. Thread (Figure 3.48).
7. Fillet (Figure 3.49).
8. Sketch, extrude (join) (Figure 3.50).
9. Hole (counterbore) (Figure 3.51).
10. Chamfer (Figure 3.52).
11. Thread (internal) (Figure 3.53).
12. Sketch, extrude (join) (Figure 3.54).

Figure 3.42 - Base revolve feature

Figure 3.43 - Extruded hexagon feature

Figure 3.44 - Extruded lug feature

Figure 3.45 - Lug feature fillets

Figure 3.46 - Threaded hole

Figure 3.47 - Circular pattern (lug, fillets, threaded hole)

Figure 3.48 - Internal threads

Figure 3.49 - Add fillet

Figure 3.50 - Extrude (join) to next surface

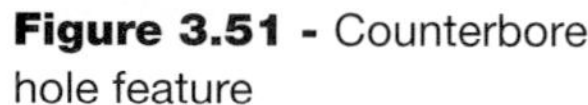

Figure 3.51 - Counterbore hole feature

Figure 3.52 - Chamfer hole

Figure 3.53 - Internal threads

Figure 3.54 - Extruded stop

■ Detailed Modeling Steps

1. Open the part file **Body.**
2. Before we start, select Tools > Application Options . . . from the menu bar. The Options dialog box opens. Click on the Display tab. On the right side under Shaded Display Modes, remove the check mark by Active > Edge Display, as shown in Figure 3.55. If necessary move the dialog box off to the side so that you can see the model. Click the Apply button. You should see that the model edges are no longer displayed, providing a more realistic shaded display. Click OK.
3. Right-click, New Sketch—select the bottom face shown in Figure 3.56.
4. Look At the sketch.
5. Project the X and Z Axes and the Center Point onto the sketch plane.
6. Use the Polygon tool to sketch a six-sided polygon with the center at the origin of the coordinate system, and a vertex coincident with the vertical Z Axis.
7. Use the General Dimension tool to add a dimension of 57 across the flats of the hexagon. See Figure 3.57.

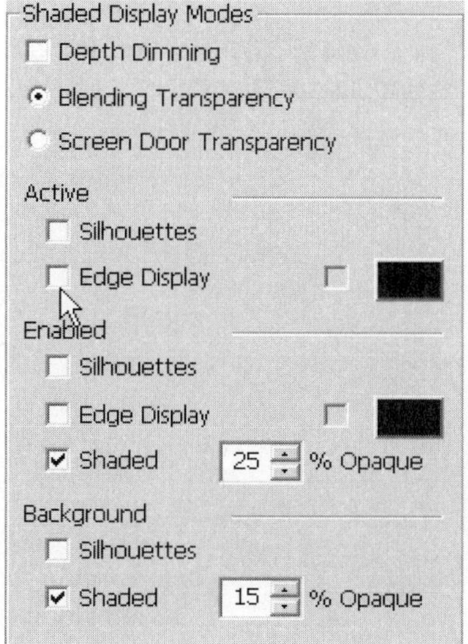

Figure 3.55 - Edge display option

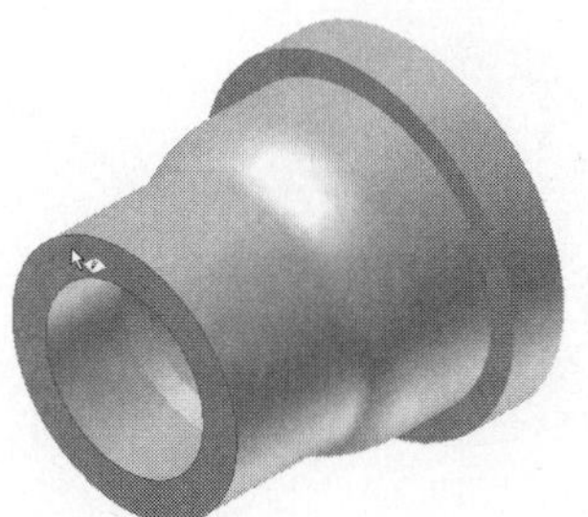

Figure 3.56 - New sketch

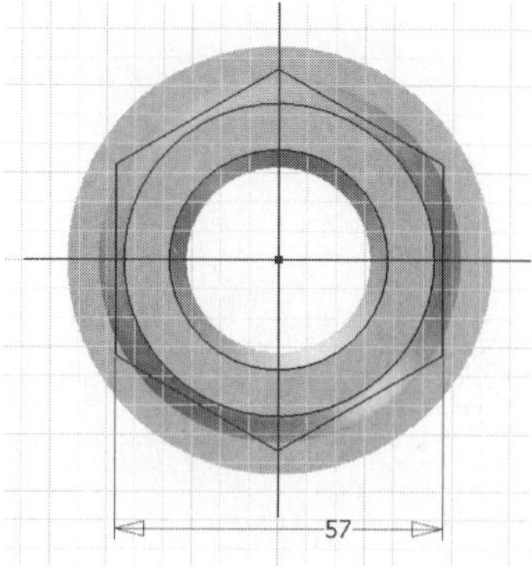

Figure 3.57 - Dimensioned polygon sketch

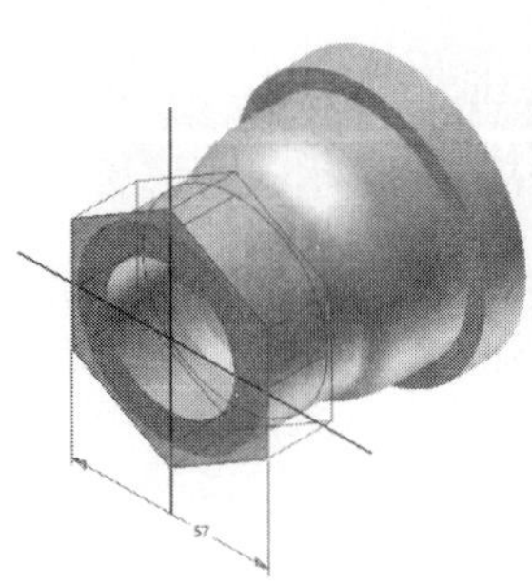

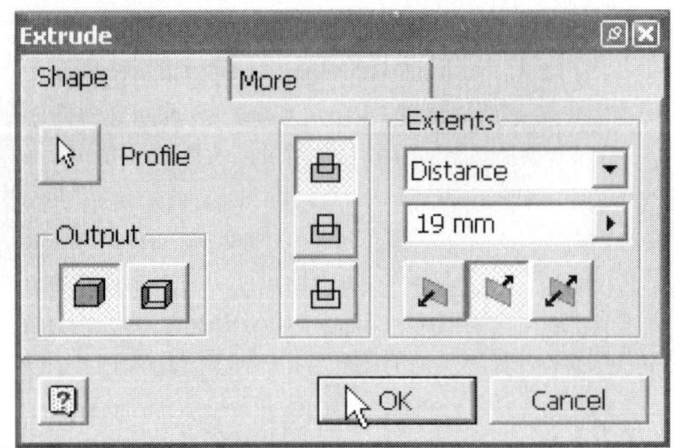

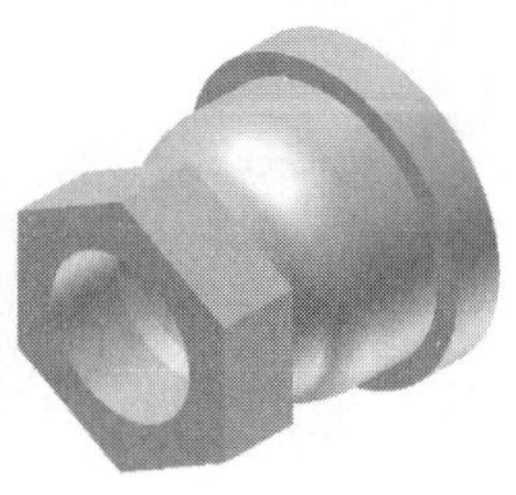

Figure 3.58 - Extruded hexagon feature construction

Figure 3.59 - Rotate view

Figure 3.60 - New sketch plane

8. Right-click, Finish Sketch.
9. Right-click, Isometric View.
10. Extrude the area between the outer edge of the face and the polygon. The extruded distance is 19, and it is a joining operation. Note the extrusion direction shown in Figure 3.58. Click OK when done.
11. Remember to save .
12. Use the 3D Rotate tool to change the view to something similar to that shown in Figure 3.59.
13. Right-click, New Sketch—select the face shown in Figure 3.60.
14. Look At the sketch.
15. Project the X and Z Axes, Center Point.
16. Use the Line tool to sketch the shape shown in Figure 3.61. The horizontal lines are coincident with the outer edge of the face; the arc is tangent to the horizontal lines.
17. Add a symmetric constraint to the two horizontal lines, symmetric about the X Axis.
18. Use the General Dimension tool to add the dimensions shown in Figure 3.62.

    TIP: The linear dimension can be made to display diameter by selecting the arc and the vertical axis, then right-clicking and choosing Linear Diameter.

19. Right-click, Finish Sketch.
20. Right-click, Isometric View.

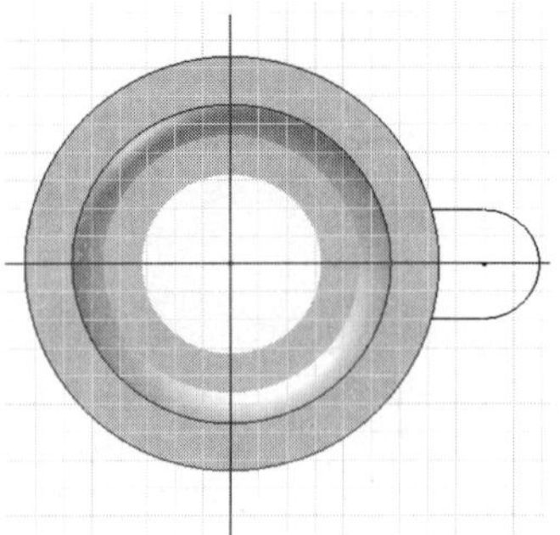

Figure 3.61 - New sketch

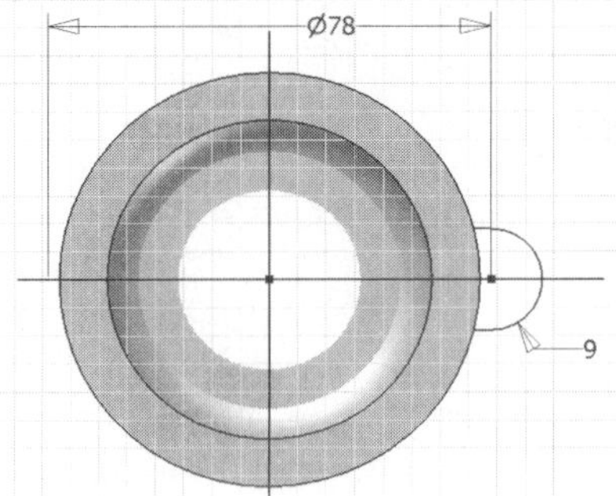

Figure 3.62 - Fully constrained sketch

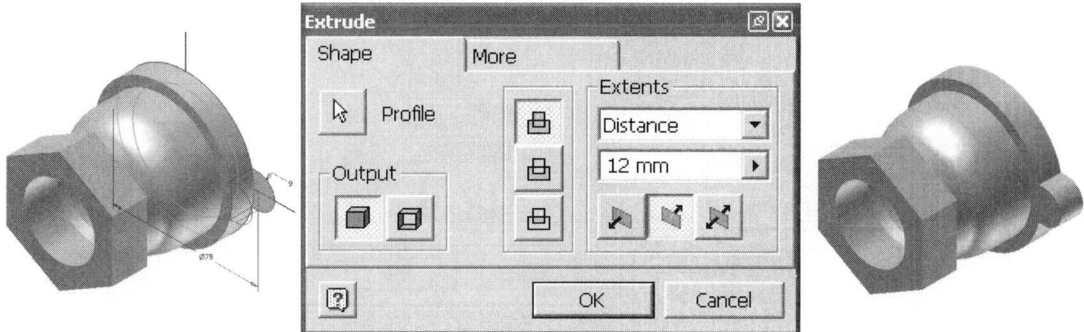

Figure 3.63 - Extrusion operation

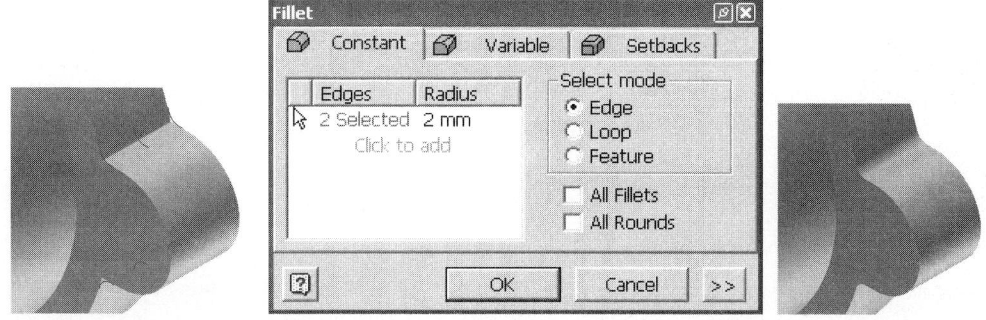

Figure 3.64 - Fillet feature operation

21. Extrude the sketch, using the parameters shown in Figure 3.63. When completed, the solid should be similar to the object at the extreme right in the figure.
22. Use the Fillet tool to fillet the edges shown in Figure 3.64 on the left. When finished, the solid should be similar to the object at the extreme right.
23. Use the 3D Rotate tool to change the view, as shown in Figure 3.65.
24. New Sketch—choose the face shown in Figure 3.66.
25. Use the Point, Hole Center tool to add a hole center coincident with the arc center from the recently created lug feature. Note that a green circle will appear when the cursor is coincident with the lug center.
26. Finish Sketch.
27. Using the Hole tool, create a threaded hole with the parameters in Figure 3.67.
28. Isometric View.
29. Use Circular Pattern to array the lug, lug fillets, and threaded hole feature. Click the Features button and then select these three features, either from the part browser or in the graphics area. Next click the Rotation Axis button and select the inside of the base revolved feature to define the rotation axis. In the Placement area set the count to 6 and the angle to 360 deg. Your model should now resemble Figure 3.68 on the left. If you still cannot see the feature array, try clicking the More << button, and set the Positioning Method to Fitted. Click OK when ready.
30. Use the Thread tool to create the internal threads shown in Figure 3.69. The part should resemble the solid shown in the lower right corner of Figure 3.69.

Figure 3.65 - Rotated view

Figure 3.66 - New sketch plane

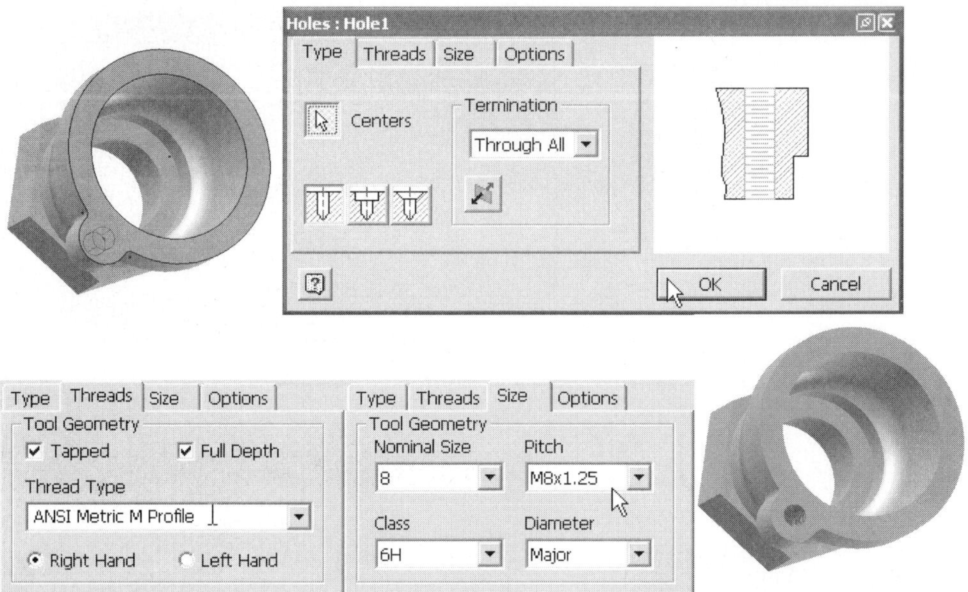

Figure 3.67 - Threaded hole feature operation

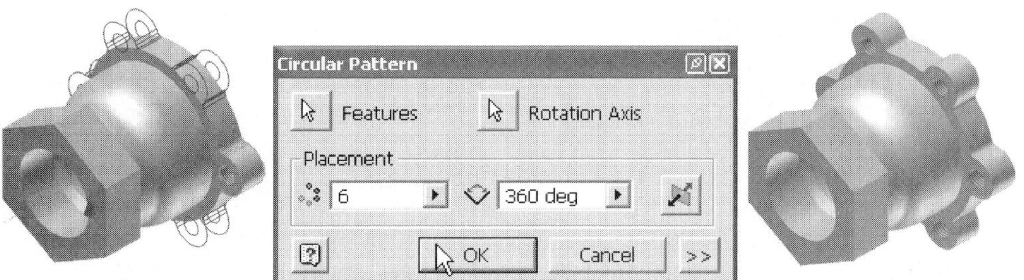

Figure 3.68 - Circular pattern operation of lug, fillets, threaded hole

31. Use the Fillet tool to fillet the edge as shown in Figure 3.70. The resulting fillet can be seen on the part on the extreme right.
32. Change to a wireframe display.
33. From the Origin folder in the part browser, turn on the visibility of the XY Plane (expand the folder, right-click on XY Plane, select Visibility). See Figure 3.71.
34. Select the Workplane tool from the Part Features menu on the panel bar. Click and drag up on the XY Plane displayed in the modeling work space. An Offset dialog box opens. Enter 44 in the dialog box, and then click the Apply (green check symbol) button (or press Enter) as shown in Figure 3.72. A new work plane is created, parallel to the XY Plane, and offset 44 units from it. Your screen should now appear as shown in Figure 3.73.
35. New Sketch. Select the offset work plane.
36. Look At the sketch.
37. Project the Y Axis onto the sketch plane.
38. Use the Circle tool to add a circle coincident with the Y Axis, and below the top edge of the part.
39. Use the General Dimension tool to add the dimensions shown in Figure 3.74.

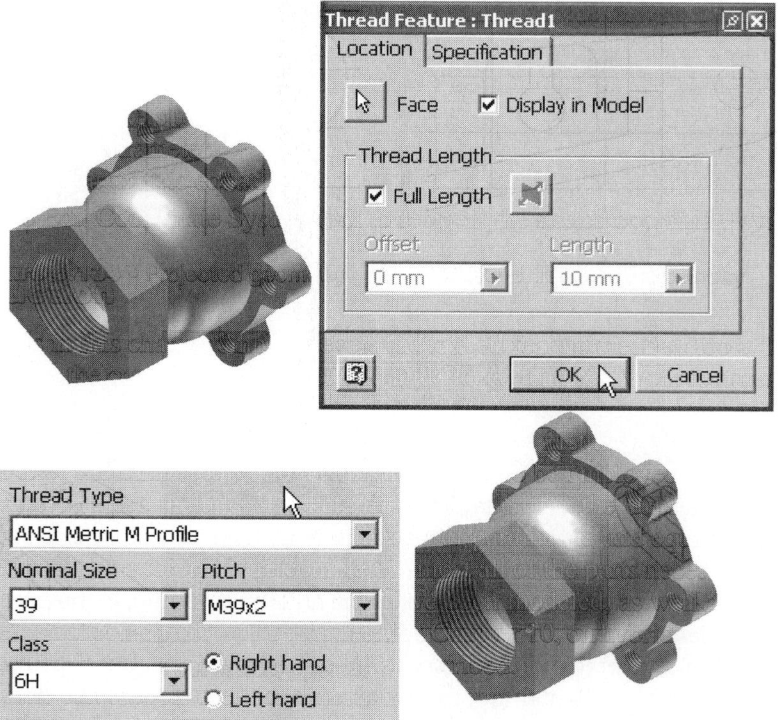

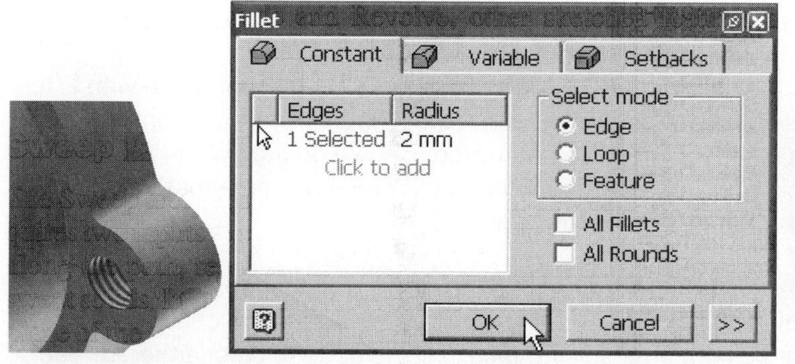

Figure 3.69 - Internal thread feature operation

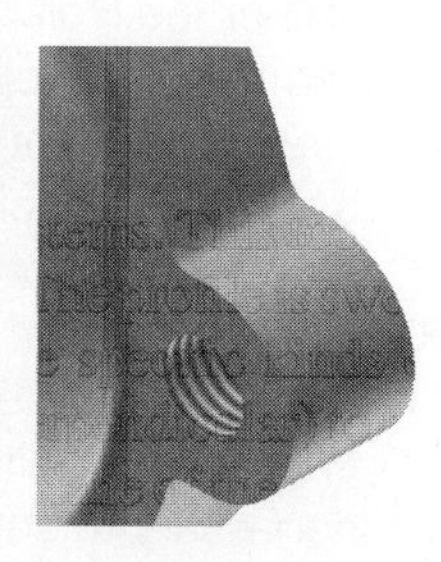

Figure 3.70 - Fillet feature operation

Figure 3.71 - Work Plane visibility

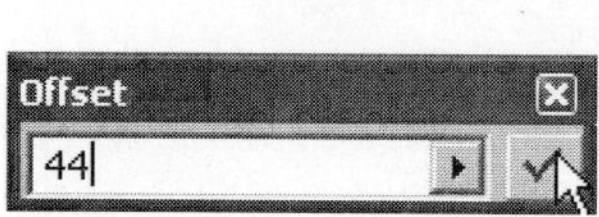

Figure 3.72 - Offset dialog box

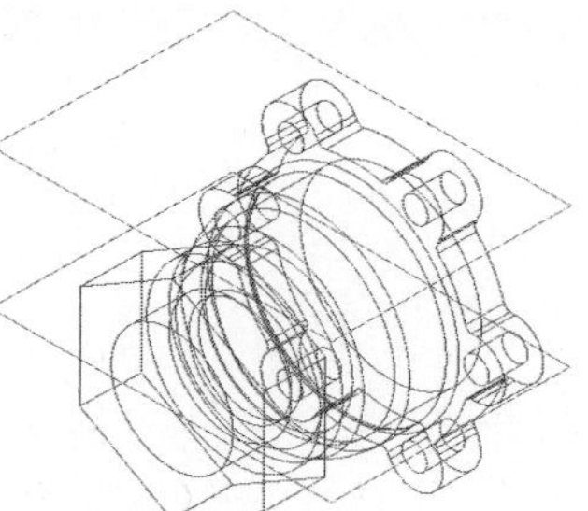

Figure 3.73 - Offset workplane

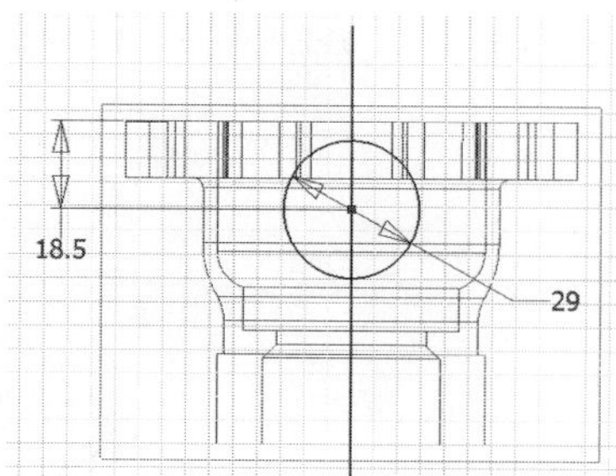

Figure 3.74 - Fully constrained sketch

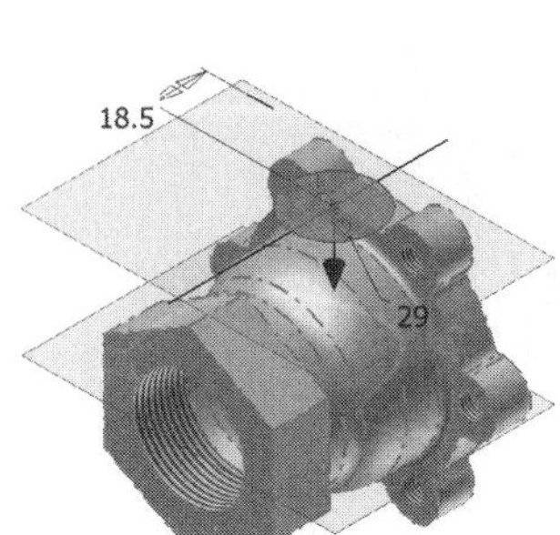

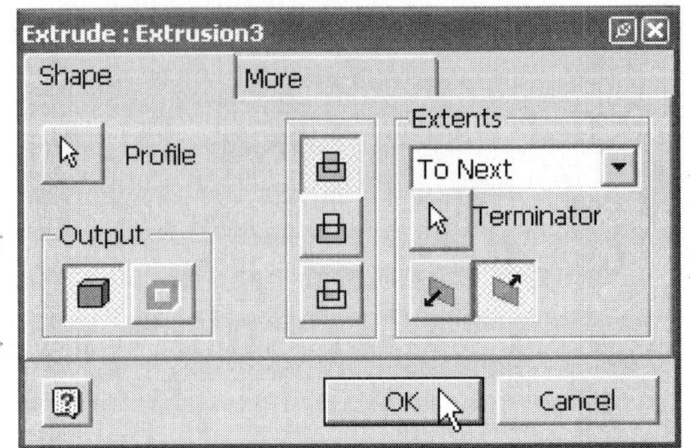

Figure 3.75 - Extrude (join) to next surface

Figure 3.76 - New sketch plane

Figure 3.77 - Rotate view

40. Finish Sketch.
41. Isometric View.
42. Change to a shaded display.
43. Extrude the circular profile to the base feature by using a Join operation, with the Extents set at To Next (Figure 3.75). Click OK.
44. Turn off the visibility of the offset work plane and the XY Plane. In the part browser, select the features, right-click, and de-select Visibility from the context menu.
45. New Sketch. Select the face shown in Figure 3.76.
46. Use the Point, Hole Center tool to add a hole center coincident with the center of the circular face.
47. Finish Sketch.
48. Use the 3D Rotate tool so that the inside of the valve body is visible. See Figure 3.77. Rotating the view in this way is helpful, since it will be necessary to select the inside surface of the valve body as the termination surface for the hole that will be created in the next step.
49. Select the Hole tool. In the Hole dialog box, select a counterbore hole with a To Termination. Select the inside curved surface of the part, as shown on the left in Figure 3.78. Enter the counterbore dimensions shown on the right in the dialog box, and then click OK. (If you encounter a problem, make sure that the "Tapped" option on the Threads tab is unchecked).
50. Isometric View.
51. Use the Chamfer tool to chamfer the inside edge of the most recently created hole. Use the distance and angle method to create the chamfer. It is necessary to select both an adjoining face and the edge to be chamfered. See Figure 3.79.

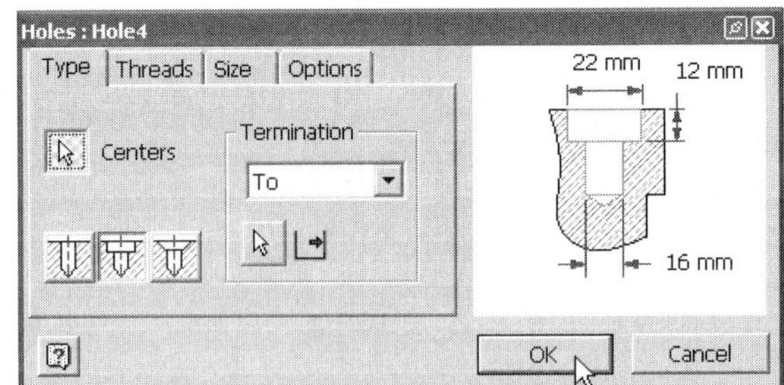

Figure 3.78 - Counterbore hole feature operation

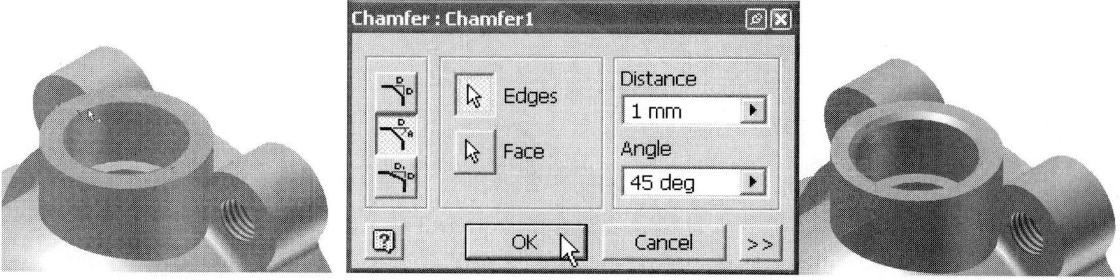

Figure 3.79 - Chamfer feature operation

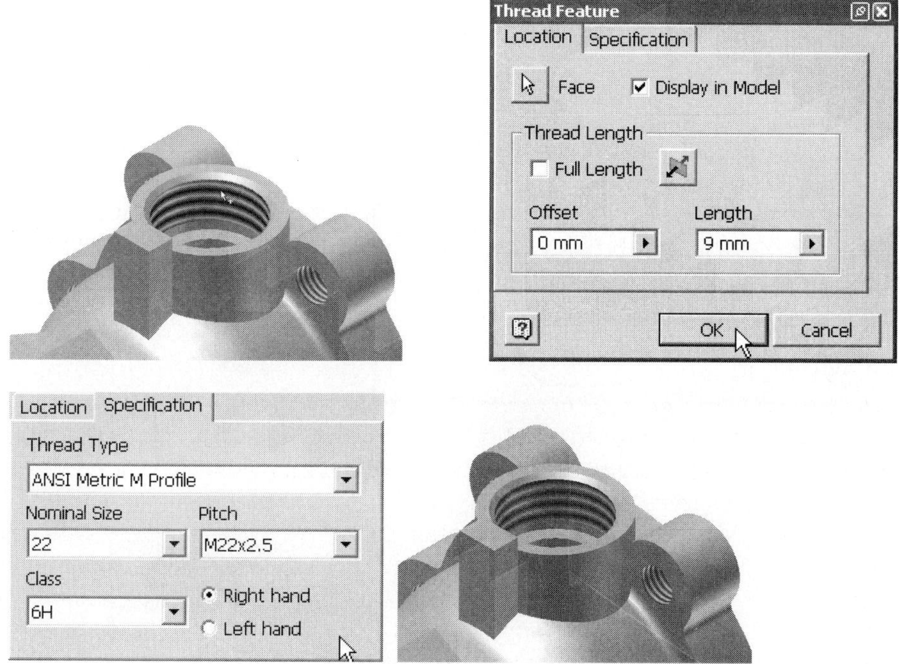

Figure 3.80 - Internal threads feature operation

52. Use the Thread tool from the Part Features menu on the panel bar to add internal threads to the counterbore hole. Use the parameters shown in Figure 3.80.

53. Change to a wireframe display.

54. Turn on the visibility of the XY Plane.

55. Select the Work Plane tool from the Part Features menu of the panel bar. Select and drag up the XY Plane. Enter an offset distance of 47 in the Offset dialog box that appears, and then click the green check mark. Your screen should be similar to Figure 3.81.

56. New Sketch. Select the offset work plane.

57. Look At the sketch.

58. Project the Y Axis on to the sketch.

59. Use the Rectangle tool to draw a rectangle over the part profile.

60. Use the General Dimension tool to add the dimensions shown in Figure 3.82.

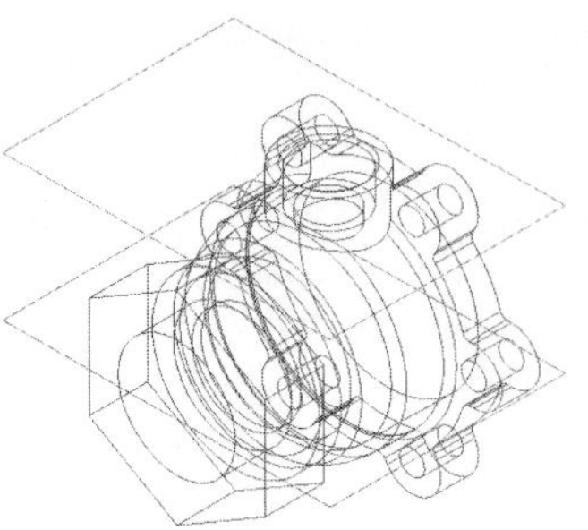

Figure 3.81 - Offset workplane

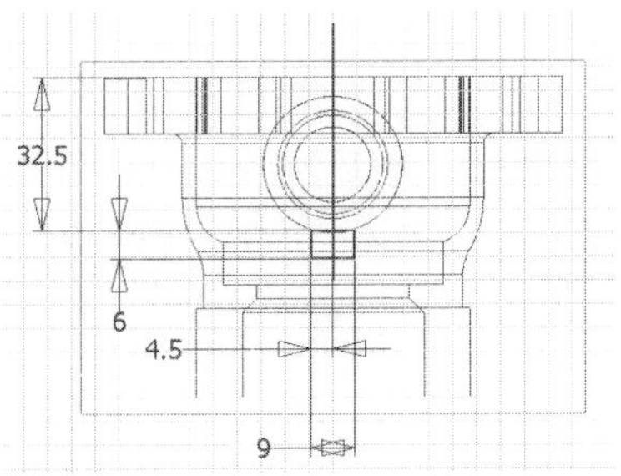

Figure 3.82 - Fully constrained sketch

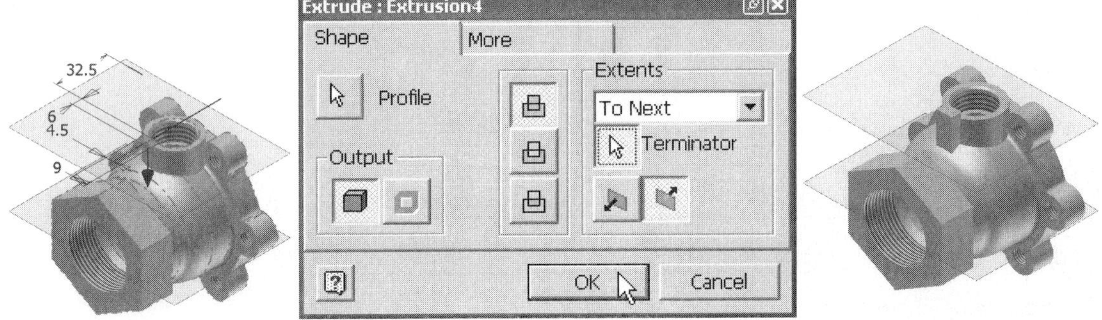

Figure 3.83 - Extrude (join) operation

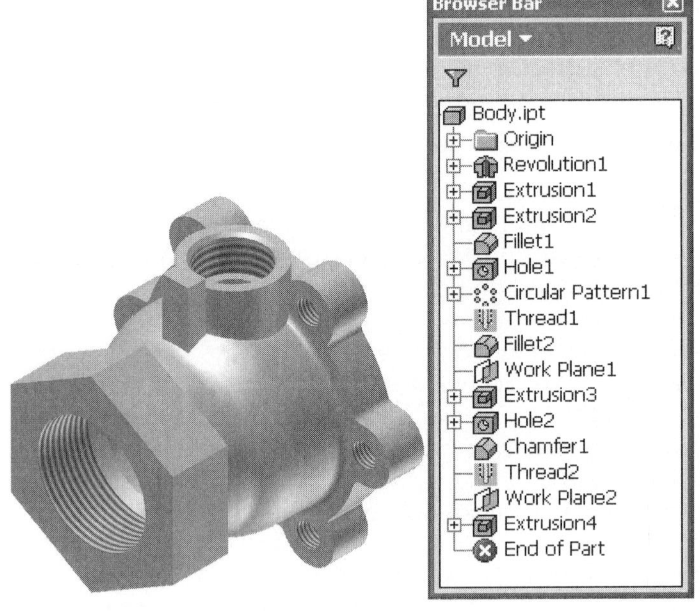

Figure 3.84 - Completed valve body

61. Finish Sketch.
62. Isometric View.
63. Change display to Shaded.
64. Extrude the rectangular profile, using the settings shown in Figure 3.83. This last feature serves as a stop for the valve handle.
65. Turn off the visibility of the work plane and XY Plane. The finished part should resemble Figure 3.84. Save the part file. This completes Tutorial 7.

QUESTIONS

1. Which is not a sketched feature?
   a. Sweep
   b. Extrude
   c. Coil
   d. Chamfer
   e. Split
2. T F When the Revolve tool is used, the revolve dialog prompts for the selection of a profile and a path.
3. When you use the output option for sketched features, the icon refers to ________ while the icon refers to ________.
4. What are the three Boolean operations?
5. When you consider termination options, the distance option will usually refer to the ________ operation while the angle option will usually refer to the ________ operation.
6. Which is not a placed feature?
   a. Hole
   b. Chamfer
   c. Rib
   d. Fillet
7. Match the work feature icons with the work feature names shown below.
   Work Point
   Work Plane
   Work Axis
8. T F The Work Plane tool useful in situations where the existing feature geometry fails to provide a suitable sketch plane for the next feature to be added.

PROBLEMS

1. Use a metric part template file to create the part shown in the drawing below. Save the part as **Pin** in the BallValve Project Workspace.

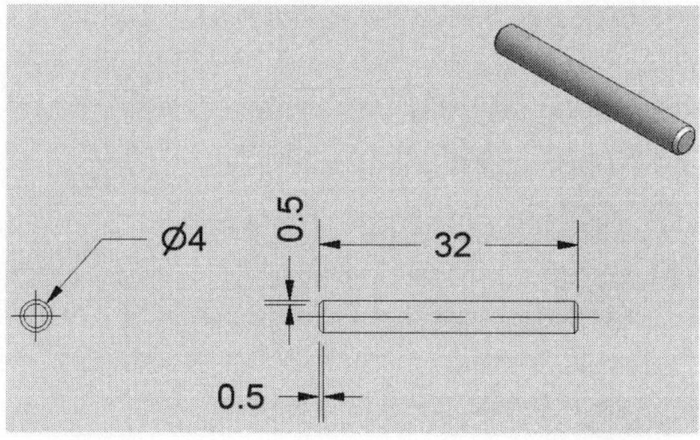

2. Complete the Body Cap part started in Chapter 2. Use the drawing shown below to add the additional features. Save the file.

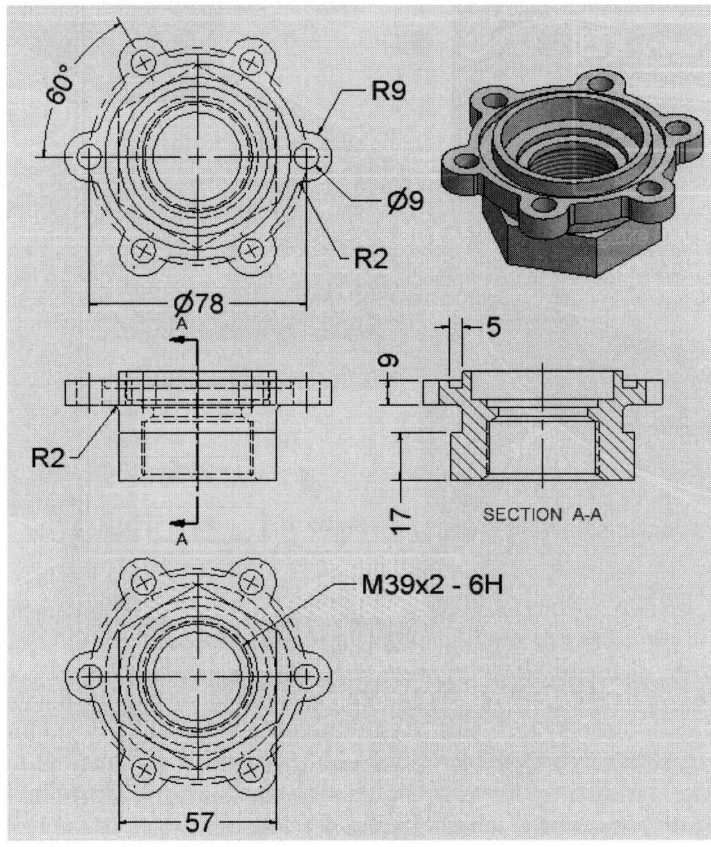

CHAPTER 4

Part Editing

LEARNING OBJECTIVES

- Rename features and sketches in the part browser
- Explain how sketches and features can be edited from the part browser
- Suppress and delete features
- Turn on/off the visibility of features and sketches
- Describe the circumstances under which it is appropriate to use a shared sketch
- Use a shared sketch to create a feature
- Explain what is meant by a parent-child relationship
- Use the Select Other tool to choose the desired geometry
- Use the Measure tool to measure lengths and distances, angles and areas
- Change the color of parts, part features, and part faces
- Name the two surface analysis styles available from the Analyze Faces tool
- Display and update the center of gravity of a part
- Use the Center point arc tool to add geometry to a sketch
- Use the Three point arc tool to add geometry to a sketch
- Use the Work Plane tool to create a work plane tangent to a cylindrical face
- Use the 3D Rotate (Common View) tool to redefine the default isometric view of a model
- Use the Mirror tool to mirror sketch geometry about an axis

Introduction

One of the most impressive aspects of a parametric modeler is the relative ease with which a part model can be quickly modified. Given this ability to easily edit parts, a designer can quickly correct mistakes, explore multiple design solutions or iterations, develop part families, etc. Most all part editing in Autodesk Inventor is handled via the browser bar (or part browser) in conjunction with context menus (right mouse button).

Part Browser

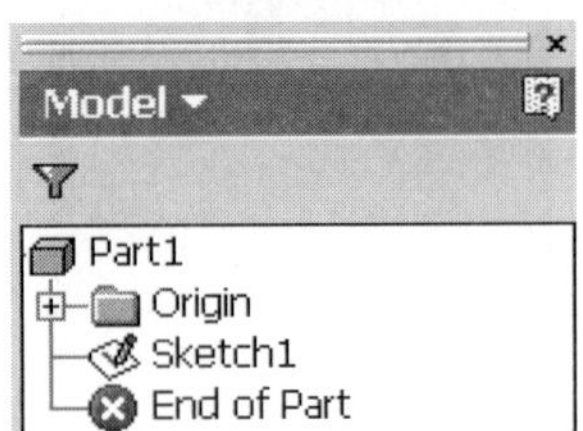

Figure 4.1 - Part browser; initial elements

Figure 4.2 - Expanding feature to reveal consumed sketch

When a new part file is started in Inventor, the first element appearing in the part browser is the Origin folder. As we saw in Chapter 1, expanding the Origin folder by clicking on the + symbol to the left of the element reveals the default reference work geometry. Similarly, clicking on the – symbol collapses the Origin folder.

If Inventor part files are set up to automatically start in sketch mode (this is how Inventor opens when launched for the first time), then an element with the name Sketch1 appears below the Origin folder in the browser, as seen in Figure 4.1.

Otherwise the Sketch1 element appears as soon as a new sketch is started. Sketches are automatically assigned indexed names (e.g., Sketch1, Sketch2, etc.) as new sketches are added to the part.

Once a base feature is created, the feature appears in the part browser. This feature will typically be an extrusion or a revolution, and will be index named accordingly (e.g., Extrusion1, Revolution1). The Sketch1 element is no longer visible. However, if the base feature is expanded, as shown in Figure 4.2, the Sketch1 element will be revealed beneath the feature. In the jargon of parametric modeling, we say that the sketch has been *consumed* by the feature.

As features (sketched, placed, or work) are added to the part, they appear in the part browser. Features are also assigned indexed names according to the kind of feature they represent (e.g., Fillet1, Hole2, and Work Plane1). Unlike sketched features, placed and work features cannot normally be expanded further, since these features do not depend upon any underlying sketch geometry.

The features are added to the part browser in the order in which they were created; consequently once the part is complete the part browser can provide a history of the part's development. The part browser is sometimes referred to as a *feature tree,* with the base feature serving as the foundation for the part. All other features are arranged in a hierarchy beneath the base feature.

Figure 4.3 shows a completed part with the accompanying part browser. Notice that if the cursor is placed over a feature name in the browser, the feature is highlighted in the graphics area.

NOTE: The part shown in Figure 4.3 is used in this chapter to demonstrate various editing techniques in Inventor. The name of this part is ch2_edit.ipt, and the file is available on the CD.

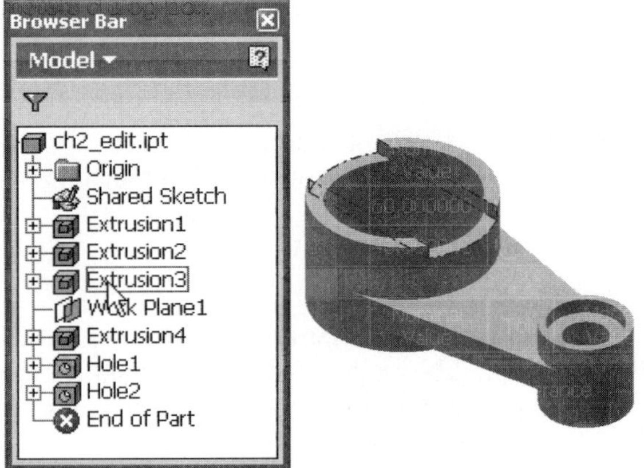

Figure 4.3 - Part with associated feature tree

It is possible to rename features. This is done by left-clicking on the text portion of the element name (not the icon). A box appears around the text. Left-click a second time inside the box, and then type in the new name. Figure 4.4 shows the part browser for the ch2_edit.ipt file with renamed features.

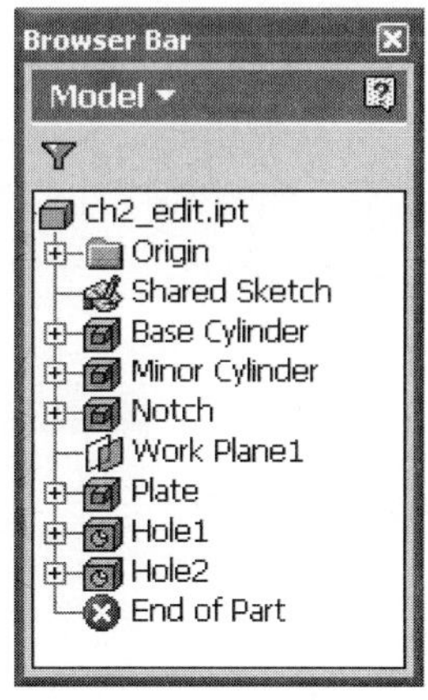

Figure 4.4 - Renaming features in the part browser

Part Browser Context Menu

Right-clicking on a feature icon in the part browser accesses the part editing commands via context menus. Assuming that a sketched feature has been selected, the Edit Sketch option will be available, as shown in Figure 4.5. Note that the options available on the context menu will vary slightly depending on the kind of feature selected. Context menus for sketch elements also present slightly different menu options.

Selecting the Edit Sketch option opens the sketch for editing. At this point the sketch can be edited by changing dimensions, constraints, or even the shape of the sketch itself. In Figure 4.6, the two dimensions are changed. Double-clicking on a dimension opens the Edit dimension dialog box, at which point the dimension can be modified. Once all the modifications have been made, right-click in the graphics area, and select Finish Sketch.

The Edit Feature option opens the Feature dialog box. In Figure 4.7, the original feature depth of 0.5 inch is changed to 2 inches.

A feature can also be deleted (by choosing Delete from the context menu), or *suppressed.* The effect of suppressing the Notch feature is shown in Figure 4.8. Although the notch no longer appears on the model, it is still listed (dimmed with a line through it) in the part browser. Right-clicking on the element and choosing Unsuppress Features will restore the notch.

The *visibility* of work features and sketches can be turned on and off from the part browser. In Figure 4.9, the visibility of a work plane is in the process of being turned on. A feature or sketch will be dimmed in the part browser if it is not visible.

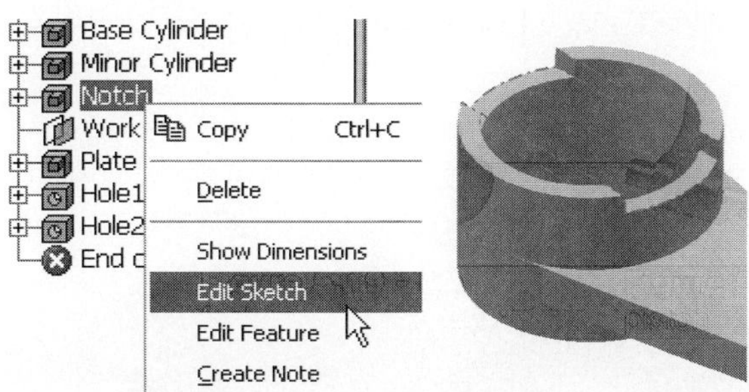

Figure 4.5 - Sketch edit access

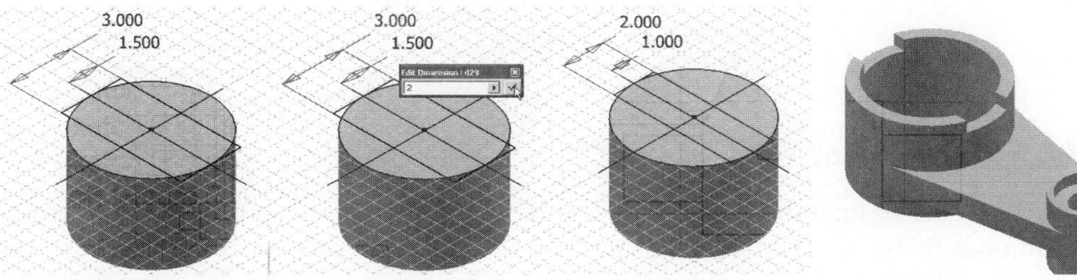

Figure 4.6 - Editing a sketch

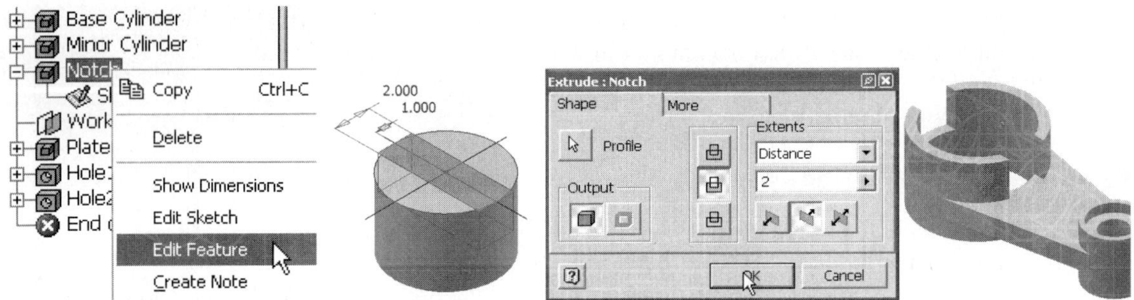

Figure 4.7 - Editing a feature

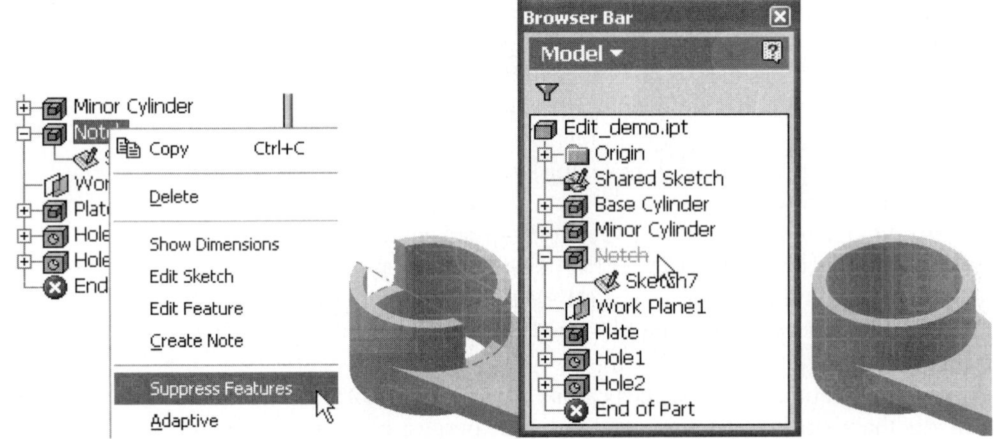

Figure 4.8 - Feature suppression

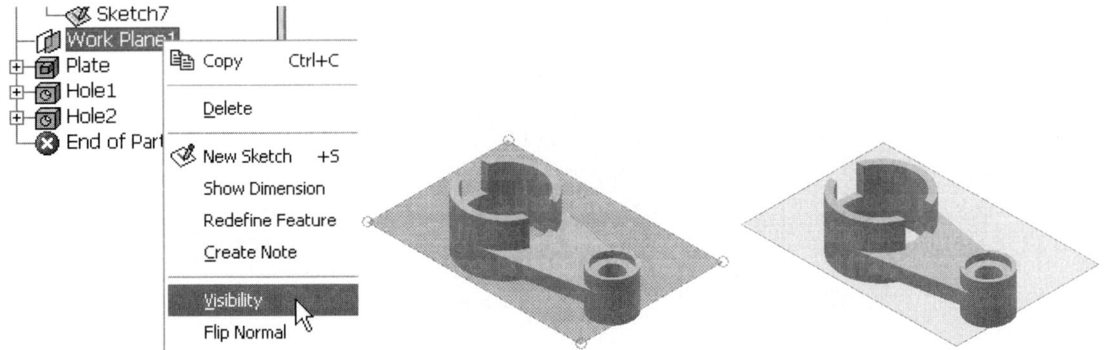

Figure 4.9 - Feature visibility

Shared Sketches

Once a sketched feature has been created, the same sketch can be used again to create additional features. This situation arises whenever two or more features share the same sketch plane. The base feature for our sample part is a large diameter cylinder. The second feature is also a cylinder, but with a smaller diameter and a different

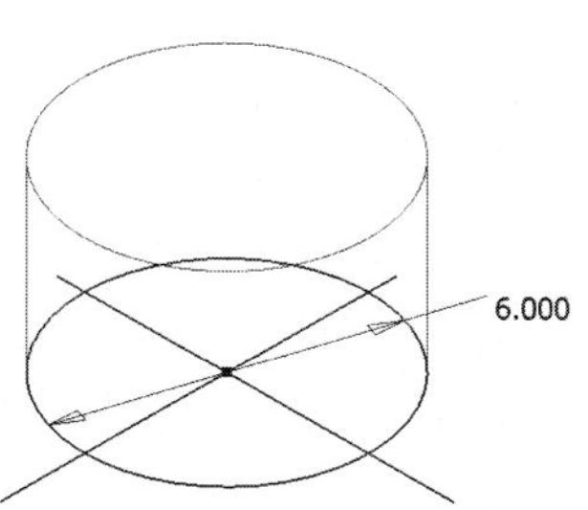

Figure 4.10 - Base feature

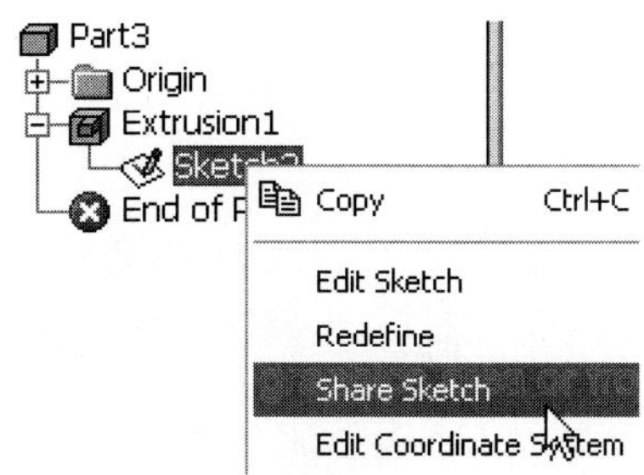

Figure 4.11 - Share sketch selection

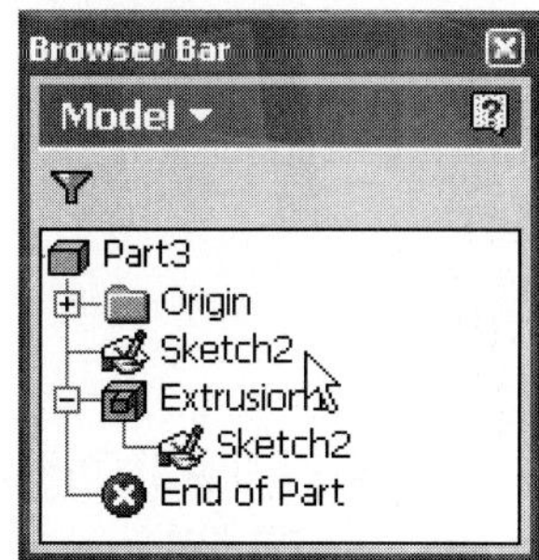

Figure 4.12 - Shared sketch in part browser

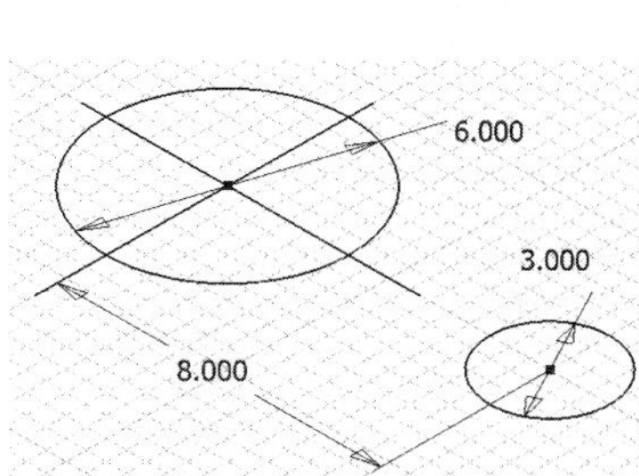

Figure 4.13 - Shared sketch

Figure 4.14 - Second extruded figure

height. Since the bottom of both cylinders originates on the same plane, the base cylinder's sketch can be shared to create the second cylinder. This is illustrated in the Figures 4.10 and 4.11. After the base feature is created (Figure 4.10), the feature is expanded in the part browser to reveal the consumed sketch.

Right-clicking on the sketch element reveals a context menu. After selecting Share Sketch (Figure 4.11), an unconsumed copy of the sketch appears in the part browser above the base feature (Figure 4.12). This is the shared sketch.

After right-clicking on the shared sketch and choosing Edit Sketch, an additional circle is sketched, sized, and positioned with respect to the original sketch (Figure 4.13). Once sketch mode is exited, the smaller circle can be extruded using the Extrude tool (Figure 4.14). Note that we could just as well have sketched and dimensioned both circles on the original sketch. In extruding the base feature, only the larger circle would have been selected. The consumed sketch could then have been shared, and a new feature created from the shared sketch.

Parent-Child Relationships

In Figure 4.13, a distance dimension of 8 was used to position the small cylinder with respect to the large cylinder. This creates a relationship between the two (cylinder) features; the position of the small cylinder feature depends upon the base feature. In parametric modeling jargon, we say that there is a *parent-child relationship* between the two cylinders; the large cylinder is the parent feature to the small cylinder child feature.

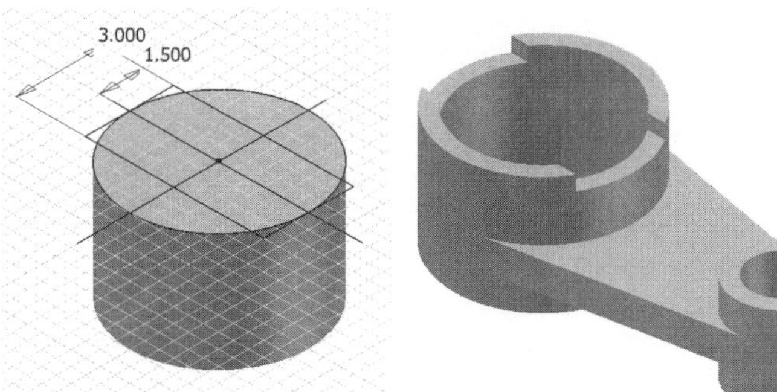

Figure 4.15 - Notch feature child to parent cylinder feature

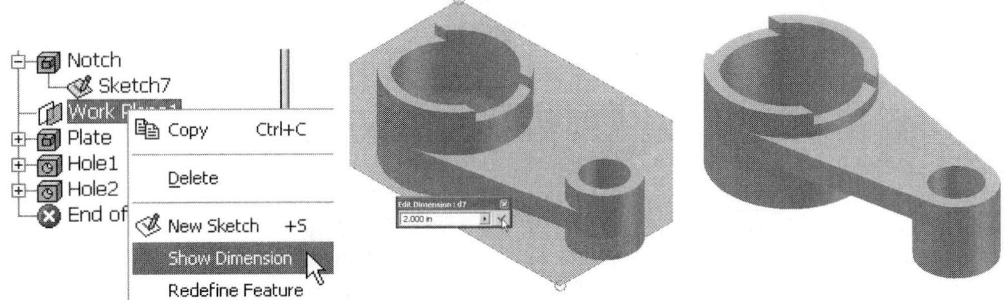

Figure 4.16 - Work plane feature parent to child plate feature

Another example is the notch. The sketch plane of the notch feature is the top face of the base feature. The notch therefore depends upon, or is a child to, the base feature. See Figure 4.15.

A third example is the plate feature. Work Plane1 was used as the sketch plane for the feature. The position of the plate depends upon (is a child to) the position of the work plane. In Figure 4.16, the height of the work plane is increased from 1 to 2. Notice the effect this has on the plate feature. Note that in order for this change to take place, it was first necessary to update the part by selecting the Update tool on the standard toolbar.

Select Other and Measure Tools

Select Other Tool

In the process of model building and editing, it is often necessary to select model geometry from the graphics window. As the cursor passes over the model, different geometric entities are briefly highlighted. If the cursor momentarily pauses over the model, the Select Other tool appears.

Since there can be a great many selectable geometric entities (i.e., parts, features, faces, work planes, edges, axes, vertices, work points) in a given region of even a simple model, the Select Other tool can be used to advantage to cycle through, identify, and select an entity of interest, all without having to change the view.

Once the Select Other tool appears, keep the cursor over the tool and then move it slightly either to the right or the left, until the corresponding right/left arrow turns green. Left-click on the arrow to cycle through the selectable geometry in the region. Once the entity of interest is identified, move the cursor back to the center of the tool until it turns green, and then left-click to select the geometry.

If the cursor is over the model in the graphics window, the Select Other tool can also be activated by right-clicking, and then selecting Select Other. . . .

Measure Tool

The Measure tool is used to take dimensions directly from the model in the graphics window. The Measure tool has four different modes; Distance, Angle, Loop, Area. The tool can be accessed either from a right mouse button context menu (right-click in the graphics window, then select Measure), or from the menu bar (Tools > Measure mode).

Upon selecting one of the Measure tool modes, the Measure box opens and the cursor changes to show a ruler, as shown in Figure 4.17.

Distance mode can be used to determine edge lengths (Figure 4.18A), the diameter and circumference of radial features (Figures 4.18B and C), and distances between two selectable points on the model (Figure 4.18D).

As shown in Figures 4.18C and 4.19A, the Select Other tool can be used in measure mode to cycle through and select the intended geometry.

Figure 4.19 shows a measure distance operation where the Select Other tool is used to select the cylindrical face of a hole (Figure 4.19A). At this point the Measure box displays the position of the hole face with respect to the origin (Figure 4.19B). In Figure 4.19C an adjacent hole has been selected, resulting in the display of the distance between the two hole centers (Figure 4.19D). To continue measuring distances, click the arrow shown in Figure 4.19D, and then select Reset from the context menu that appears (Figure 4.19E).

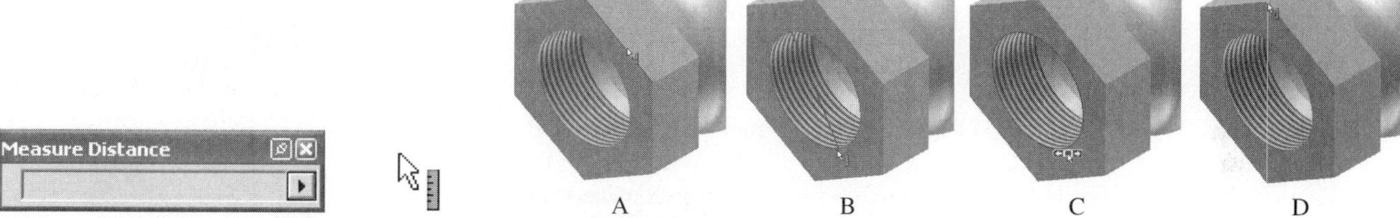

Figure 4.17 - Measure box and measure mode cursor icon

Figure 4.18 - Distance measure: edge, diameter, circumference, and distance between two points

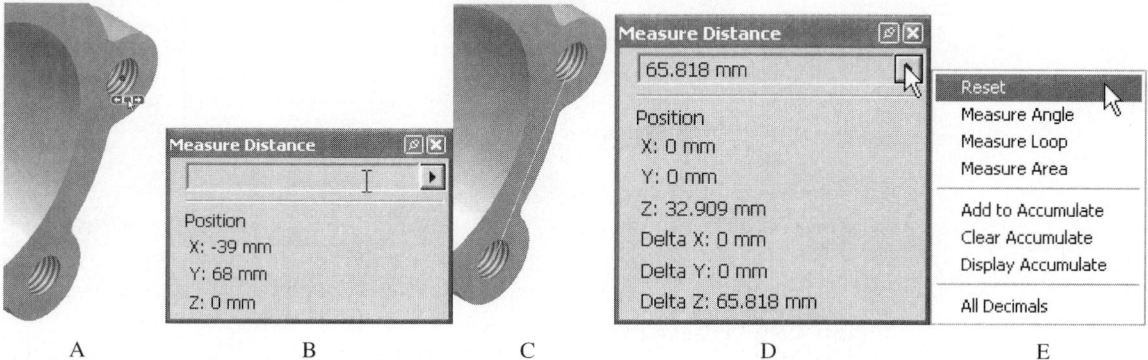

Figure 4.19 - Measure distance (between two points) operation

Using the Measure tool to measure angles, loops, or areas is similar to measuring distances. Finally, note (in Figure 4.19D) that it is possible to accumulate measurements. For example, the total surface area of a part can be calculated by using the Accumulate options.

Part Color: Part, Feature, Face

Model color can be controlled at the part, feature, and face levels. To change the color of a part, use the material drop-down list from the standard toolbar, as shown in Figure 4.20. Note that many of the colors are associated with materials, and that some colors are translucent, while others apply a texture mapping to the part. The option "As Material" assigns color based upon the material assigned to the part.[1]

Different part features can be assigned colors distinct from the global part color. This is accomplished from the part browser. Right-click on the feature, and select Properties from the context menu (Figure 4.21A). The Feature Properties dialog box appears. Select a feature color from the Feature Color Style drop-down list, as shown in Figure 4.21B. The result is shown in Figure 4.21C.

New to Inventor Release 6 is the ability to change the color of individual part faces. To change a face color, highlight the face in the graphics window, right-click, and select Properties from the context menu. The Face Properties dialog box, shown in

Figure 4.20 - Material drop-down list

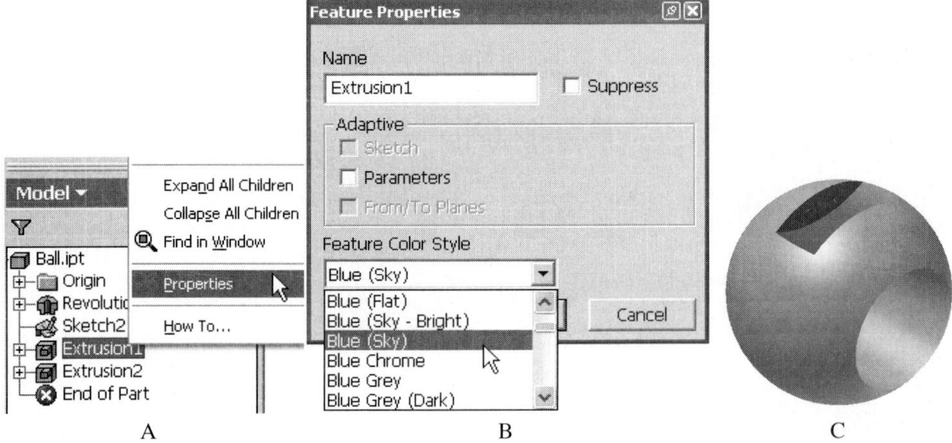

Figure 4.21 - Feature color selection

[1]Materials are covered in Chapter 8.

Figure 4.22, appears. Select a color from the Face Color Style drop-down list, and then click OK. The result of this operation is shown on the right in Figure 4.22.

Analyze Faces

Analyze Faces, another Release 6 addition, is used to evaluate surface quality. This tool is accessed from the menu bar (Tools > Analyze Faces). This opens the Analyze Faces dialog box, shown in Figure 4.23. Analyze Faces can also be toggled on and off from the standard toolbar.

Two styles are available, Zebra pattern and Draft analysis. Zebra pattern is used to evaluate surface continuity by projecting parallel lines onto the model. Parallel stripes indicate flat areas on the model, whereas jagged stripes are an indication that the surface curvature is not constant.

Draft analysis is helpful in determining whether a model can be manufactured by casting. The color spectrum indicates how the draft angle changes within a specific range.

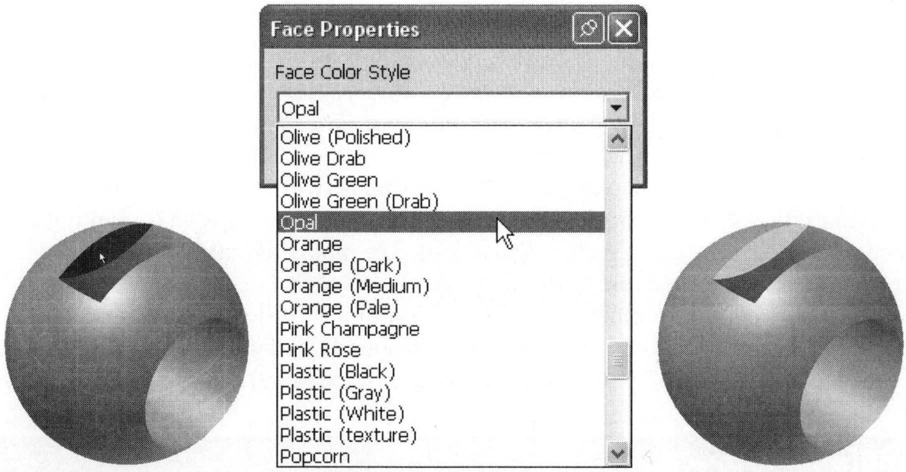

Figure 4.22 - Face color operation

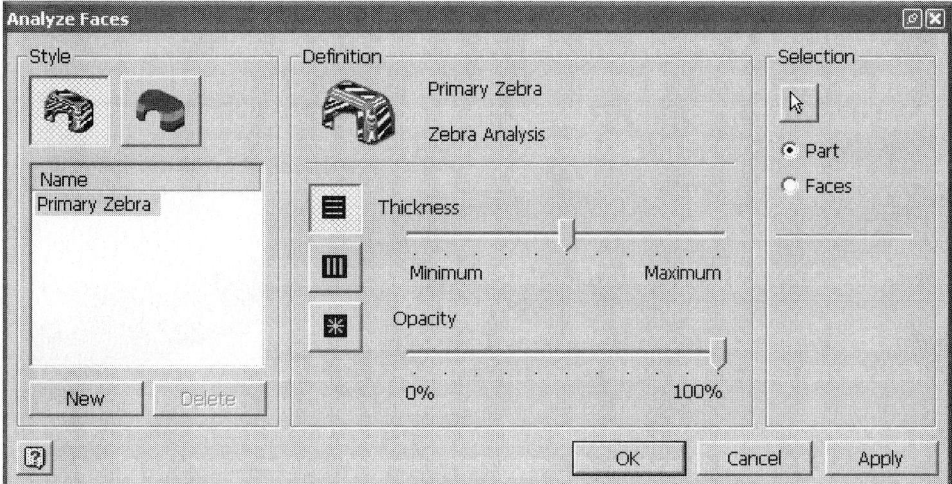

Figure 4.23 - Analyze Faces dialog box

Figure 4.24 - Analyze Faces; from left to right, zebra pattern analysis and draft analysis

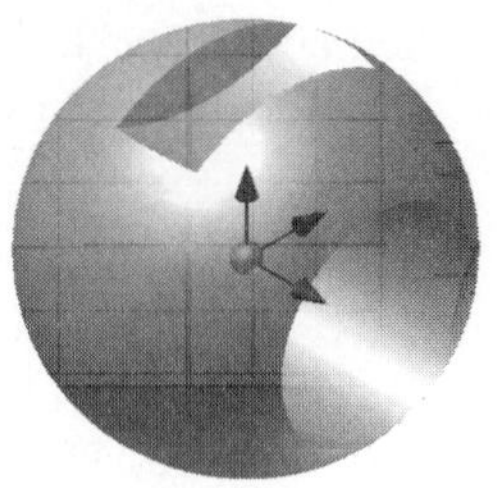

Figure 4.25 - Center of gravity icon displayed

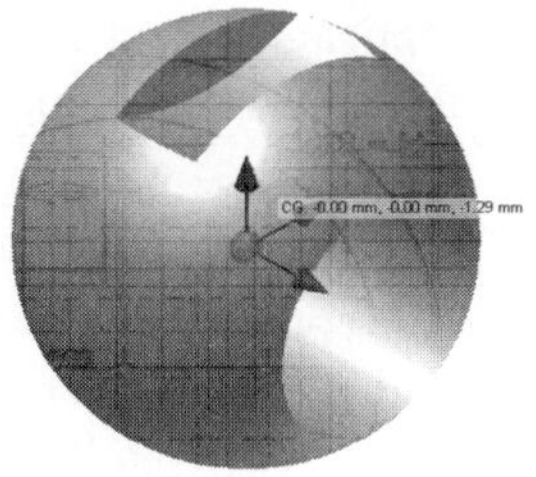

Figure 4.26 - Center of gravity coordinates displayed

Both styles contain settings that can be adjusted from the Analyze Faces dialog box. It is also possible to define new styles based on these settings. Otherwise the default styles, primary zebra and primary draft, are used. Examples of both the zebra pattern and draft analysis are shown in Figure 4.24.

Center of Gravity

Another tool new to Inventor Release 6 allows for the display of the center of gravity of both parts and assemblies. The Center of Gravity (COG) tool is accessed from the menu bar (View > Center of Gravity). If this tool is turned on, the center of gravity symbol, shown in Figure 4.25, is displayed.

It is also possible to display the COG coordinates within the graphics window, as shown in Figure 4.26. To do this, either rotate the view so that the COG symbol can be directly selected, or use the Select Other tool (i.e., place the cursor over the symbol, then use Select Other to cycle to the COG symbol).

If the part geometry changes, the COG symbol will be dimmed. To update the mass properties, select Tools > Update Mass Properties from the menu bar. See Chapter 8 on advanced assembly modeling for more information on the COG tool.

TUTORIAL 5 Handle (continued)

In Chapter 2 the base feature shown in Figure 4.27 had been created and saved. We will now add two additional features in order to complete the part.

Build Strategy

1. Share sketch, extrude (Figure 4.28).
2. Sketch, extrude (Figure 4.29).

Figure 4.27 - Handle base feature

Figure 4.28 - Extrude shared sketch

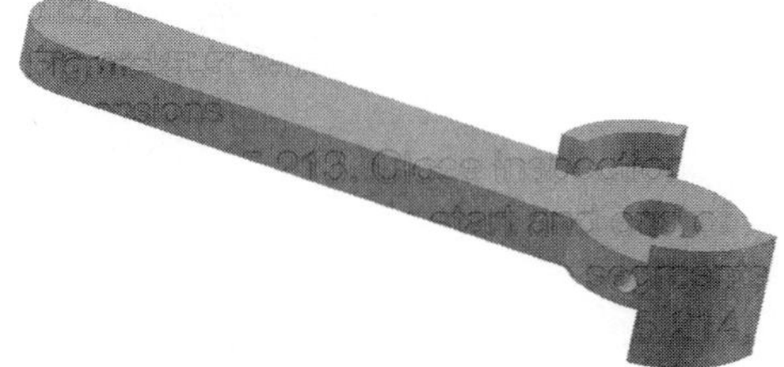

Figure 4.29 - Completed handle

Detailed Modeling Steps

1. Open the part file **Handle.** Recall that the handle base feature was created by extruding two closed areas in both directions from the sketch plane. However, other enclosed areas exist on this sketch. We will now re-use (i.e., share) this sketch, extruding these unused areas to create an additional feature. See Figure 4.30.
2. In the Browser Bar, expand Extrusion1 by clicking on the + sign. The consumed (Sketch1) should now be visible in the browser below Extrusion1. Right-click on Sketch1, and then select Share Sketch as shown in Figure 4.31. The sketch dimensions should now be visible on the screen.[2] Also, in the browser Sketch1 should appear both as a consumed sketch below Extrusion1, and as a shared sketch above Extrusion1. See Figure 4.32.

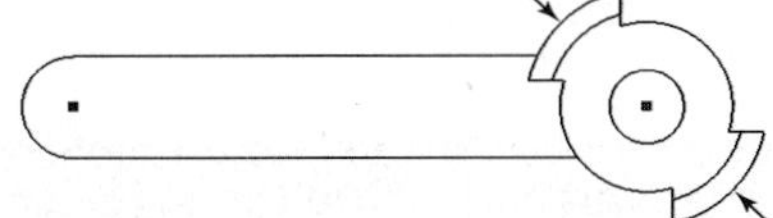

Figure 4.30 - Base feature sketch showing separate areas

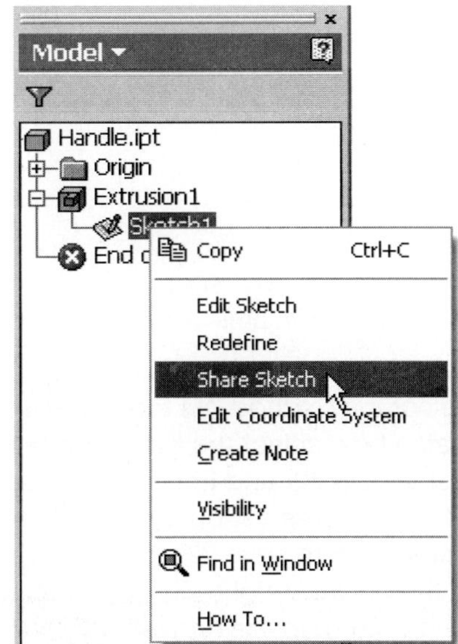

Figure 4.31 - Accessing share sketch tool

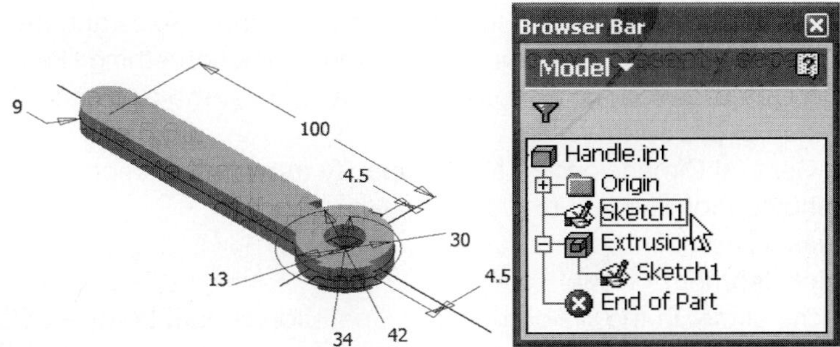

Figure 4.32 - Shared sketch

[2]When a sketch is shared, the visibility of that sketch is turned on.

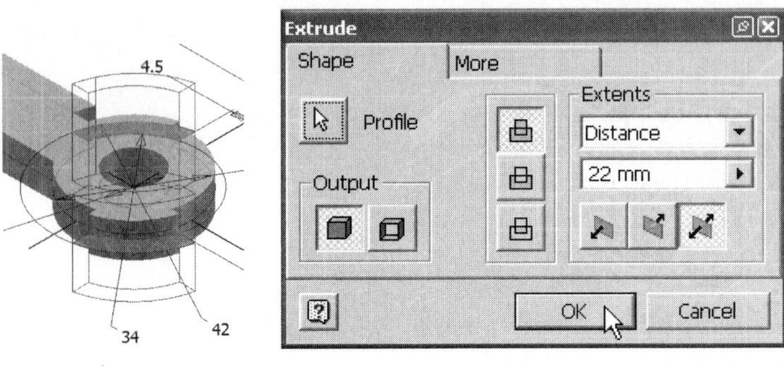

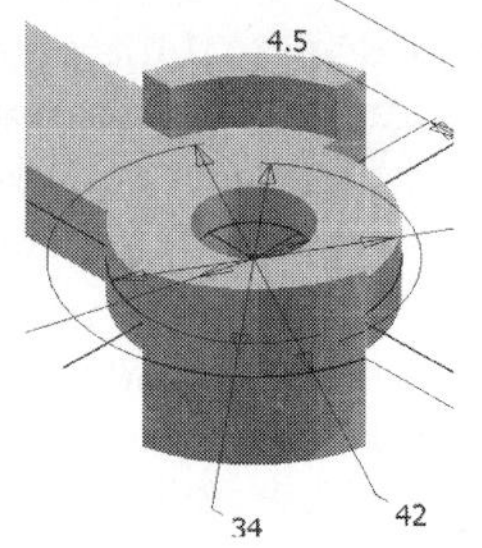

Figure 4.33 - Extrude shared sketch operation

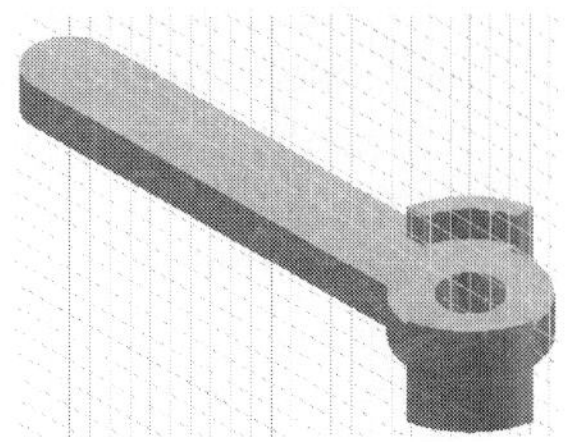

Figure 4.34 - New sketch (default XZ) plane

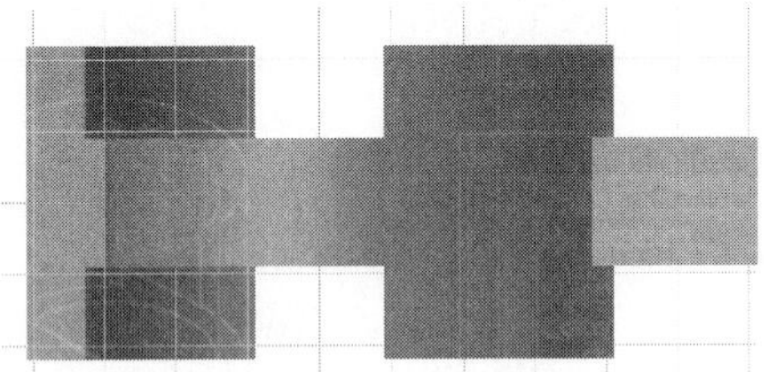

Figure 4.35 - View of handle seen with line of sight normal to XZ plane

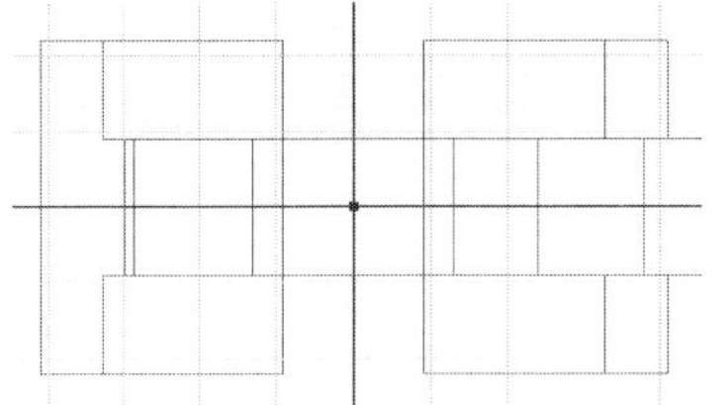

Figure 4.36 - Wireframe display with projected axes

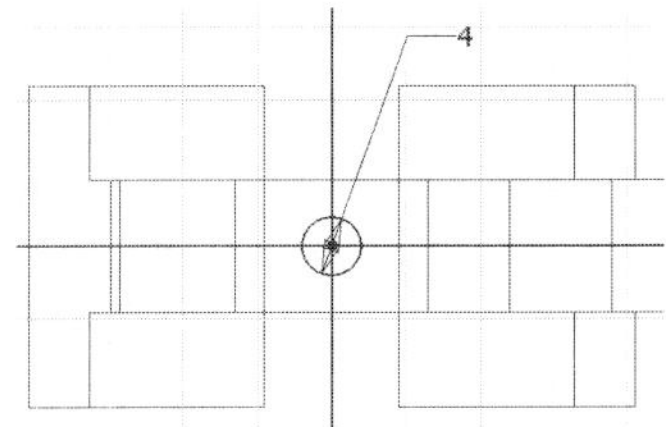

Figure 4.37 - Fully constrained sketch

3. Select the Extrude tool. After selecting the Profile button in the Extrude dialog box, select the two outer profiles shown in Figure 4.33. The Extrusion distance is 22, and the extrusion takes place from the mid-plane. After clicking OK, your screen should resemble the image on the extreme right.
4. Remember to save.
5. In the browser, right-click on Sketch1, and then de-select Visibility. Expand the Origin folder in the browser by clicking on the + preceding the Origin folder icon. Right-click on the XZ Plane, and then select New Sketch. Your screen should resemble Figure 4.34.
6. From the standard toolbar, click on the Look At icon, and then click on the XZ Plane in the browser. The screen should look something like Figure 4.35.
7. Again from the standard toolbar, change to a Wireframe display.
8. Using the Project Geometry tool, project the X and Z Axes and the Center Point onto the sketch plane. Your screen should look something like Figure 4.36.
9. Using the Center point circle tool, place a circle at the projected Center Point (intersection of projected X and Z Axes).
10. Use the General Dimension tool to specify the circle diameter as 4. Your screen should look like Figure 4.37.
11. Right-click, Finish Sketch.
12. Right-click, Isometric View.
13. Extrude the circle. In the dialog box, the Operation should be set to Cut, the Extents to All, and the direction should be from the mid-plane in both directions. See Figure 4.38.
14. After changing back to a Shaded display and rotating your view slightly, your model should resemble Figure 4.39.
15. Save the part file. This completes Tutorial 5.

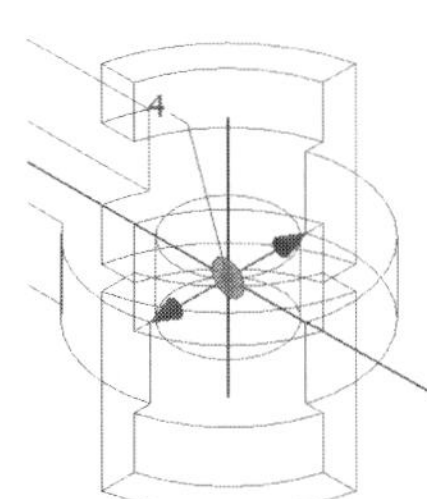

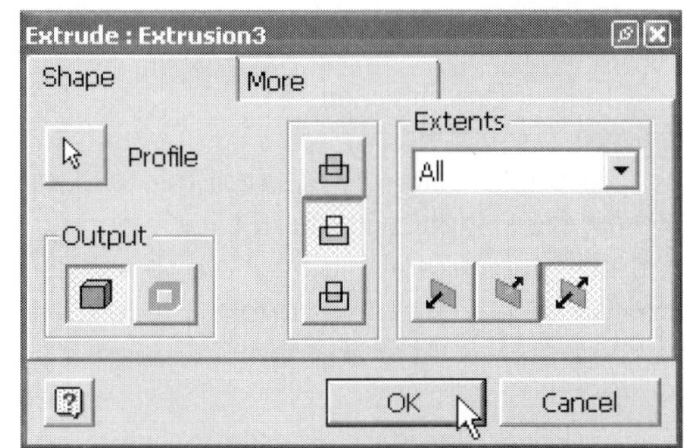

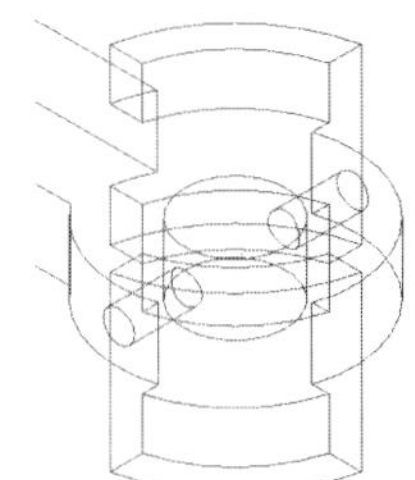

Figure 4.38 - Extruded cut operation

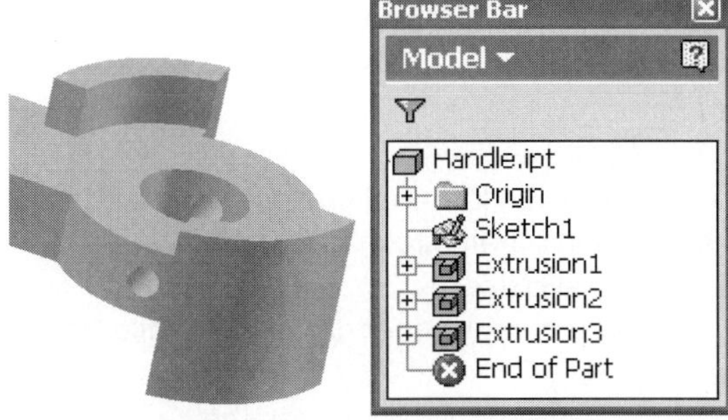

Figure 4.39 - Detail of recently added extruded cut feature

Ball — Tutorial 8

Build Strategy

1. Sketch (Figure 4.40).
2. Revolve (Figure 4.41).
3. Sketch, then extrude (cut) from mid-plane (Figure 4.42).
4. Share sketch, sketch, then extrude (cut) (Figure 4.43).

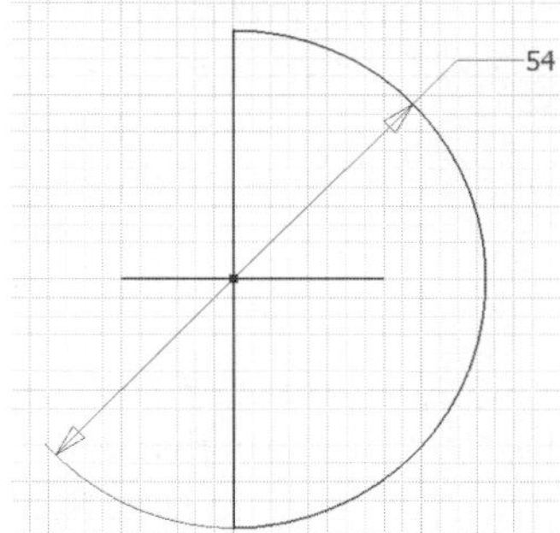

Figure 4.40 - Base feature sketch

Figure 4.41 - Base revolved feature

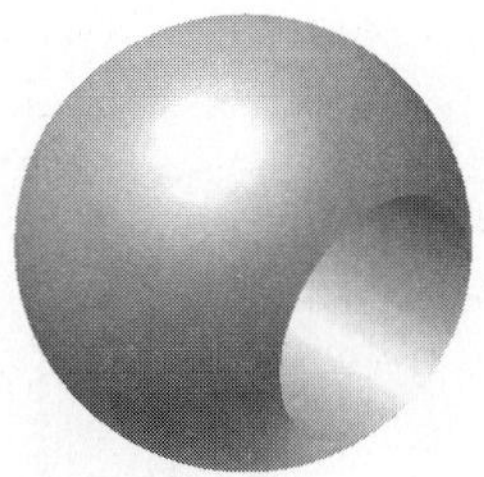

Figure 4.42 - Extruded (cut) feature

Figure 4.43 - Completed ball part

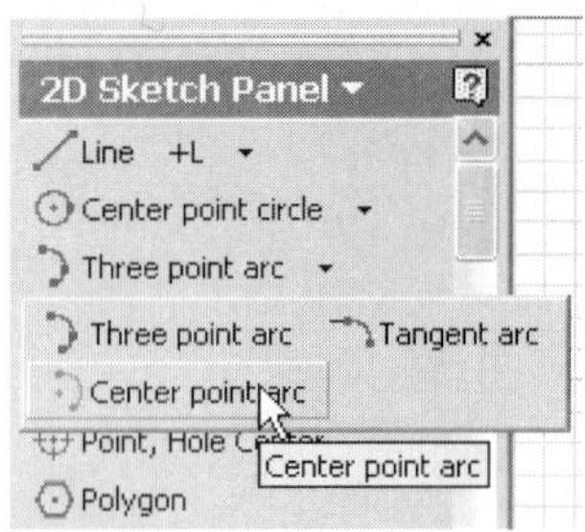

Figure 4.44 - Access to center point arc tool

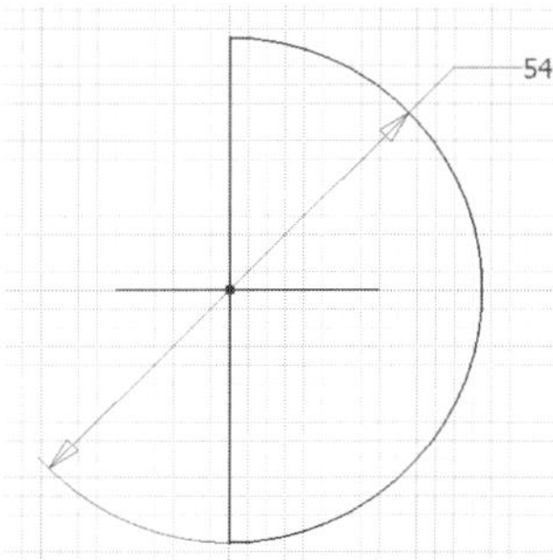

Figure 4.45 - Fully constrained base feature sketch

Figure 4.46 - Base revolved feature

Detailed Modeling Steps

1. Start a new metric part file.
2. Project X and Y Axes, Center Point.
3. Change the Vertical Y Axis to a centerline (select the axis, click the down arrow beside the Style field on the standard toolbar, click Centerline).
4. Use the Center point arc tool (see Figure 4.44) to sketch an arc with its center at the projected coordinate system origin. Place the arc endpoints coincident with the vertical Y Axis and the 180 degree arc to the right of the vertical centerline.
5. Use the Line tool to sketch a vertical line connecting (i.e.,) the two arc endpoints.
6. Use the General Dimension tool to specify the diameter (54) of the arc. The sketch should be similar to Figure 4.45. Note that to change from a radius to a diameter dimension, right-click, and then select Diameter while in the process of creating the dimension.
7. Right-click, Finish Sketch.
8. Right-click, Isometric View.
9. Revolve . See Figure 4.46. Note that while one's first inclination may be to revolve a circle (rather than a half circle), this cannot be done because the profile cannot cross the axis of revolution.
10. Save the file as **Ball.** It is recommended that the file be regularly saved every five to ten minutes.
11. Expand the Origin folder. After right-clicking on the YZ Plane, choose New Sketch.
12. Choose Look At , and then select the YZ Plane in the browser.
13. Change to a Wireframe Display.
14. Use the Project Geometry tool to project the Y and Z Axes, as well as the Center Point.
15. Use the Center point circle tool to place a circle coincident with the projected origin.
16. Use the General Dimension tool to add a diameter of 32 to the circle. The sketch should be similar to Figure 4.47.
17. Right-click, Finish Sketch.
18. Right-click, Isometric View.
19. Extrude the circle; specify a cut operation, All Extents, from the mid-plane (Figure 4.48). Click OK when finished. Note that it may be necessary to select

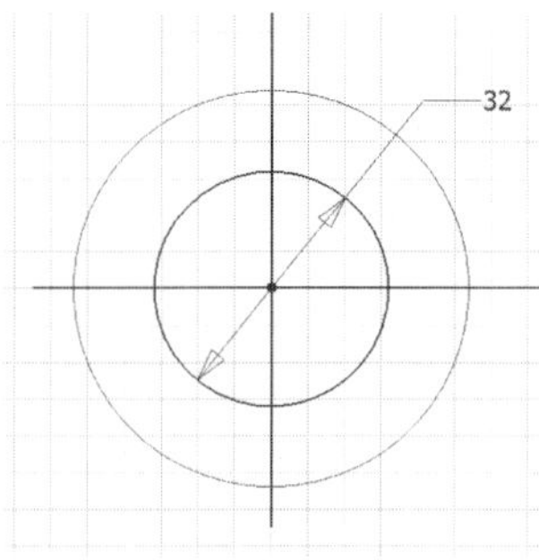

Figure 4.47 - Fully constrained sketch

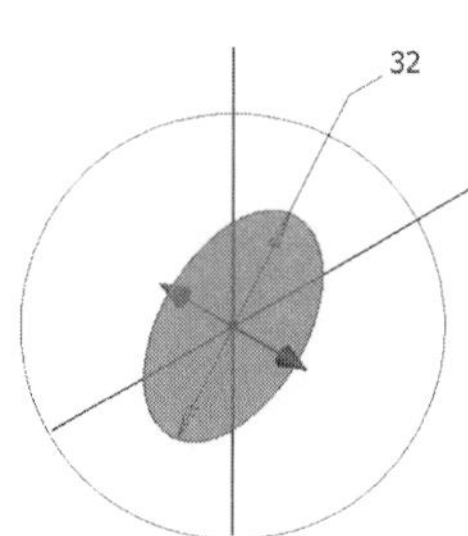

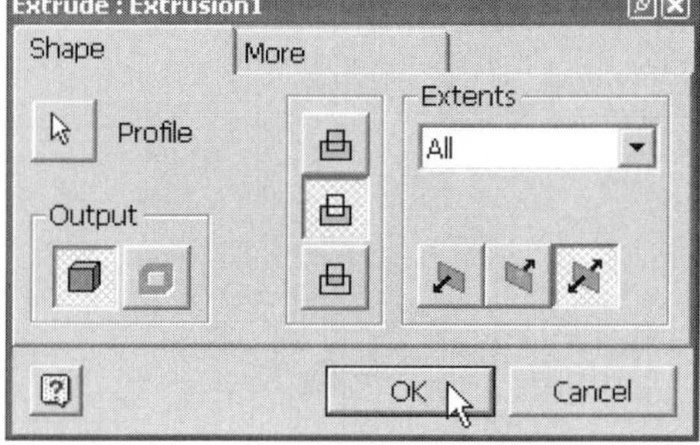

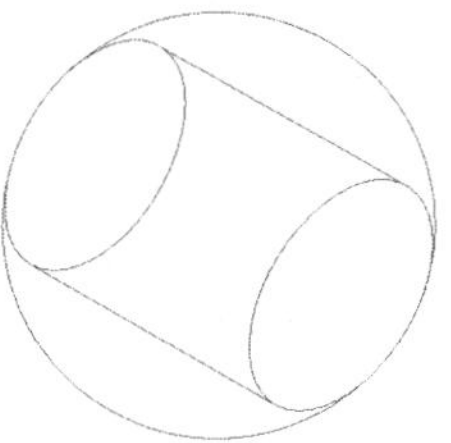

Figure 4.48 - Extrude (cut) operation

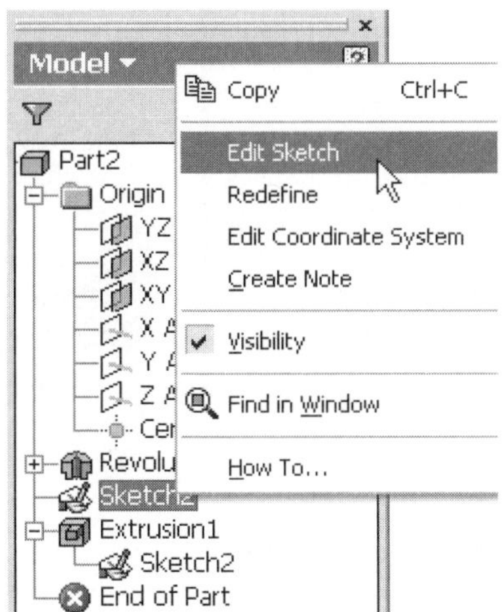

Figure 4.49 - Edit shared sketch access

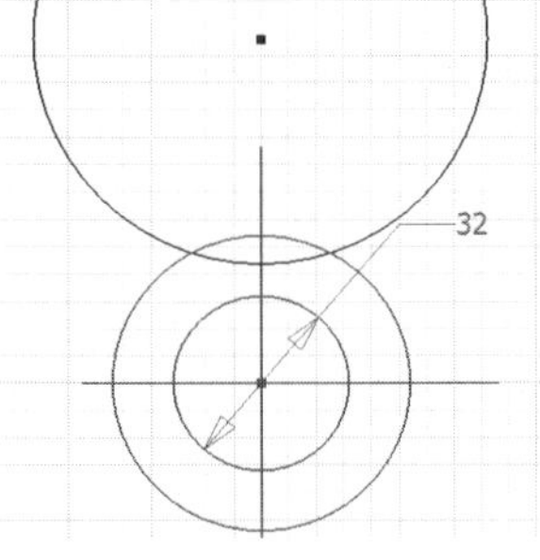

Figure 4.50 - New circle sketched on existing (shared) sketch

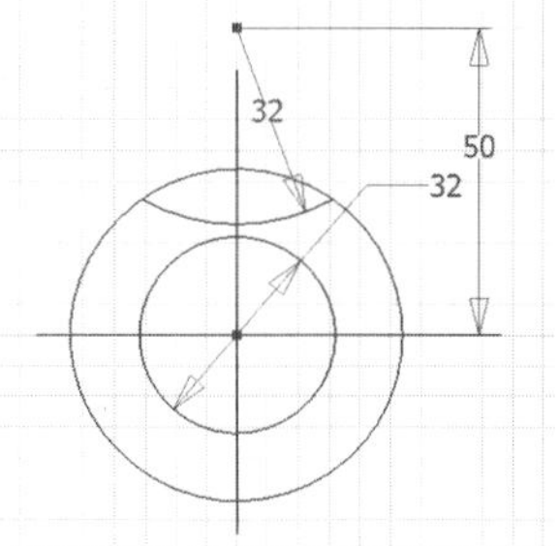

Figure 4.51 - Fully constrained sketch

the profile to be extruded; to do so, you must first select the Profile button on the Extrude dialog box.

20. Change the display to Shaded. Since the next feature is also extruded from the default YZ plane, we can once again use the Share Sketch tool.

21. Expand Extrusion1 in the part browser, right-click on the consumed sketch, and choose Share Sketch (see Figure 4.31).

22. Right-click on the unconsumed shared sketch (Sketch2 in my file) in the browser, and select Edit Sketch, as shown in Figure 4.49.

23. Choose Look At, and then select the shared sketch in the browser.

24. Change to a Wireframe Display.

25. Use the Project Geometry tool to project the sphere (i.e., outer diameter circle) onto the sketch plane.

26. Use the Circle tool to sketch a circle whose center is on (i.e.,) the vertical Z Axis, well above the sphere. The circle should intersect the projection of the sphere. In addition, the circle should not extend below the horizontal Y Axis. Your sketch should resemble Figure 4.50.

27. Use the Trim tool to trim the circle with the projection of the sphere. Only the portion of the circle inside the sphere should remain. Note that the sphere must be projected in order for this step to work properly; see step 25.

28. Use General Dimension to add the dimensions shown in Figure 4.51.

29. Right-click Finish Sketch.

30. Right-click Isometric View.

31. Change the display to Shaded.

32. Extrude the closed profile (eye lid shape) from the mid-plane, using a cut operation, with a distance of 13 mm (Figure 4.52).

33. Use the part browser to turn off the visibility of the shared sketch. The completed part is shown in Figure 4.53.

34. Save the part file. This completes Tutorial 8.

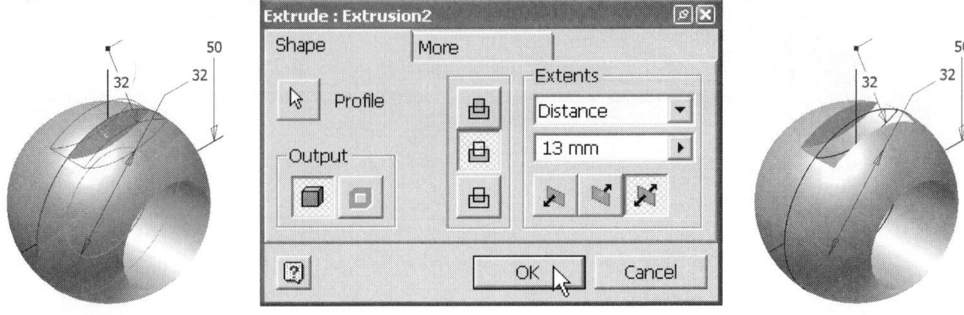

Figure 4.52 - Extruded cut operation

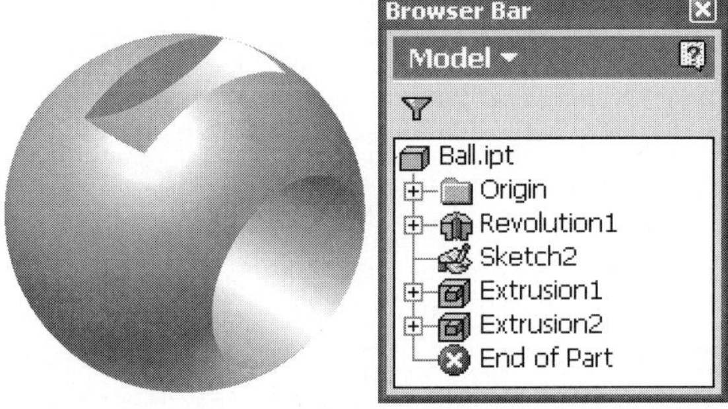

Figure 4.53 - Completed part

TUTORIAL 9 Stem

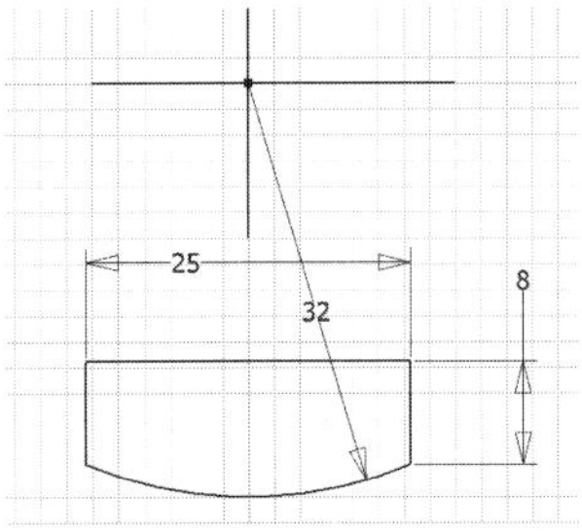

Build Strategy

1. Sketch (Figure 4.54).
2. Extrude (Figure 4.55).
3. Revolve (intersect) (Figure 4.56).
4. Create work plane(s), sketch, extrude (join) (Figure 4.57).
5. Redefine isometric, extrude (cut) (Figure 4.58).

Figure 4.54 - Fully constrained base feature sketch

Figure 4.55 - Base feature

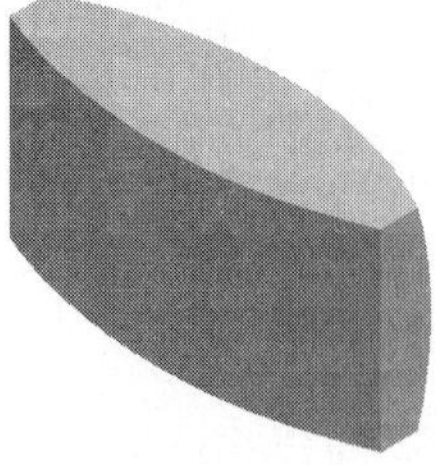

Figure 4.56 - Revolved feature

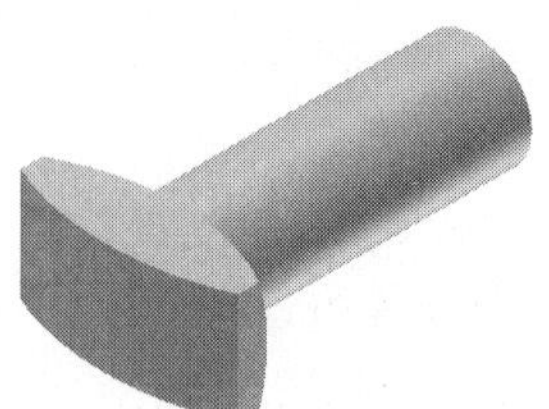

Figure 4.57 - Extruded join feature

Figure 4.58 - Completed Stem part

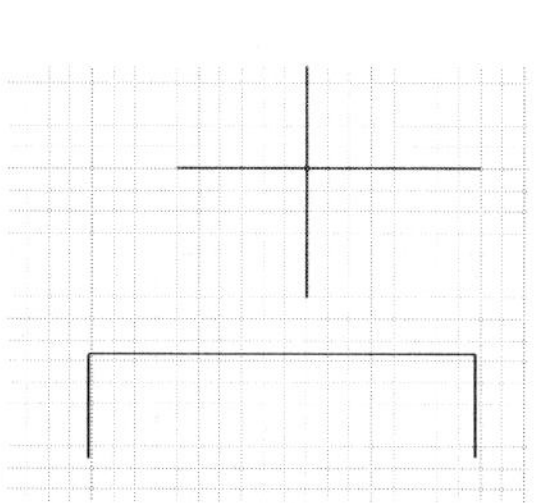

Figure 4.59 - Start of base feature sketch

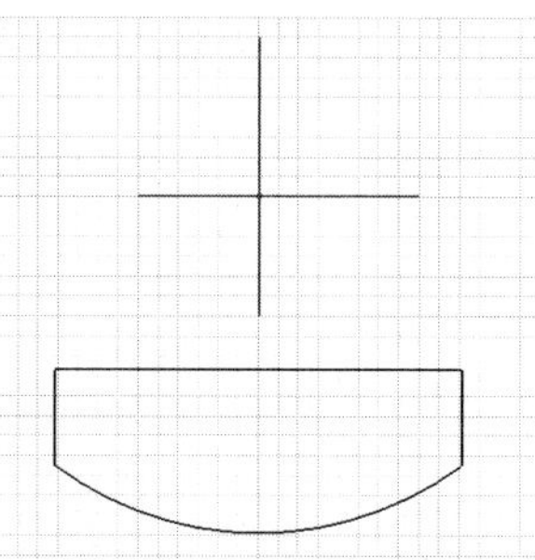

Figure 4.60 - Base feature sketch without dimensions

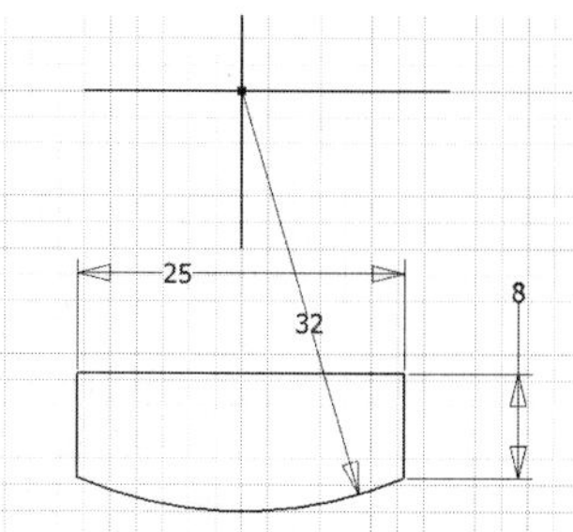

Figure 4.61 - Fully constrained base feature sketch

■ Detailed Modeling Instructions

1. Start a new metric part file.
2. Project the X and Y Axes and Center Point onto the sketch plane.
3. Use the Line tool to sketch Figure 4.59.
4. Add Equal and Symmetric constraints to the vertical lines. The line of symmetry is the vertical Y Axis.
5. Use the Center point arc tool to draw an arc whose center is at the projected Center Point. Also, the arc endpoints should be coincident with the lower endpoints of the previously sketched vertical lines. The sketch should be similar to Figure 4.60.
6. Use the General Dimension tool to add the dimensions shown in Figure 4.61.
7. Right-click, Finish Sketch.
8. Right-click, Isometric View.
9. Extrude the profile a distance of 13, from the mid-plane. Click OK when done. The resulting shape should be similar to that shown on the extreme right of Figure 4.62.
10. Save the file as **Stem.** It is recommended that the file be regularly saved every five to ten minutes.
11. Since the next feature is also extruded from the default (XY) plane, we can use the Share Sketch tool. In the browser, expand Extrusion1 by clicking on the leading +. Sketch1 should now be visible. Right-click on Sketch1, then choose

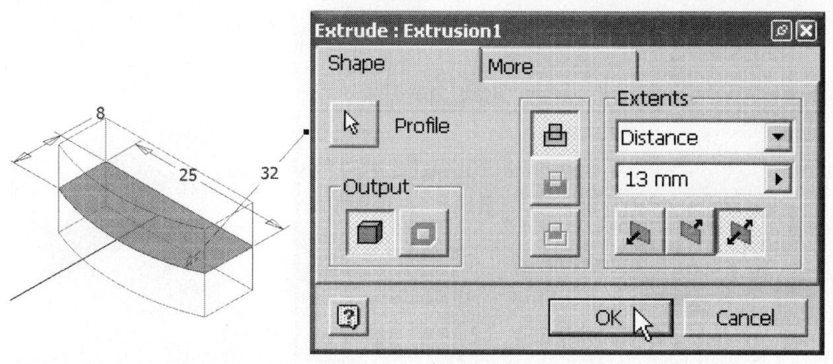

Figure 4.62 - Base extrude feature operation

Share Sketch. Another Sketch1 appears in the browser, above Extrusion1. Also the sketch is now visible in the work area (see Figure 4.31, if necessary).

12. Right-click on Sketch1 in the browser, and then choose Edit Sketch (see Figure 4.49, if needed).
13. Look At the sketch.
14. Use the Center point arc tool to sketch an arc similar to that shown in Figure 4.63. The arc endpoints and the center point should all be on (i.e.,) the projected vertical Y Axis.
15. Use the Line tool to draw a line connecting the two arc endpoints.
16. Use the General Dimension tool to add the dimensions shown in Figure 4.64.
17. Right-click, Finish Sketch.
18. Right-click, Isometric View
19. Revolve the half circle profile (Figure 4.65). Use the vertical axis as an axis of revolution. Note that the Boolean operation is *intersection.* Also note that upon assembly, this Stem part will sit inside the slot feature (see Figure 4.52) on the Ball part created in the previous tutorial.
20. Turn off the visibility of the shared sketch. Figure 4.66 shows the resulting solid.
21. Select the Work Plane tool from the (Part Features) panel bar. Select the XZ Plane (in the Origin folder of part browser), and then select the cylindrical face of the solid, as shown in Figure 4.67. A work plane parallel to the XZ Plane and tangent to the cylindrical face has been created.

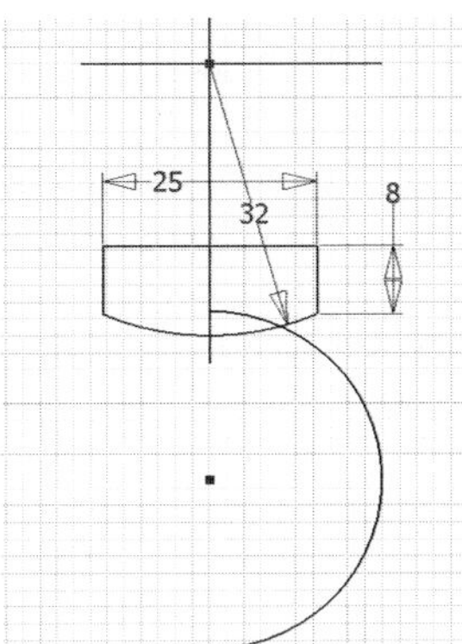

Figure 4.63 - Arc sketched on existing (shared) sketch

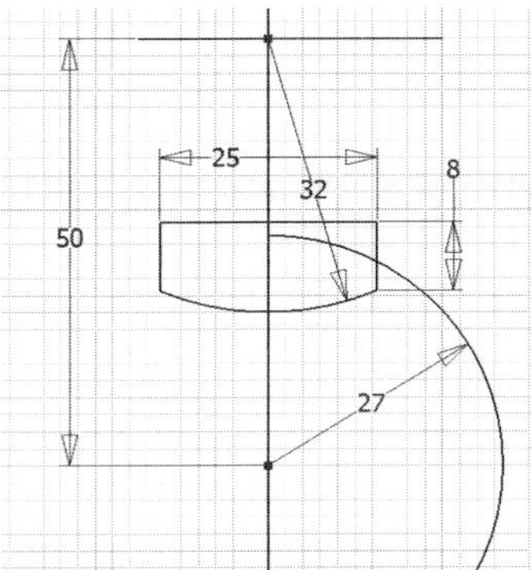

Figure 4.64 - Fully constrained sketch

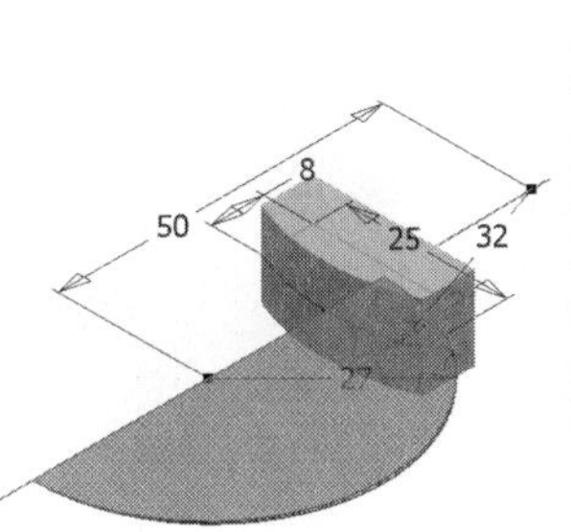

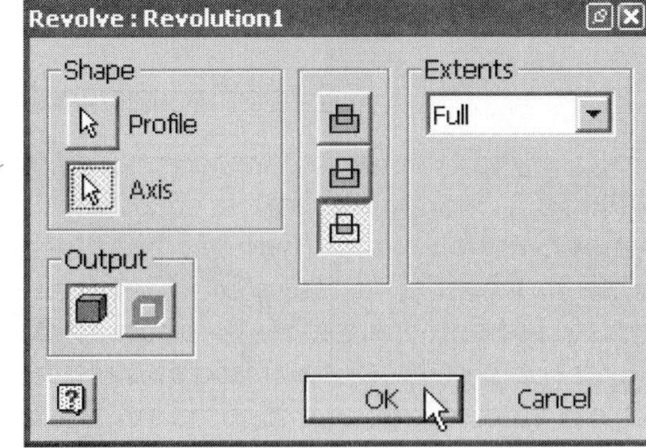

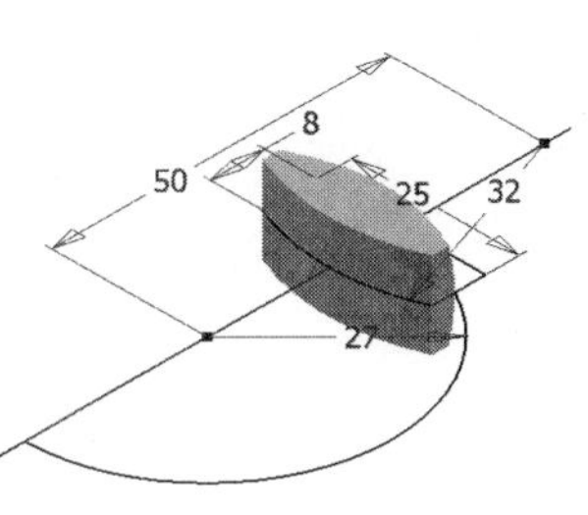

Figure 4.65 - Revolved feature intersection operation

Figure 4.66 - Revolved (intersect) feature

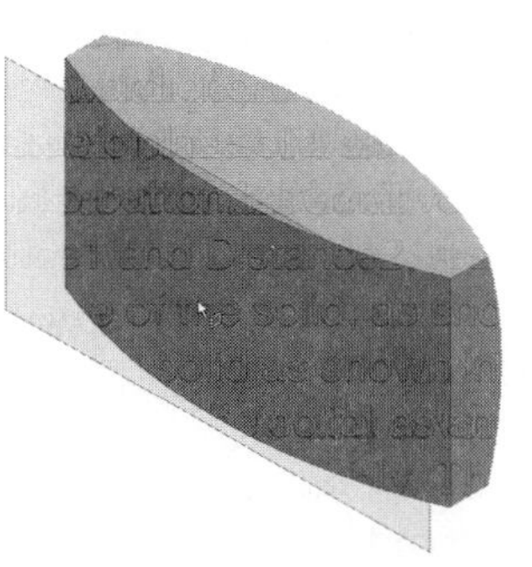

Figure 4.67 - Tangent work plane feature

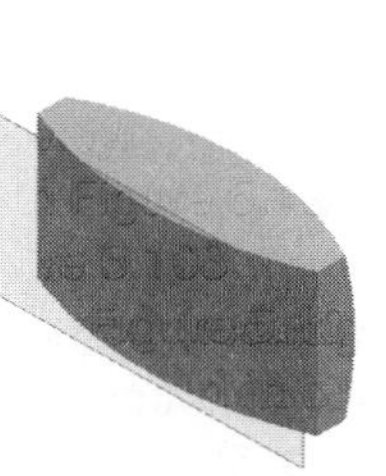

Figure 4.68 - Offset work plane feature

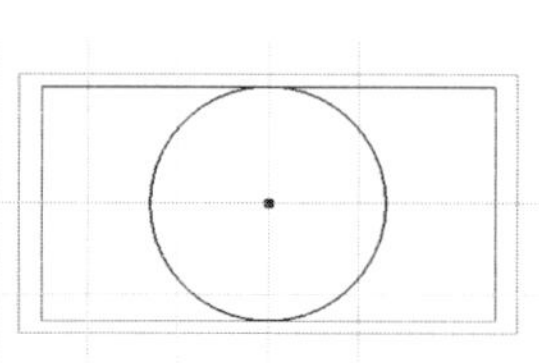

Figure 4.69 - Fully constrained sketch

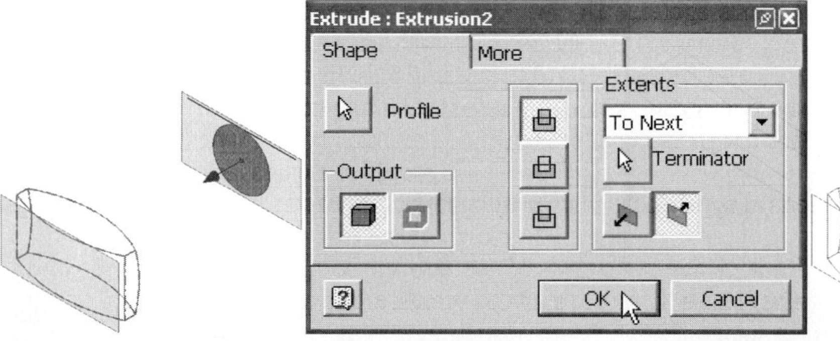

Figure 4.70 - Extruded join feature (from work plane)

22. Select the Work Plane tool again. This time click and drag the tangent work plane up and to the right. The Offset dialog box appears. Enter 42 as an offset distance. After Zooming All, the screen should be similar to Figure 4.68.
23. Right-click, New Sketch, then select the new offset work plane.
24. Look At the new sketch.
25. Change the display to wireframe.
26. Project the Center Point onto the sketch.
27. Use the Center point circle tool to draw a circle whose center is at the projected center point.
28. Add a Tangent constraint between the circle and the long edge of the solid. Your screen should resemble Figure 4.69.
29. Right-click Finish sketch.
30. Right-click Isometric View.
31. Extrude to the next surface, using a join operation. Click OK when done (Figure 4.70).
32. Change to a shaded display.
33. Turn off the visibility of the two work planes. The part should be similar to that shown in Figure 4.71.
34. It is at times useful to change the default isometric view. Use 3D Rotate/ Common View to accomplish this. NOTE: See Figures 1.19, 1.20, and 1.21 and the accompanying discussion in Chapter 1 for an overview.

    TIP: In 3D Rotate mode, the Space Bar acts as a toggle between Free Rotate and Common View.

Figure 4.71 - Extruded (join) feature

Select 3D Rotate, then right-click and select Common View. As shown in Figure 4.72, from left to right, select the arrows in such a way that the view on the extreme right results. Now right-click and select Redefine Isometric.

35. Change the display to wireframe.
36. Right-click, New Sketch. Select the YZ Plane as the sketch plane.
37. Look At the sketch plane.
38. Zoom All to center the view.
39. Project the Y Axis onto the sketch plane.
40. Use the Center point circle tool to sketch a circle on the Y Axis (see Figure 4.73).
41. Use the General Dimension tool to add the dimensions shown in Figure 4.73.
42. Right-click, Finish Sketch.
43. Right-click, Isometric View.
44. Extrude with a cut operation, all the way through, from the mid-plane (Figure 4.74).
45. Use the color control on the standard toolbar to change the color of the stem to Beige (Light), as shown in Figure 4.75.
46. After changing to a shade display the part and browser should be similar to Figure 4.76.
47. Save the file. This completes Tutorial 9.

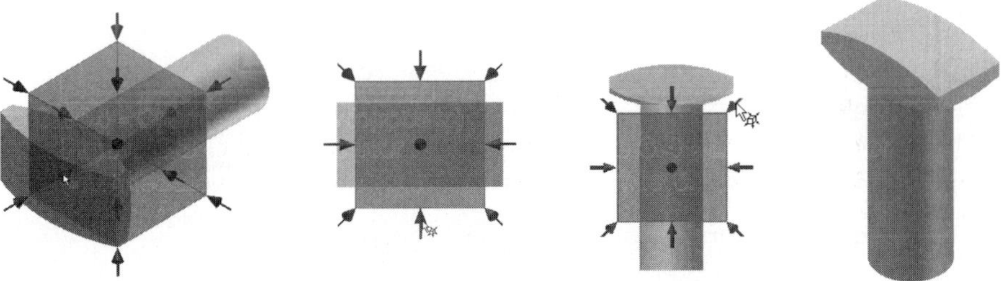

Figure 4.72 - Using Common View to redefine the default isometric view

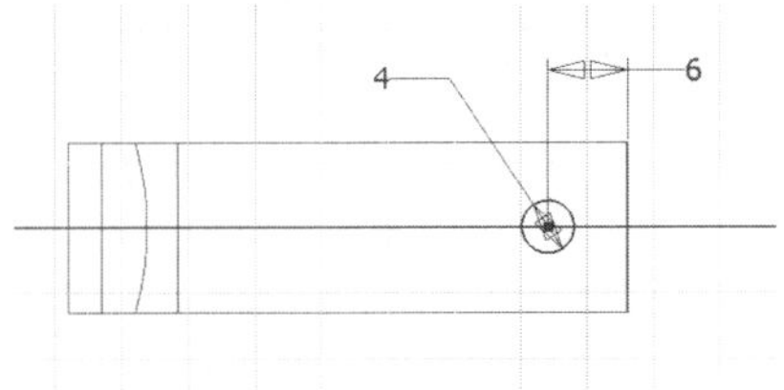

Figure 4.73 - Fully constrained sketch

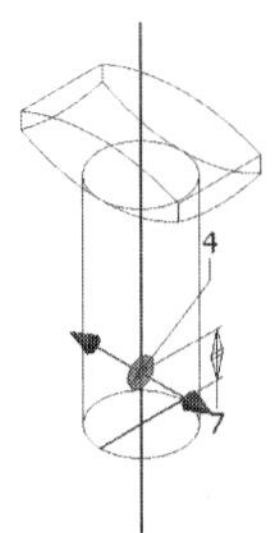

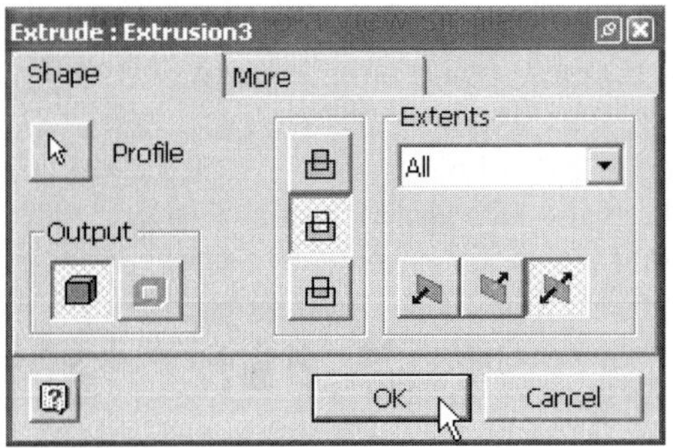

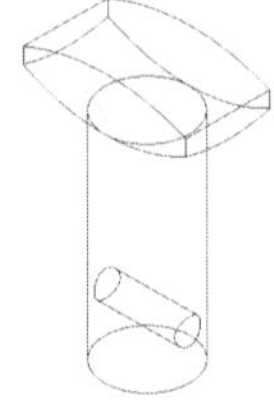

Figure 4.74 - Extruded (cut) feature operation

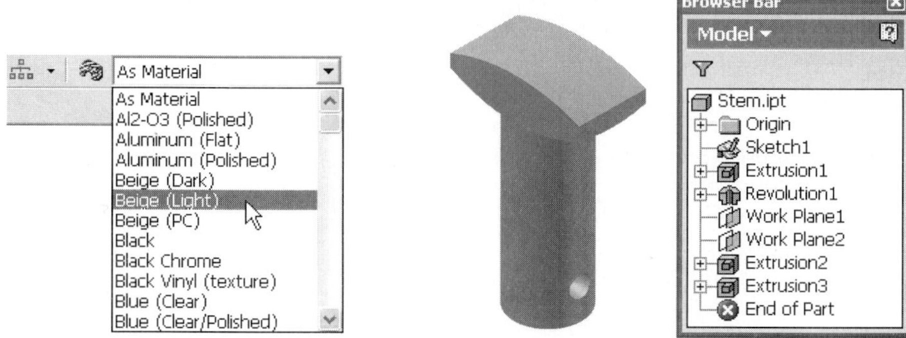

Figure 4.75 - Changing part color

Figure 4.76 - Completed part

Creating a New Project (Garlic Press) TUTORIAL 10

The next part to be modeled is not part of the ball valve assembly. For this reason we will now create a new project.

1. Open the Project Editor by selecting File > Projects . . . from the menu bar.
2. Click the New button at the bottom of the dialog box.
3. From the Inventor project wizard dialog box, click the Next button (the Personal Workspace for Group Project radio button should be selected).
4. Modify the Project File Name and Location in the Inventor project wizard dialog box as shown below, and then click the Finish Button (Figure 4.77).
5. Click the OK button when prompted to create a new project path.
6. With the Garlic Press project highlighted, click the Apply button.

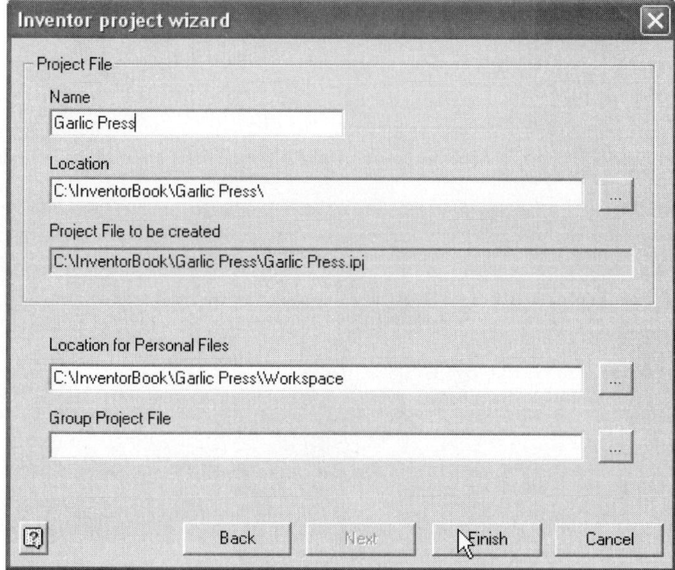

Figure 4.77 - New project dialog box

TUTORIAL 11 Piston (Garlic Press)

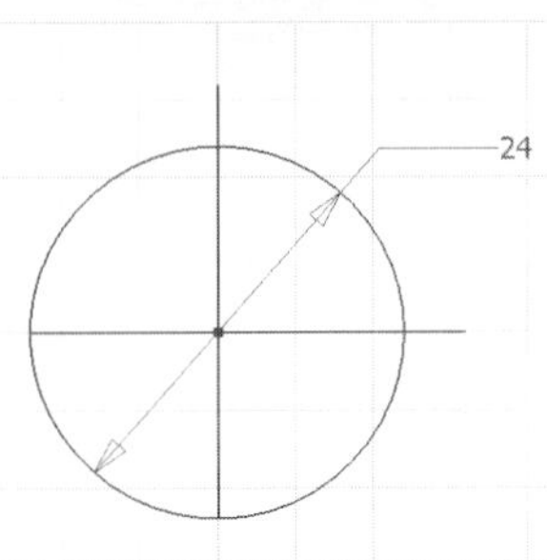

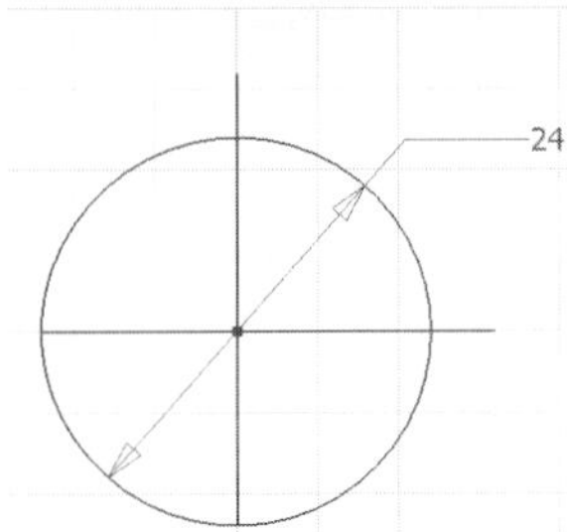

Figure 4.78 - Base feature sketch

Figure 4.79 - Base extruded feature

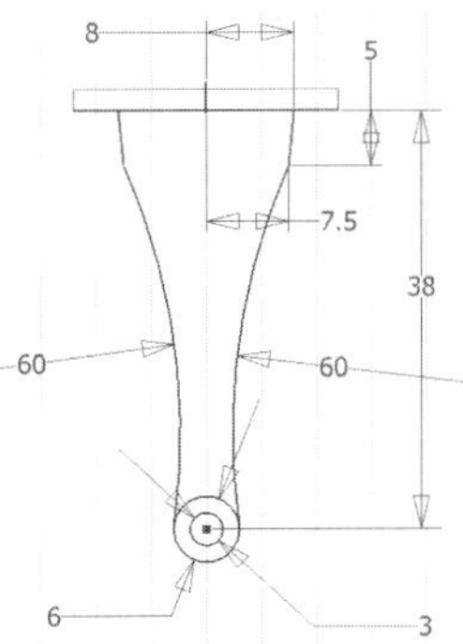

Figure 4.80 - Fully constrained sketch

Build Strategy

1. Sketch (Figure 4.78).
2. Extrude (Figure 4.79).
3. Sketch (Figure 4.80).
4. Extrude (join) (Figure 4.81).
5. Share sketch.
6. Extrude (join) (Figure 4.82).
7. Extrude (join) (Figure 4.83).

Detailed Modeling Steps

1. Start a new metric part file.
2. Project the X and Y Axes, and the Center Point.
3. Use the Circle tool to sketch circle at origin.
4. Use General Dimension tool (see Figure 4.84).
5. Finish Sketch.
6. Isometric View.
7. Extrude the sketch, as shown in Figure 4.85.
8. Save the file as **Piston.** It is recommended that the file be regularly saved every five to ten minutes.
9. Wireframe display.
10. New sketch—YZ Plane.
11. Look At the sketch.
12. Project the Z Axis and the bottom edge of Extrusion1 onto the sketch plane, as shown in Figure 4.86.
13. Draw two concentric circles, coincident with the Z Axis, and then dimension as shown in Figure 4.87.
14. Sketch an inclined line as shown in Figure 4.88, coincident with the bottom edge of the extrusion. Dimension as shown in Figure 4.88.
15. Use the Three point arc tool to sketch the arc shown in Figure 4.89. The arc is coincident with both the inclined line and large diameter circle. Add the R60 dimension. Note that when using Three point arc, the two endpoints are selected first, and then a third point on the arc.
16. Add a tangent constraint between the arc and the circle, as shown in Figure 4.90.

Figure 4.81 - Extruded join feature

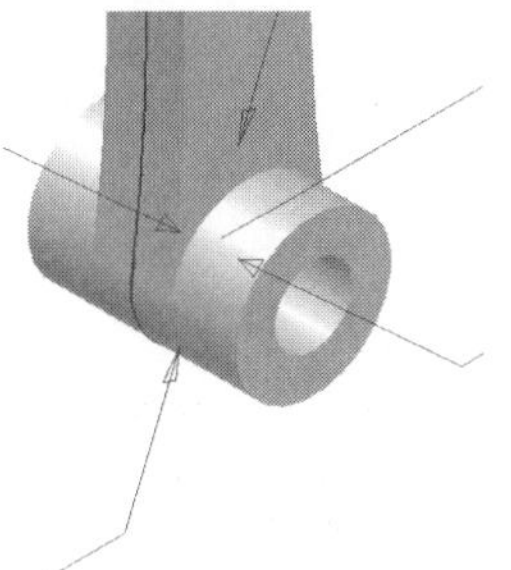

Figure 4.82 - Extrude (join) shared sketch

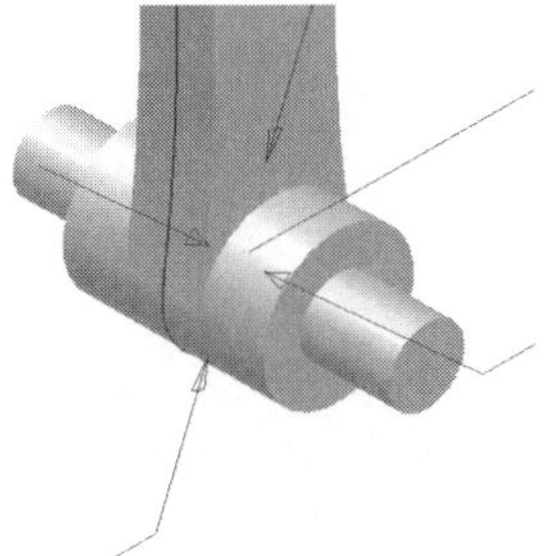

Figure 4.83 - Extrude (join) shared sketch

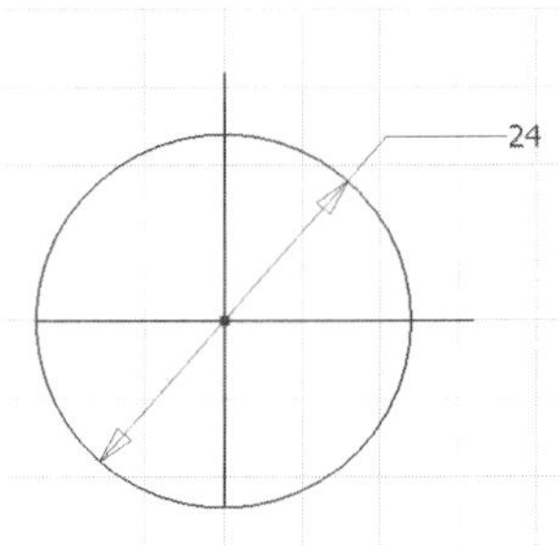

Figure 4.84 - Base feature sketch

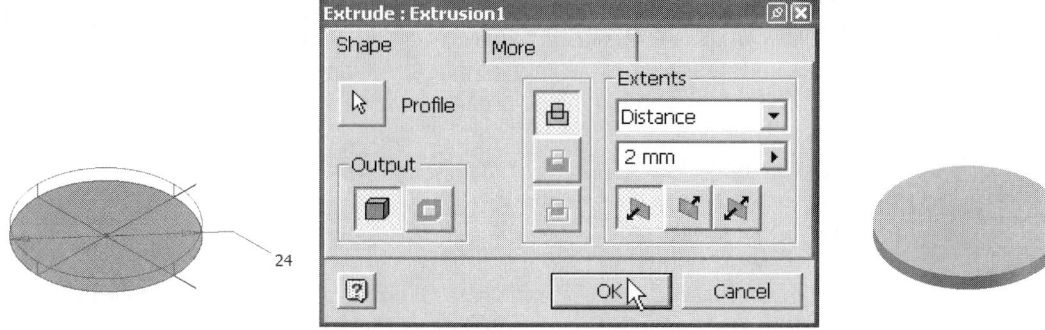

Figure 4.85 - Extruded base feature operation

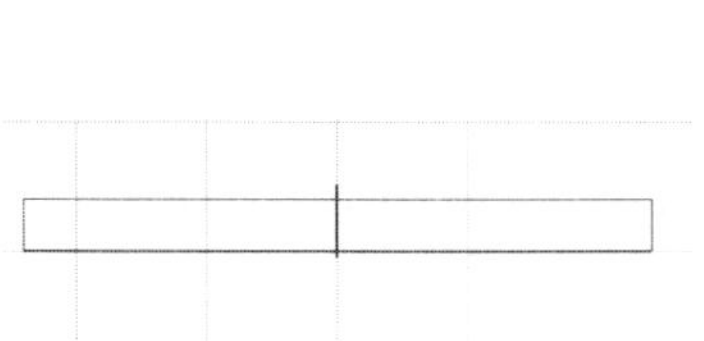

Figure 4.86 - Projected geometry

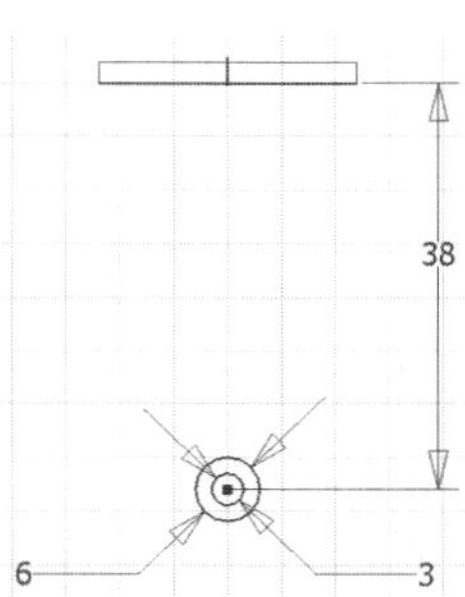

Figure 4.87 - Dimensioned sketch geometry

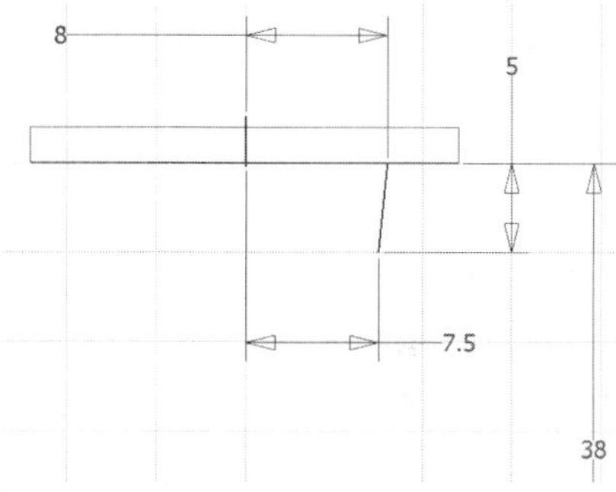

Figure 4.88 - Additional sketch geometry

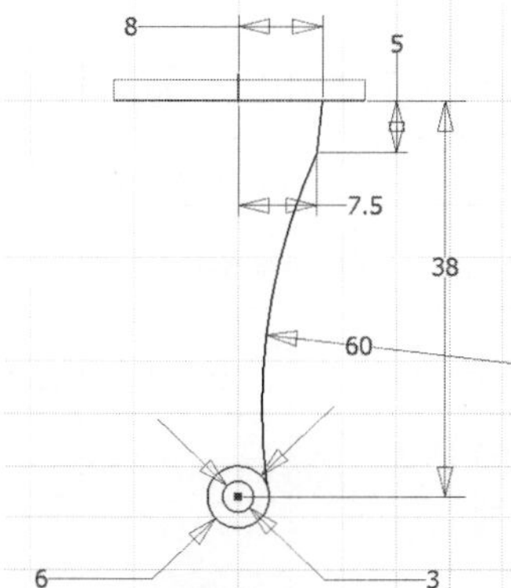

Figure 4.89 - More sketch geometry

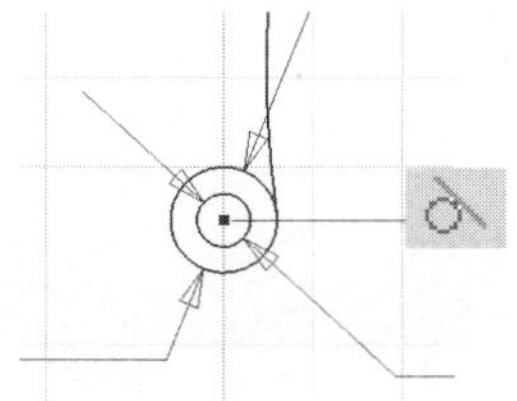

Figure 4.90 - Add tangent constraint

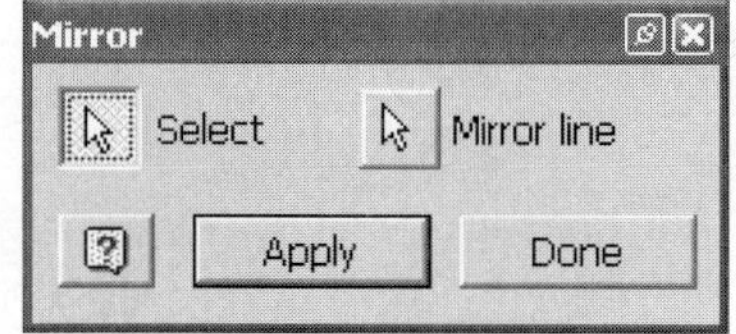

Figure 4.91 - Mirror dialog box

17. Use the Mirror tool to mirror the inclined line about the projected Z Axis. The Mirror dialog box is shown in Figure 4.91. First select the inclined line, then click on the Mirror line button and select the vertical axis, then click on the Apply button, then select Done.

18. Again use the Three point arc tool to create an arc coincident with the endpoint of the mirrored inclined line and the large diameter circle.

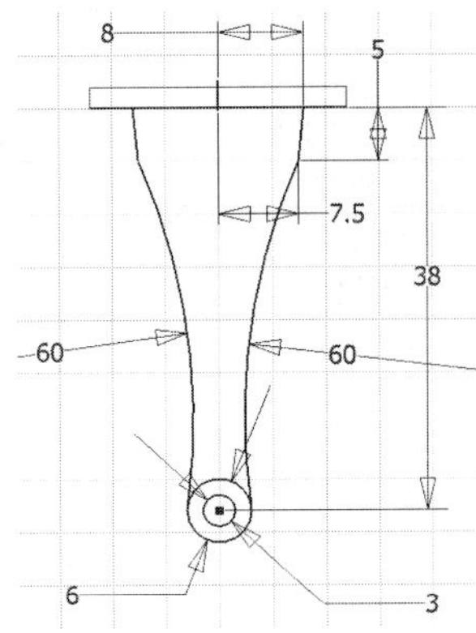

Figure 4.92 - Fully constrained sketch

19. Dimension the arc with a radius of 60, and add a tangent constraint between the arc and the large diameter circle.
20. Use the Line tool to sketch a horizontal line coincident with the upper endpoints of the inclined lines. The sketch should now resemble Figure 4.92.
21. Finish Sketch.
22. Isometric View.
23. Extrude as shown in Figure 4.93.
24. Shaded display.
25. Expand Extrusion2 (i.e., the second extrusion) in the part browser, right-click on Sketch2, and then select Share Sketch, as shown in Figure 4.94.
26. Extrude the area between the inner and outer circles of the shared sketch. See Figure 4.95.
27. Switch to Wireframe display.
28. Extrude the area of the inner diameter circle, as shown in Figure 4.96.
29. Shaded display.
30. Turn off visibility of shared sketch.
31. Isometric view. The part should now appear like Figure 4.97.
32. Save the part. Other part features will be added in Chapter 5.

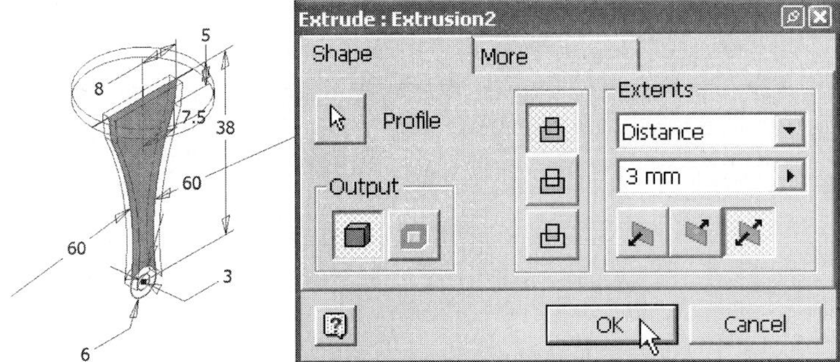

Figure 4.93 - Extruded feature operation

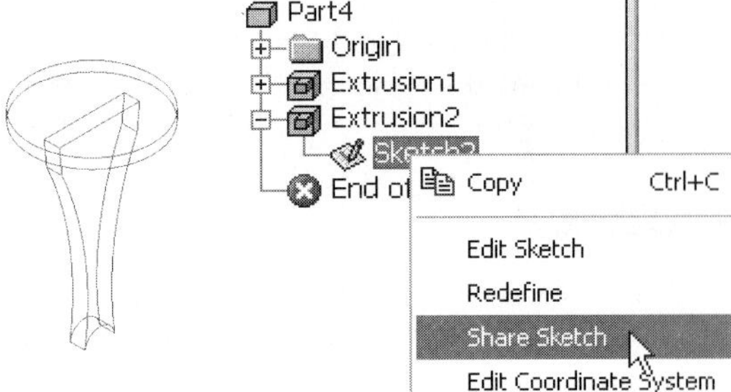

Figure 4.94 - Share sketch

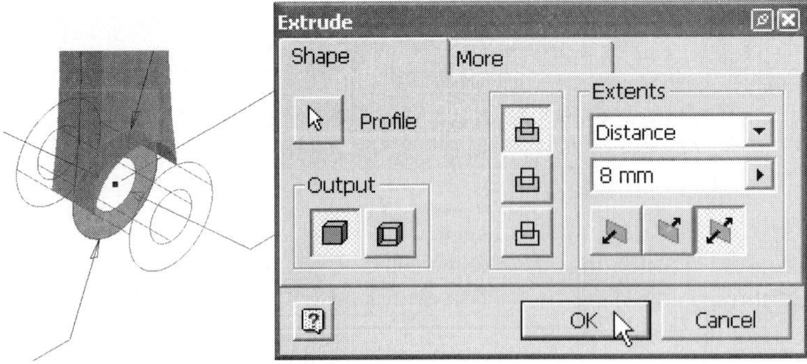

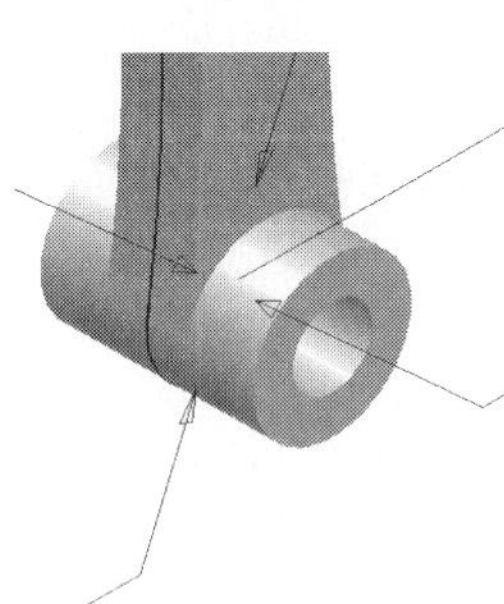

Figure 4.95 - Extruded (join) feature operation

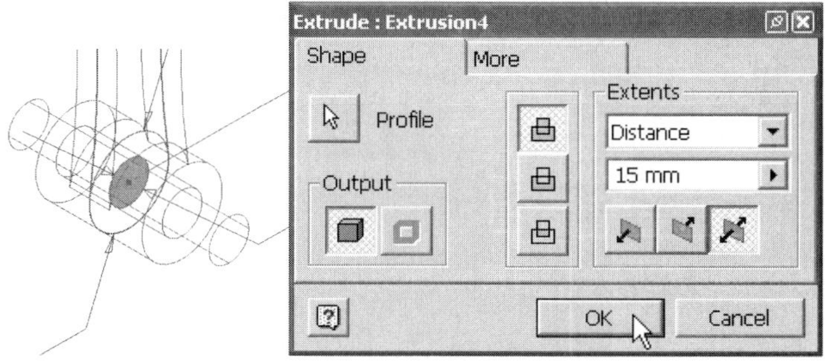

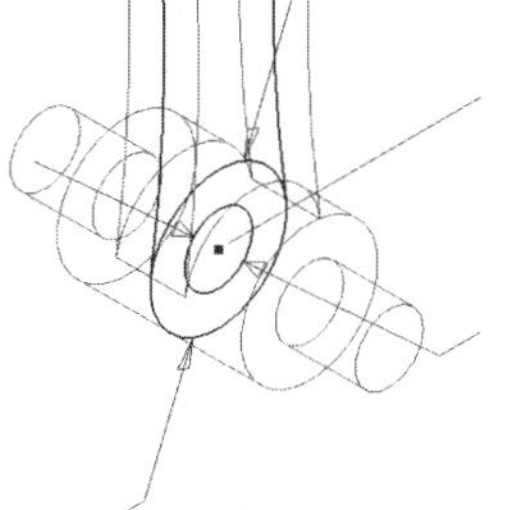

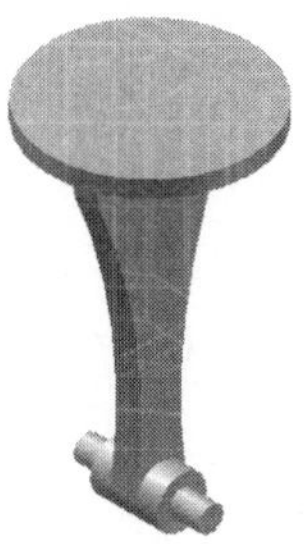

Figure 4.96 - Extruded (join) feature operation

Figure 4.97 - Completed part

QUESTIONS

1. T F Once a feature has been created the name appearing in the part browser cannot be modified.
2. Which one of these actions cannot be accomplished from the context menu in the part browser:
   a. Delete a feature
   b. Suppress a feature
   c. Rotate a feature
   d. Toggle the visibility of a feature
3. T F By default in Inventor, all sketches are considered to be shared sketches.
4. T F Parent-Child relationships describe how one feature is derived from another.

PROBLEM

1. Open the part file Arm on the CD. A drawing of this file is shown in Figure A below. Edit the file so that it conforms to the drawing shown in Figure B below.

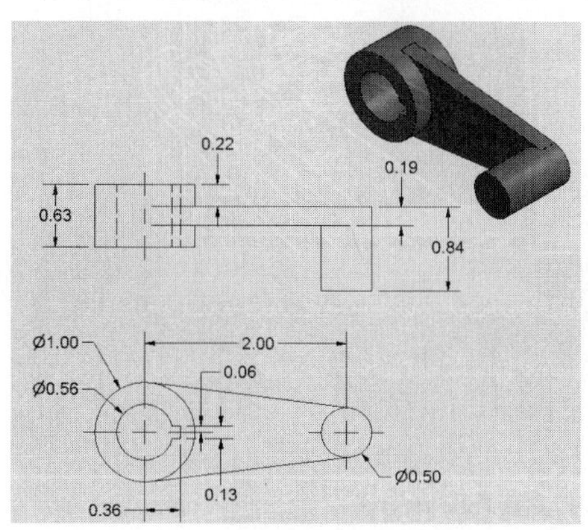

A

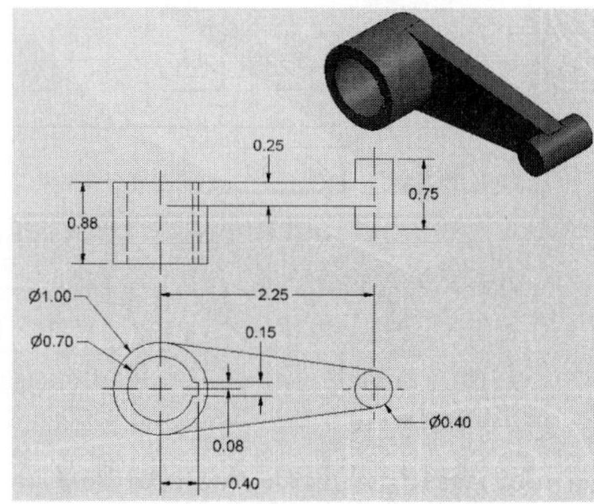

B

5 CHAPTER

Advanced Part Modeling

LEARNING OBJECTIVES

- List two inputs required to create a coil feature
- List the required inputs in order to create a loft feature
- List the two inputs required to create a sweep feature
- List the two kinds of supports that can be created with the Rib tool
- Describe two ways in which the Split tool can be used
- Identify some possible inputs needed to create a shelled feature
- Identify several inputs that must be specified in order to create a face draft
- Provide an example of a situation where a 3D sketch is necessary
- List two uses of the parameters dialog box
- Use the Project Geometry tool to project existing feature geometry onto a sketch
- Use the Rectangular Pattern tool to array sketch geometry in a rectangular pattern
- Use the Rib tool to create a rib feature
- Use the Mirror Feature tool to mirror a feature about a plane
- Use the Shell tool to shell an existing solid feature
- Use the Loft tool to create a lofted feature with parallel cross sections
- Use the Fillet tool to create a variable radius feature
- Use the Split tool to spilt the face of a part
- Use the Face Draft tool to add a draft angle to the face of a part
- Use the Include Geometry tool to add 2D sketch geometry to a 3D sketch
- Use the 3D sketch environment to create a 3D sketch
- Use the Edit Dimension tool to write parametric equations
- Control the bend radius used on a 3D sketch
- Use the Sweep tool to create a swept solid along a 3D path
- Use the Loft tool to create a lofted feature with nonparallel cross sections
- Use the Coil tool to create a coil feature
- Use the Work Axis tool to create a work axis feature.

- Use the Work Plane tool to create:
  - A work plane through two lines
  - An angled work plane
- Use the Parameters tool to:
  - Rename parameters
  - Write parametric equations
- Use the Edit Coordinate System tool to reorient the sketch coordinate system

Introduction

The tutorials in this chapter employ several advanced feature creation tools. These include tools for the creation of both sketched and placed features. Work features are also used in different ways to assist in part creation. Other feature duplication techniques, Mirror Feature and Rectangular Pattern, are used for the first time.

Some new sketching tools and techniques will also be employed for the first time in this chapter. These include 3D sketches, tools used for the duplication of sketch geometry (e.g., mirror and pattern), and the use of parameters and equations.

Once the chapter tutorials are completed, most all of the parts necessary to assemble the ball valve and the garlic press will have been modeled, as well as a toast rack part model. Additional parts will be modeled in Chapter 10, on hybrid modeling, where new tools introduced with Release 6 will be described.

All of the part models used for demonstration purposes in this chapter are included on the CD in the back of the book.

Sketched Features

In addition to Extrude and Revolve, other sketched features available in Inventor include Sweep, Coil, Loft, Rib, and Split. All of these sketched feature tools will be used in the tutorials appearing at the end of this chapter.

Sweep

The Sweep tool is useful, for example, in the modeling of piping systems. This tool requires two inputs to create a solid feature, a closed profile and a path. The profile is swept along the path, resulting in a swept solid. Extrude and Revolve are specific kinds of swept solids. In an extrusion, the path is a straight line normal (i.e., perpendicular) to the plane of the profile. In a revolution, the path is a circle normal to the plane of the profile.

As with the Extrude and Revolve tools, it is now possible in Release 6 to sweep open profiles to create a swept surface feature. Examples of this are provided in Chapter 10.

Two sketches are required to create a sweep; one sketch is the profile, the other the path. In Figure 5.1, an Ellipse (accessed from the Circle flyout on the panel bar) is swept along a Spline path (accessed from the Line flyout on the panel bar).

Since only two of the three (X, Y, and Z) coordinates in a Spline vary, the sweep path is two-dimensional. Later we will see how to create 3D sketches. This will allow for the creation of sweep paths that are three-dimensional.

Coil

The Coil tool is most commonly used to model helical features like springs. A single sketch is required as input. The sketch should include a closed profile as well as an axis

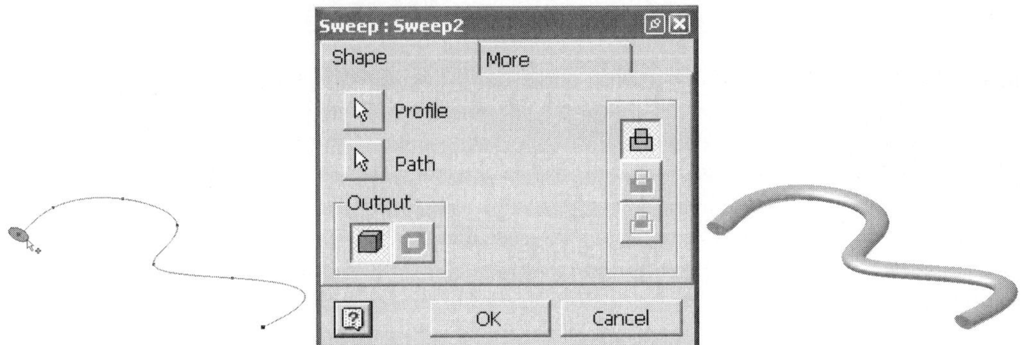

Figure 5.1 - Sweep feature operation

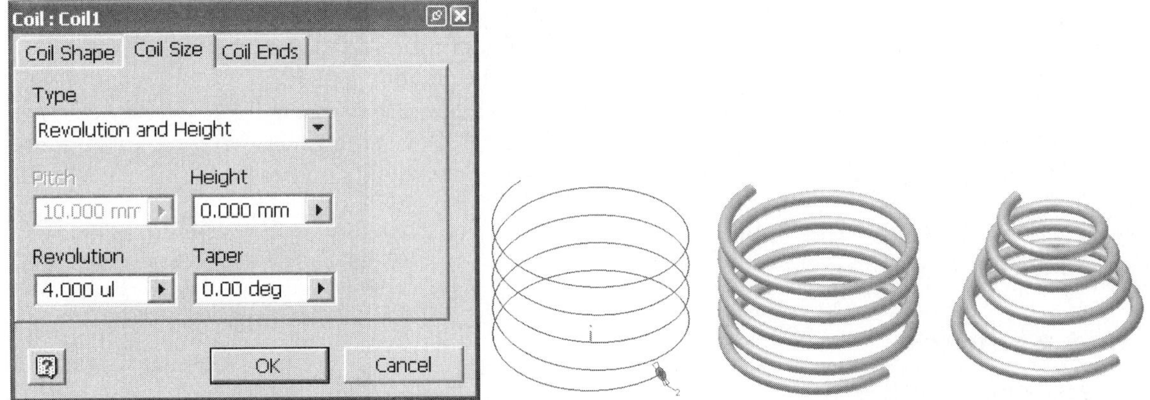

Figure 5.2 - Coil feature operation

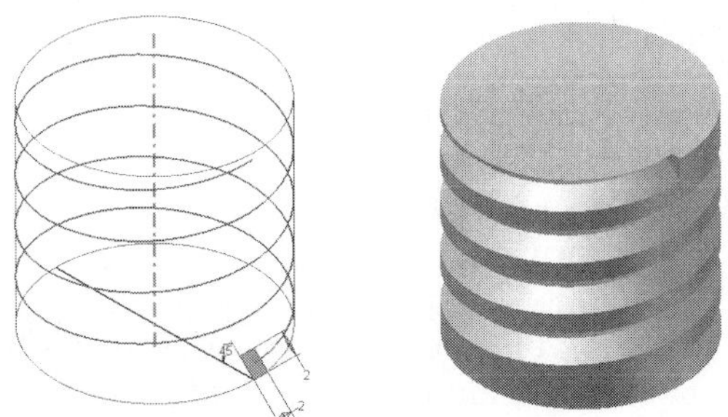

Figure 5.3 - Solid feature threads cut from cylinder

of revolution. The Coil dialog box includes three different tabs: Coil Shape, Coil Size, and Coil Ends. Figure 5.2 shows the Coil tool in operation, as well as two different coils produced from the same sketch.

Although threads are most commonly represented *cosmetically,* it is possible to model threads as an actual solid feature, as shown in Figure 5.3. Here a second coil feature with a diamond-shaped cross section (created with the Three point rectangle tool) is used to cut material from a base cylinder.

Loft

Lofted shapes are frequently encountered in the aerospace and marine industries. Examples include the fuselage and wings of an aircraft, and the hull of a boat. Shapes with varying cross section are candidates for lofting. A minimum of two sketches are required to create a lofted feature with the Loft tool. In the lofting process, additional work planes will typically need to be created for use as sketch planes. Figure 5.4 shows the sequence of steps employed in creating a lofted solid of a half hull.

Although the sketch planes used to create a lofted solid are frequently parallel to one another, this need not be the case, as shown in Figure 5.5.

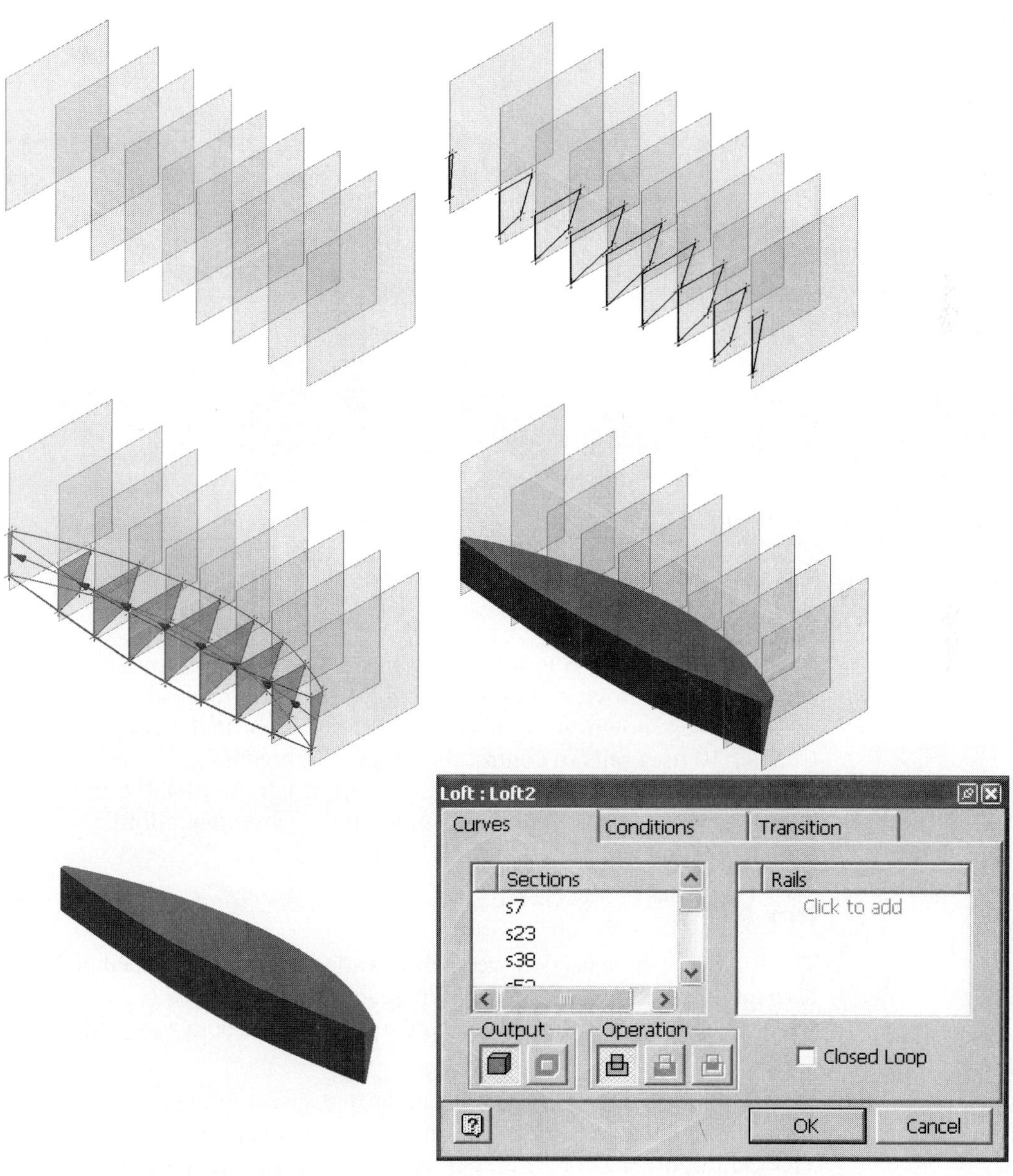

Figure 5.4 - Lofted half hull

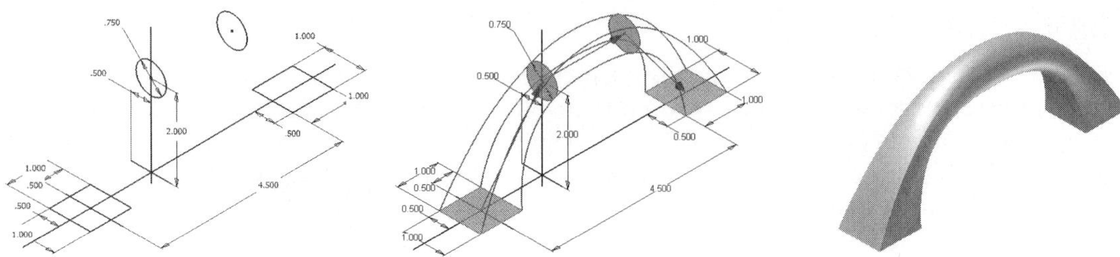

Figure 5.5 - Lofted solid with non-parallel sketch planes

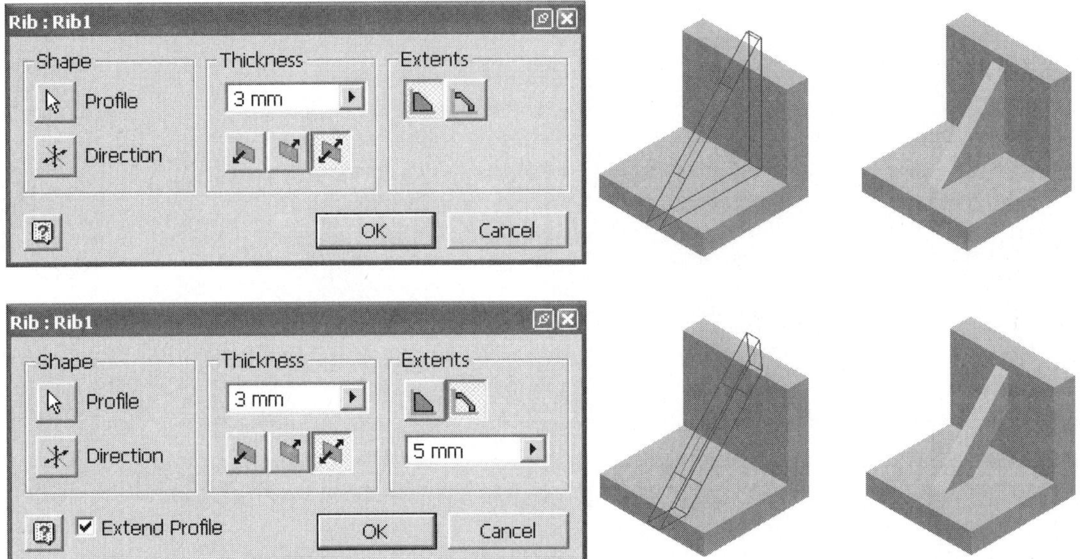

Figure 5.6 - Rib feature operation for rib and web

New to Release 6 is the ability to define lofts with *rails* as well as with cross sections. Rails provide additional control over the resulting lofted shape. Although the examples shown in Figures 5.4 and 5.5 do not make use of rails, Tutorial 30 in Chapter 10 uses rails to control the shape of a creamer.

As with Extrude, Revolve, and Sweep, it is now possible in Release 6 to create lofted surfaces, as well as lofted solids. In the event that a lofted surface is to be created, both open and closed profiles can be used.

Rib

The Rib tool can be used to create thin-walled supports. An open profile sketch is used to create these supports; the supports can either be closed (i.e., a rib) or open (i.e., a web). Examples showing the Rib tool used to create both a rib and a web are shown in Figure 5.6.

Split

The Split tool is used both to remove a portion of a part (split part), as well as to split a part (split face) so that a taper can be applied to part faces on either side of

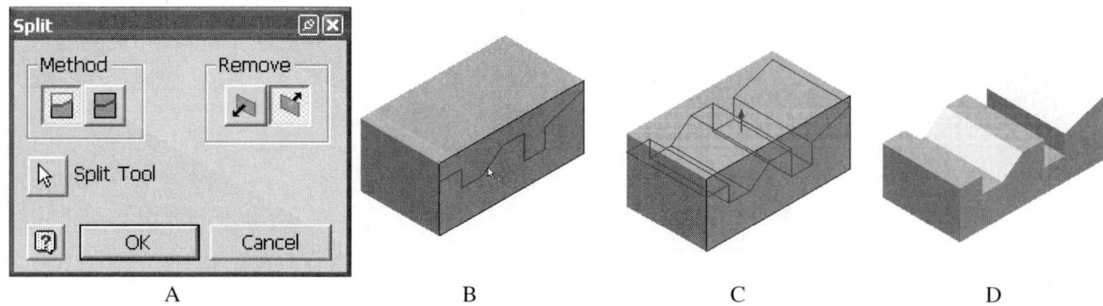

Figure 5.7 - Split part

Figure 5.8 - Split face

the split or *parting line*. Either a parting line sketched on a part face or work plane can be used to split a part. This means that Split sometimes acts like a sketched feature (e.g., sketched parting line) while at other times it is more like a placed feature (e.g., work plane used to split).

Figure 5.7B shows a parting line sketched on the face of a box. In Figure 5.7, this parting line sketch is used in conjunction with the Split tool to remove that part of the box above the line. Although it is not possible to directly save both sides of a split part, this can still be accomplished by using the Save Copy As command to create a second copy of the part. If there is a parting line sketch, this should be created before saving. Opening each file in turn, use the Split tool to save different sides of the part in two different part files.

Figure 5.8 shows the Split tool used to split part faces above and below the parting line. A draft (i.e., taper) angle can now be applied to the different faces using the Face Draft tool. See Figure 5.11.

Placed Features

Thus far the Hole, Fillet, Chamfer, and Thread placed feature tools have been introduced. In the remaining tutorials the Shell and Face Draft tools will also be used.

Shell

The Shell feature tool is used to remove material from the inside of a part, leaving a hollow cavity with walls of a specified thickness. It is possible to specify faces that are

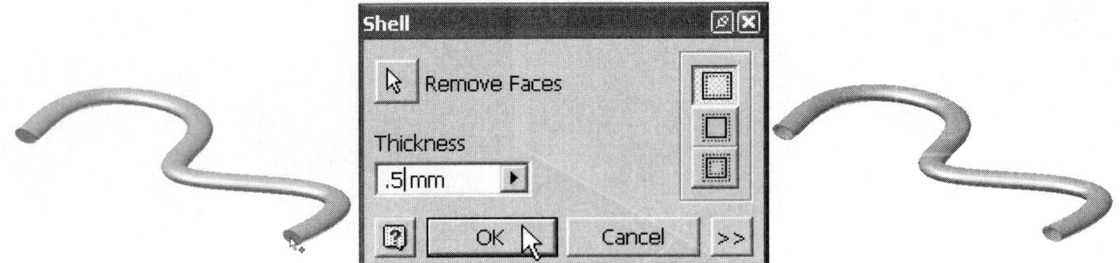

Figure 5.9 - Shell feature

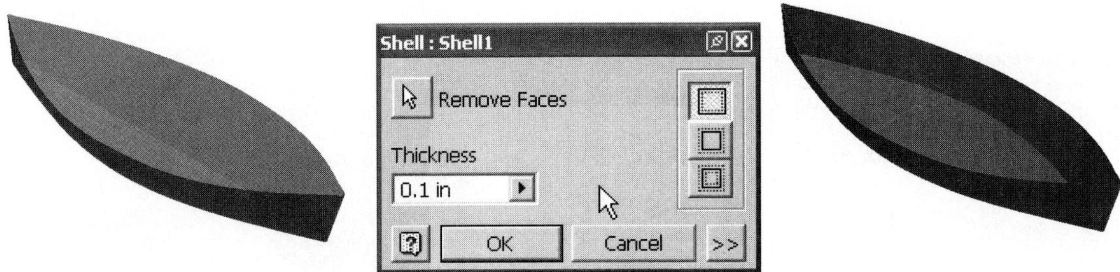

Figure 5.10 - Hull form shelled

removed from the shell. Faces that are not removed form the shell walls. In Figure 5.9, the swept solid is shelled, with both end faces removed. The result is a hollow tube.

In another example, a lofted hull form is shelled. Only the top face is removed, resulting in Figure 5.10.

Face Draft

The Face Draft tool applies a taper angle to specified faces of a part. This taper angle, or *draft,* allows molded parts to be easily extracted from the mold. A draft angle can either be specified at the time that an extrusion is created, or added later to an existing part using Face Draft. The effect of the Face Draft tool is determined by several factors. First, a *pull direction* must be specified. This refers to the direction that the mold is pulled in order to extract the part. Next the face to which the taper is applied is selected, along with the *fixed edge.* This is the edge of the tapered face where the taper begins. No material is added/removed from the fixed edge. Finally, the *taper angle* must be specified.

Face Draft is often used in conjunction with the Split tool. In Figure 5.11, Face Draft is applied to a part face that had previously been split (see Figure 5.8). After selecting Face Draft from the panel bar, the Face Draft dialog box opens (Figure 5.11A). It is first necessary to indicate a pull direction. This is done by selecting the top face. A vertical arrow extending from the top face and indicating the direction that the mold will be pulled is now visible (Figure 5.11B, top face). This direction can be flipped, if necessary. Next the face to be tapered and the fixed edge of this face are selected. In Figure 5.11B the parting line passing along the frontal face is highlighted. The icon that looks like a check mark with an attached arrow indicates both the fixed edge and the direction of the taper. In this case, the inclined (parting) line is the fixed edge, and material will be removed from the part (since the arrow points into the part). The result of applying this draft to a single face is shown in Figure 5.11C. Notice that the taper is only applied up to the parting line previously created with the Split tool.

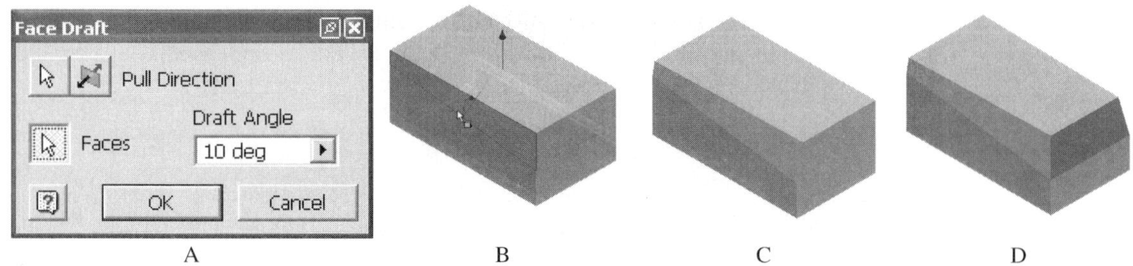

Figure 5.11 - Face draft operation

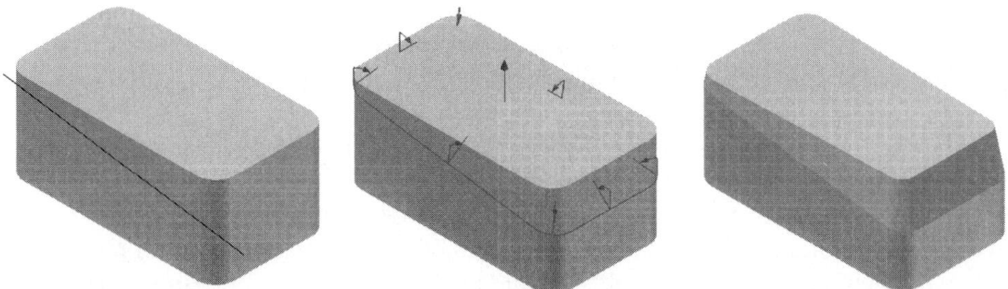

Figure 5.12 - Face draft with tangent faces

It is possible to taper several faces within the same Face Draft feature. This has been done in Figure 5.11D.

If the tapered face consists of several tangent faces, the taper is applied to all faces, as shown in Figure 5.12.

3D Sketch

The routing of pipes, cables, ducts, etc., can be modeled using sweep features. The sweep paths for these features will typically not lie (exclusively) in a two-dimensional plane, however. For this reason, Inventor includes a 3D sketch environment.

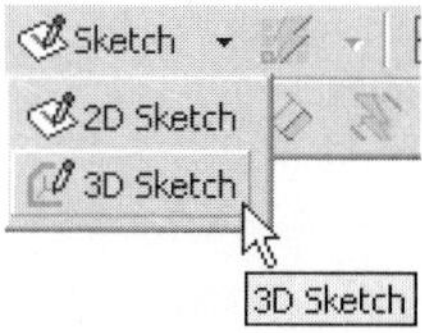

Figure 5.13 - 3D sketch access

The 3D Sketch environment is accessed either by clicking the right mouse button and selecting New 3D Sketch, or from the standard toolbar, as shown in Figure 5.13.

Once the 3D Sketch environment is entered, the 3D Sketch tools appear in the panel bar, as shown in Figure 5.14.

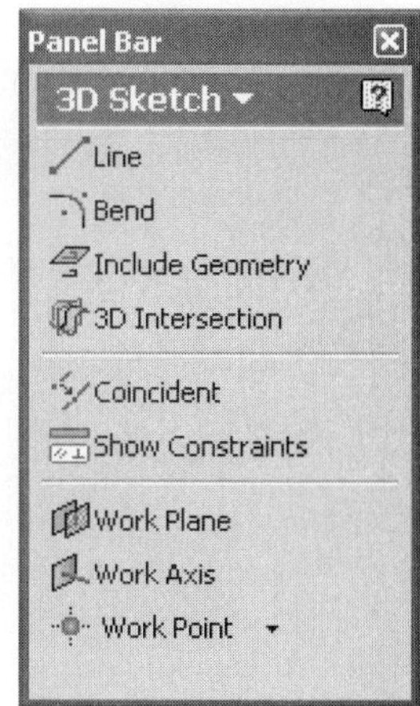

Figure 5.14 - 3D Sketch tools

In Inventor it is not possible to sketch geometry independent of a 2D sketch plane. This means that in order to construct a three-dimensional path in Inventor, 2D segments of the path must first be created on sketch planes. The segments are then incorporated into the 3D sketch using the Include Geometry tool. 3D line segments can then be included in the 3D sketch with the 3D Line tool. 3D lines can only be drawn to the endpoints of included geometry, to the vertices of existing geometry, or to Work Points.

Bend radii can either be included automatically between 3D line segments, or afterwards by using the 3D Bend tool. The Auto-Bend Radius applied to a 3D sketch path is set from the Sketch tab of the Document Settings dialog box, accessed by selecting Tools > Document Settings . . . from the menu bar. To determine whether an Auto-Bend radius will be applied, right-click when the 3D Line tool is in operation. If Auto-Bend is checked, a radius will automatically be applied between joined line segments. See the Toast Rack tutorial in this chapter for an example of creating a 3D sketch, and then using it to create a 3D sweep.

In Chapter 10, on hybrid modeling, some new 3D sketch environment tools will be discussed and used in the tutorials at the end of the chapter. These include the 3D Intersection tool and the Grounded Work Point tool.

Parameters and Equations

In Chapter 2 a *parameter* was defined as a named quantity whose value can change. In Inventor parameters are used to define the size and shape of features, as well as to position parts with respect to one another in an assembly (see Chapter 6). Parameters are added to Inventor files whenever dimensions are added to a sketch, when values are entered into feature dialog boxes, and when assembly constraints are created. In each of these cases, values entered from the keyboard are assigned an indexed name (e.g., d0, d1). In the case of sketch dimensions, the parameter name can be seen by the user at the top of the Edit Dimension dialog box, as seen in Figure 5.15. For this particular parameter the name is d5, and the value is 60.

Figure 5.15 - Edit dimension dialog box

The Parameters tool available from the panel bar gives access to the Parameters dialog box, shown in Figure 5.16.

For this particular part file, two model parameters have been created, d0 and d1. From the Parameters dialog box parameters can be renamed (to something more meaningful, like length) and values can be changed. This is done by clicking on the cell to be edited, and then typing in the information. It is also possible to write equations relating parameters. In Figure 5.17, the parameters have been renamed, the value of the

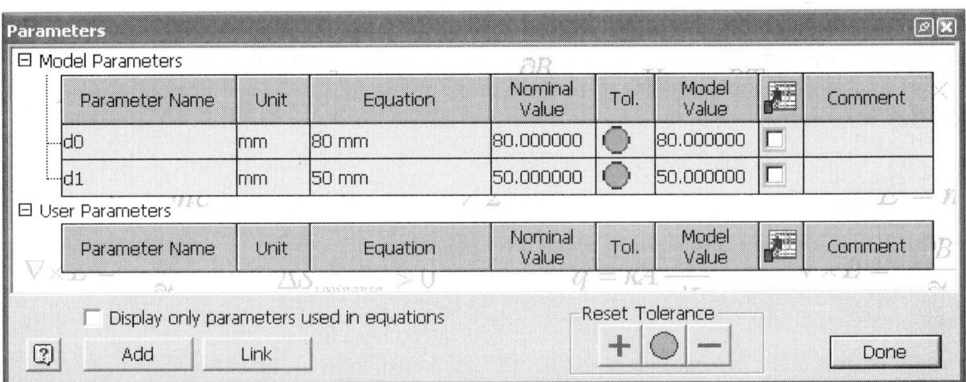

Figure 5.16 - Parameters dialog box

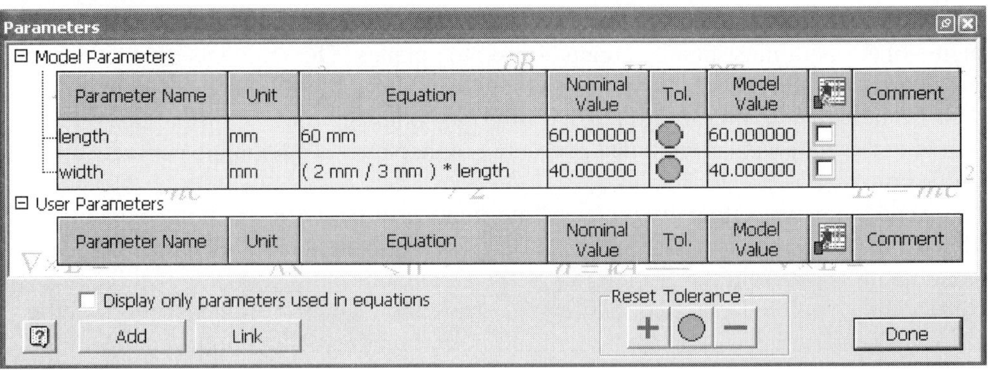

Figure 5.17 - Modified parameters

length parameter has been changed, and the width has been set equal to two-thirds of the length. In the tutorials at the end of this chapter, equations will be written in the Edit Dimension dialog box. In this case, equations need to be preceded by an equal (=) sign.

Piston (Garlic Press) (continued) TUTORIAL 11

In Chapter 4 the piston shown in Figure 5.18 had been created and saved. We will now add a couple of additional features in order to complete the part.

Build Strategy

1. Rib (Figure 5.19).
2. Mirror rib feature (Figure 5.20).

Detailed Modeling Steps

1. Open the part file Piston in the Garlic Press project folder.
2. New Sketch on the XZ Plane.
3. Look At sketch.
4. Wireframe display.
5. Project Geometry—*limiting element* of outer cylinder and vertical edge of Extrusion2 (i.e., the second extrusion in the part browser), as shown in Figure 5.21.
6. Zoom in on cylinders (see Figure 5.22).

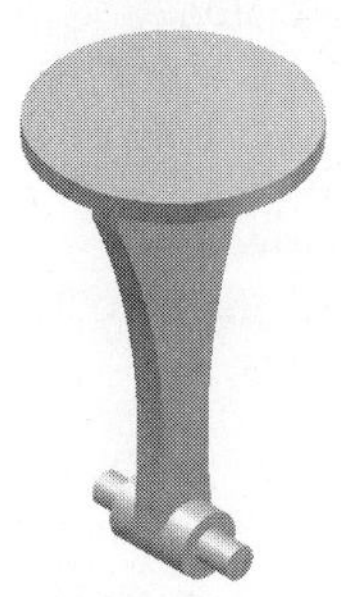

Figure 5.18 - Piston part

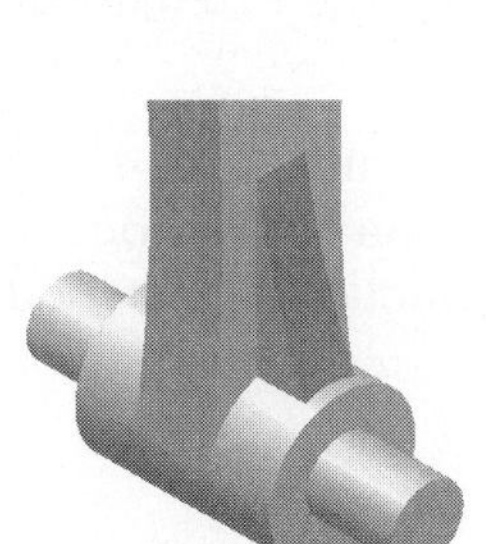

Figure 5.19 - Rib feature

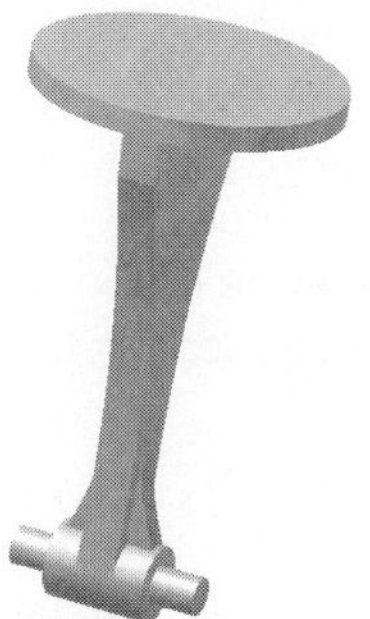

Figure 5.20 - Completed piston

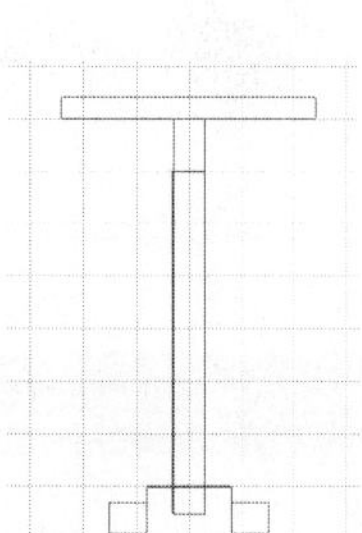

Figure 5.21 - Projected geometry

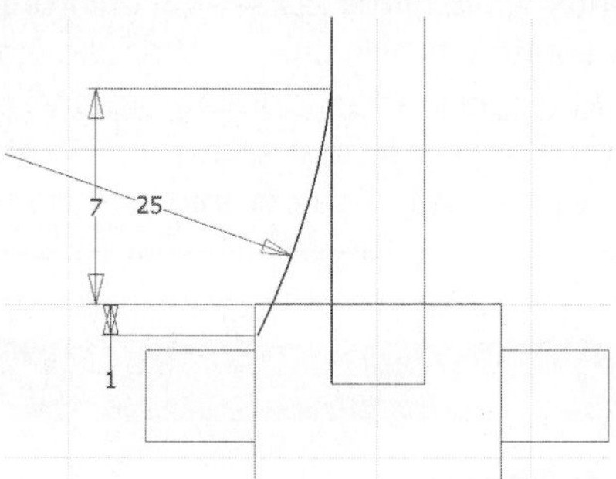

Figure 5.22 - Rib feature sketch

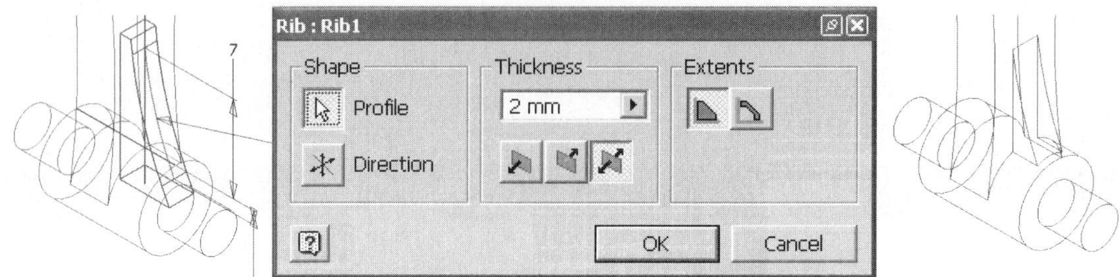

Figure 5.23 - Rib feature operation

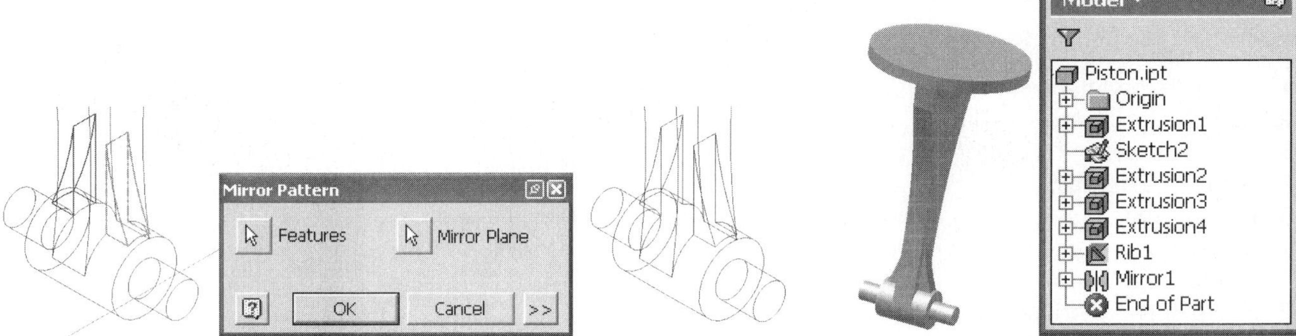

Figure 5.24 - Mirror feature operation

Figure 5.25 - Completed piston

7. Use the Three point arc tool to create an arc with one end coincident with the vertical edge of Extrusion2, and the other endpoint terminating below the limiting element of the large diameter cylinder. Add the dimensions, as seen in Figure 5.22.
8. Finish Sketch.
9. Isometric View.
10. Zoom in on the rib sketch.
11. Use the Rib tool, as shown in Figure 5.23.
12. Use the Mirror Feature tool to mirror the rib feature, as seen in Figure 5.24. Use the YZ Plane as the mirror plane. Note that it is also possible to mirror sketch geometry while in the 2D sketch environment, using the Mirror tool. The Mirror (sketch) tool is used in the following tutorial (12), steps 28 and 29.
13. Shaded display. After zooming out and rotating the view, the completed piston should be similar to that shown in Figure 5.25.
14. Save the part. This completes Tutorial 11.

TUTORIAL 12 Cylinder Handle

Build Strategy

1. Sketch (Figure 5.26).
2. Extrude (Figure 5.27).

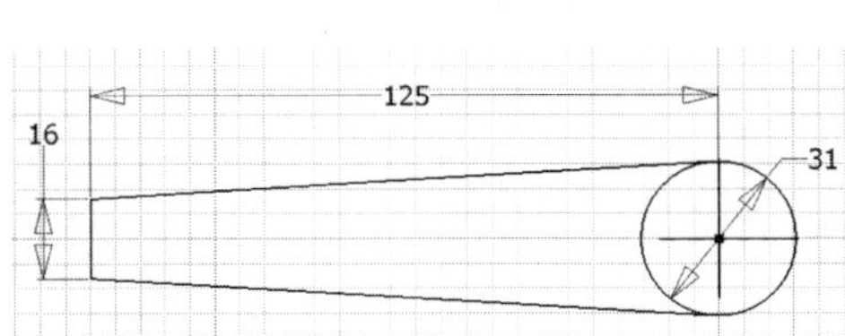

Figure 5.26 - Base feature sketch

Figure 5.27 - Extruded base feature

Figure 5.28 - Shelled base feature

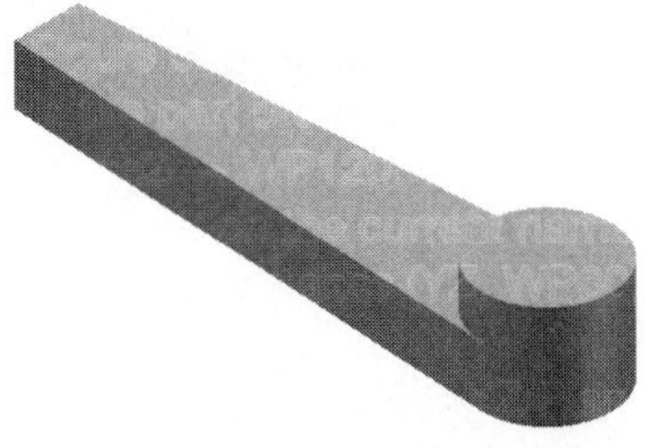

Figure 5.29 - Extrude shared sketch

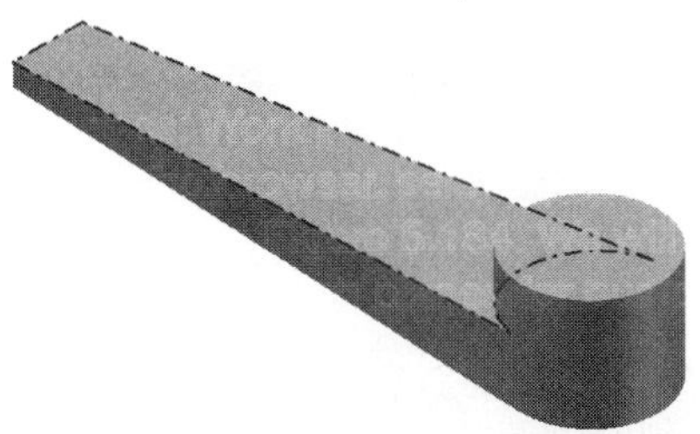

Figure 5.30 - Chamfer

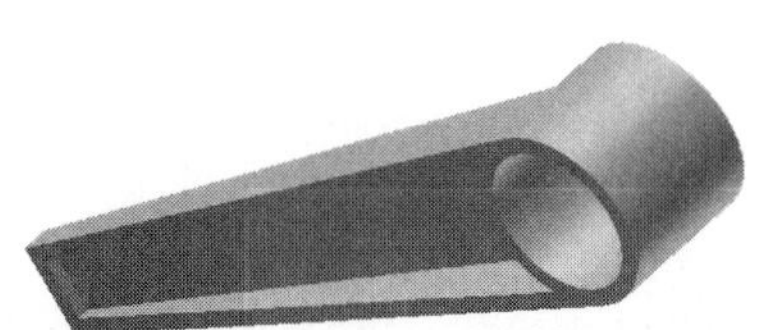

Figure 5.31 - Shell feature

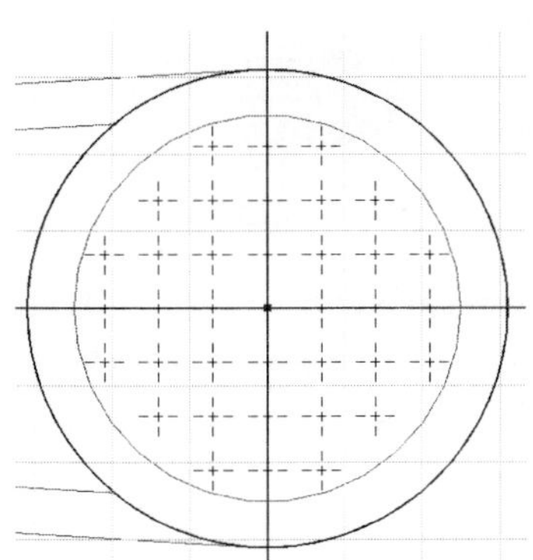

Figure 5.32 - Hole pattern

Figure 5.33 - Holes added

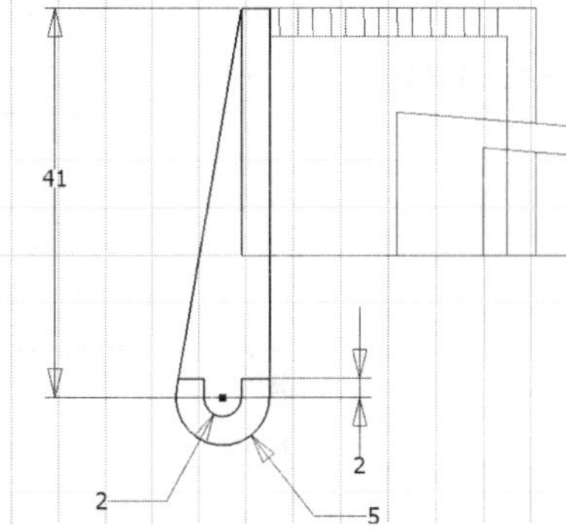

Figure 5.34 - Fully constrained sketch

3. Shell (Figure 5.28).
4. Extrude (join - shared sketch) (Figure 5.29).
5. Chamfer (2 distances) (Figure 5.30).
6. Shell (Figure 5.31).
7. Sketch (Point, Hole Center, Rectangular Pattern, Mirror (twice)) (Figure 5.32).
8. Hole (Figure 5.33).
9. Sketch (Figure 5.34).
10. Extrude (join, portion of the previous sketch) (Figure 5.35).
11. Loft (join) (Figure 5.36).
12. Extrude (join, shared sketch) (Figure 5.37).
13. Fillet (Variable radius, two symmetrical edges) (Figure 5.38).
14. Fillet (linear, several edges) (Figure 5.39).

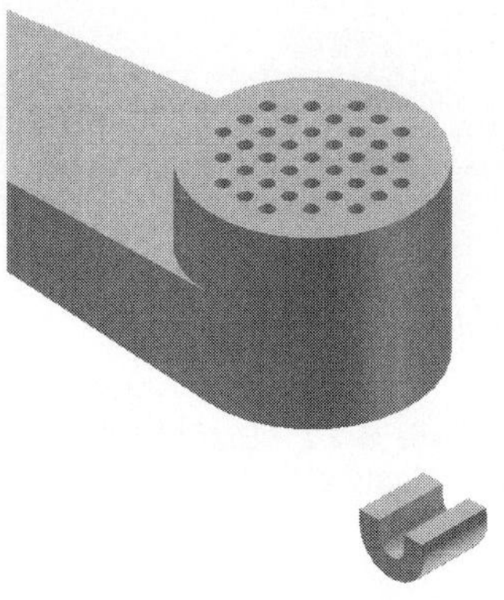

Figure 5.35 - Extrude (join)

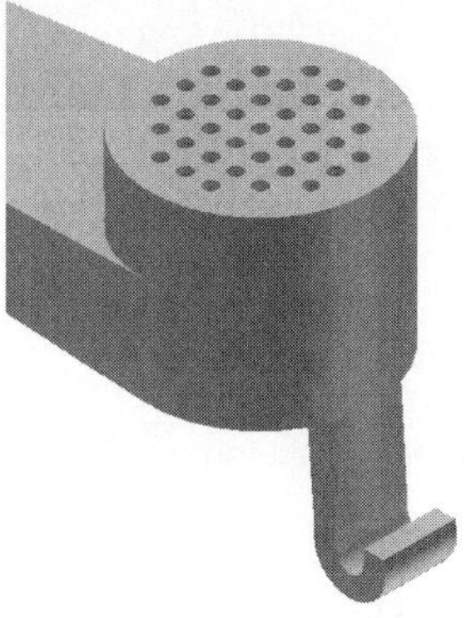

Figure 5.36 - Loft (join) feature

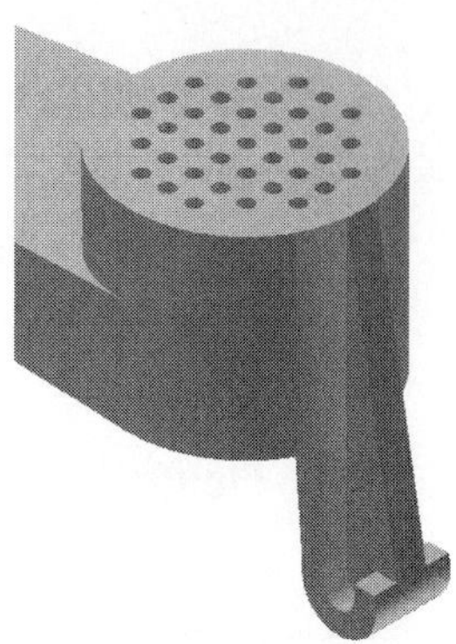

Figure 5.37 - Extrude (join) shared sketch

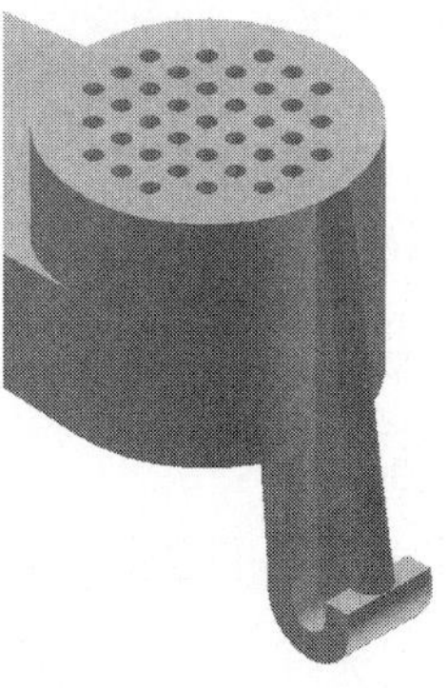

Figure 5.38 - Variable radius features

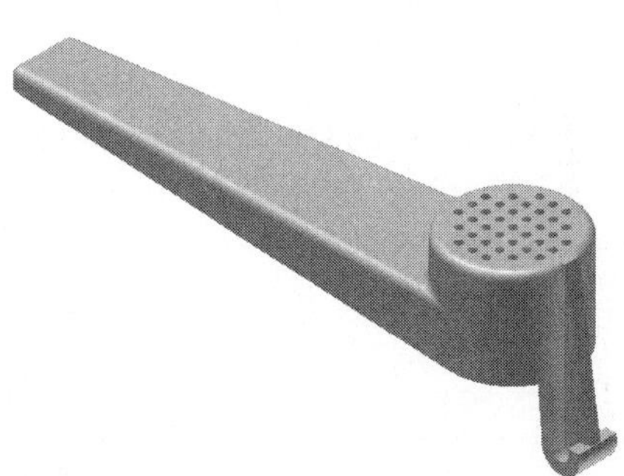

Figure 5.39 - Completed part

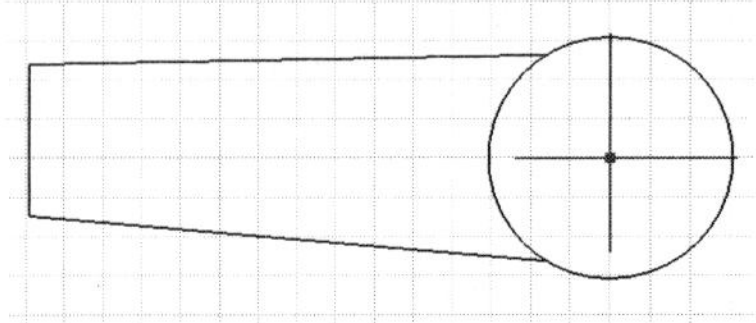

Figure 5.40 - Rough sketch

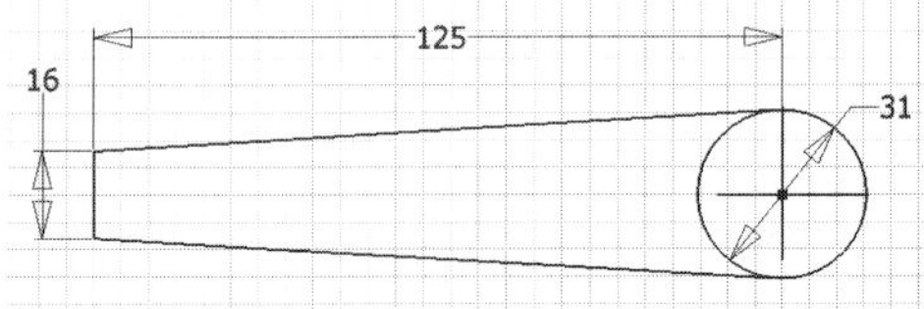

Figure 5.41 - Base feature sketch

Detailed Modeling Steps

1. Start a new metric file.
2. Project X and Y Axes and Center Point.
3. Sketch the diagram in Figure 5.40. The inclined lines are coincident with the circle.
4. Add a Symmetric constraint to the inclined lines, about the X Axis.
5. Add a Tangent constraint to an inclined line and the circle.
6. Dimension the sketch as shown in Figure 5.41.

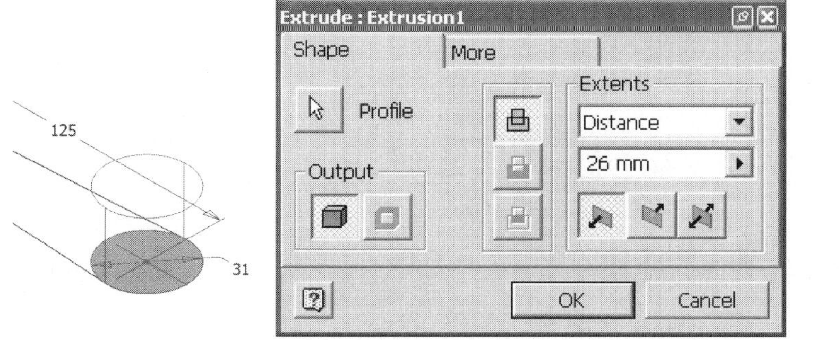

Figure 5.42 - Extruded base feature

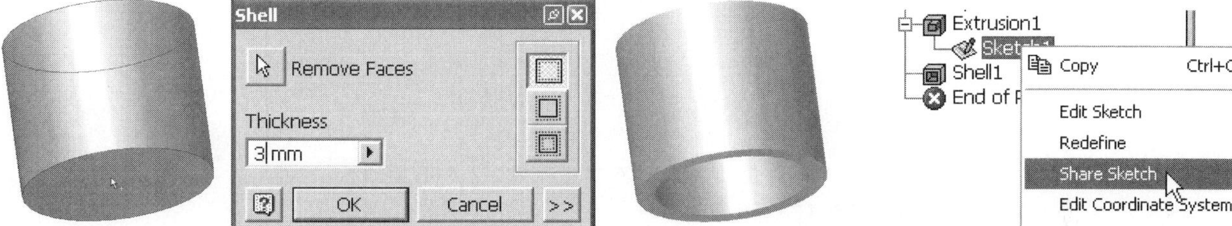

Figure 5.43 - Shell feature operation

Figure 5.44 - Share sketch

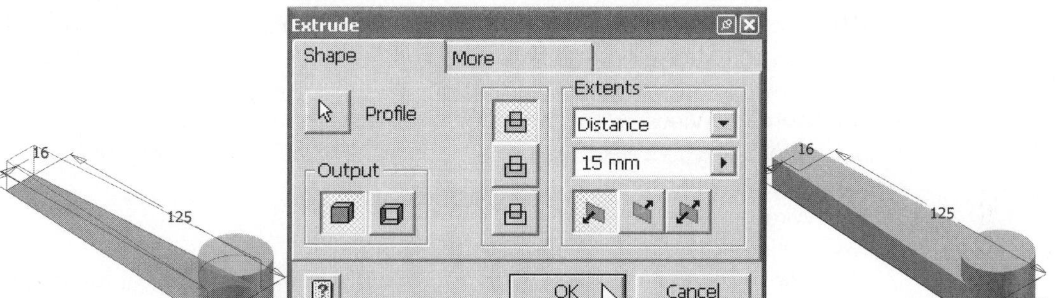

Figure 5.45 - Extrude shared sketch

7. Finish Sketch.
8. Isometric View.
9. Extrude the circle. The completed extrusion appears on the right in Figure 5.42.
10. Save the file as **Cylinder_Handle.** It is recommended that the file be regularly saved every five to ten minutes.
11. Use the Shell tool on the cylinder, as shown in Figure 5.43. Rotate the cylinder to aid in the selection of the cylinder bottom as a removed face.
12. Share the consumed sketch of the extruded cylinder (Figure 5.44).
13. Isometric View.
14. Extrude (Figure 5.45).
15. Turn off visibility of shared sketch.
16. Wireframe display.

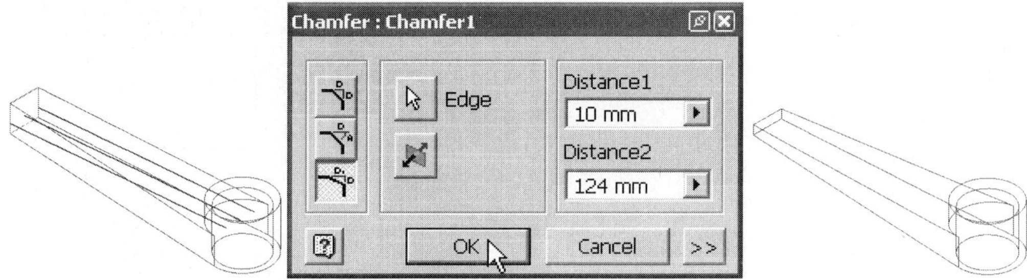

Figure 5.46 - Chamfer (two distances)

Figure 5.47 - Shell feature

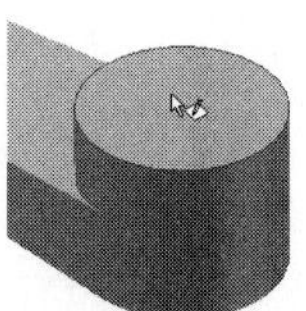

Figure 5.48 - New sketch plane

17. Chamfer the second extrusion, as shown in Figure 5.46. The resulting part should resemble the image on the far right in Figure 5.46.
18. Shaded display.
19. Use the Shell tool a second time, as shown in Figure 5.47. Use the 3D Rotate tool so that the bottom face can be selected for removal. The resulting solid is shown in the image on the far right in Figure 5.47.
20. Isometric View.
21. New Sketch - select the top cylindrical face, as shown in Figure 5.48.
22. Look At sketch plane.
23. Wireframe display.
24. Project X and Y Axes and Center Point.
25. Use Point, Hole Center tool to add a single hole center at the projected center point.
26. Use the Rectangular Pattern tool with the settings shown in Figure 5.49. First select the Geometry button, then select the hole center; next select the Direction 1 Arrow button, select the X Axis (in the graphics area), enter the Count (4) and Spacing (3.5 mm) as shown; now select the Direction 2 Arrow button, select the Y Axis (in the graphics area), enter the Count (4) and Spacing (3.5 mm). Now click the More Info >> button. Deselect the Associative check box. Click OK.
27. Because the rectangular pattern is no longer associative, it is possible to edit the pattern. While holding down the Shift key, select the three hole centers in the upper right corner, right-click, and select Delete. The resulting pattern should appear as shown in Figure 5.50. Note that the display is set to wireframe. This allows us to see the cylindrical wall thickness.
28. Using the Mirror tool, first select the hole centers that do not lie on the Y Axis, select the Mirror line button, then select the Y Axis, and then select the Apply button (Figure 5.51).

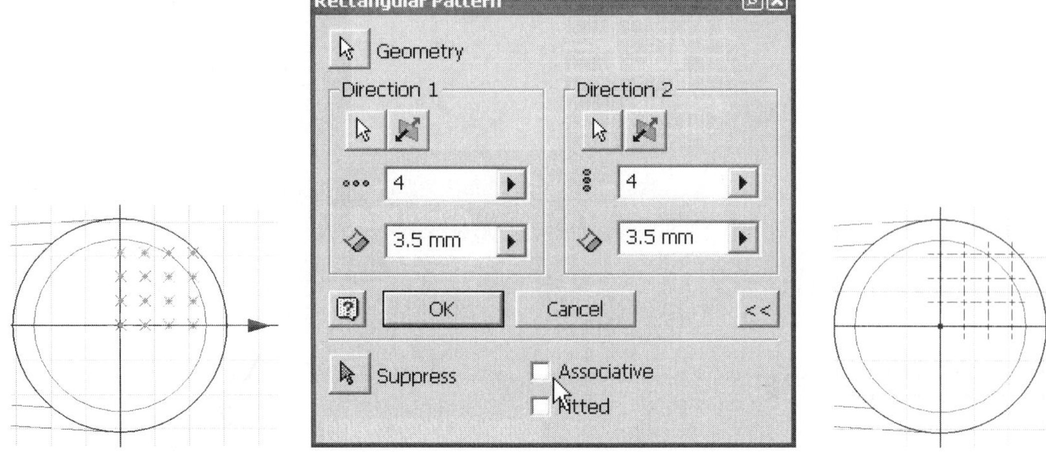

Figure 5.49 - Pattern hole centers

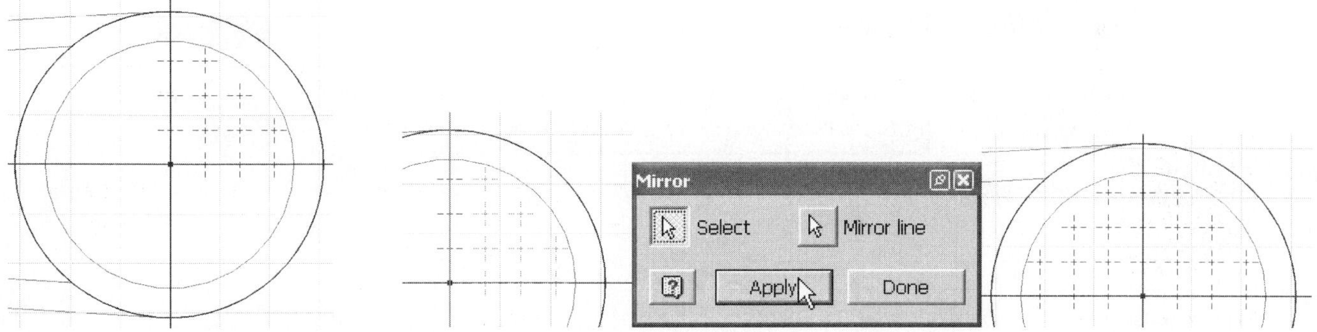

Figure 5.50 - Edited hole pattern

Figure 5.51 - Mirror hole pattern about Y Axis

29. The Mirror dialog box should still be active (if not, activate it). Select all hole centers that do not lie on the X Axis, select the Mirror line button, choose the X Axis as the mirror line, and click the Apply button, then click the Done button. The resulting sketch should be similar to Figure 5.52.

30. Finish Sketch.

31. Isometric View.

32. Shaded Display.

33. Use the Hole tool, as shown in Figure 5.53. The resulting part should be similar to the right-hand side of the figure.

34. New Sketch, on the XZ Plane.

35. Look At the new sketch.

36. Wireframe display.

37. Use the Project Geometry tool to project both the inside and outside limiting elements of the cylinder onto the sketch plane, as shown in Figure 5.54.

38. Use the Line tool to duplicate the sketch shown in Figure 5.55.

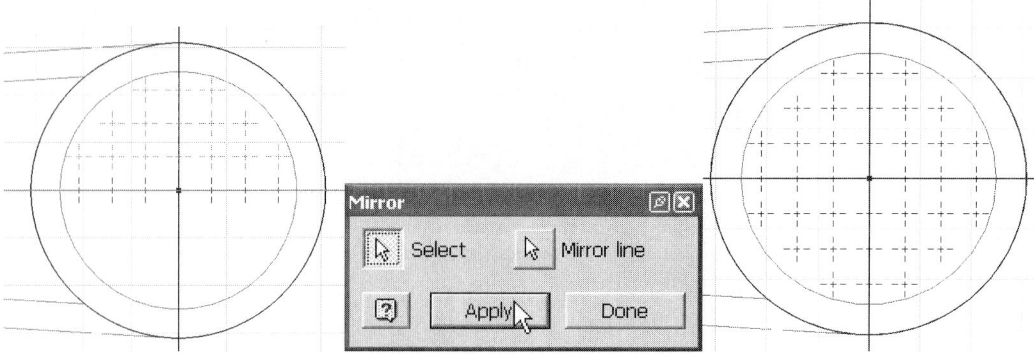

Figure 5.52 - Mirror hole pattern about X Axis

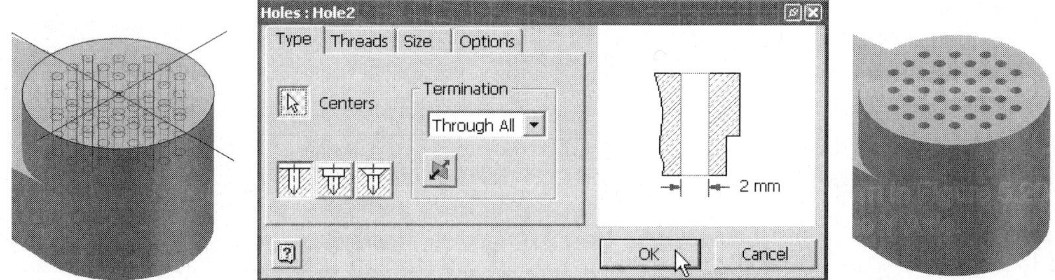

Figure 5.53 - Hole feature operation

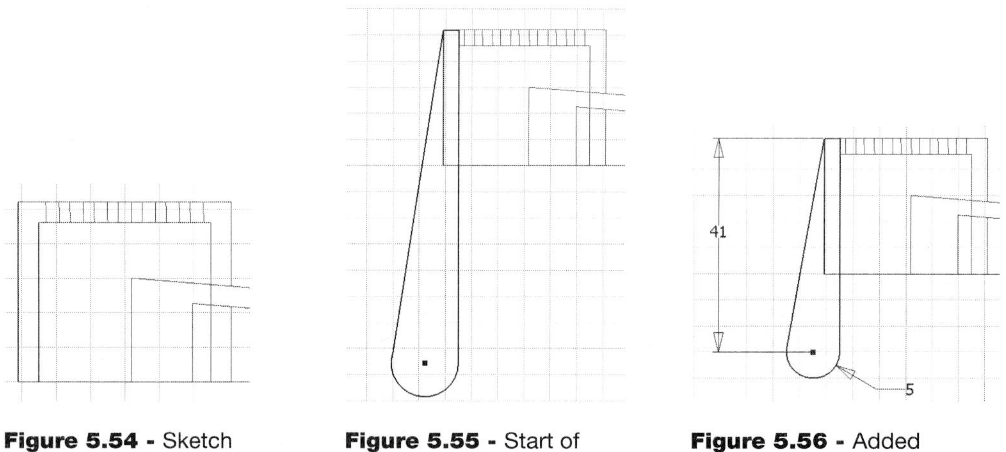

Figure 5.54 - Sketch plane with projected edges

Figure 5.55 - Start of new sketch

Figure 5.56 - Added dimensions

39. Add Tangent constraints as needed between the arc and the two line segments.
40. Add the dimensions shown in Figure 5.56.
41. Use the Line tool to add the lines and arc shown in Figure 5.57. Note that the two arcs are concentric ◎, and that the horizontal line segments are collinear.
42. Add dimensions to complete the sketch, as shown in Figure 5.58.

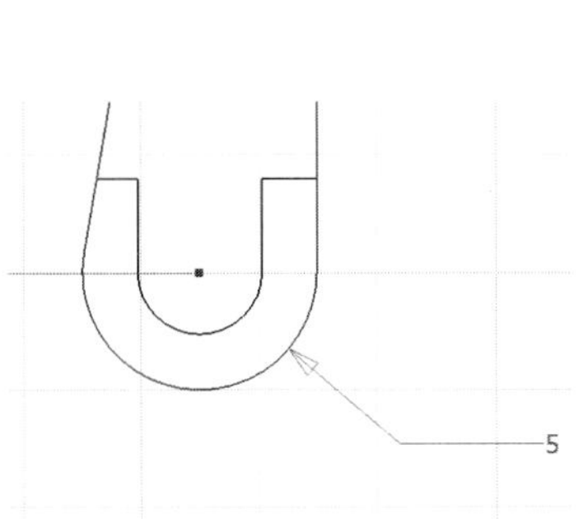

Figure 5.57 - Add to sketch

Figure 5.58 - Fully constrained sketch

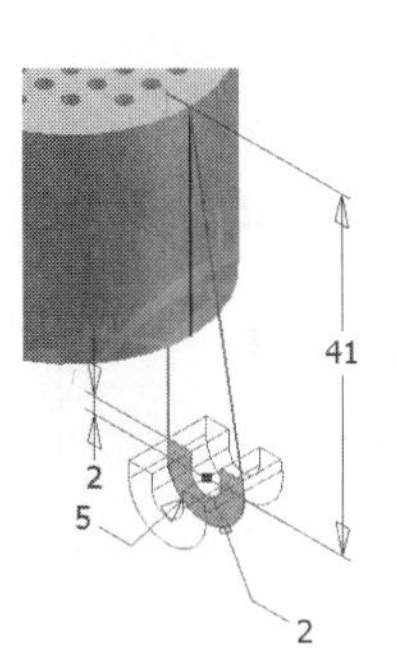

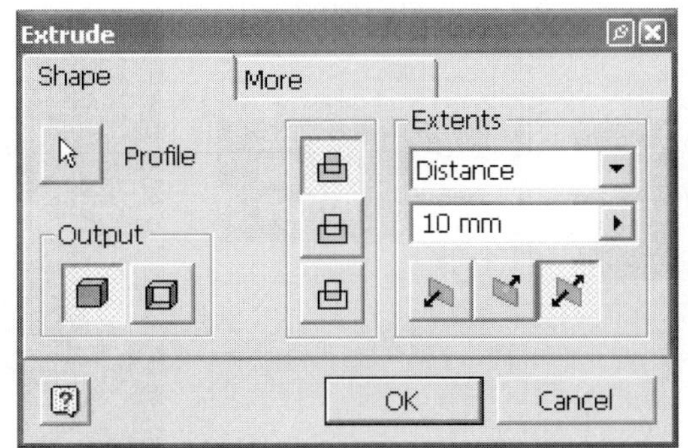

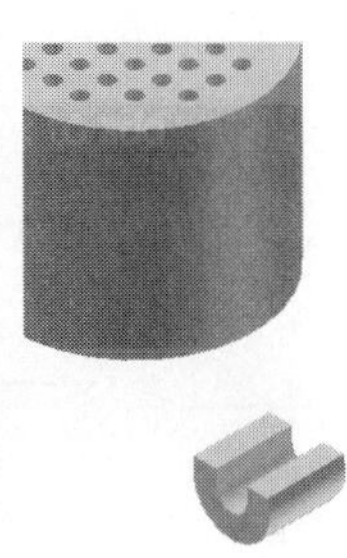

Figure 5.59 - Extrude (join)

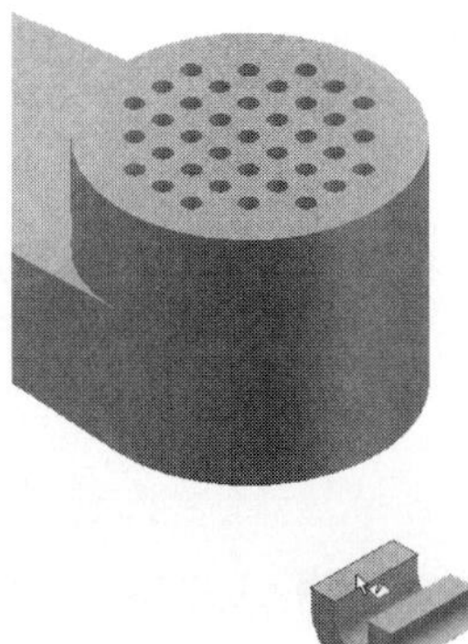

Figure 5.60 - New sketch

43. Finish Sketch.
44. Isometric View.
45. Shaded Display.
46. Zoom in on the sketch.
47. Extrude a portion of the sketch, as shown in Figure 5.59. The right side of the figure shows the resulting extrusion.
48. We will now create a loft feature to connect the two presently separated solids shown on the right in Figure 5.59. Start by creating a new sketch, using the face shown in Figure 5.60.
49. Finish Sketch. Note that when you create a new sketch plane on an existing face in Inventor, the profile of the face is automatically projected onto the sketch plane.
50. Create another new sketch, using the face shown in Figure 5.61. Use 3D Rotate to compose the view.
51. Look At the sketch.
52. Wireframe display.
53. Project the X Axis onto the sketch plane, as shown in Figure 5.62.
54. Use the Line tool to sketch the two horizontal lines shown in Figure 5.63. Both lines should be coincident with the two concentric circles.

Figure 5.61 - Another new sketch

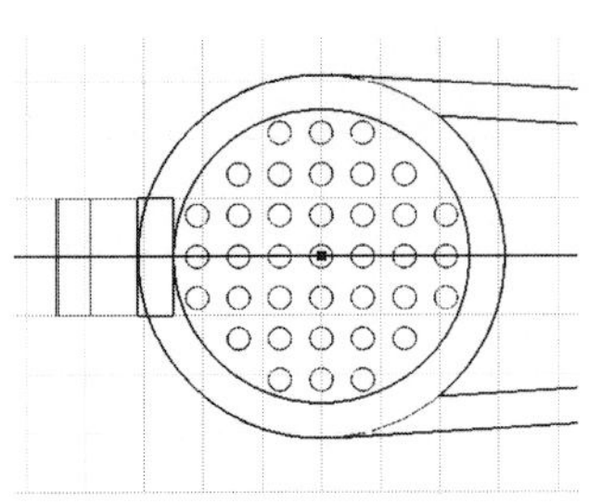

Figure 5.62 - Project X Axis onto sketch plane

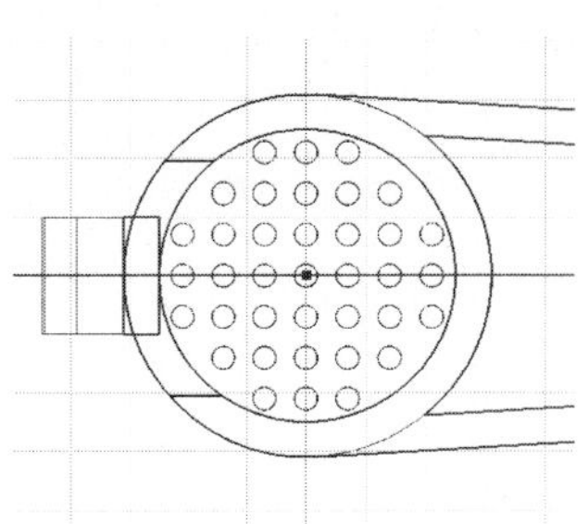

Figure 5.63 - Sketch two horizontal lines

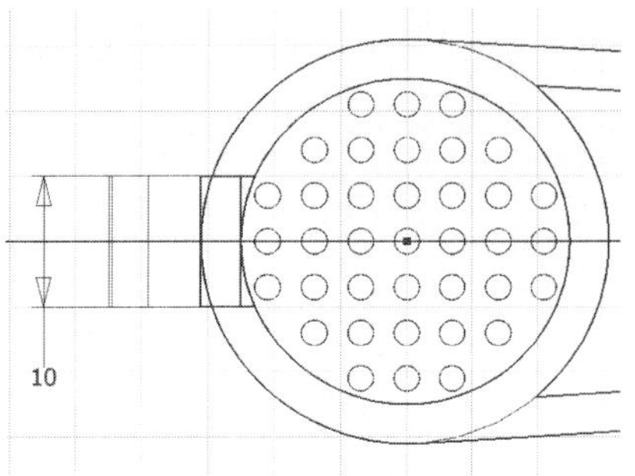

Figure 5.64 - Fully constrained sketch

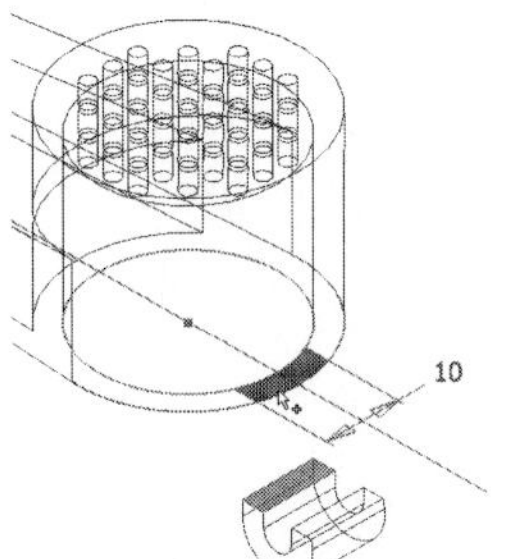

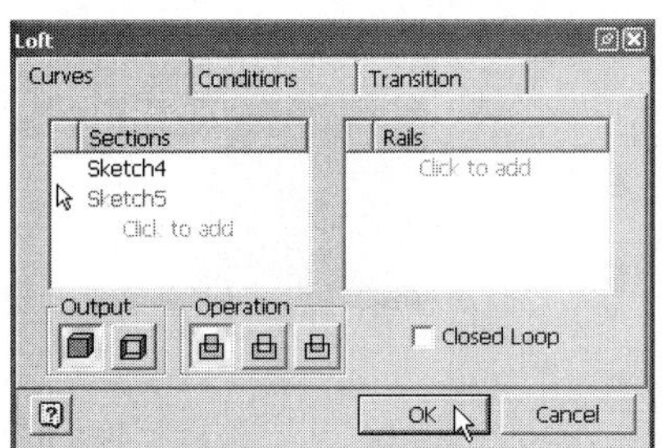

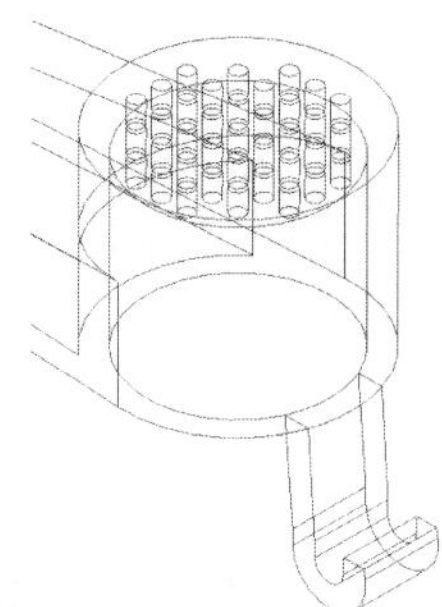

Figure 5.65 - Loft (join) feature

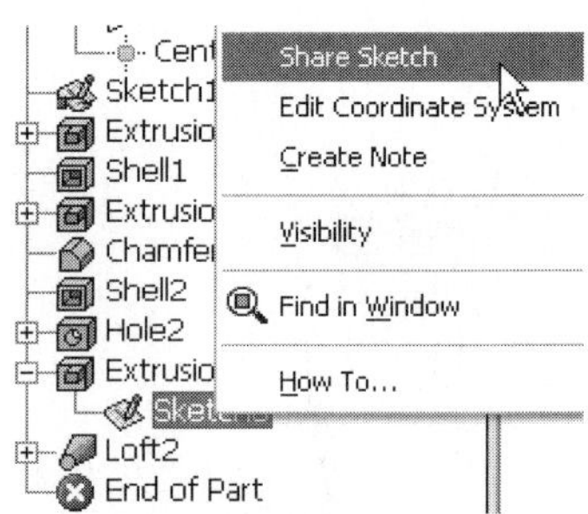

Figure 5.66 - Share sketch

55. Add a symmetry constraint to the two horizontal lines, with the X Axis as the line of symmetry.
56. Add the dimension shown in Figure 5.64.
57. Finish Sketch.
58. Isometric View.
59. Use the Loft tool as shown in Figure 5.65. First select the rectangular sketch (top face of the U-shape solid feature). Because the second sketch makes use of geometry from a previous sketch and consequently includes two different closed profiles, it is necessary to select this sketch twice. The first click will highlight the entire sketch. For the second selection, click in the vicinity of the smaller enclosed area, as shown in Figure 5.65 on the left.
60. Change to a shaded display.
61. Share the sketch of the most recently created extrude feature, as shown in Figure 5.66.
62. Extrude the shared sketch as shown in Figure 5.67 (select both areas).
63. Hide the visibility of the shared sketch.
64. Use the Fillet tool to create a *variable radius fillet.* After selecting Fillet in the panel bar, select the Variable tab. Select the edge shown in Figure 5.68 (on the left). The upper point radius is 0; the lower point radius is 2. After the start point radius is correct, click to the left of End in the dialog box to set the other radius. A Blue circle appears at the edge point being currently set. Click OK when done. Now duplicate the variable radius fillet on the opposite symmetrical edge.

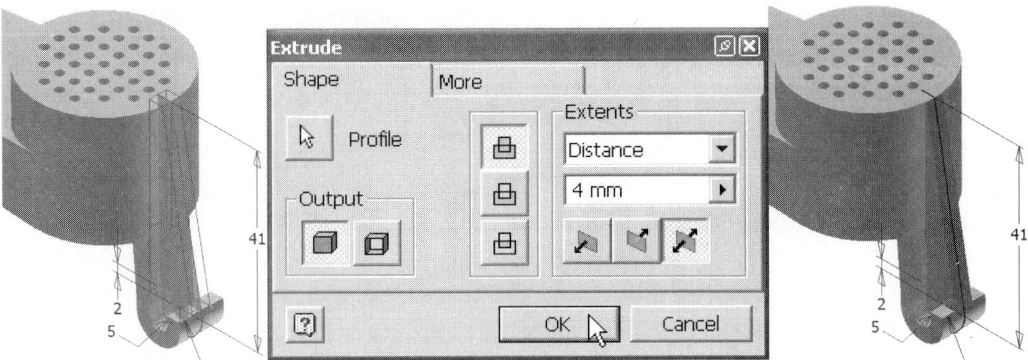

Figure 5.67 - Extrude (join) shared sketch

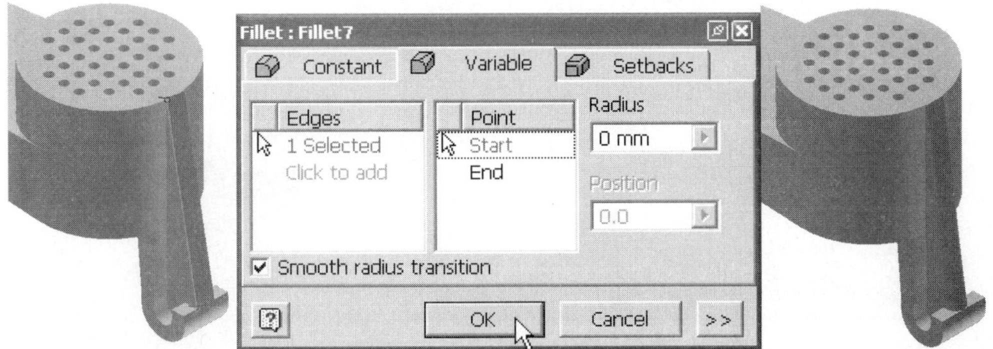

Figure 5.68 - Variable radius feature

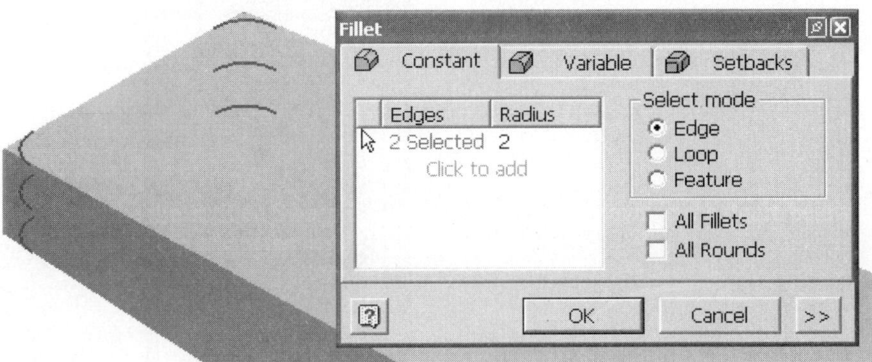

Figure 5.69 - Fillet feature

Figure 5.70 - Change view

65. Fillet the edges shown in Figure 5.69.
66. Use Common View to change the isometric view to that shown in Figure 5.70.
67. Fillet the edge as shown in Figure 5.71.
68. Fillet the edge as shown in Figure 5.72.
69. Isometric View. The completed part appears in Figure 5.73.
70. Save the part. This completes Tutorial 12.

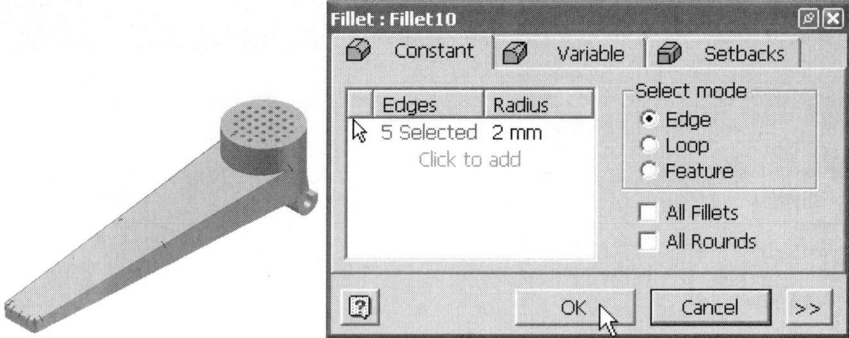

Figure 5.71 - Add fillet feature

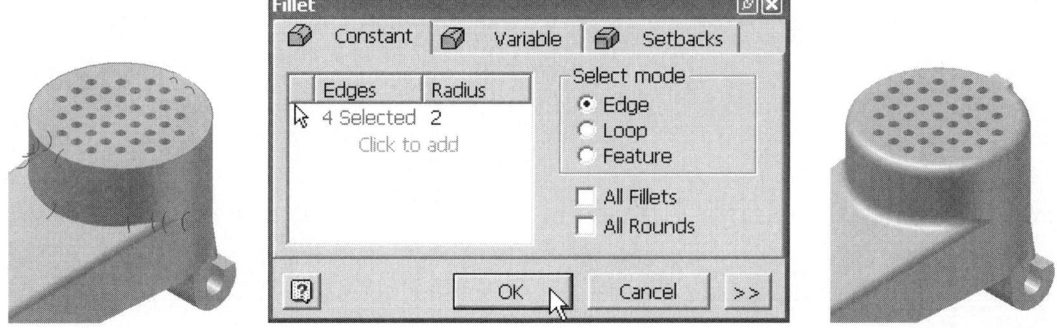

Figure 5.72 - Additional fillet features

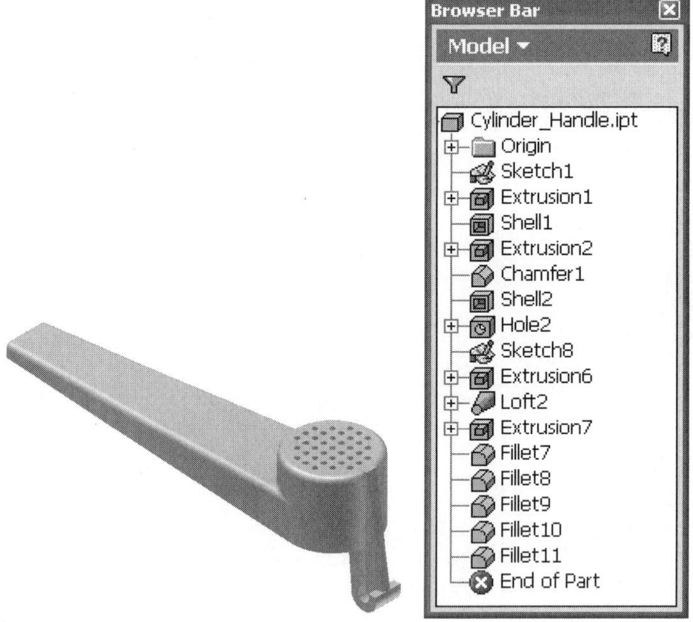

Figure 5.73 - Completed part

Handle (Garlic Press) — Tutorial 13

Build Strategy

1. Sketch (Figure 5.74).
2. Extrude (Figure 5.75).
3. Chamfer (two distances) (Figure 5.76).
4. Split (all faces, YZ Plane) (Figure 5.77).
5. Chamfer (two distances) (Figure 5.78).
6. Chamfer twice (two distances) (Figure 5.79).
7. Chamfer twice (two distances) (Figure 5.80).
8. Chamfer (two distances) (Figure 5.81).
9. Face Draft (2 faces, 1 side) (Figure 5.82).

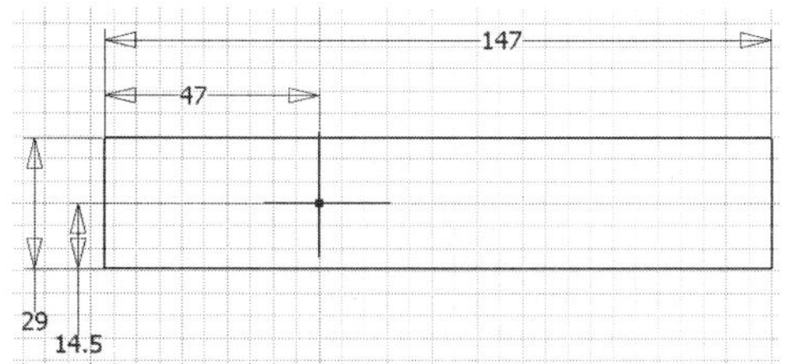

Figure 5.74 - Base feature sketch

Figure 5.75 - Extrude base feature

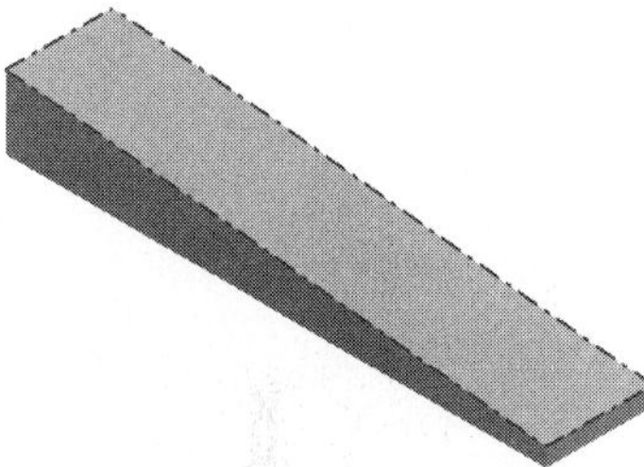

Figure 5.76 - Chamfer feature

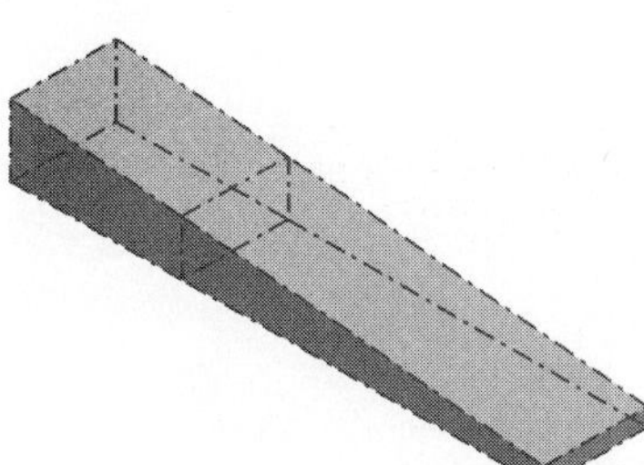

Figure 5.77 - Split feature: all faces, about YZ plane

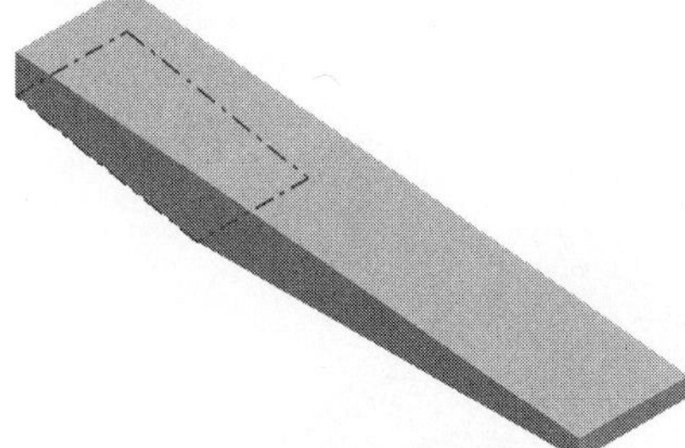

Figure 5.78 - Chamfer feature (bottom forward end)

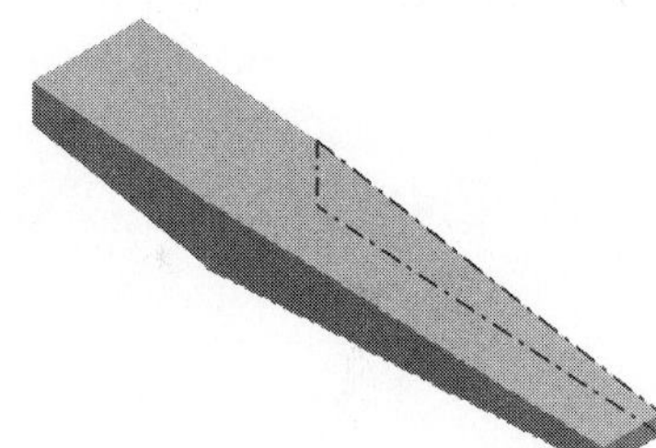

Figure 5.79 - Chamfer both forward sides

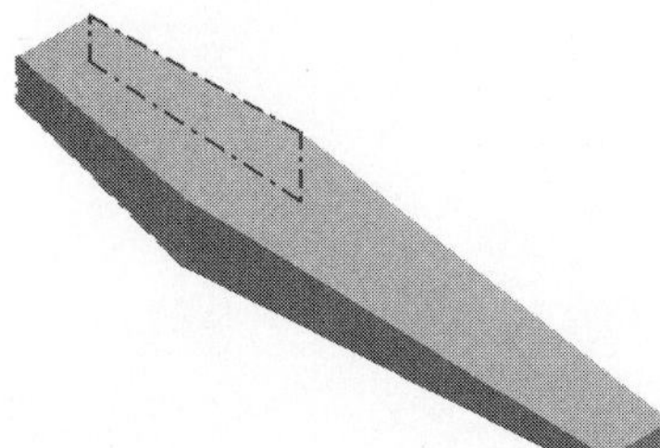

Figure 5.80 - Chamfer both rear sides

Figure 5.81 - Chamfer forward end

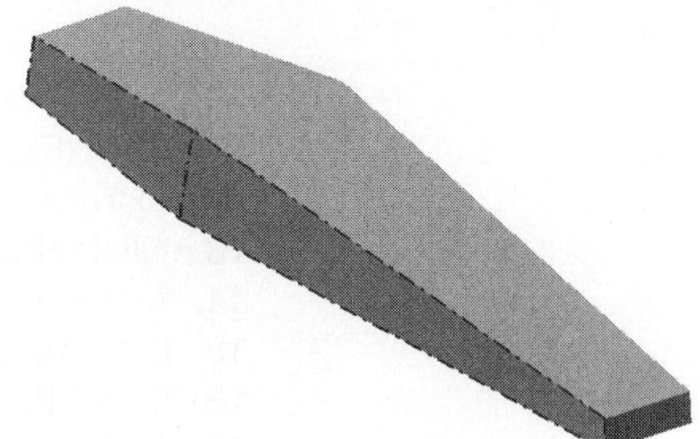

Figure 5.82 - Face draft

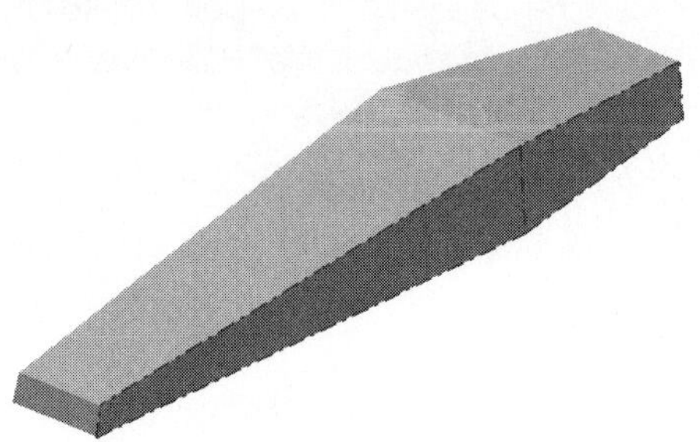

Figure 5.83 - Face draft, opposite side

Figure 5.84 - Shell

Figure 5.85 - Extrude feature

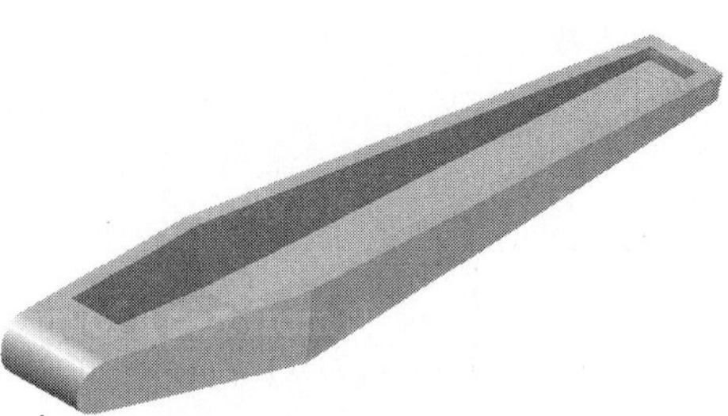

Figure 5.86 - Round end (two fillets)

Figure 5.87 - Extrude (cut) feature

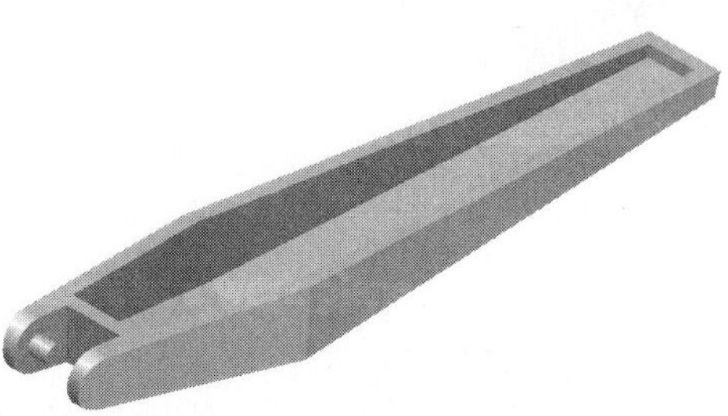

Figure 5.88 - Extrude feature

10. Face Draft (2 faces, 1 side) (Figure 5.83).
11. Shell (Figure 5.84).
12. Extrude (join) (Figure 5.85).
13. Fillet (Figure 5.86).
14. Extrude (cut) (Figure 5.87).
15. Extrude (join) (Figure 5.88).
16. Mirror Feature (Figure 5.89).
17. Extrude (cut) (Figure 5.90).
18. Extrude (join) (Figure 5.91).
19. Fillet (Figure 5.92).

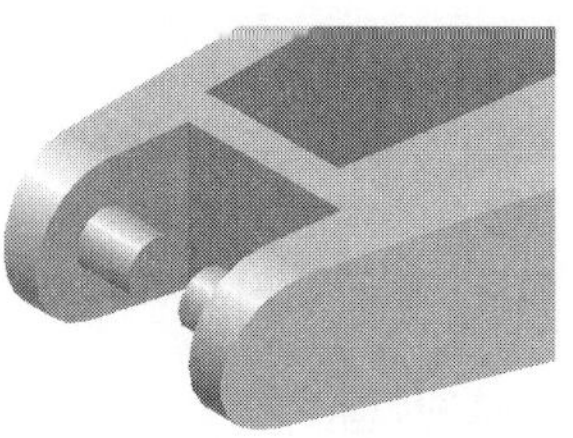

Figure 5.89 - Mirror feature

Figure 5.90 - Extruded (cut) feature

Figure 5.91 - Extruded feature

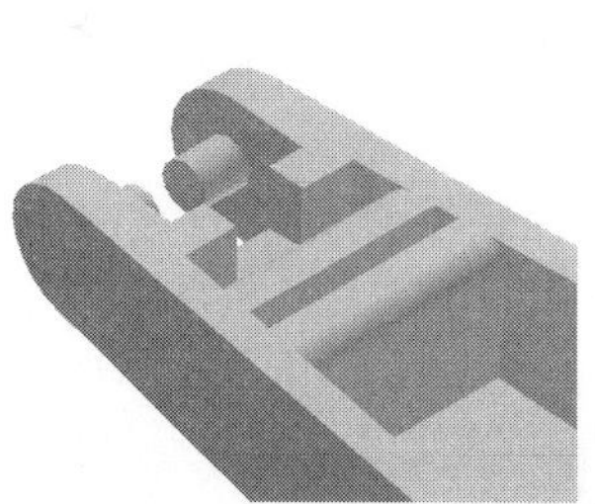

Figure 5.92 - Fillet feature

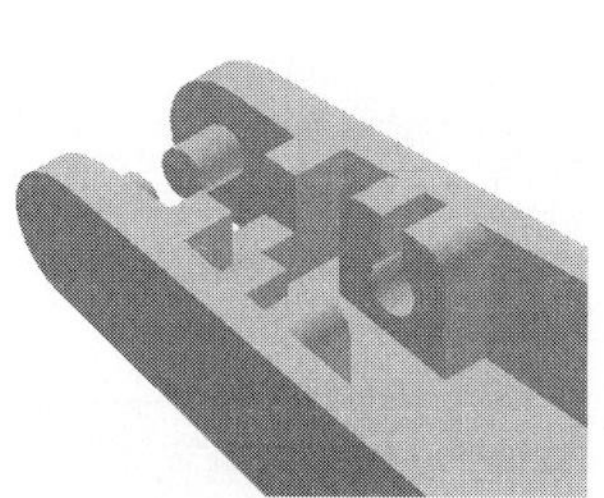

Figure 5.93 - Extruded cut feature

Figure 5.94 - Completed part

20. Extrude (cut) (Figure 5.93).
21. Completed part (Figure 5.94).

■ Detailed Modeling Steps

1. Start a new metric part file.
2. Project X and Y Axes and Center Point on sketch.
3. Use Rectangle tool and then add the dimensions shown in Figure 5.95.
4. Change the dimension display by right-clicking in the graphics area; now select Dimension Display > Name.
5. Add the next dimension as an equation, as shown in Figure 5.96. The new dimension should be half the height. Be sure to use the parameter name corresponding to the height in *your* sketch. It may be different from the one shown in the figure, depending upon the order in which the dimensions were added.

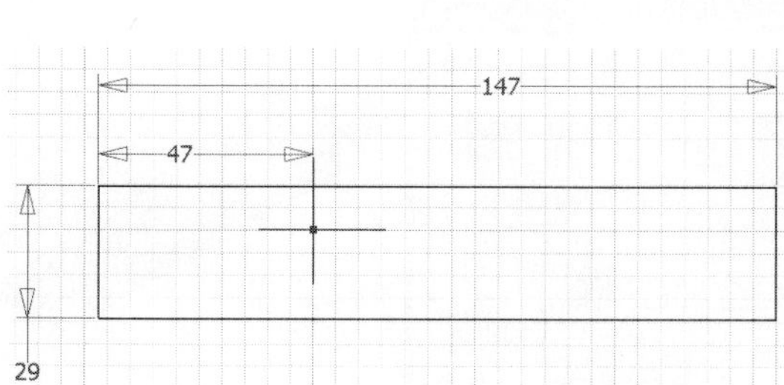

Figure 5.95 - Preliminary sketch

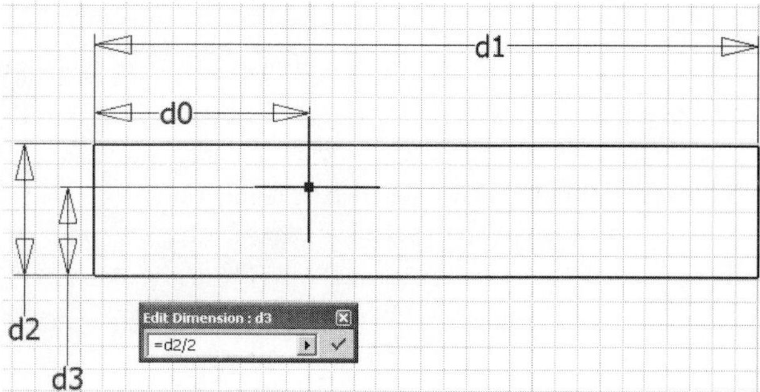

Figure 5.96 - Entering a dimension as an equation

6. Change the dimension display by right-clicking in the graphics area; now select Dimension Display > Value.
7. Finish Sketch.
8. Isometric View.
9. Extrude the sketch, as shown in Figure 5.97. The resulting solid should be similar to the figure on the right.
10. Save the file as **Handle.** It is recommended that the file be regularly saved every five to ten minutes.
11. Chamfer the solid using two distances, as shown in Figure 5.98. If necessary, use the Flip button to reverse the distances on the model. The resulting solid should resemble the figure on the right.
12. Use the Split tool to split all of the part faces about the YZ Plane (available from the Origin folder on the part browser), as shown in Figure 5.99.

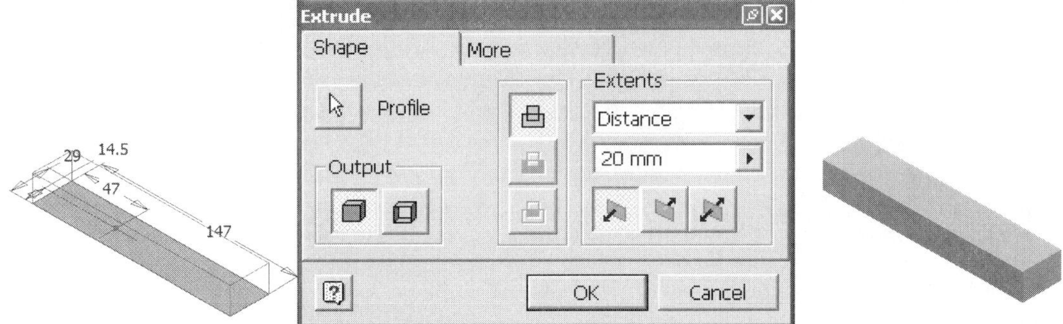

Figure 5.97 - Extrude base feature operation

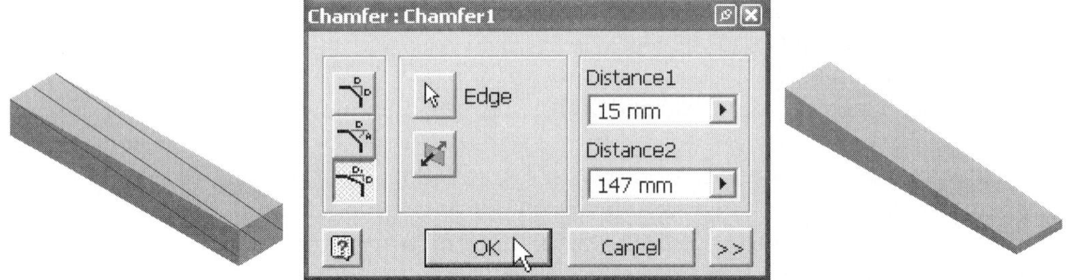

Figure 5.98 - Chamfer feature operation

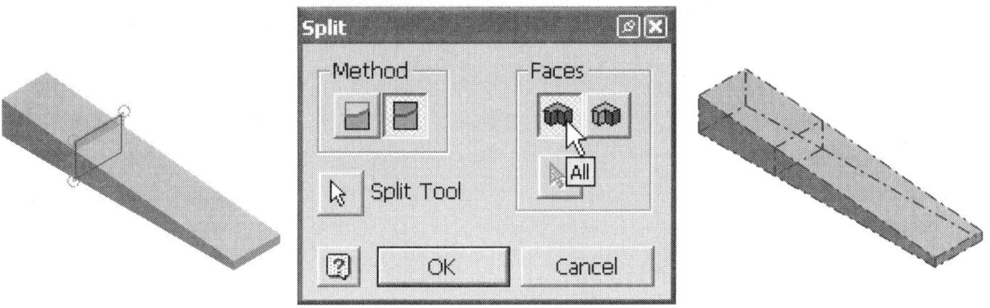

Figure 5.99 - Split feature operation

13. Change to a Wireframe display.
14. Chamfer the bottom, forward end of the solid part, as shown in Figure 5.100.
15. Chamfer the forward side of the solid as shown in Figure 5.101. Note that it was necessary to use the Flip button. Alternatively, the 6.5 and 100 dimensions could be switched for Distance1 and Distance2, respectively.
16. Chamfer the opposite side of the solid, as shown in Figure 5.102.
17. Chamfer the rear side of the solid as shown in Figure 5.103.
18. Chamfer the opposite edge of the solid, as shown in Figure 5.104. Note that it was necessary to use the flip button.
19. Zoom in on the right end of the solid.
20. Chamfer the forward end of the solid as shown in Figure 5.105.
21. Isometric View.

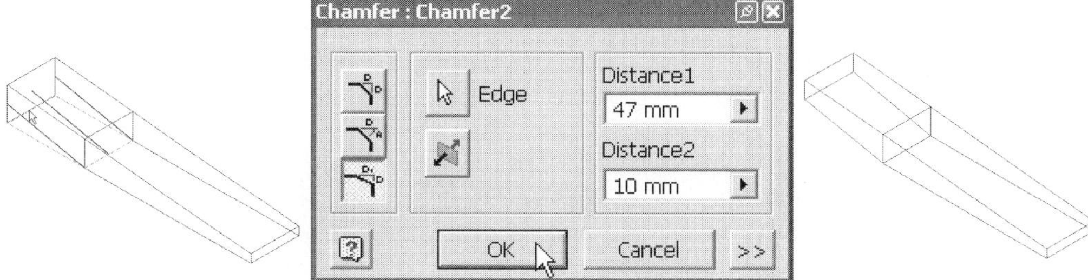

Figure 5.100 - Chamfer feature operation

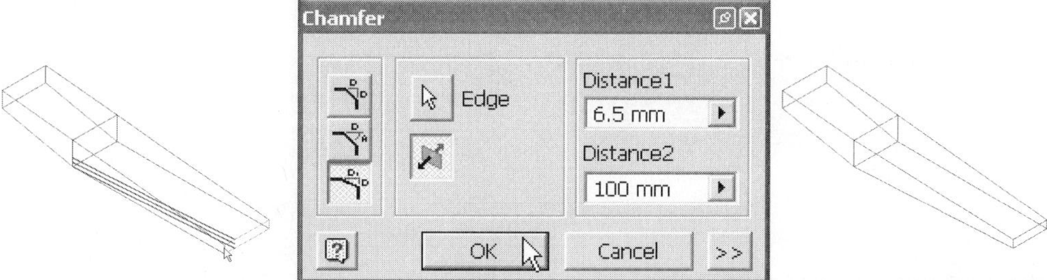

Figure 5.101 - Chamfer forward side

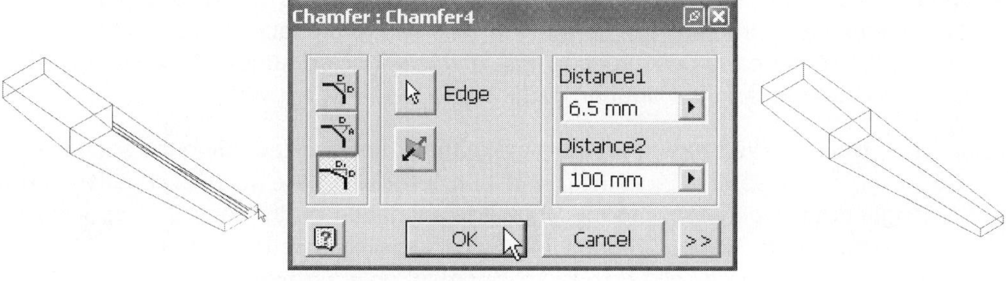

Figure 5.102 - Chamfer forward side (opposite)

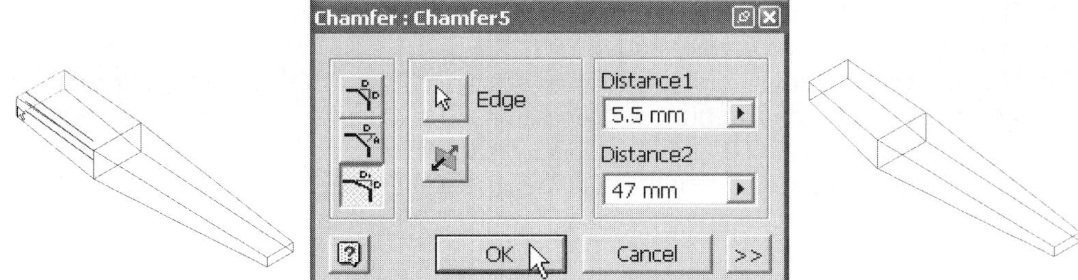

Figure 5.103 - Chamfer rear side

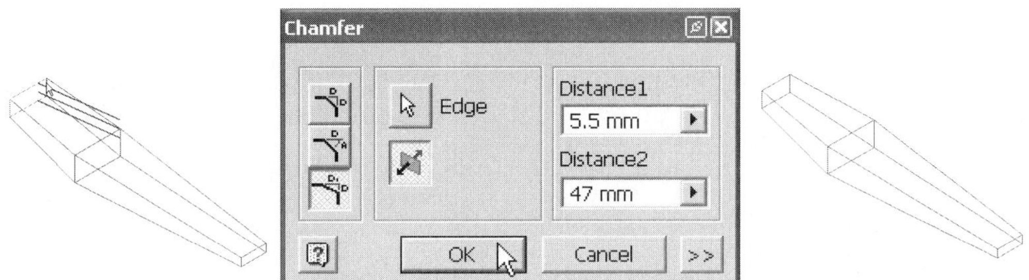

Figure 5.104 - Chamfer rear side (opposite)

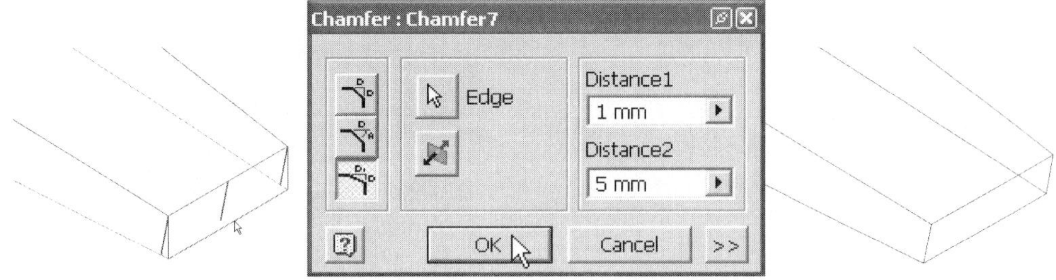

Figure 5.105 - Chamfer forward end

22. Shaded display.
23. Use the 3D Rotate to change to a view similar to that shown in Figure 5.106.
24. Use the Face Draft tool to add a draft angle to the faces shown in Figure 5.107. Note the pull direction (up), the (lower) fixed edges of the two faces to be tapered, and the draft angle.

    TIP: The face draft tool requires some getting used to. Reviewing the section earlier in this chapter should help with the terminology (e.g., pull direction, fixed edge, draft angle). Viewing the Show Me animation on the face draft tool is also recommended (Visual Syllabus, Part Modeling palette, Face Draft icon). Best of all, create a solid box and experiment on your own.

Figure 5.106 - Change view

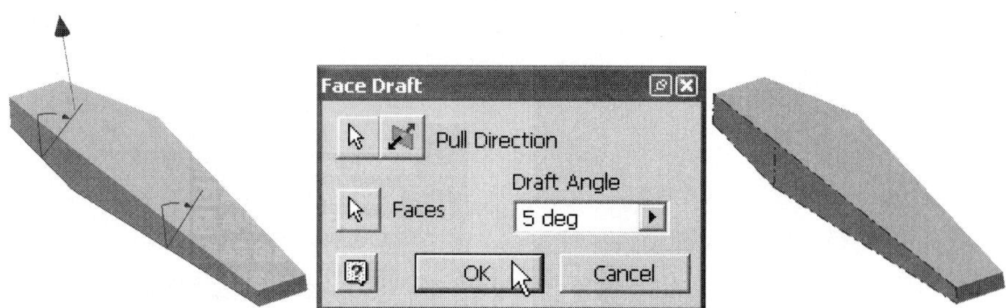

Figure 5.107 - Face draft operation (two faces)

Figure 5.108 - Change view

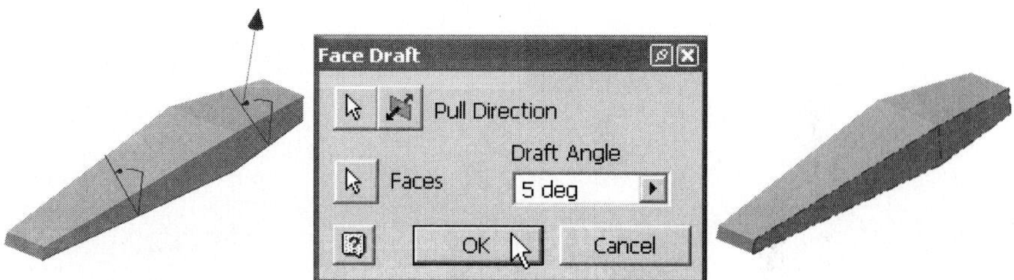

Figure 5.109 - Face draft operation (two faces, opposite)

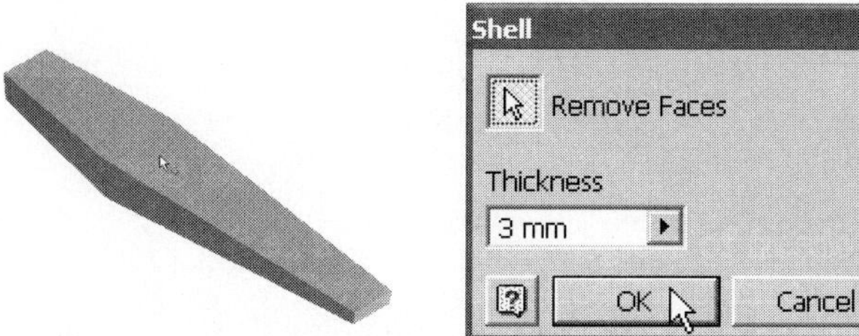

Figure 5.110 - Shell feature operation

25. Use the 3D Rotate tool (Common View) change to a view similar to that shown in Figure 5.108.
26. Again use the Face Draft tool to add the draft angles to the faces shown in Figure 5.109.
27. Isometric View.
28. Use the Shell tool as shown in Figure 5.110.
29. New Sketch on the XZ Plane.
30. Look At sketch.
31. Wireframe display.
32. Project edges shown in Figure 5.111.

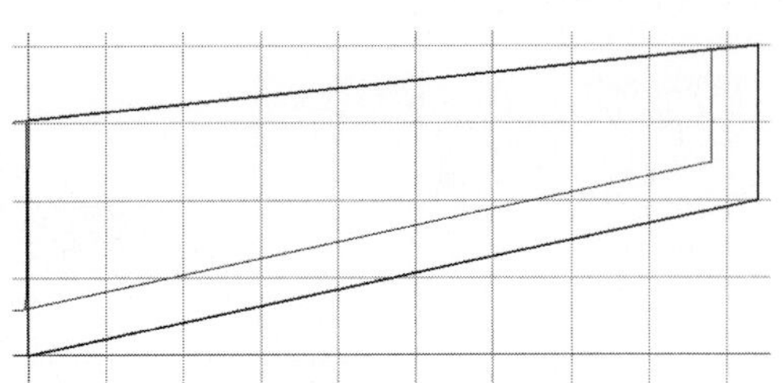

Figure 5.111 - Project edges

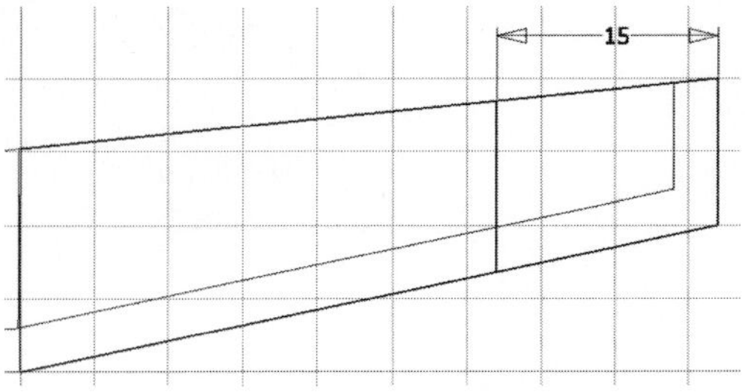

Figure 5.112 - Fully constrained sketch

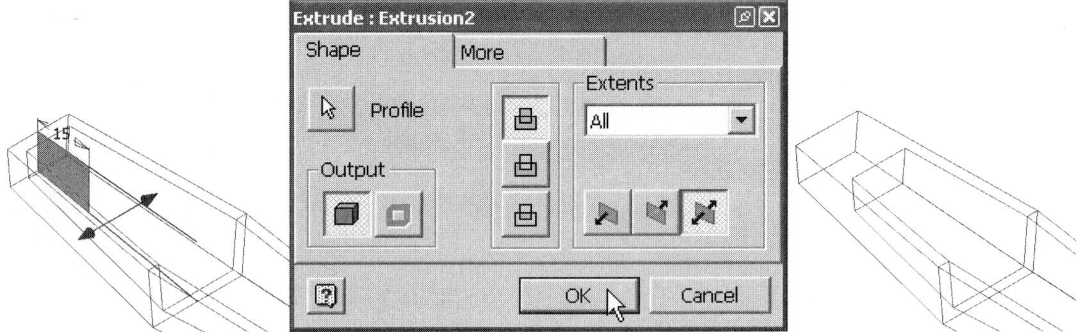

Figure 5.113 - Extrude (join) feature operation

Figure 5.114 - Change view

33. Use the Line tool to sketch the vertical line, coincident with the two projected lines. Next add the dimension, as shown in Figure 5.112.
34. Finish Sketch.
35. Isometric View.
36. Use the Extrude tool as shown in Figure 5.113.
37. Shaded display.
38. Use 3D Rotate, Common View mode to change to the isometric view shown in Figure 5.114.
39. Use the Fillet tool to add two rounds as shown in Figure 5.115.
40. New Sketch - select the XY Plane.
41. Look At the sketch plane.
42. Zoom In.
43. Wireframe Display.

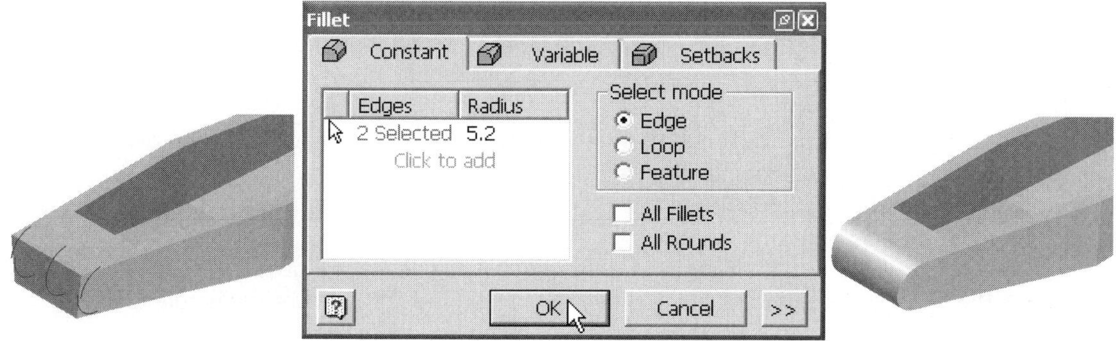

Figure 5.115 - Fillet feature operation

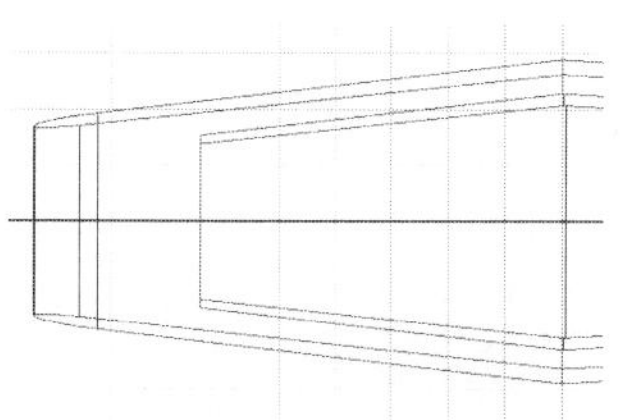

Figure 5.116 - Projected geometry

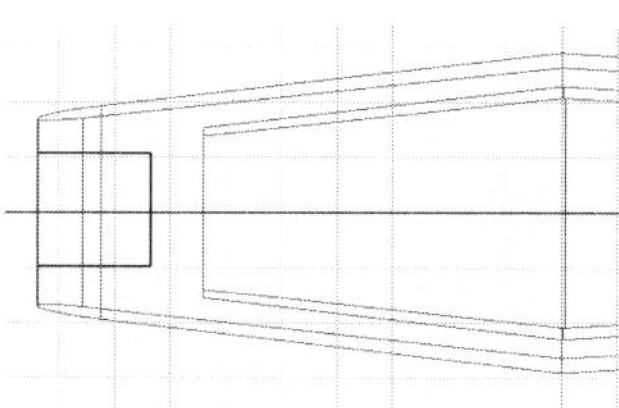

Figure 5.117 - Sketch rectangle

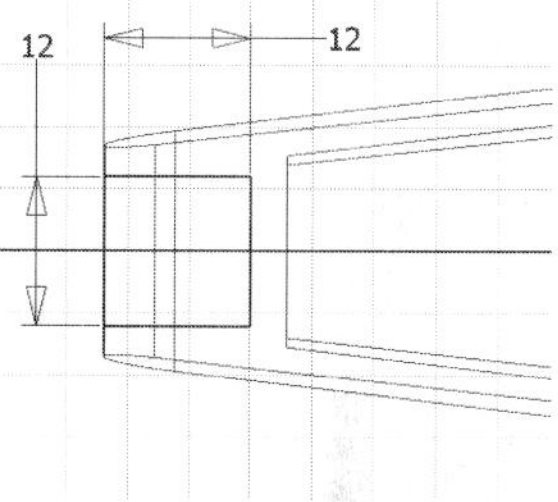

Figure 5.118 - Fully constrained sketch

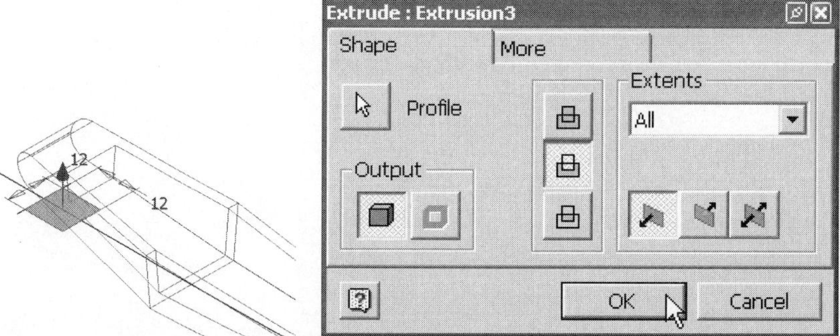

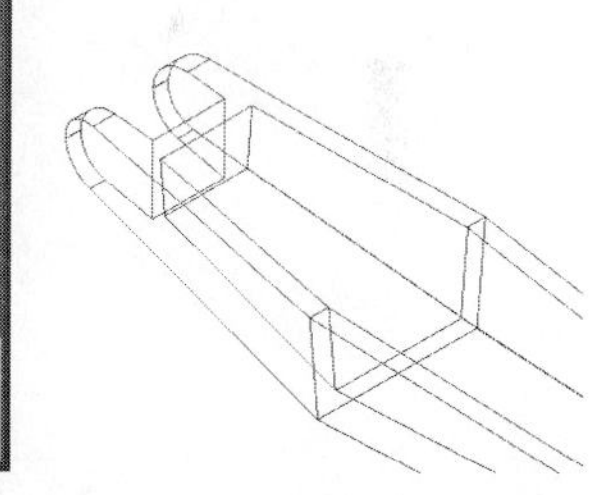

Figure 5.119 - Extrude (cut) feature operation

44. Project the X Axis and the leading edge of the solid, as shown in Figure 5.116.
45. Use the Rectangle tool as shown in Figure 5.117. One corner (and therefore one side) of the rectangle is coincident with the left edge of the existing solid.
46. Add a symmetry constraint to the two horizontal edges (about the X Axis) of the rectangle.
47. Add the dimensions shown in Figure 5.118.
48. Finish Sketch.
49. Isometric View.
50. Extrude (cut) as shown in Figure 5.119.

51. Shaded Display.
52. New Sketch - select the face shown in Figure 5.120.
53. Use the Center point circle tool to add a circle (coincident with the existing center point).
54. Use Common View to modify the isometric view, and then dimension the circle as shown in Figure 5.121.
55. Finish Sketch.
56. Extrude as shown in Figure 5.122.
57. Use the Mirror Feature tool as shown in Figure 5.123. Use the XZ Plane as the mirror plane. Figure 5.123 on the right shows the resulting solid.
58. New Sketch, using the face shown in Figure 5.124.
59. Look At the sketch.
60. Wireframe display.
61. Project the Z Axis onto the sketch plane.

Figure 5.120 - New sketch

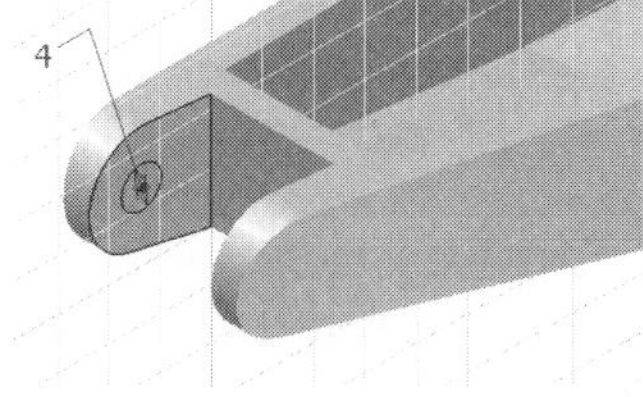

Figure 5.121 - Fully constrained sketch

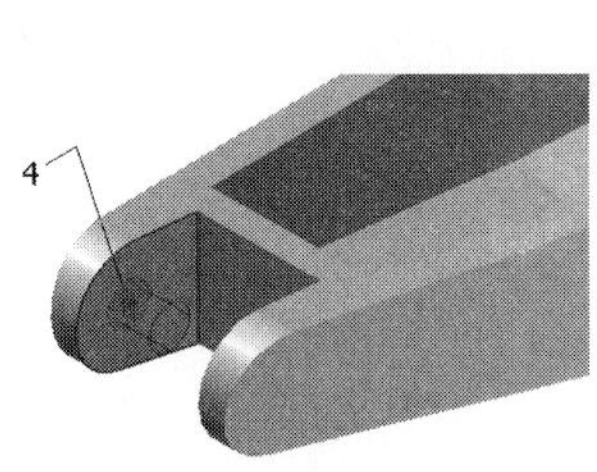

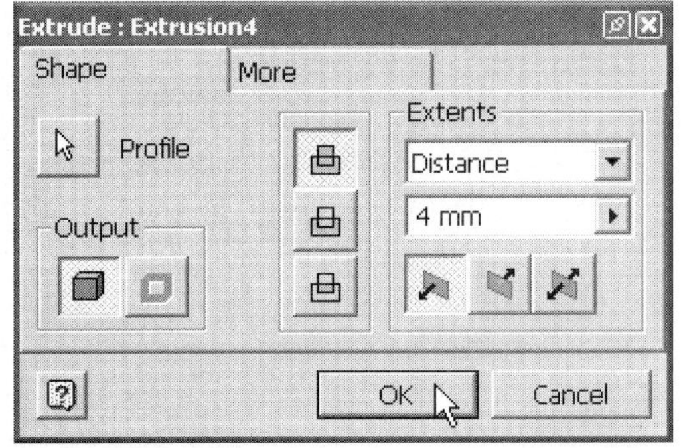

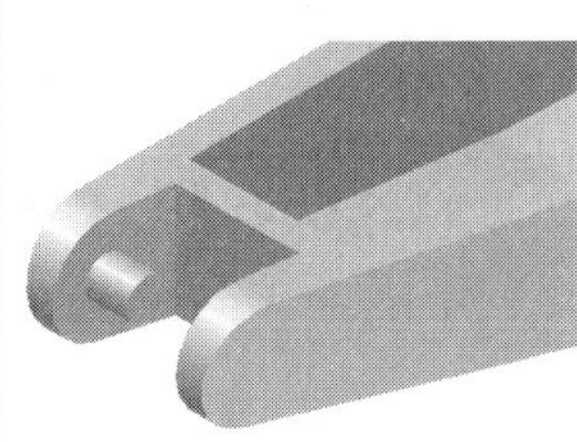

Figure 5.122 - Extrude feature operation

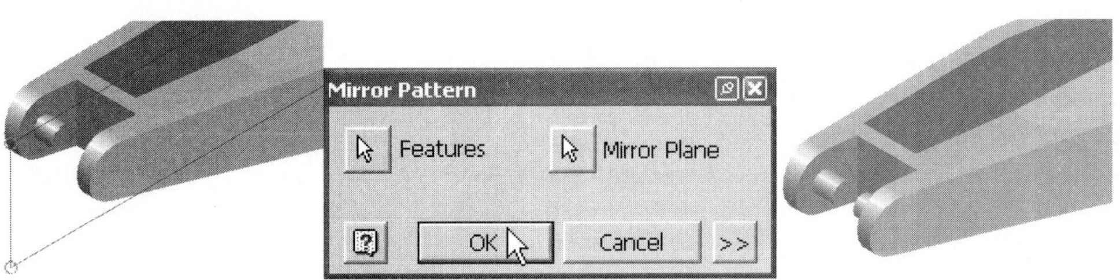

Figure 5.123 - Mirror feature operation

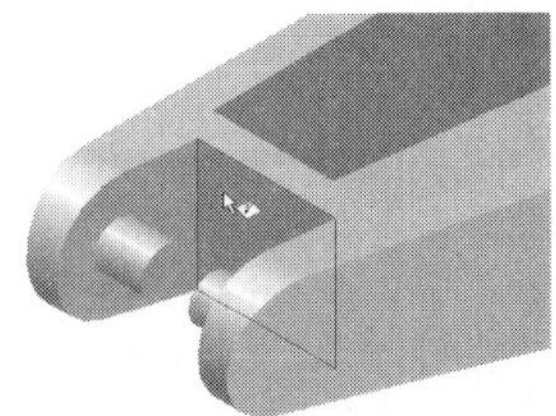

Figure 5.124 - New sketch

62. Add a rectangle to the sketch, as shown in Figure 5.125. The right side of the rectangle is coincident with the existing vertical edge.
63. Add a symmetry constraint to the horizontal sides of the rectangle (about the Z Axis).
64. Add the dimensions shown in Figure 5.126.
65. Finish Sketch.
66. Isometric View.
67. Use the Extrude tool as shown in Figure 5.127. Note the cut operation.
68. Shaded Display.
69. New Sketch on the XZ Plane.
70. Look At the sketch plane.
71. Zoom in.
72. Wireframe display.
73. Project Geometry onto the sketch plane, as shown in Figure 5.128.
74. Use the Center point circle and General Dimension tools to add the circle with dimensions shown in Figure 5.129.
75. Use the Line and Trim tool to modify the sketch as shown in Figure 5.130. Note that all four vertical lines are coincident (on both ends) with existing geometry.
76. Add the dimensions as shown in Figure 5.131.
77. Finish Sketch.
78. Isometric View.
79. Extrude the profile as shown in Figure 5.132.
80. Shaded Display.
81. Use the Fillet tool to add the rounds shown in Figure 5.133.

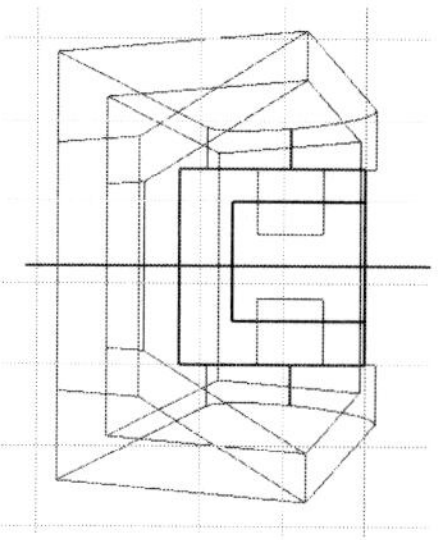

Figure 5.125 - Rectangle sketch

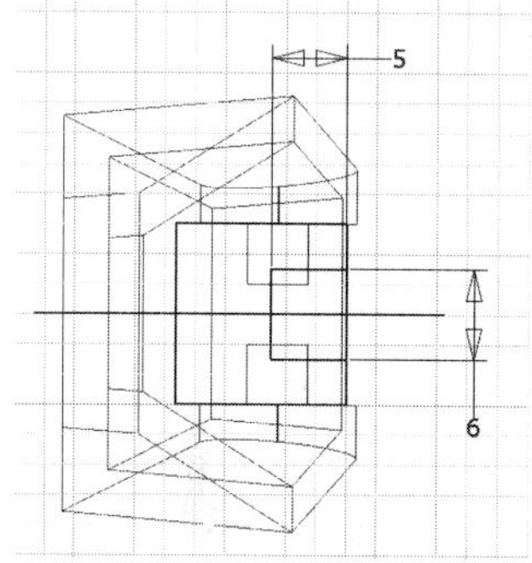

Figure 5.126 - Fully constrained sketch

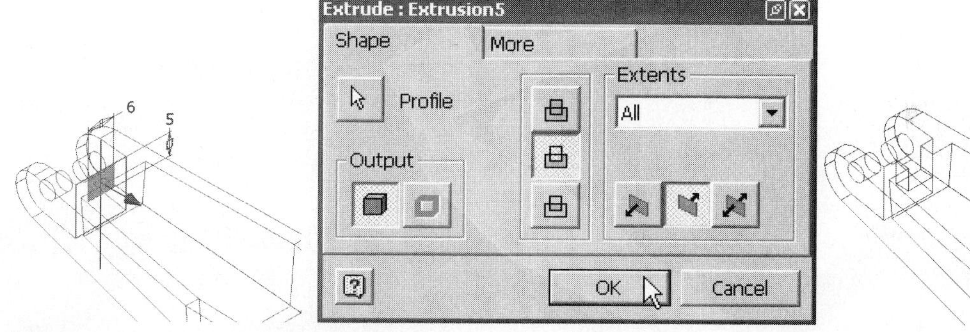

Figure 5.127 - Extruded feature operation

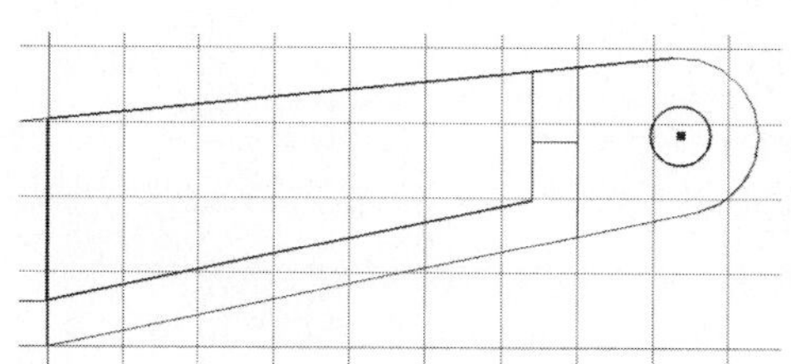

Figure 5.128 - Projected geometry

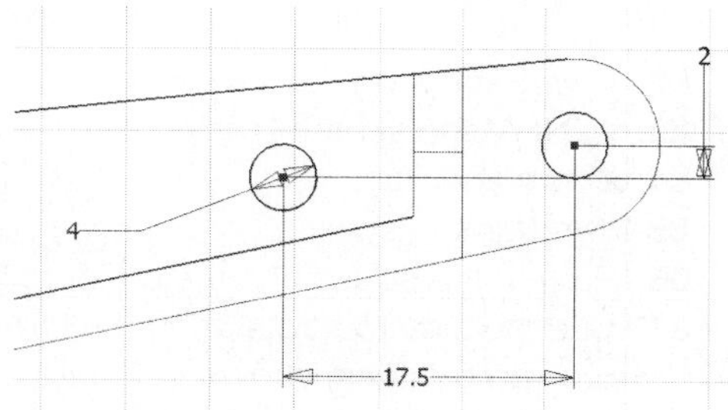

Figure 5.129 - Sketch geometry

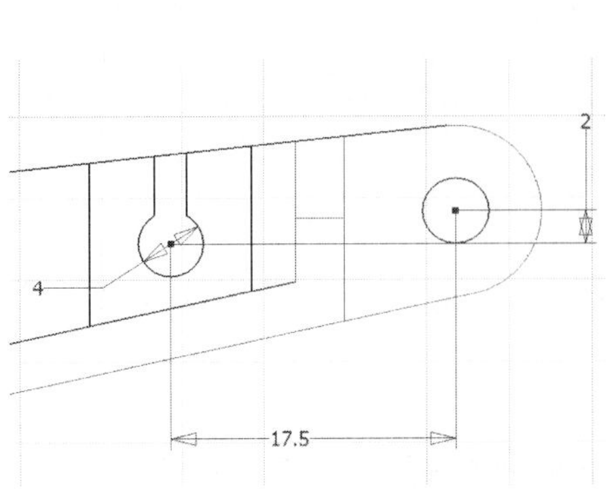

Figure 5.130 - More sketch geometry

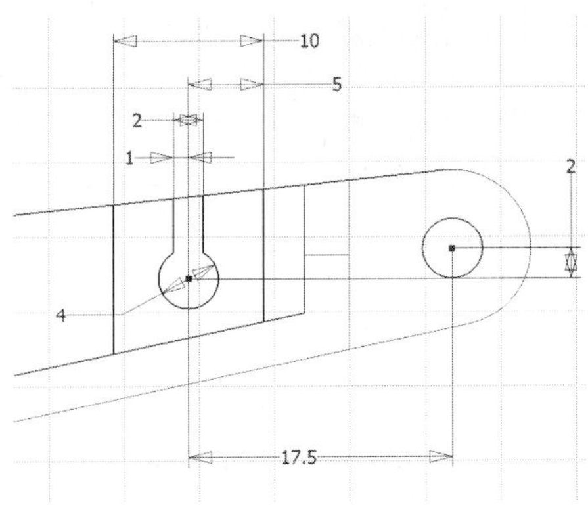

Figure 5.131 - Fully constrained sketch

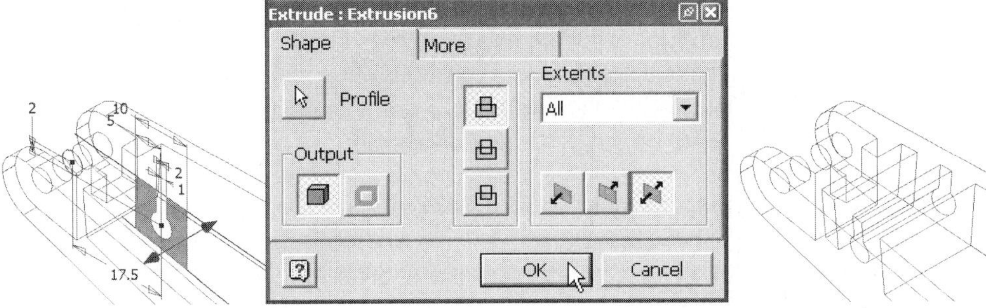

Figure 5.132 - Extruded feature operation

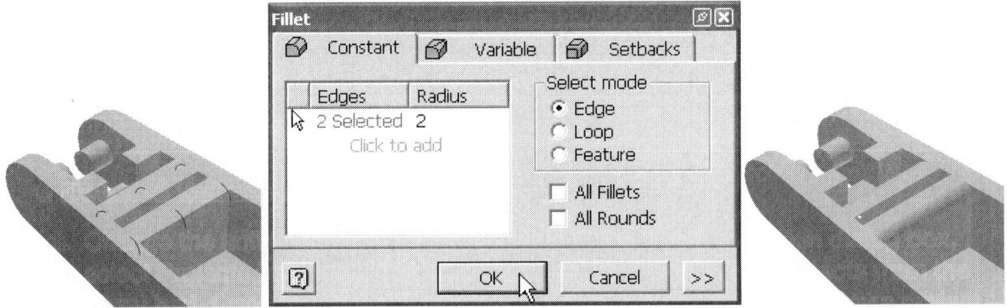

Figure 5.133 - Fillet feature operation

82. New Sketch on the XZ Plane.
83. Look At the sketch plane.
84. Zoom and Pan.
85. Wireframe display.
86. Project the edges shown in Figure 5.134.
87. Use the Line tool to sketch the two inclined lines (both coincident with the projected edges) shown in Figure 5.135. These additional lines serve to form a closed profile which can now be extruded.
88. Finish Sketch.
89. Isometric View.

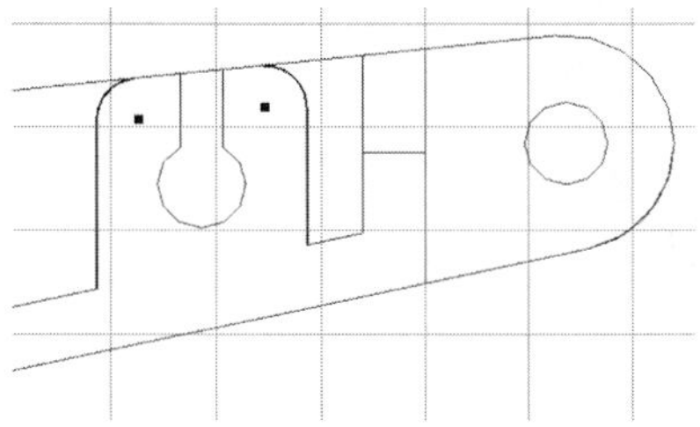

Figure 5.134 - Projected geometry

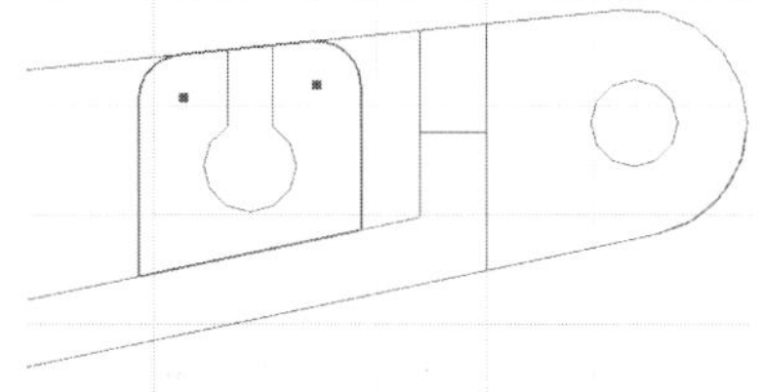

Figure 5.135 - Add geometry

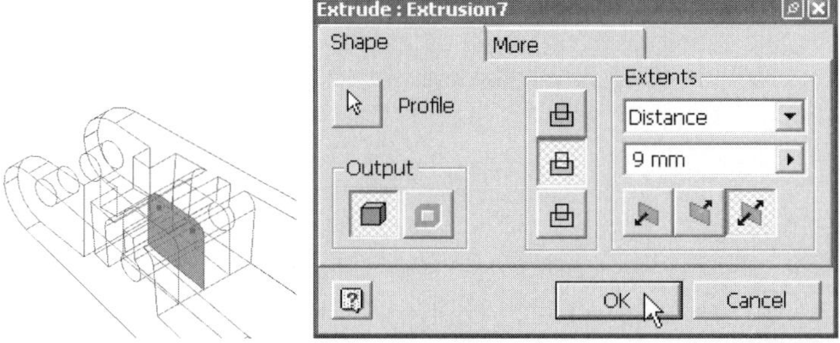

Figure 5.136 - Extruded cut feature operation

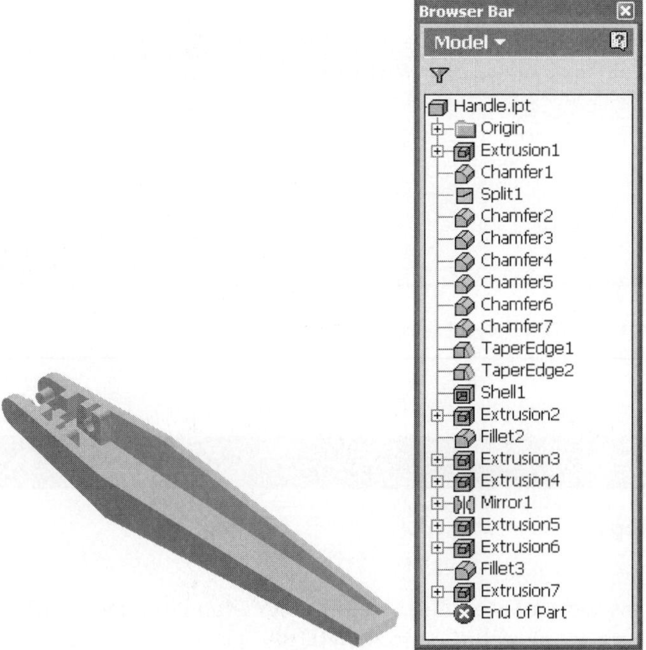

Figure 5.137 - Completed Handle part

90. Extrude the profile as shown in Figure 5.136. Note the cut operation.
91. Shaded display.
92. Isometric View. Figure 5.137 shows the completed part.
93. Save the file. This completes Tutorial 13. Note that this also completes all of the garlic press parts.

TUTORIAL 14 Creating a New Project (Toast Rack)

1. Open the Project Editor by selecting File > Projects . . . from the menu bar.
2. Click the New button.
3. From the Inventor project wizard dialog box, click the Next button (the Personal Workspace for Group Project radio button should be selected).
4. Modify the Project File Name and Location in the Inventor project wizard dialog box as shown in Figure 5.138, then click the Finish Button.
5. Click the OK button when prompted to create a new project path.
6. With the Toast Rack project highlighted, click the Apply button.

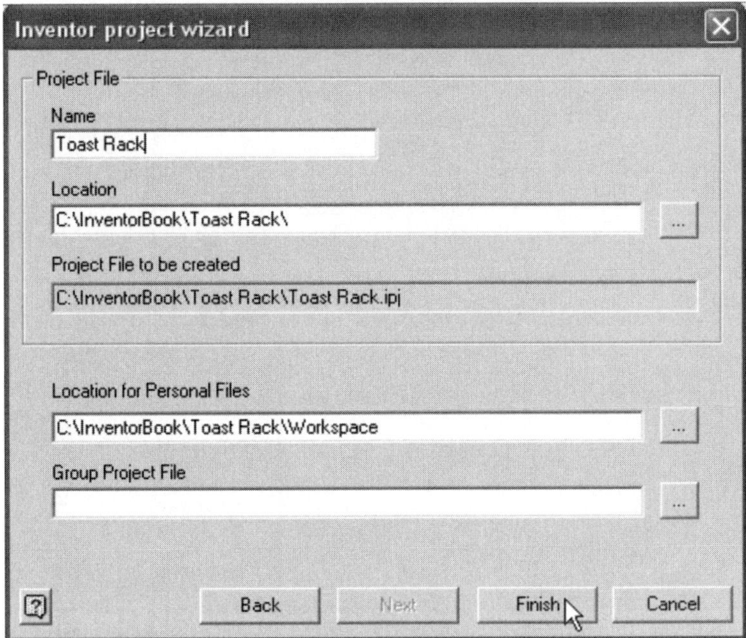

Figure 5.138 - Project wizard dialog box

TUTORIAL 15 Toast Rack

Build Strategy

1. Sketch (Figure 5.139).
2. Create a new work plane, offset from the XY Plane (Figure 5.140).
3. Project geometry, sketch (Figure 5.141).
4. Create a 3D sketch, using Include Geometry and the Line tools (Figure 5.142).
5. On the XZ Plane, project the leg from the 3D sketch, then sketch a circle at the point of projection (Figure 5.143).
6. Sweep (Figure 5.144).
7. Set up for Coil (Figure 5.145).
8. Coil (Figure 5.146).

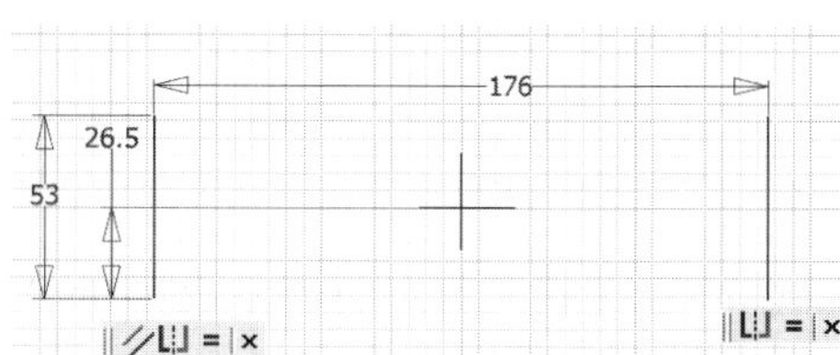

Figure 5.139 - Fully constrained sketch

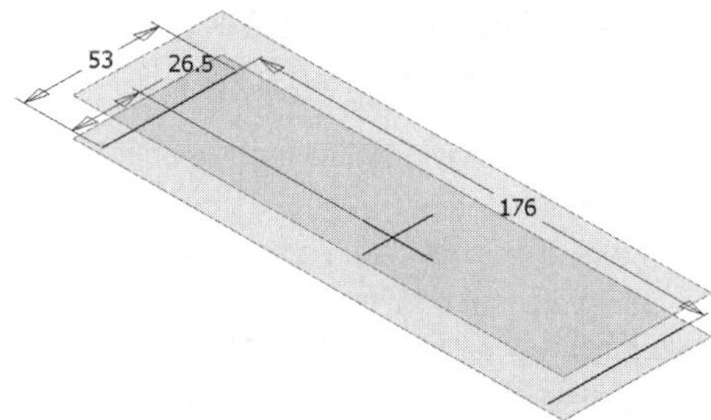

Figure 5.140 - Offset work plane

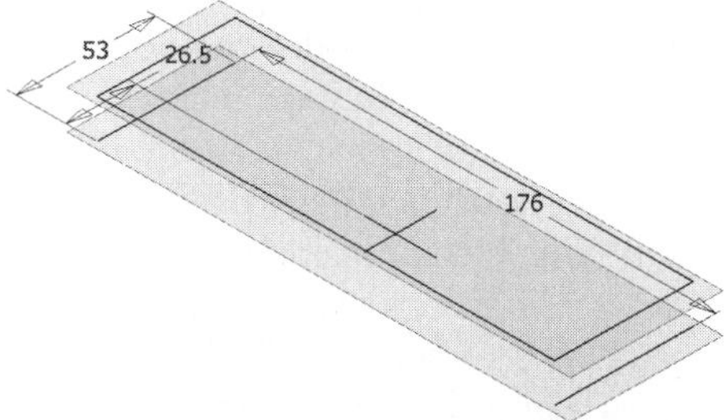

Figure 5.141 - Another sketch

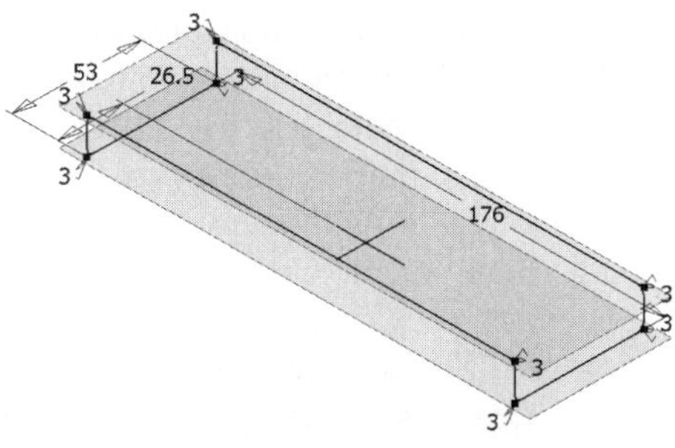

Figure 5.142 - 3D sketch path

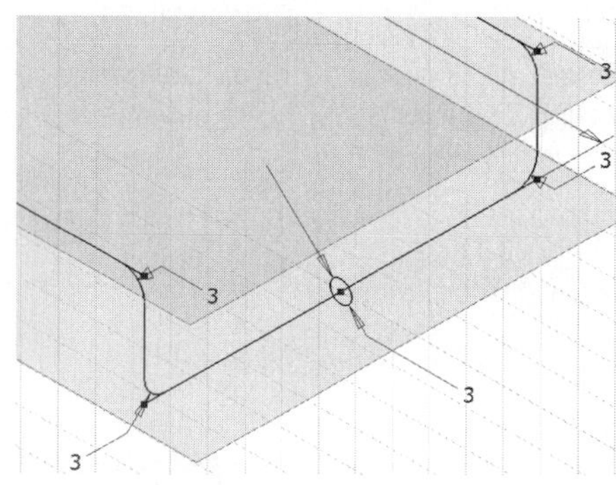

Figure 5.143 - Circle profile

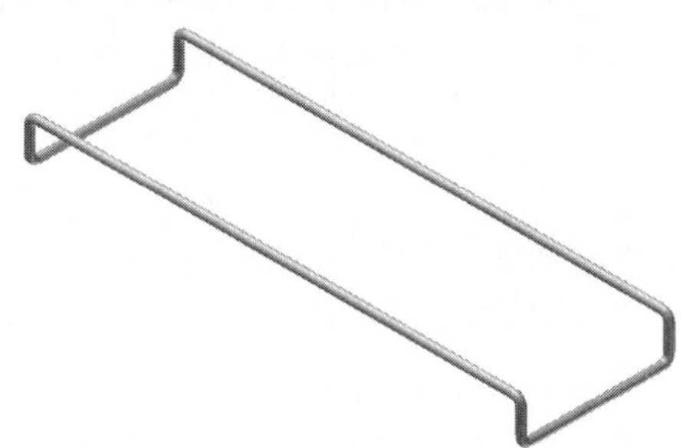

Figure 5.144 - Sweep feature

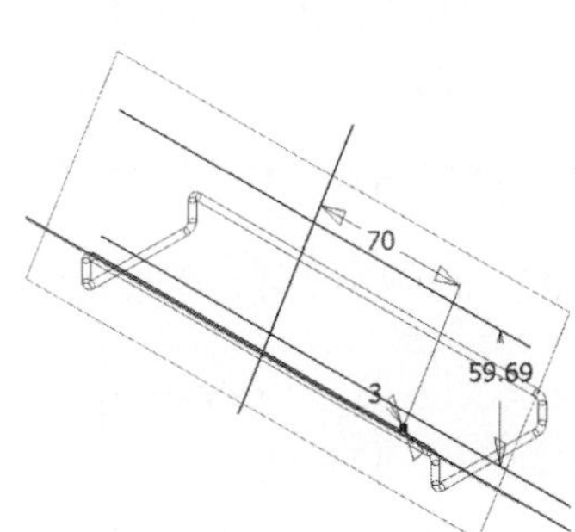

Figure 5.145 - Coil set up

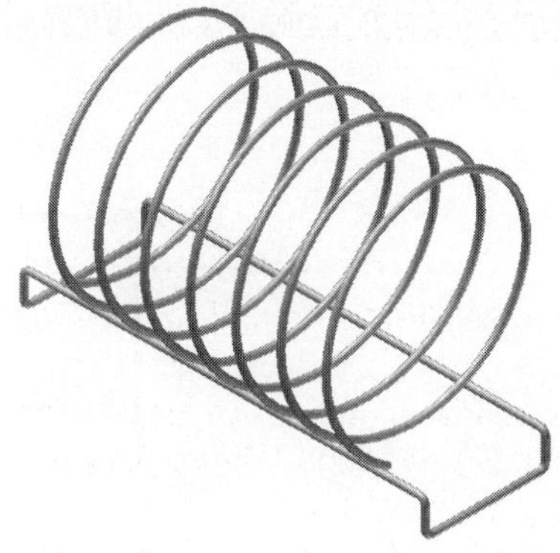

Figure 5.146 - Completed part

Detailed Modeling Steps

1. Start a new metric part file.
2. Use the Project Geometry tool to project the X and Y Axes onto the sketch plane.
3. Use the Line tool to sketch a single "vertical" line to the left of the origin (Figure 5.147).
4. Use the Mirror tool to mirror the "vertical" line about the vertical Y Axis. First select the "vertical" line, then click on the Mirror line button and select the vertical axis, then click on the Apply button, then select Done.
5. Add an Equal constraint to the two "vertical" lines. Note that in using the Mirror tool to create the second "vertical" line, a symmetric constraint has been added to the mirrored line. How do I know this?
6. Use the General Dimension tool to add the dimensions shown in Figure 5.148.
7. After right-clicking in the graphics area, select Dimension Display > Name. The dimension names are now displayed, not their values, as shown in Figure 5.149.
8. Again using the General Dimension tool, enter an expression that will center the "vertical" line so that the line's endpoints will be equidistant from the horizontal axis, as shown in Figure 5.150. The sketch is now fully constrained. Recall that geometric constraints (i.e., symmetric, equal) have also been applied.
9. To once again display the dimensions as values, right-click in the graphics area and select Dimension Display > Value (Figure 5.151).
10. Finish Sketch.
11. Isometric View.
12. Turn on the Visibility of the XY Plane.
13. Use the Work Plane tool to create an offset work plane 15 mm above the default XY Plane. After selecting the Work Plane tool, in the graphics area click on

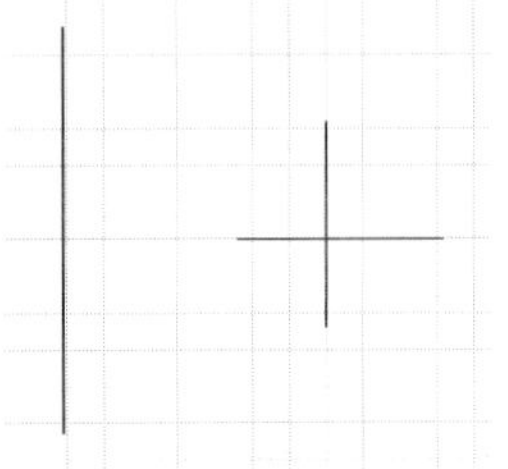

Figure 5.147 - Preliminary sketch

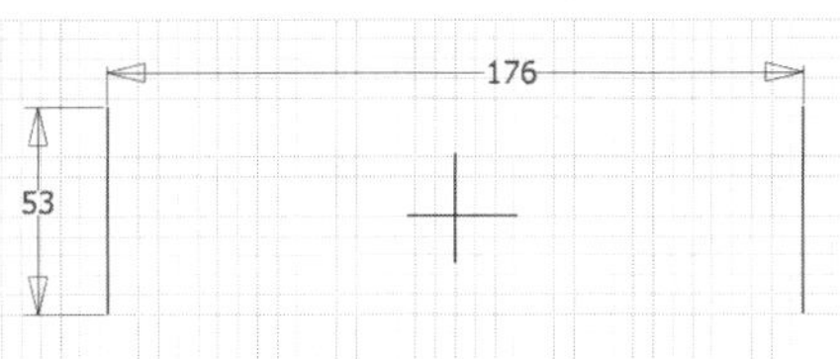

Figure 5.148 - Partial sketch with some dimensions

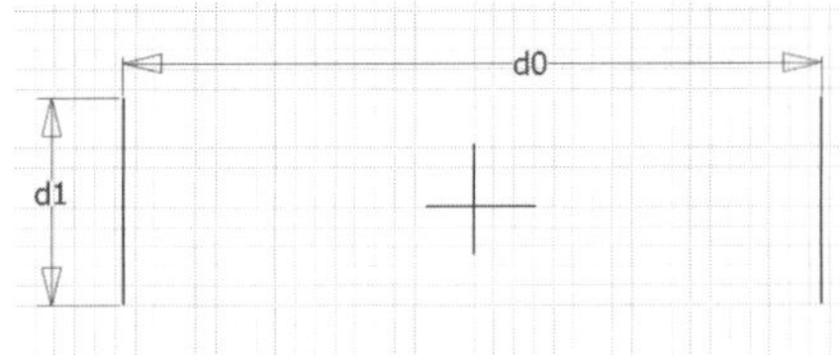

Figure 5.149 - Parametric dimension names displayed

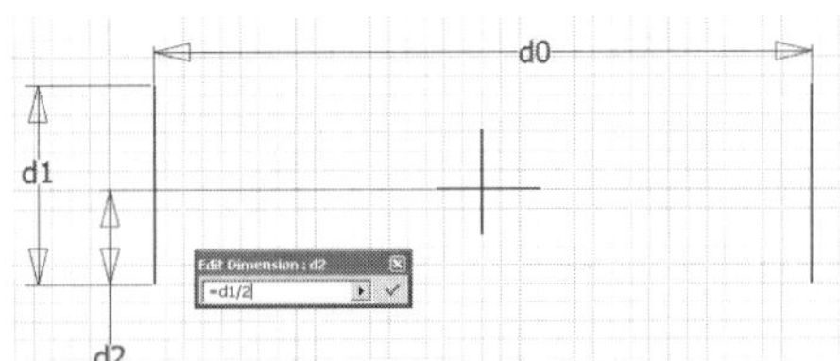

Figure 5.150 - Dimension entered as equation

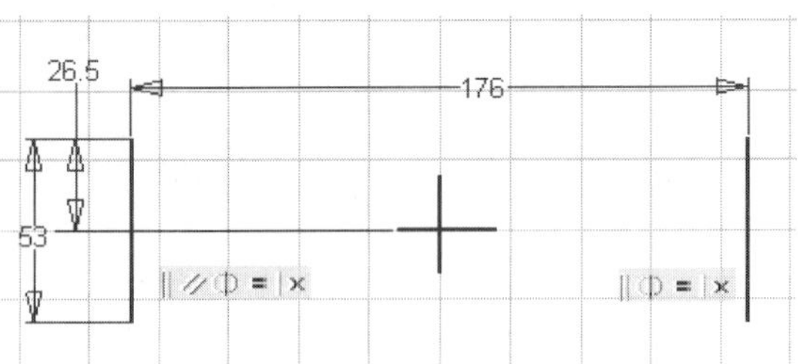

Figure 5.151 - Fully constrained sketch with constraints displayed

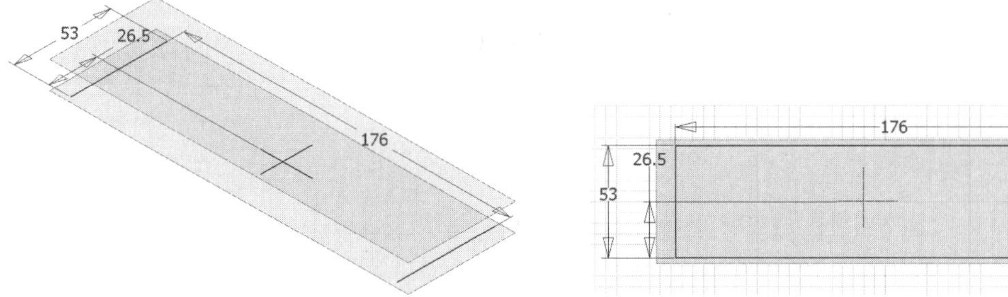

Figure 5.152 - Offset work plane

Figure 5.153 - Completed sketch

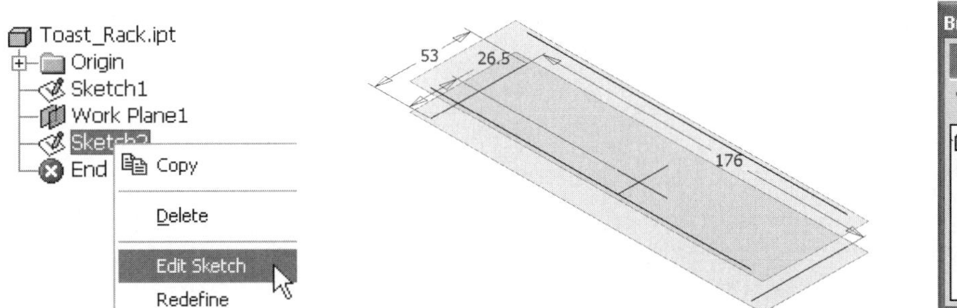

Figure 5.154 - Edit sketch access

Figure 5.155 - Reference lines deleted

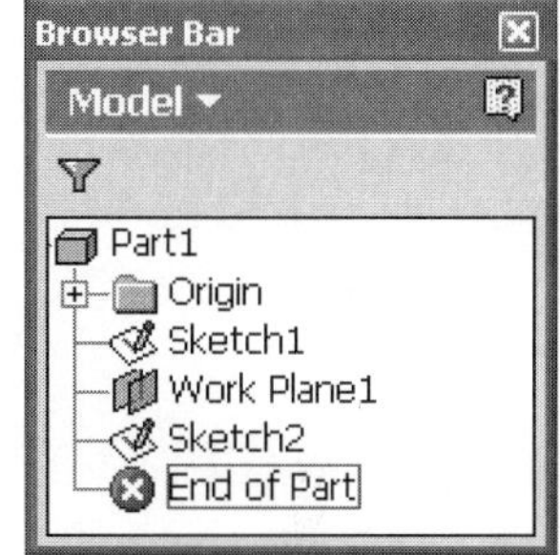

Figure 5.156 - Part browser

the XY Plane and drag up. The Offset dialog box appears. Enter **15**, and then click Apply (green check mark). Your screen should be similar to Figure 5.152.

14. New Sketch. Select the offset work plane.
15. Look At the new sketch.
16. Use the Project Geometry tool to project the "vertical" lines from the previous sketch onto the new sketch plane.
17. Use the Line tool to sketch two parallel horizontal lines whose endpoints are coincident with the projected "vertical" lines (Figure 5.153). Note that the projected lines on this sketch have a reference linetype, while the sketched lines have a normal linetype. How can this be verified?
18. Finish Sketch.
19. Save the file as **Toast_Rack.** It is recommended that the file be regularly saved every five to ten minutes.
20. Isometric View
21. Before continuing, we will delete the projected reference geometry on the second sketch on the offset work plane. In the part browser, right-click on Sketch2 and select Edit Sketch, as shown in Figure 5.154.
22. Select a length 53 reference line, and then hit the Delete key. Do the same for the remaining length 53 reference line.
23. Finish Sketch. The screen should now resemble Figure 5.155.
24. In the part browser there should be two sketches and a work plane, as shown in Figure 5.156. We will now use these sketches to help create a 3D sketch.
25. Right click, and then select New 3D Sketch. The panel bar now displays the 3D Sketch tools, as shown in Figure 5.157.

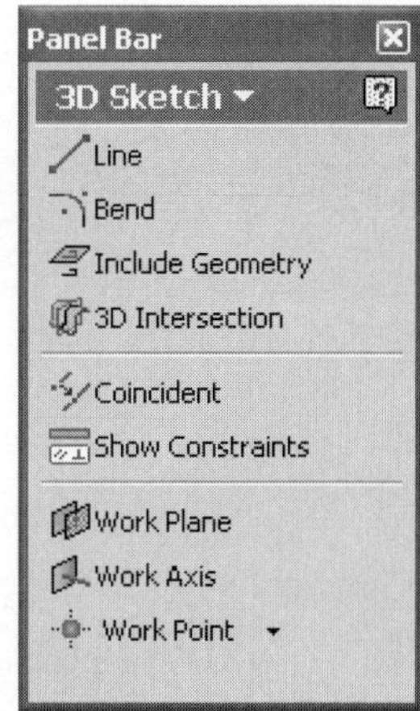

Figure 5.157 - 3D Sketch tools

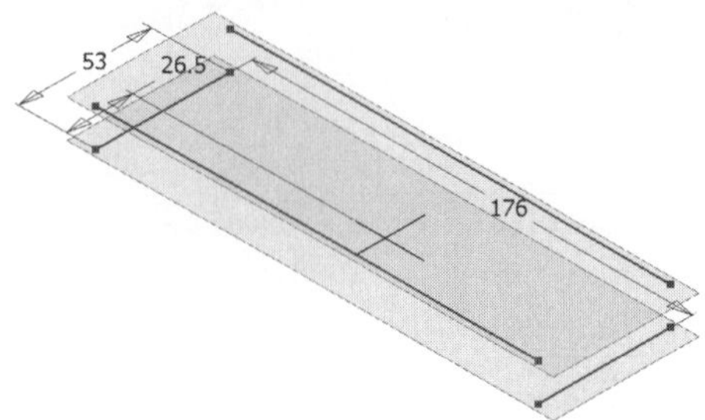

Figure 5.158 - Include 2D sketch geometry in 3D sketch

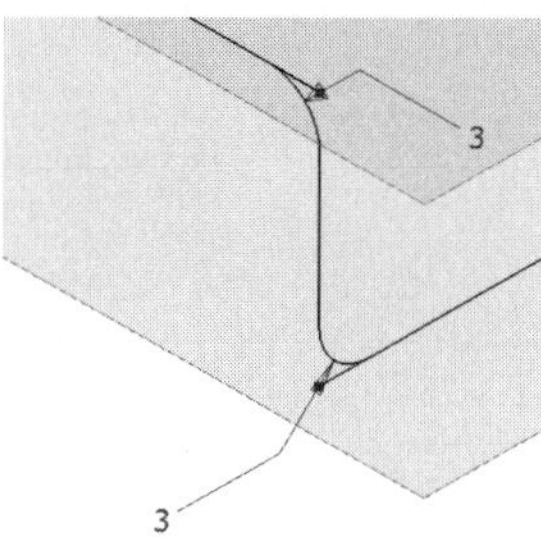

Figure 5.159 - 3D line with bend radius

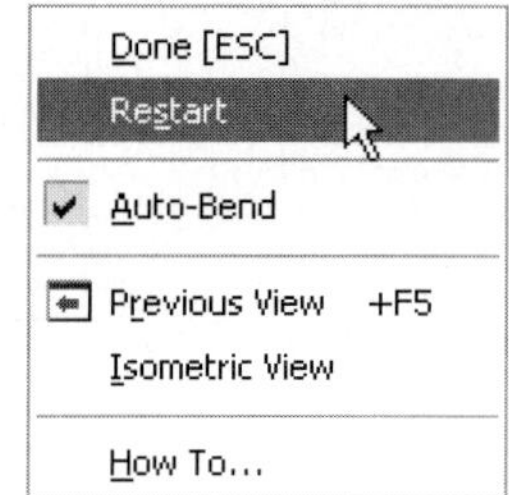

Figure 5.160 - Context menu with option to restart the 3D line tool

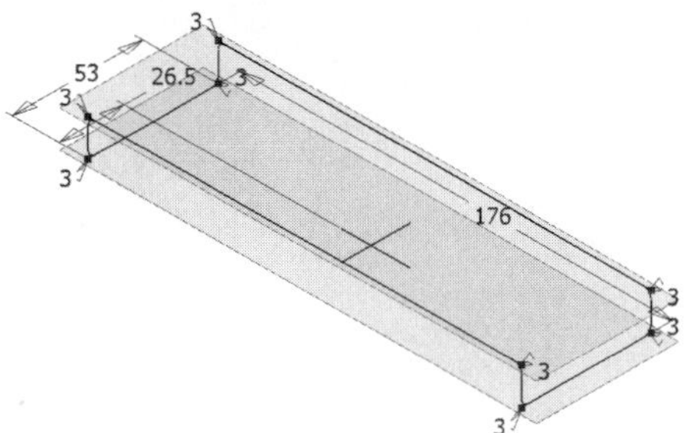

Figure 5.161 - Completed 3D sketch path

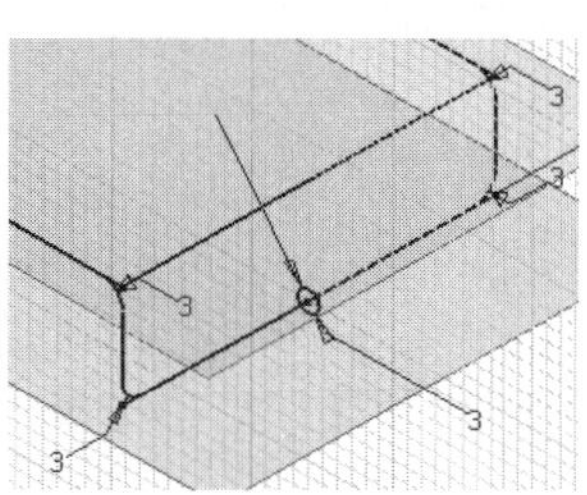

Figure 5.162 - Circle profile

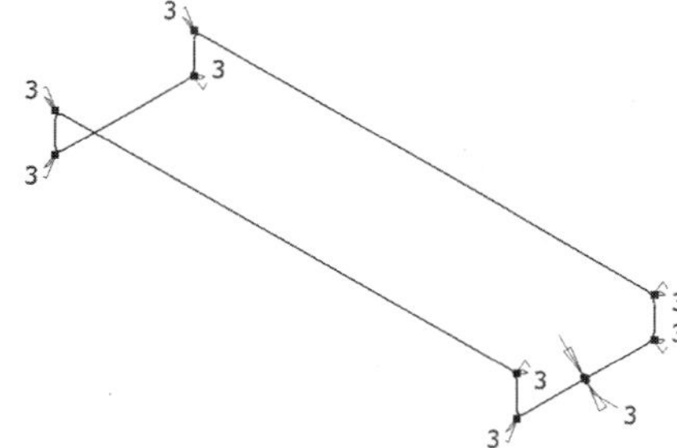

Figure 5.163 - Sweep path and profile

26. Select the Include Geometry tool from the 3D Sketch panel bar. Select the four lines (two length = 53 lines from Sketch1, and the two length = 176 lines from Sketch2), as shown in Figure 5.158. Right-click and then select Done.
27. Select Tools > Document Settings . . . from the menu bar. Click on the Sketch Tab. Change the Auto-Bend Radius to **3**, and then click OK.
28. We are now ready to create a 3D path for the sweep. Click on the Line tool from the panel bar. First click on an endpoint of a line in the XY Plane, then on the endpoint of the line directly above this point. A vertical line is immediately created connecting the two lines, with a bend radius of 3 applied at both ends. This can be seen in Figure 5.159.
29. Right-click, and then select Restart, as shown in Figure 5.160. This allows us to continue drawing separate 3D line segments at the three other corners.
30. Add the remaining vertical line segments at the corners of the 3D sketch. The result should be similar to Figure 5.161.
31. Right-click, Finish 3D Sketch. We will now create the profile to be swept along the 3D path.
32. New Sketch. Select the XZ Plane.
33. Use the Project Geometry tool to project the length = 53 line segment on the lower front side of the screen onto the sketch plane. The projected line appears as a point.
34. Use the Circle tool to sketch a circle centered on the projected line.
35. Use the General Dimension tool to specify a diameter of **3** for the circle, as shown in Figure 5.162.
36. Finish Sketch.
37. Turn off the visibility of the first two 2D sketches, as well as the visibility of all work planes (Figure 5.163).

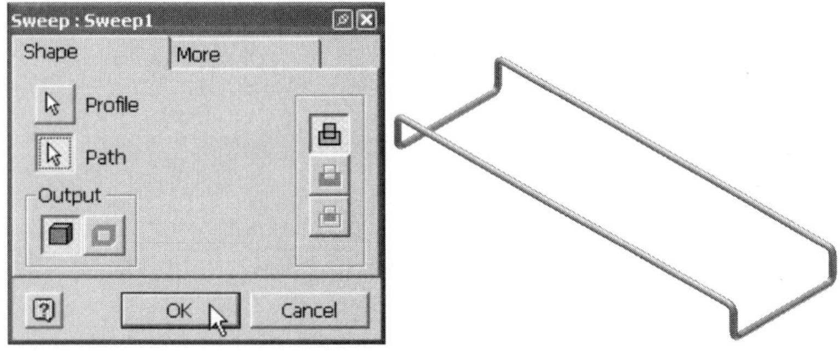

Figure 5.164 - Sweep feature operation

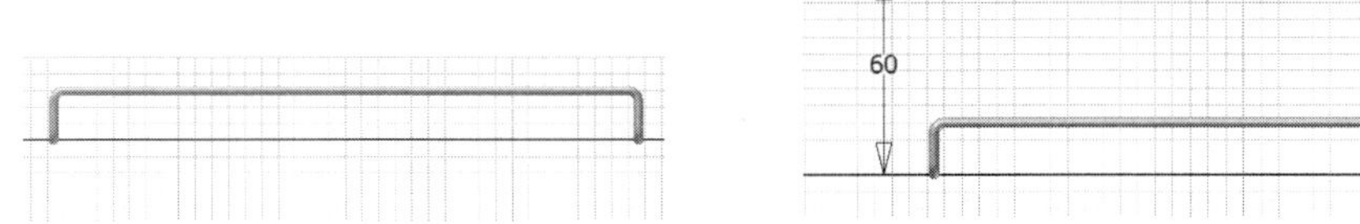

Figure 5.165 - New sketch with projected axis

Figure 5.166 - Completed sketch

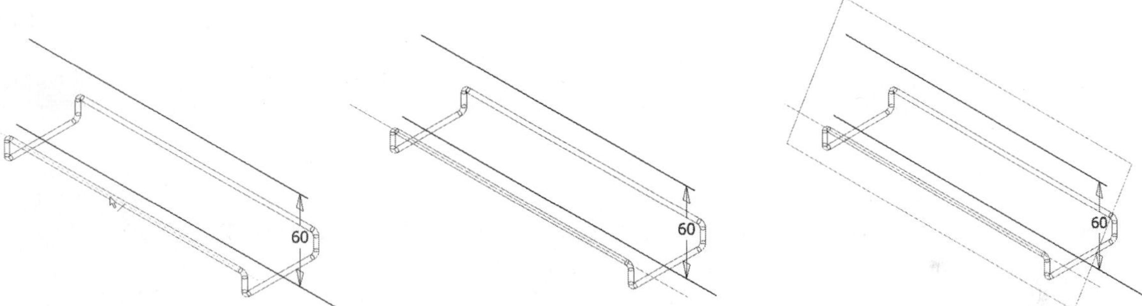

Figure 5.167 - Work axis feature

Figure 5.168 - Angled work plane feature

38. Select the Sweep tool. With the Profile button active select the circle sketch, if necessary. Then click on the Path button and select the 3D sketch, as shown in Figure 5.164. Click OK. The resulting swept solid should resemble Figure 5.164.
39. New Sketch. Select the XZ Plane.
40. Look At the new sketch.
41. Project the X Axis onto the sketch plane. Your screen should resemble Figure 5.165.
42. Use the Line tool to sketch a horizontal line above the base feature.
43. Use the General Dimension tool to add the dimension shown in Figure 5.166.
44. Finish Sketch.
45. Isometric View.
46. Change to a wireframe display.
47. Select the Work Axis tool from the panel bar, and then select the long leg of the sweep feature in the foreground of the screen as seen in Figure 5.167.
48. Select the Work Plane tool from the panel bar, and then select the work axis and the horizontal line above the base feature. A work plane is created that passes through the two lines, as shown in Figure 5.168.

49. New Sketch. Select the work plane.
50. Look At the sketch.
51. Use the Project Geometry tool to project the X and Z Axes onto the sketch plane.
52. Zoom in on the right half of the sketch. Project the upper limiting edge of the sweep, as shown in Figure 5.169.
53. Use the Circle tool to draw a small circle. Add the dimensions shown in Figure 5.170, and use a tangent constraint to constrain the circle to the projected sweep limiting edge.
54. Finish Sketch.
55. Isometric View. The screen should now resemble Figure 5.171.
56. Select the Coil tool from the Panel Bar. The Coil dialog box appears. The Profile (i.e., the circle sketch) should be automatically selected, and the Axis button should be depressed. Select the horizontal line 60 mm above the XY Plane, as shown in Figure 5.172.
57. Click on the Coil Size tab, and enter the inputs shown in Figure 5.173, then click OK.
58. After changing to a shaded display, and turning off the visibility of the work plane and work axis, the screen should resemble Figure 5.174.
59. Save the part file. This completes Tutorial 15.

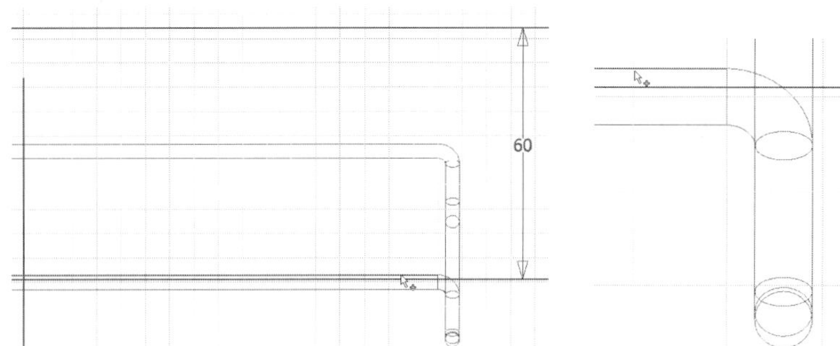

Figure 5.169 - Projected limiting element of sweep

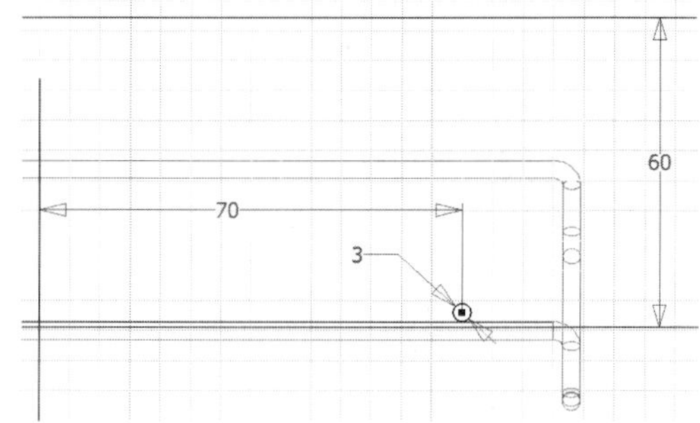

Figure 5.170 - Fully constrained sketch

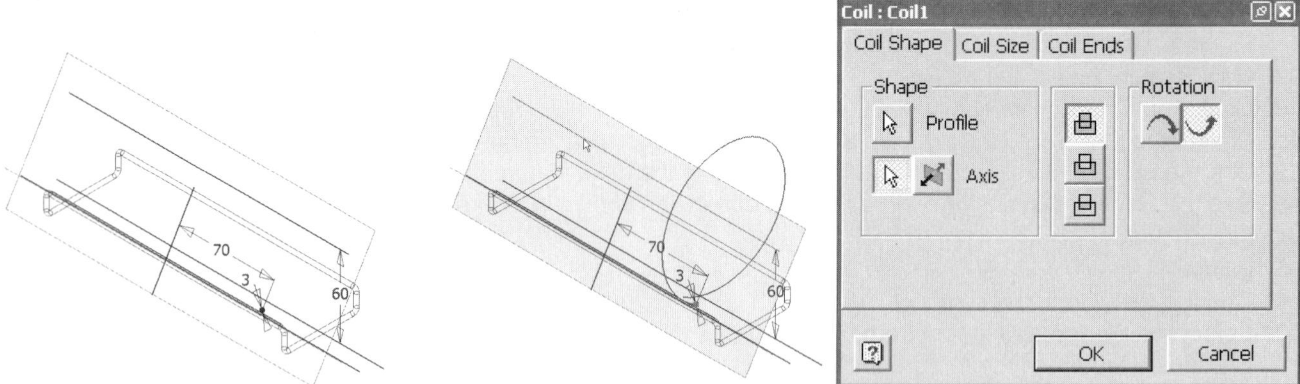

Figure 5.171 - Coil set up

Figure 5.172 - Coil feature operation 1

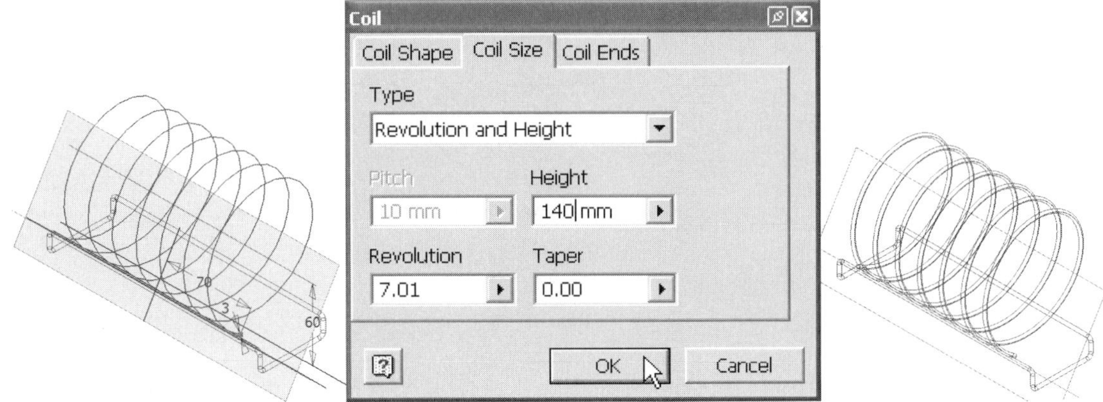

Figure 5.173 - Coil feature operation 2

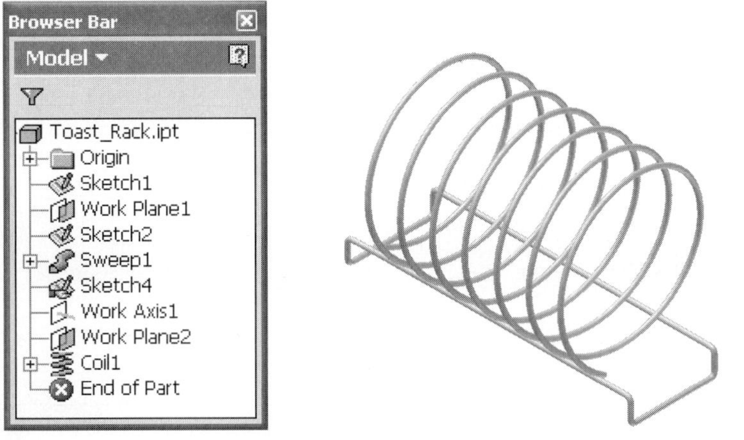

Figure 5.174 - Completed Toast Rack

Spring (Ball Valve Project Folder) — TUTORIAL 16

Build Strategy

1. Create angled work planes (Figure 5.175).
2. Sketch (Figure 5.176).
3. Sketch (Figure 5.177).
4. Sketch (Figure 5.178).
5. Sketch (Figure 5.179).
6. Sketch (Figure 5.180).
7. Sketch (Figure 5.181).
8. Loft (Figure 5.182).

Detailed Modeling Steps

This spring part (shown in Figure 5.182) is modeled with a single loft feature. The loft is comprised of six sketches. Each sketch plane is rotated 60 degrees with respect to

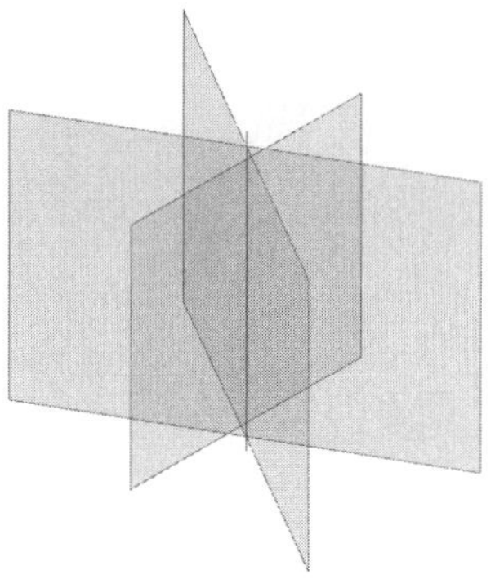

Figure 5.175 - 3 angled work planes

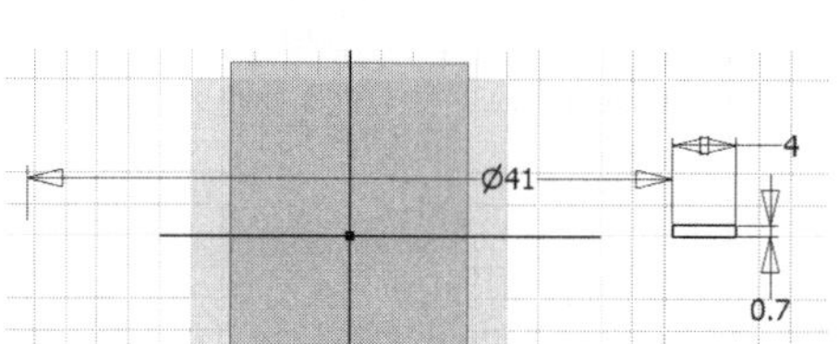

Figure 5.176 - First sketch

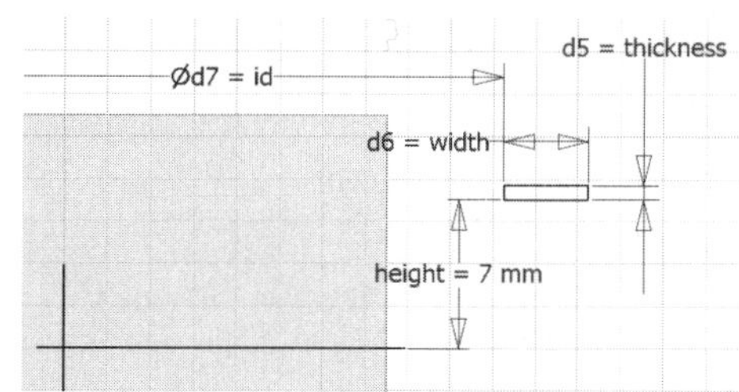

Figure 5.177 - Second sketch

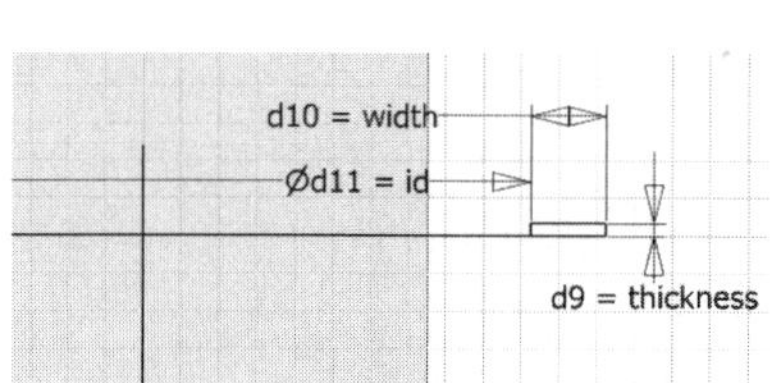

Figure 5.178 - Third sketch

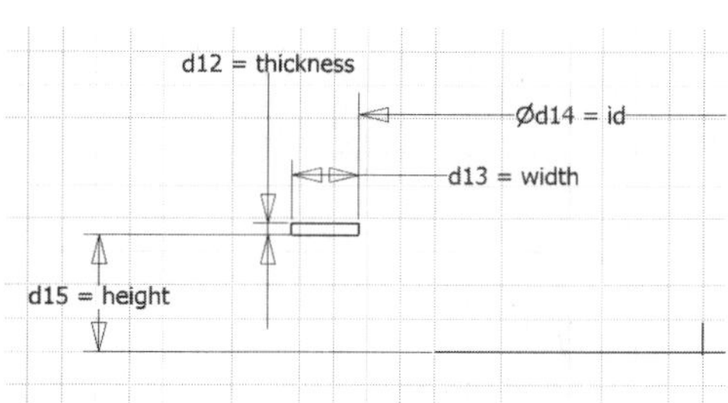

Figure 5.179 - Fourth sketch

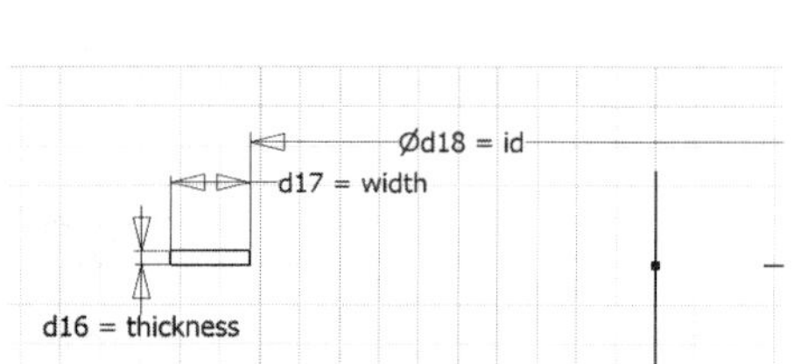

Figure 5.180 - Fifth sketch

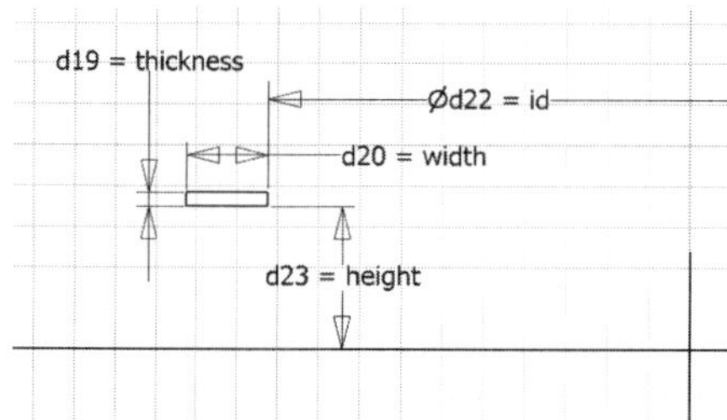

Figure 5.181 - Sixth sketch

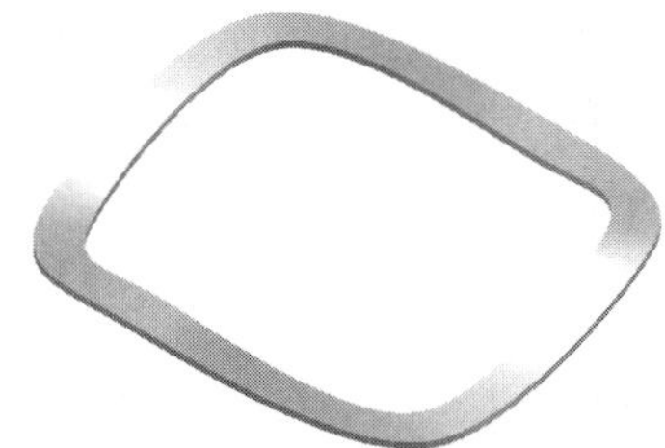

Figure 5.182 - Completed part

the previous sketch plane. The profile, a rectangle, is the same for all six sketches. Apart from the 60 degree sketch plane rotation angle, the only other path variable is that the rectangle moves alternately up and down by a fixed distance from one sketch to the next.

1. Change the active projects folder back to Ball Valve. From the Open dialog box, select Projects from the What To Do column (or select File > Projects from the menu bar). Once the Projects window is open, click on the Ball Valve project name; then click the Apply button. The Ball Valve Workspace is now the active folder.

2. Start a new metric part file.
3. Isometric View.
4. Finish Sketch.
5. Delete Sketch1 in the part browser.
6. From the Origin folder of the part browser, turn on the visibility of both the YZ Plane and the Z Axis (right-click on the feature, then select Visibility).
7. Use the Work Plane tool to create an angled work plane. First select the Z Axis, and then select the YZ Plane (in either the graphics area or from the browser). Enter 60 in the Angle dialog box to create a new work plane passing through the Z Axis and rotated 60 degrees with respect to the YZ plane.
8. Repeat the process in step 7, this time creating a work plane that is rotated 120 degrees with respect to the YZ Plane. Your screen should now resemble Figure 5.183.
9. In the part browser rename the work planes: Work Plane1 → WP60, Work Plane2 → WP120. To rename a feature in the browser, select, pause, and then left-click on the current name, then modify. See Figure 5.184. We will use these three work planes (YZ, WP60, and WP120) as sketch planes for our lofted solid spring part.
10. New Sketch - on the YZ Plane.
11. Look At the Sketch.
12. In the browser, right-click on the YZ Plane and de-select Auto-Resize. This will prevent the work plane from re-sizing as we sketch.
13. Project the Y and Z Axes and the Center Point.
14. Use the Rectangle tool to sketch a rectangle as shown in Figure 5.185. One corner of the rectangle should be coincident with the horizontal Y Axis.
15. Add the dimensions shown in Figure 5.186. Note that to add the diameter dimension, start by clicking the vertical axis, then the left edge of the rectangle, then right-click and select Linear Diameter.
16. Right-click in the graphics area, then select Dimension Display > Name to display the parametric names as seen in Figure 5.187.
17. Click on the Parameters icon panel bar. Rename the parameters, as shown in Figure 5.188, then click Done.
18. Finish Sketch.
19. Save the file as **Spring.** It is recommended that the file be regularly saved every five to ten minutes.
20. Rename the first sketch "Sketch1" in the part browser (see step 9 above). Do this for future sketches (e.g., Sketch2, Sketch3, etc.), six in total.
21. Isometric View. Your screen should now resemble Figure 5.189. We will proceed counterclockwise around the Z Axis, creating a new sketch every 60 degrees. The next sketch will be on WP60.

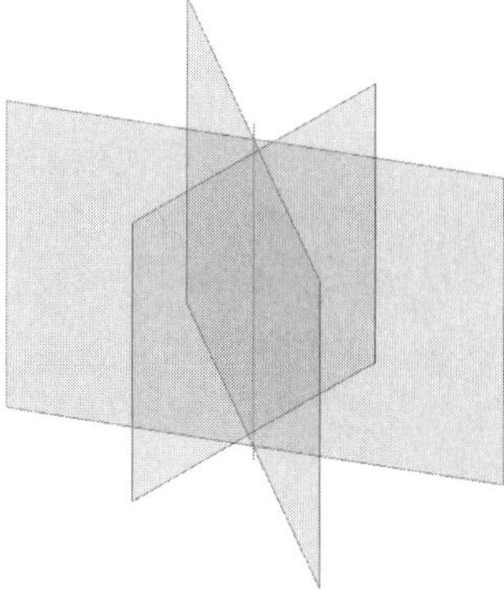

Figure 5.183 - Three angled work planes (60 degree rotations)

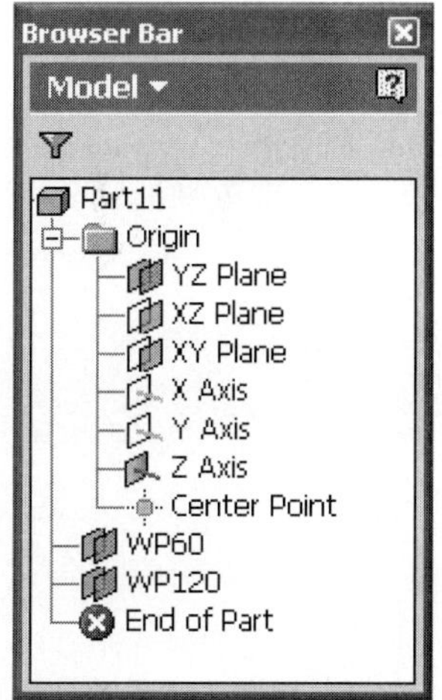

Figure 5.184 - Part browser with renamed work planes

Figure 5.185 - Preliminary sketch

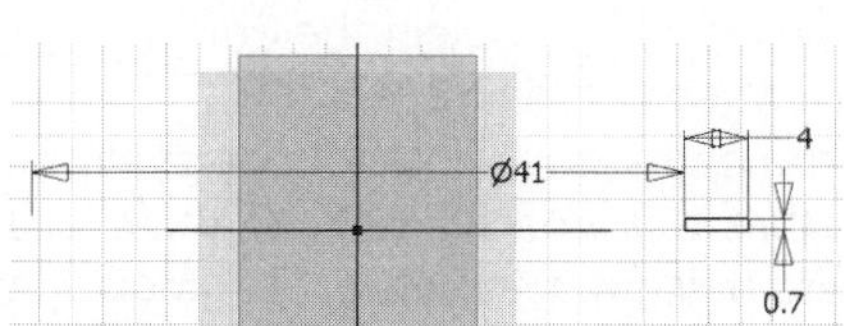

Figure 5.186 - Dimensioned first sketch

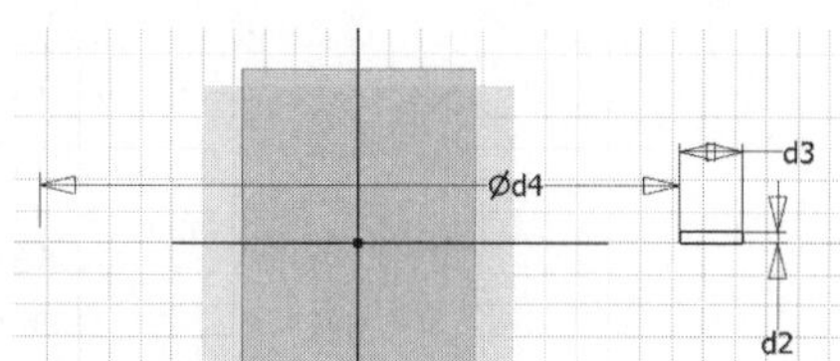

Figure 5.187 - First sketch with parameter names

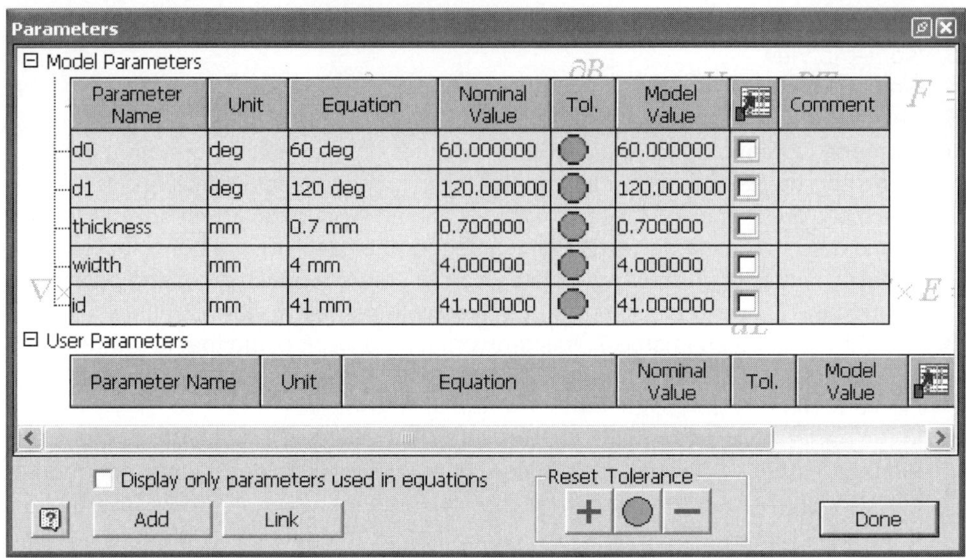

Parameters

Model Parameters

Parameter Name	Unit	Equation	Nominal Value	Tol.	Model Value		Comment
d0	deg	60 deg	60.000000		60.000000		
d1	deg	120 deg	120.000000		120.000000		
thickness	mm	0.7 mm	0.700000		0.700000		
width	mm	4 mm	4.000000		4.000000		
id	mm	41 mm	41.000000		41.000000		

User Parameters

Parameter Name	Unit	Equation	Nominal Value	Tol.	Model Value

Display only parameters used in equations

Add | Link | Reset Tolerance | Done

Figure 5.188 - Parameters dialog box

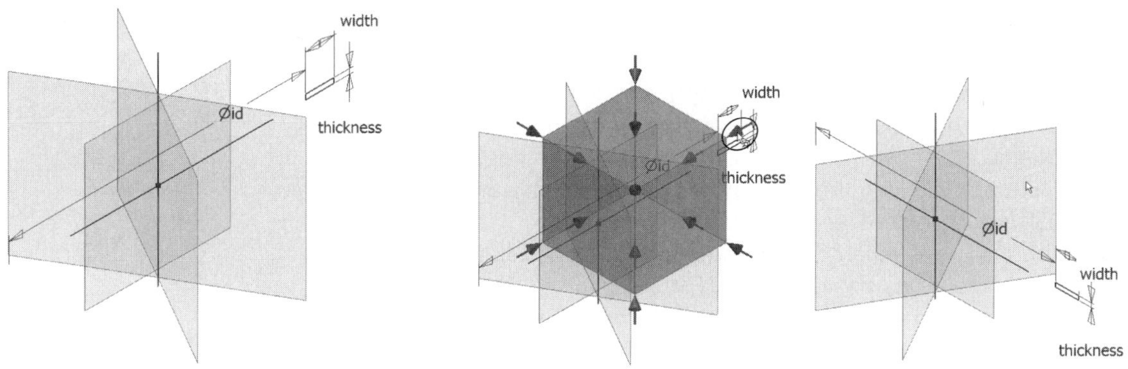

Figure 5.189 - Three work planes, one sketch

Figure 5.190 - Change view

22. To get a better look at the next sketch plane, use Common View as shown in Figure 5.190 to rotate the view counterclockwise by 90 degrees.
23. New Sketch - on WP60.
24. Turn off the visibility of Sketch1, WP120, the YZ Plane and the Z Axis.
25. Before using the Look At tool, we need to modify the sketch plane coordinate system. We can do this with the Edit Coordinate System tool, available on the 2D Sketch panel bar. Select the Edit Coordinate System tool. The screen should now resemble Figure 5.191.

    As you can see, the coordinate system does not agree with what we are used to: that is, X positive to the right, Y positive up. To change the sketch coordinate system, first select the green Y Axis arrow on the sketch plane, and then select the Z Axis in the origin folder. Your screen should now appear as shown in Figure 5.192. Right-click, then select Done. The positive Y Axis is now vertical, with the X Axis to the right.

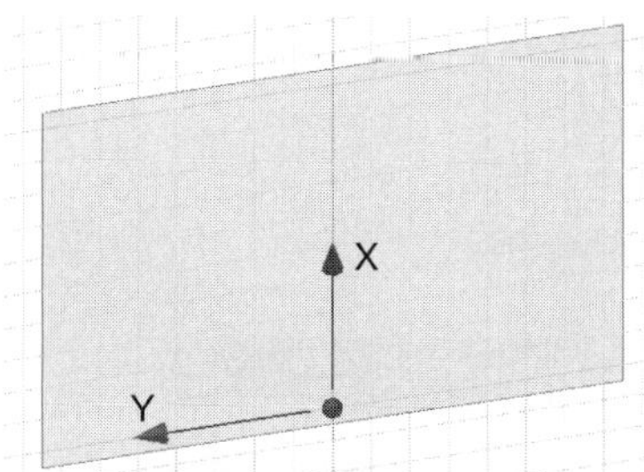

Figure 5.191 - Unfriendly sketch coordinate system

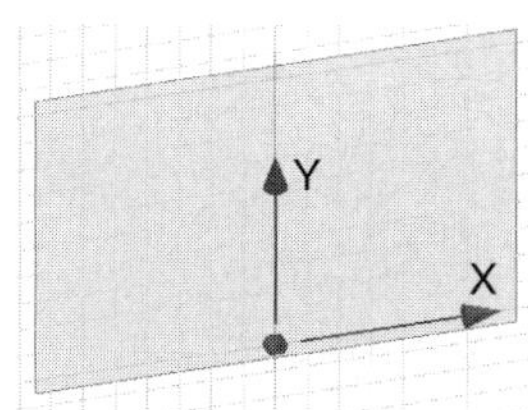

Figure 5.192 - Modified sketch coordinate system

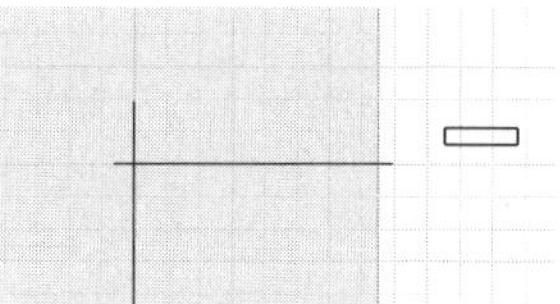

Figure 5.193 - Preliminary second sketch

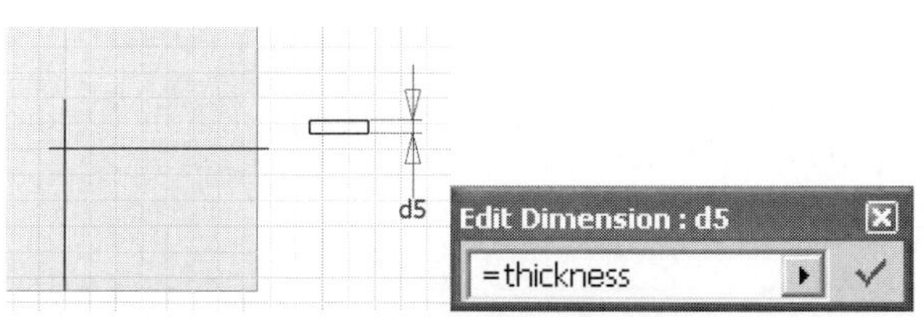

Figure 5.194 - Enter dimension as an equation

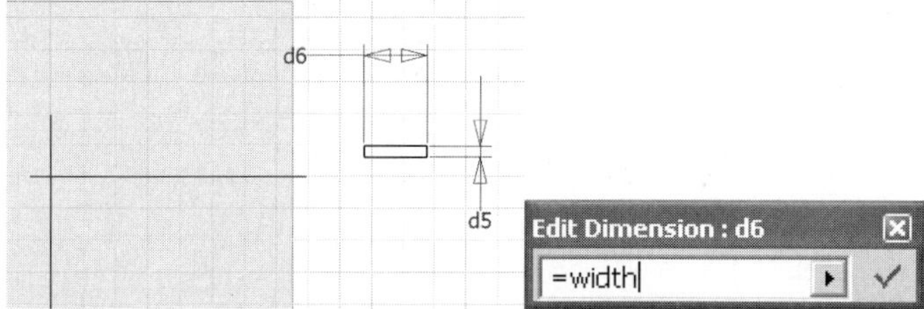

Figure 5.195 - Enter another dimension as an equation

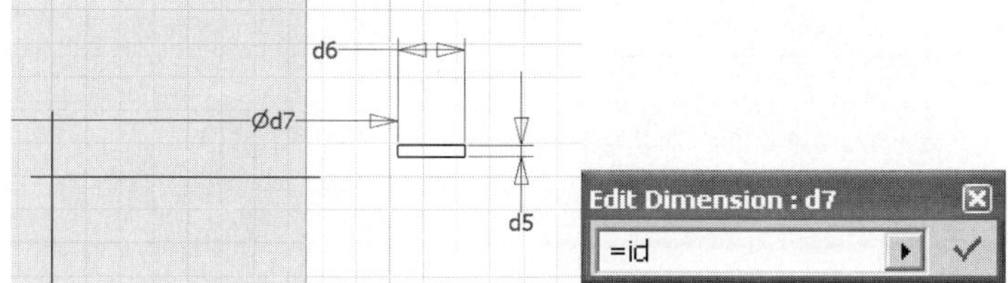

Figure 5.196 - Another dimension as an equation

26. Look At the sketch plane.
27. Project the Y and Z Axes onto the sketch plane.
28. Use the Rectangle tool to sketch the rectangle shown in Figure 5.193.
29. Add the parametric dimension entered as an equation, as shown in Figure 5.194.
30. Add the parametric dimension entered as an equation, as shown in Figure 5.195.
31. Add the parametric dimension entered as an equation, as shown in Figure 5.196.
32. Add the dimension shown in Figure 5.197. Note that this dimension is from the *top* of the rectangle to the horizontal centerline.
33. Click on the Parameters icon on the panel bar. Give the most recently created parameter, d8, the name "height" as shown in Figure 5.198, then click Done.
34. Right-click in the graphics area, then select Dimension Display > Expression. The sketch should be similar to Figure 5.199.
35. Finish Sketch.
36. Isometric View.
37. Turn off the visibility of both WP60 and the second sketch. Turn on the visibility of WP120.

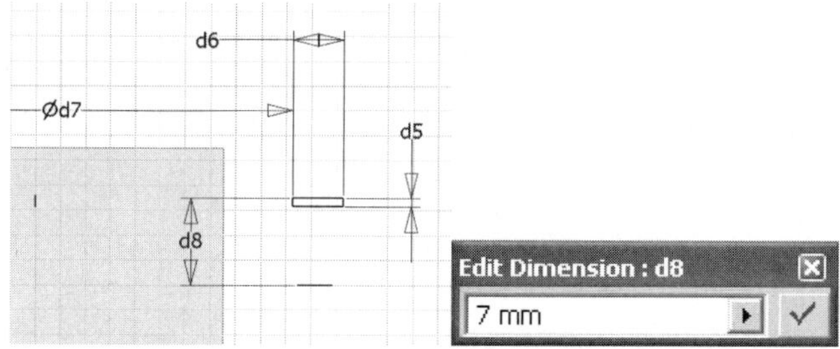

Figure 5.197 - Add another dimension

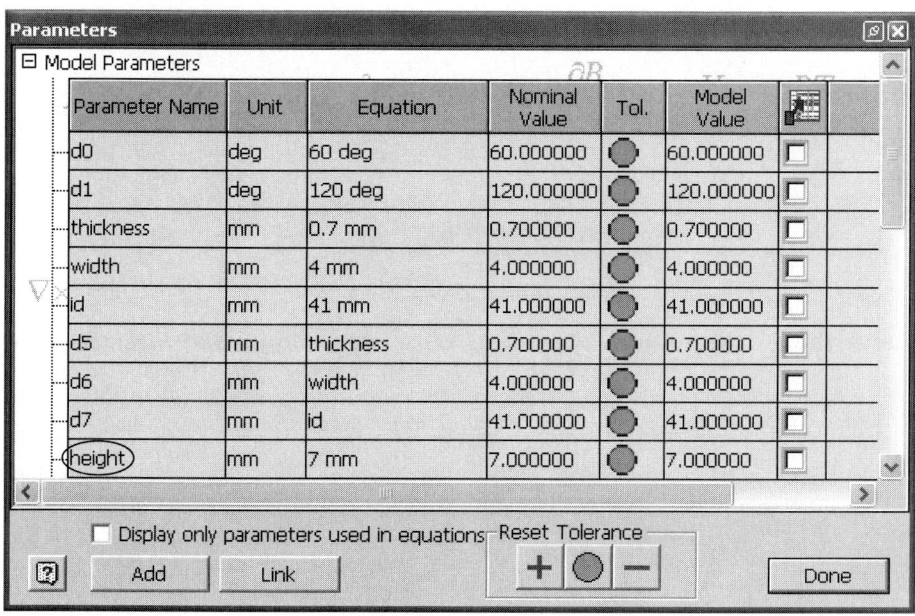

Parameters

Model Parameters

Parameter Name	Unit	Equation	Nominal Value	Tol.	Model Value	
d0	deg	60 deg	60.000000		60.000000	
d1	deg	120 deg	120.000000		120.000000	
thickness	mm	0.7 mm	0.700000		0.700000	
width	mm	4 mm	4.000000		4.000000	
id	mm	41 mm	41.000000		41.000000	
d5	mm	thickness	0.700000		0.700000	
d6	mm	width	4.000000		4.000000	
d7	mm	id	41.000000		41.000000	
height	mm	7 mm	7.000000		7.000000	

Display only parameters used in equations

Reset Tolerance

Add Link Done

Figure 5.198 - Set parameter name d8 = height

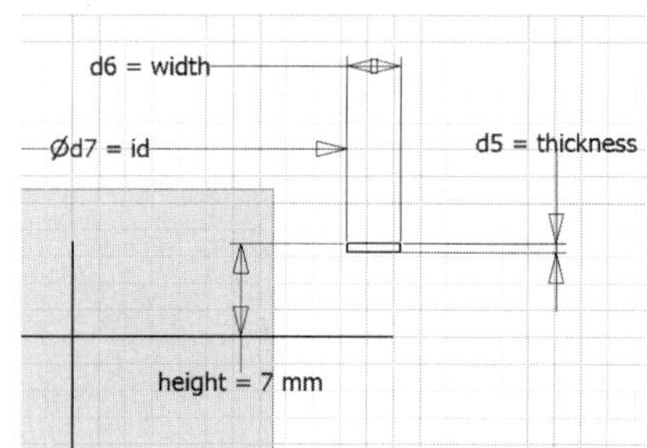

Figure 5.199 - Fully dimensioned second sketch

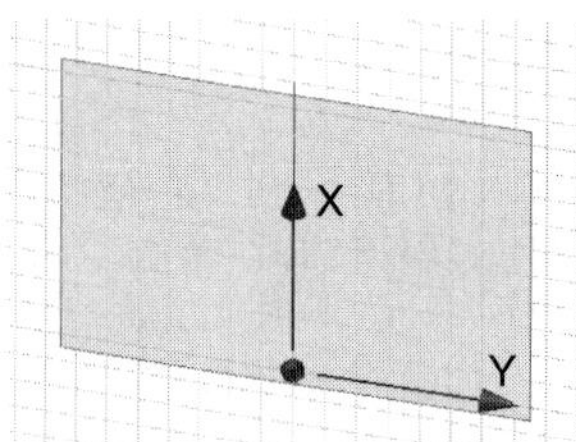

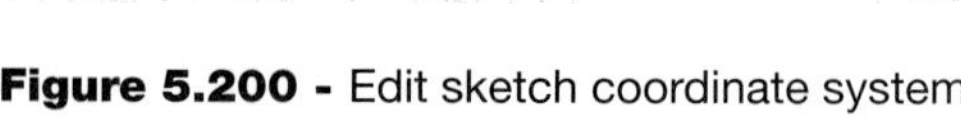

Figure 5.200 - Edit sketch coordinate system

38. New Sketch - select WP120.
39. Edit Coordinate System. Select the Y Axis arrow, and then select the Z Axis icon in the Origin folder to re-orient the coordinate system. See Figure 5.200 for a before and after comparison. Note that although the X Axis appears to point to the left, this is only because we are seeing the sketch plane from behind.

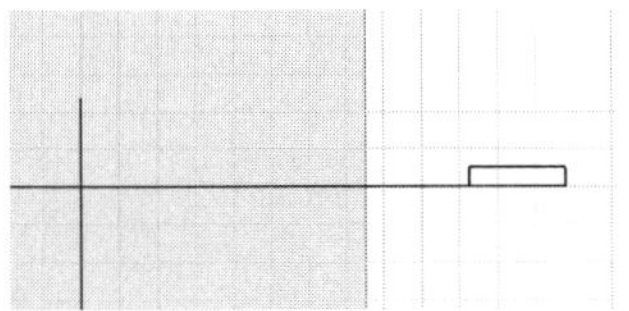

Figure 5.201 - Preliminary sketch 3

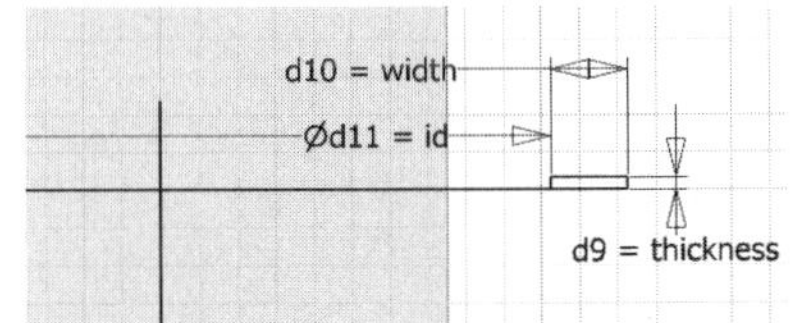

Figure 5.202 - Fully dimensioned third sketch

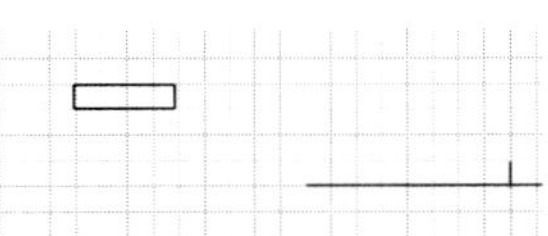

Figure 5.203 - Preliminary sketch 4

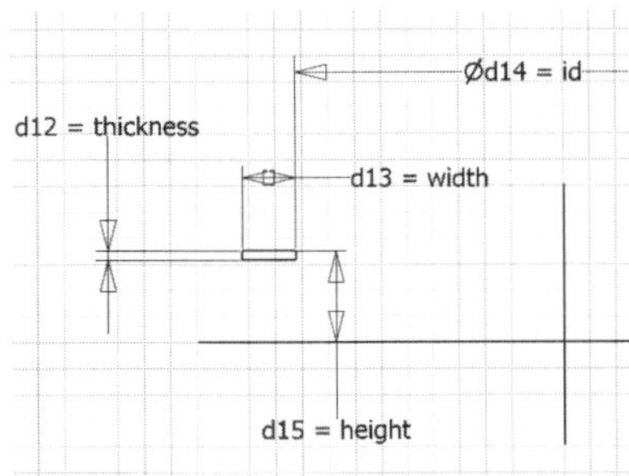

Figure 5.204 - Fully constrained sketch 4

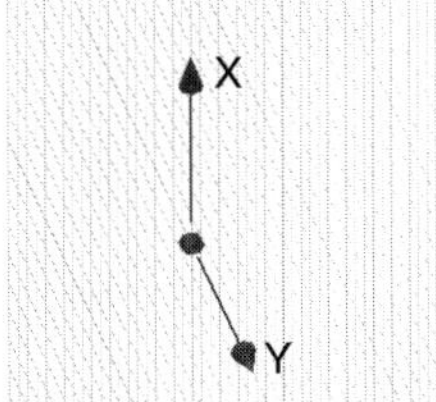

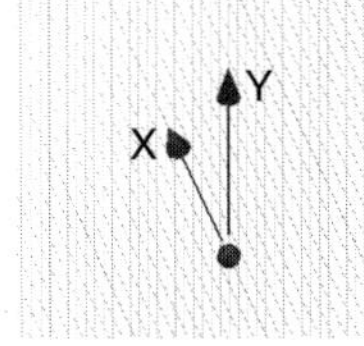

Figure 5.205 - Edit sketch coordinate system

40. Look At the sketch. Select Edit Coordinate System to verify that positive X is to the right, and that positive Y is up. Right-click, Done to exit.
41. Project the Y and Z Axes onto the sketch plane.
42. Use the Rectangle tool to sketch the rectangle shown in Figure 5.201. One corner of the rectangle should be coincident with the horizontal Y Axis.
43. Add the dimensions shown in Figure 5.202. Use equations as in steps 29–31 (e.g., for d9, enter "= thickness").
44. Finish Sketch.
45. Isometric View.
46. Turn off the visibility of the third sketch and WP120.
47. New Sketch - select the YZ Plane.
48. Look At the sketch.
49. Project the Y and Z Axes onto the sketch plane.
50. Use the Rectangle tool to sketch the rectangle shown in Figure 5.203.
51. Add the dimensions shown in Figure 5.204.
52. Finish Sketch.
53. Isometric View.
54. Turn off the visibility of the sketch.
55. New Sketch - select WP60.
56. Use the Edit Coordinate System tool to modify the sketch coordinate system, as shown in Figure 5.205. To re-orient, first select the green Y Axis arrow, then select the Z Axis in the Origin folder.
57. Look At the sketch. Select Edit Coordinate System to verify that positive X is to the right, and that positive Y is up. Right-click, Done to exit.
58. Project the Y and Z Axes onto the sketch plane.

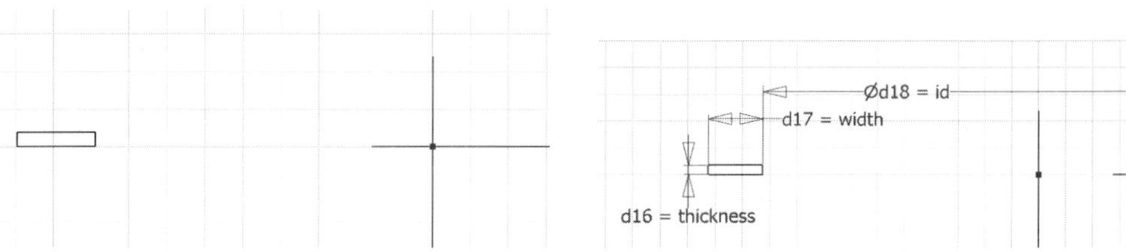

Figure 5.206 - Preliminary sketch 5

Figure 5.207 - Fully dimensioned sketch 5

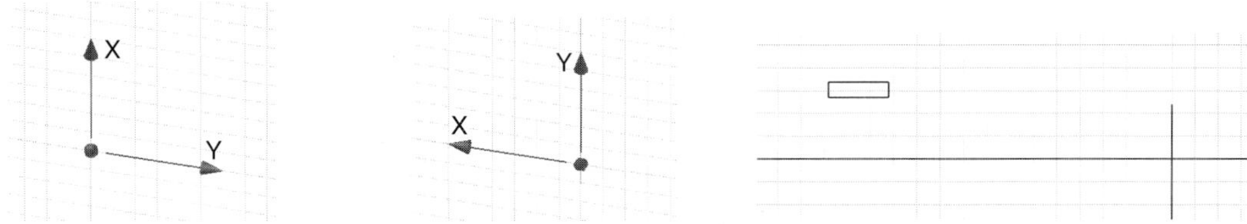

Figure 5.208 - Edit sketch coordinate system

Figure 5.209 - Preliminary sketch 6

59. Use the Rectangle tool to sketch the rectangle shown in Figure 5.206. One corner of the rectangle should be coincident with the Y Axis.
60. Add the dimensions shown in Figure 5.207.
61. Finish Sketch.
62. Isometric View.
63. Turn off the visibility of the sketch.
64. New Sketch on WP120.
65. Use the Edit Coordinate System tool to modify the sketch coordinate system, as shown in Figure 5.208. To re-orient, first select the green Y Axis arrow, then select the Z Axis in the Origin folder. Right-click and select Done to exit.
66. Look At the sketch. Select Edit Coordinate System to verify that positive X is to the right, and that positive Y is up. Right-click Done to exit.
67. Project the Y and Z Axes onto the sketch plane.
68. Use the Rectangle tool to sketch the rectangle shown in Figure 5.209.
69. Add the dimensions shown in Figure 5.210.
70. Finish Sketch.
71. Isometric View.
72. Turn on the visibility of all six sketches. The screen should resemble Figure 5.211.
73. Use the Loft tool to create a lofted solid, as shown in Figure 5.212. Starting with Sketch1, select the six sketches in the order in which they were created. Note that the Closed Loop box is checked.
74. The resulting lofted feature should resemble Figure 5.213. Close inspection of the loft in the vicinity of the first sketch plane reveals that the start and end of the loft do not blend smoothly. This problem will now be addressed.
75. Use the Common View option of the 3D Rotate tool, as shown in Figure 5.214, to obtain the principal view shown on the extreme right.
76. We can correct this discontinuity by editing the loft. Right-click on the Loft icon in the part browser, then select Edit Feature. Click on the Conditions tab of the

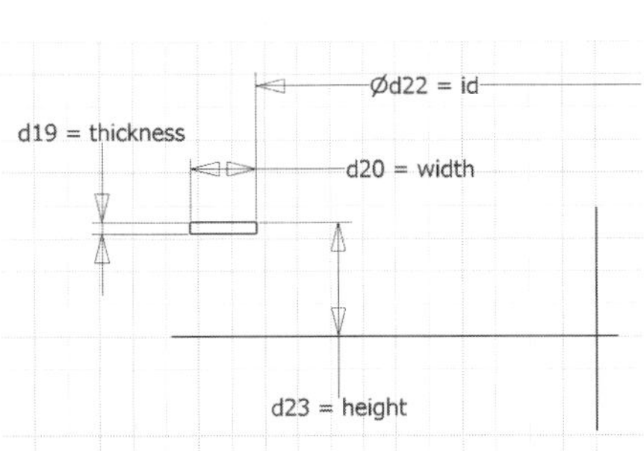

Figure 5.210 - Fully dimensioned sketch 6

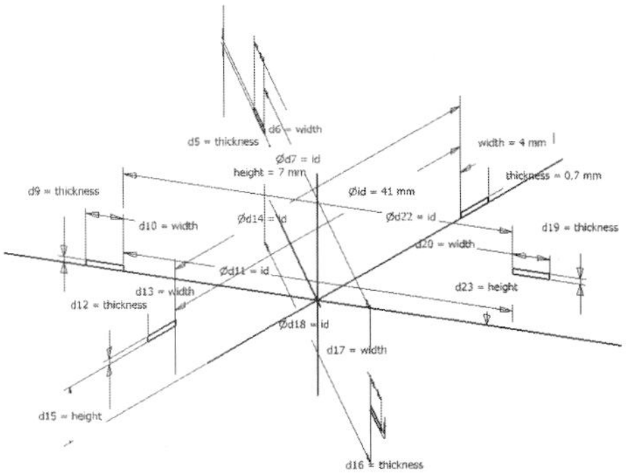

Figure 5.211 - Six sketches

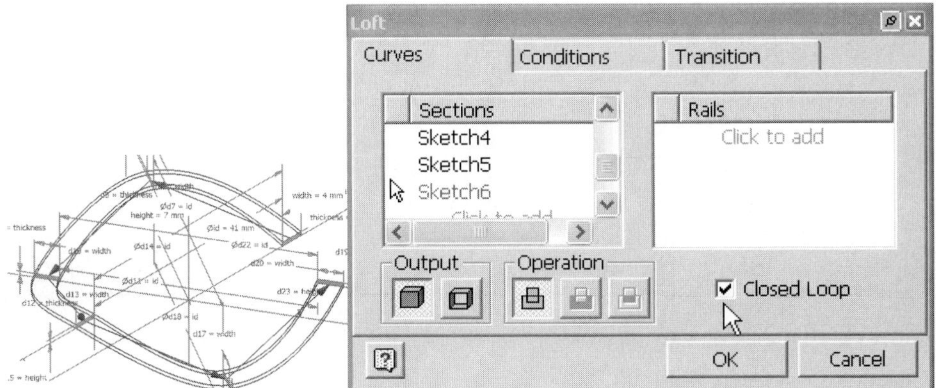

Figure 5.212 - Lofted feature operation

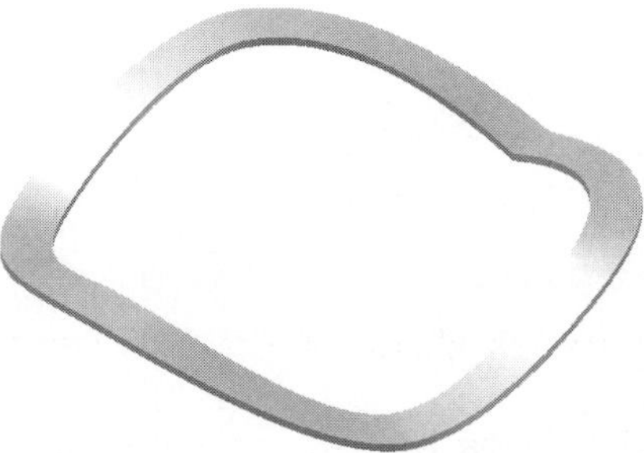

Figure 5.213 - Lofted solid

Loft dialog box. Modify both the first and the last sketch in the loft by selecting the Direction Condition button, then giving them both a Weight of 10. Note that this value was determined through experimentation. When finished, click OK (Figure 5.215). The resulting lofted feature should now be smooth throughout, as shown in Figure 5.216.

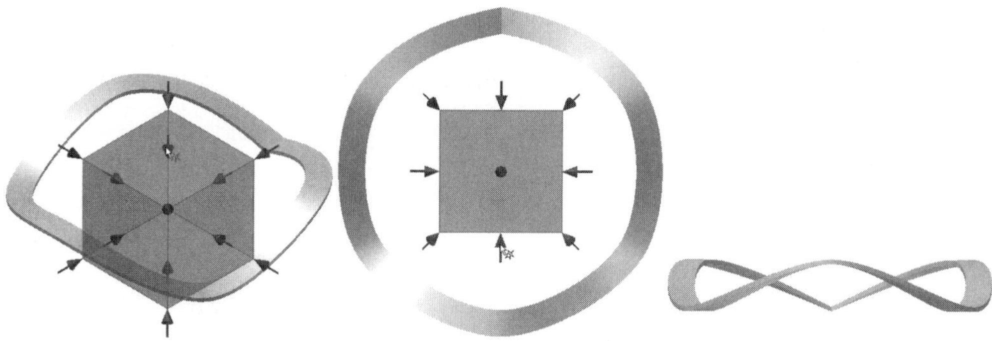

Figure 5.214 - Change view

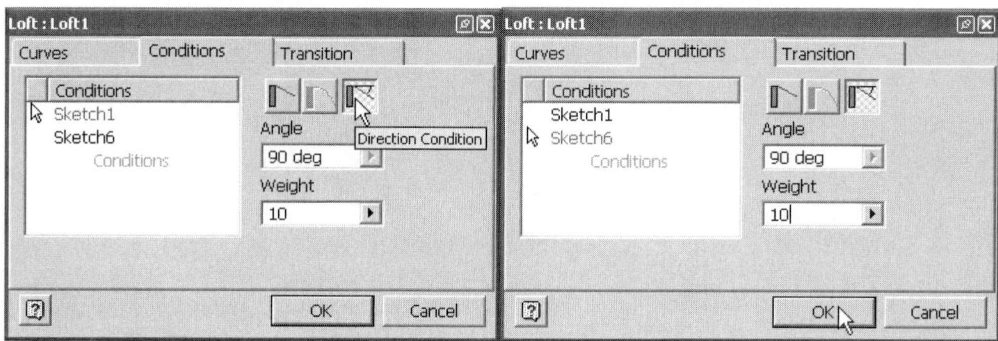

Figure 5.215 - Conditions tab, loft dialog box

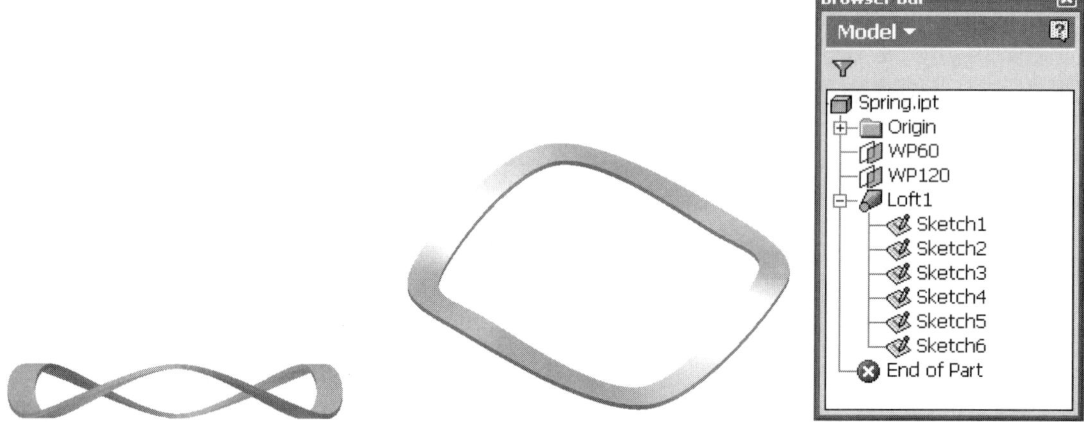

Figure 5.216 - Smoothed transition

Figure 5.217 - Completed Spring part

77. Isometric View.
78. Figure 5.217 shows the Spring part. In the browser the Loft feature has been expanded, revealing the six consumed sketches. Note that it was necessary to rename all of the sketches (see step 9 above).
79. Save the file. This completes Tutorial 16.

QUESTIONS

1. Which of the following input(s) is/are required to create a solid feature using the Sweep tool
   - **a.** Closed profile
   - **b.** Axis
   - **c.** Path
   - **d.** Distance
   - **e.** a and b
   - **f.** a and c
   - **g.** a and d
2. Which of the following input(s) is/are required to create a solid feature using the coil tool
   - **a.** Closed profile
   - **b.** Axis
   - **c.** Path
   - **d.** Distance
   - **e.** a and b
   - **f.** a and c
   - **g.** a and d
3. T F The minimum number of sketches required to create a solid using the Loft tool is three.
4. T F The Rib tool can be used not only to create a rib, but also to create a web.
5. Which of the following can the Split tool not be used for:
   - **a.** Split a part
   - **b.** Split a sketch
   - **c.** Split a face
6. T F Using the shell tool, material can be removed from the inside of a part, leaving a cavity with walls of variable thickness.
7. T F The Face Draft tool allows for the development of a face in sketch mode prior to extruding.
8. T F In the 3D sketch environment, it is possible to sketch an arc or spline independent of a 2D sketch plane.
9. Which of the following icons is the parameter tool?
   - **a.**
   - **b.**
   - **c.** f_x
   - **d.**

PROBLEMS

1. Open the file entitled SweepEx.ipt with the setup as shown in the figure below. Use the Sweep tool to create the resulting swept feature.

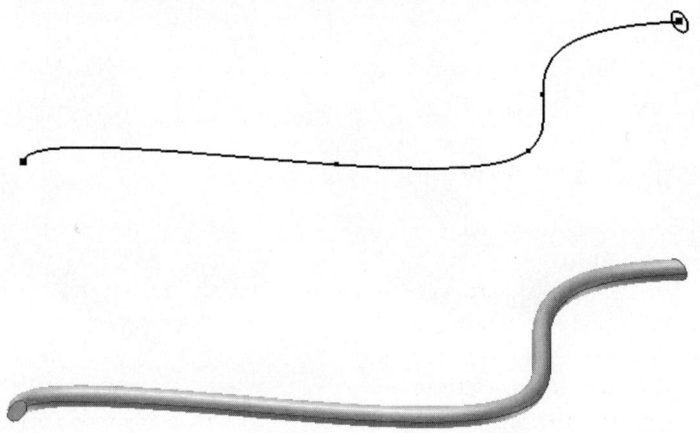

2. Open the file entitled CoilEx.ipt with the setup as shown on the left in the figure below. Use the Coil tool to create the resulting coil feature shown on the right.

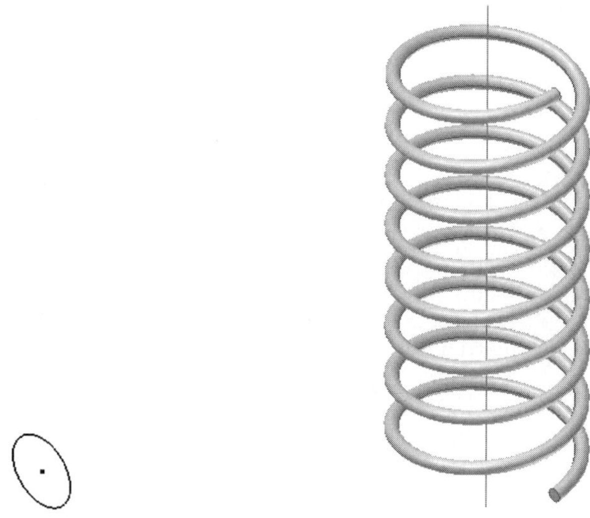

3. Open the file entitled LoftEx.ipt with the setup shown in the figure on the left below. Use the Loft tool to create the lofted feature shown in the middle figure. Finally, use the Mirror Feature tool to mirror the lofted feature about the XZ plane, to end up with the jet ski hull shown on the right.

4. Open the file entitled RibEx.ipt with the setup as shown in the figure on the left below. Use the Rib tool to create the resulting ribbed and webbed features shown respectively in the center and on the right.

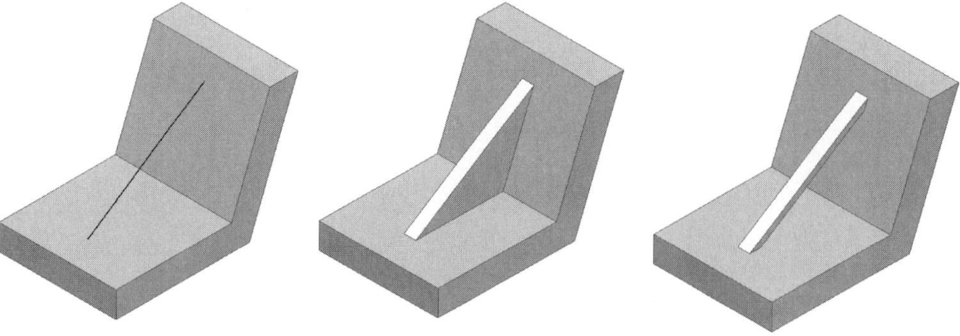

5. Open the file entitled SplitEx.ipt with the setup as shown in the figure on the left. Use the Split tool to reveal the hidden feature, as shown on the right.

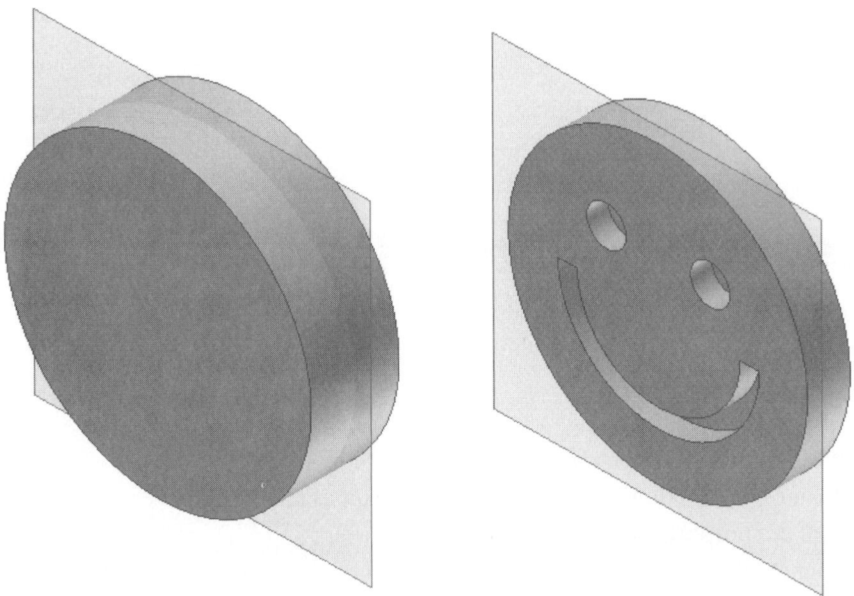

6. Open the file entitled ShellEx.ipt with the setup as shown in the figure on the left. Use the Shell tool to create the resulting shelled feature on the right.

6 CHAPTER

Assembly Modeling

LEARNING OBJECTIVES

- Identify similarities between part and assembly modeling
- List four ways a component can be added to an assembly file
- Define the purpose for assembly constraints
- Recognize the number of degrees of freedom of a part from the degree-of-freedom symbol
- List and describe the four principal types of assembly constraints
- Identify two methods for manipulating individual components within the assembly environment
- Toggle between Modeling and Position Views in the assembly browser
- List two ways to access the part environment from within an assembly file
- List three ways to return from part to assembly mode in an assembly file
- Change the color of a component in an assembly
- Use the Move Component tool to translate a single component in the assembly environment
- Use the Place Constraint tool to:
  - Add a mate constraint between a part face and a work plane
  - Add a mate constraint between two points
  - Mate (i.e., align) the axes of two parts
  - Add an insert constraint between two parts
  - Add a mate constraint between two edges
  - Add an angle constraint between two part faces
  - Add a tangent constraint between a face and a cylindrical surface
- Use the Place Component tool to add:
  - Parts to an assembly file
  - A subassembly to an assembly
- Use the Rotate Component tool to rotate a single component in the assembly environment
- Use the Section View tool to create a quarter, half, or three-quarter section view in an assembly environment
- Use the Pattern Component tool to array a part in an assembly

- Change the name of an assembly constraint in the assembly browser
- Suppress an assembly constraint in the assembly browser
- Use the assembly browser to enable/disable components in an assembly
- Edit a part from within an assembly
- Use the content library to add a part to an assembly

Introduction

Most engineered products are typically composed of several parts. The individual parts are assembled to form a finished product. In the case of even moderately complex products like a bicycle, individual parts are organized into subassemblies (e.g., rear derailleur). The different subassemblies are then combined to form the finished product.

Similarly, parametric assembly modeling is used to combine virtual components in order to create a parametric assembly model. A *component* is either a part or a subassembly of parts. Assembly modeling has been in common usage since the 1990's and was in large part developed within the aerospace and automotive industries.

In this chapter we will look at the assembly modeling capabilities within Autodesk Inventor. We will see that there are assembly modeling tools for 1) adding parts, 2) positioning (i.e., constraining) parts, 3) editing, and 4) display. We will also see that Inventor assembly and part files are associatively linked. This means that if a part is modified, it is automatically updated within the assembly. At the end of the chapter tutorials detailing the creation of the ball valve and garlic press assembly models are provided.

Part and Assembly Modeling Similarities

Most parametric modelers, Inventor included, provide separate environments for part and assembly modeling. Until now we have worked exclusively in a part modeling environment. All of the data we have created has been stored in Inventor part files with an .ipt extension. In this chapter we will use Inventor assembly files with an .iam extension.

Table 6.1 provides a summary of some of the similarities between part and assembly modeling. Both environments use a browser or tree structure providing a hierarchy of the part or assembly. In the part browser different features are seen, the sum of which is

Table 6.1 Part and assembly modeling comparison

	Part Modeling	Assembly Modeling
Tree structure	Features → Parts	Components → Assembly
Base	Feature	Component
Constraints	Geometric and dimensional	Assembly
Parent/child relationships	Between features	Between components
Editing	Sketch, feature	Component, part
Parameters	Dimensions, features	Constraints
Parametric equations	Local	Global
Documentation	Part drawings	Assembly drawings

a part. In the assembly browser, different components (e.g., parts, subassemblies) add up to a finished assembly, or product.

Just as the first feature of a part model has a special status and name (i.e., base), so too does the first part added to an assembly. Both part and assembly models make use of constraints. Much of the time spent in this chapter will be used applying assembly constraints between different parts.

We have seen that in part modeling, features are often positioned with respect to features created earlier in the part building process. This sets up dependencies between features known as parent-child relationships. Similarly, components added later in the assembly process are often dependent upon previously added parts.

Just as part sketches and features can be edited from within the part browser, so too can part files be accessed and modified from within the assembly modeling environment.

In a part file, parameters are created whenever dimensions are added to a sketch, or when values are entered in a dialog box to create features. We will see that assembly constraints are parametric, as well. Within a part file it is possible to relate parameters within a part using equations. It is also possible to use equations in an assembly file such that changing a parameter on one part changes a dimension on another part. In Chapter 7 we will see that we can extract drawings from part models. Likewise in Chapter 9 we will see that we can document our assemblies, for example by creating an exploded view drawing with a Bill of Materials.

Adding Parts to an Assembly

Components can be added to an assembly file by placing an existing part in a file, by building the part within an assembly file, by patterning a part already in the assembly, or by importing a standard part from the Autodesk Inventor Content Library. In this chapter the most common method used to add parts to an assembly will be with the Place Component tool.

Place Component

This tool allows the user to add Inventor part (.ipt) and assembly (.iam), as well as other files, to another assembly file. Multiple *instances* (or *occurrences*) of the file can be added, simply by clicking in the graphics area.

The Place Component tool is used in *bottom-up assembly modeling*. In bottom-up design all of the parts have been designed and simply need to be placed in an assembly and constrained to complete the assembly process.

Create Component

The Create Component tool allows for the creation of parts from within an assembly file. This approach is used in *top-down assembly modeling,* where part geometries are as yet undefined, although certain design criteria may be known. A common top-down design approach is to create a 2D layout of the assembly parts, constraining and then testing their motion characteristics, and only then turning the 2D part sketches into 3D features. It is also possible to employ a *middle-out design* approach, where some parts do exist and are brought into an assembly file. Other parts are created in their assembled position directly in the assembly model. In modeling these parts, it is possible to take advantage of the geometry of existing parts. Examples of this include using the face of an existing part as a sketch plane for another part, or projecting geometry from an existing part onto the sketch plane of a new part. These are more advanced topics, and are consequently not covered in this book.

Pattern Component

The Pattern Component tool is used to add several arrayed (rectangular or circular) occurrences of an existing part to an assembly. This tool is used in one of the tutorials at the end of this chapter to add a circular pattern of parts to an assembly.

Content Library

Autodesk Inventor includes a library of standard parts and steel shapes that can be dropped into assembly files. This content library consists of catalogs arranged into folders and subfolders. Catalogs contain data (template files) used to insert a particular standard part in an assembly file. An .ipt file is created for the inserted part.

The content library browser is used to access catalog content. To open the content library browser, click the down arrow next to Model in the title bar of the assembly browser, and select Library, as shown in Figure 6.1.

The top level of the library browser displaying the registered catalogs is shown in Figure 6.2.

The catalog hierarchy is navigated by double-clicking on items in the library browser window. Intermediate levels of the library browser are used to categorize content. An example intermediate level page is shown in Figure 6.3.

At the lowest level, the preview picture and the parameters for a part are displayed, as shown in Figure 6.4.

At this level Autodesk's I-drop technology can be used to add standard parts to an assembly file. An example of this is provided in the ball valve assembly tutorial at the end of this chapter.

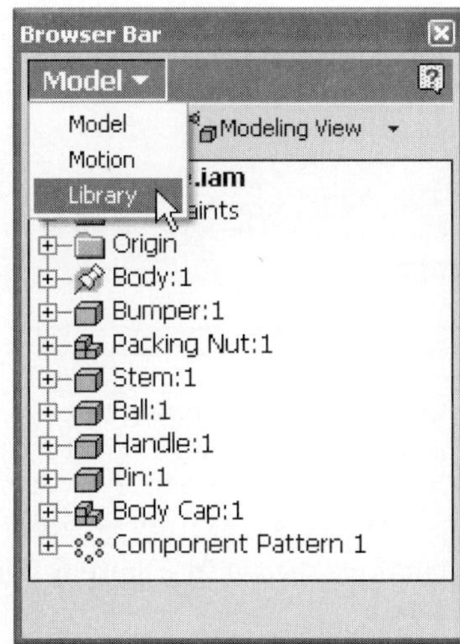

Figure 6.1 - Library browser access

Figure 6.2 - Library browser (top level)

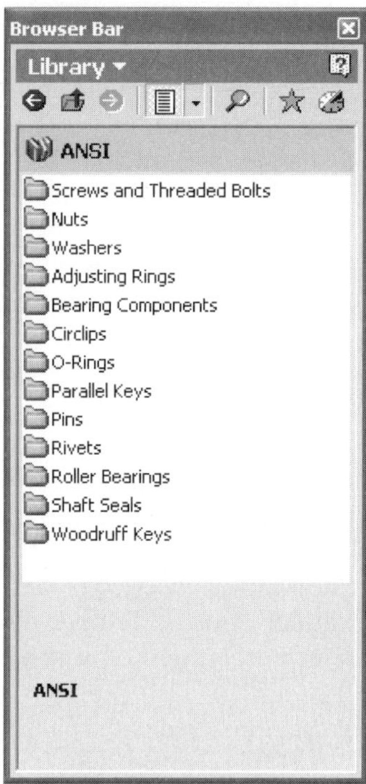

Figure 6.3 - Library browser (intermediate level)

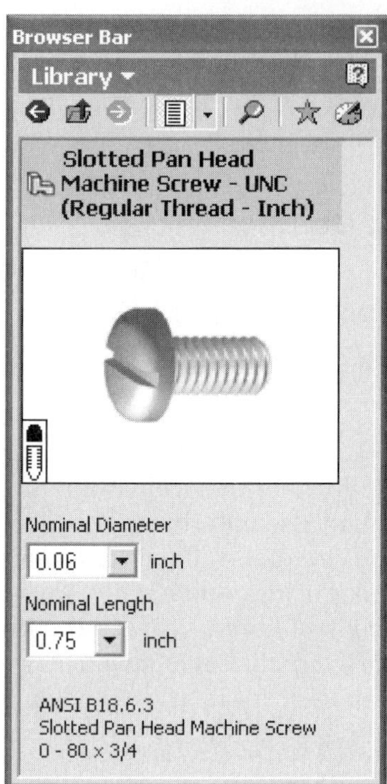

Figure 6.4 - Library browser (bottom level)

Figure 6.5 - Library browser with standard parts and steel shapes catalogs

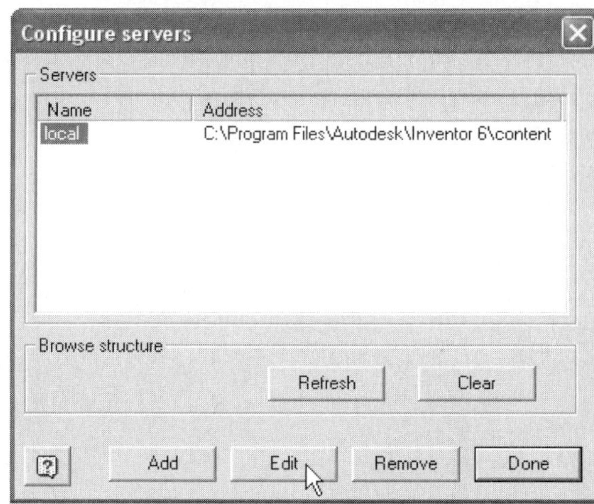

Figure 6.6 - Configure servers dialog box

Note that in order to display the steel shapes catalog, shown in Figure 6.5, it may be necessary to configure the library server.

To do so, click on the Configure button. This opens the Configure servers dialog box, shown in Figure 6.6. In the Servers list, highlight the server name, then click the Edit button.

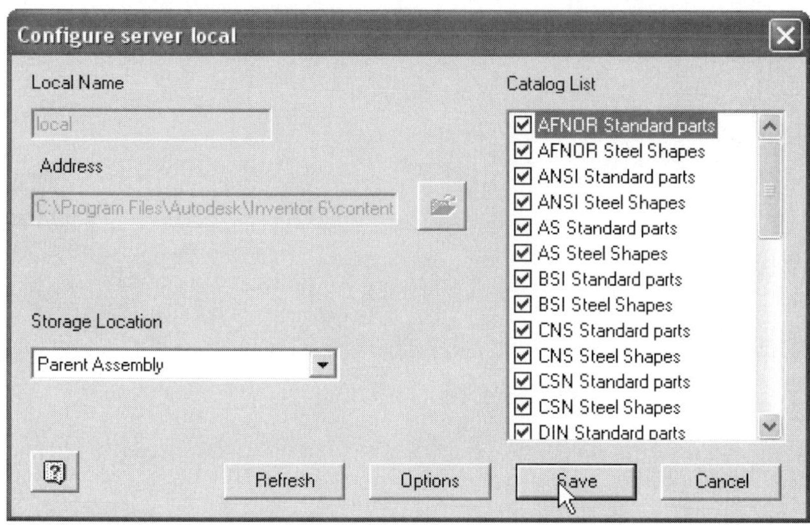

Figure 6.7 - Configure server local dialog box

In the Configure server local dialog box (Figure 6.7), place a check mark by the parts in the Catalog list that you wish to include. Click Save, Refresh, and finally, Done. A Steel Shapes catalog should now appear in the top level of the library browser.

Assembly Constraints

Assembly constraints are used to position components relative to one another. As assembly constraints are applied to a component, the component's freedom to move within the assembly is restricted.

Degrees of Freedom

A rigid body (i.e., a part) has six degrees of freedom (DOF), three in translation and three in rotation. The degrees-of-freedom symbol is shown in Figure 6.8.

As assembly constraints are applied to a component, the component's degrees of freedom are removed, and the ability of the component to move within the assembly is restricted. It is possible to simulate motion within an assembly by properly constraining the components. A non-moving part should have all DOF's removed, while a moving part should still have some DOF's remaining in the direction(s) of movement. The position of the first component placed in an assembly is fixed, or *grounded*. This means that all six DOF's have been removed. This is indicated in the assembly browser by the pushpin icon.

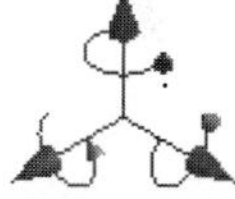

Figure 6.8 - Degrees of freedom symbol

TIP: Any component within an assembly can be grounded (ungrounded) by right-clicking on the component in the assembly browser and selecting (de-selecting) grounded.

Constraint Types

There are four principal types of assembly constraints available in Inventor; Mate, Angle, Tangent, and Insert. In order to apply a constraint between two components, the constraint type is first specified. Geometry (e.g., face or surface, axis or edge, point) is

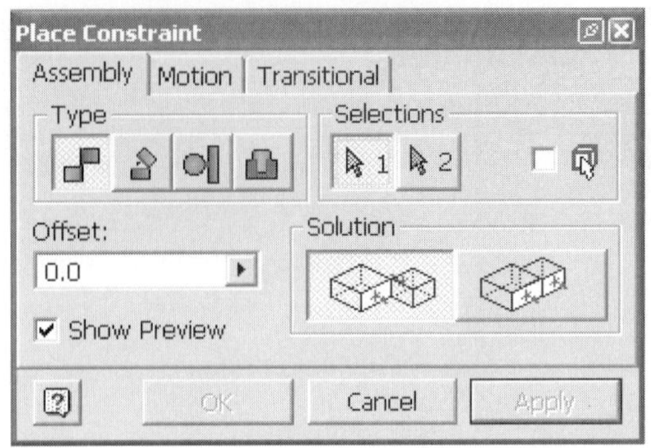

Figure 6.9 - Place constraint dialog box

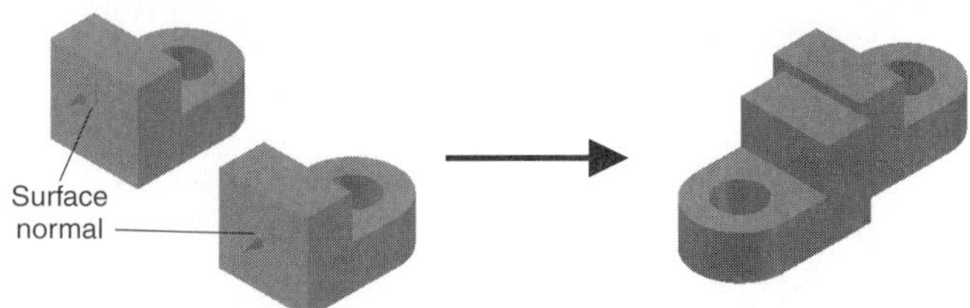

Figure 6.10 - Mate (face - face)

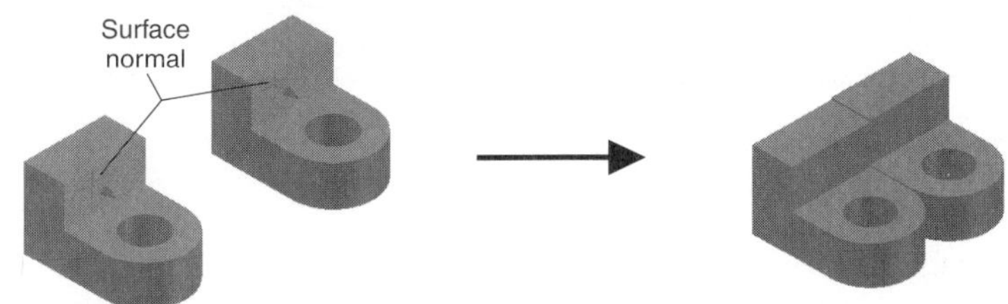

Figure 6.11 - Flush (face - face)

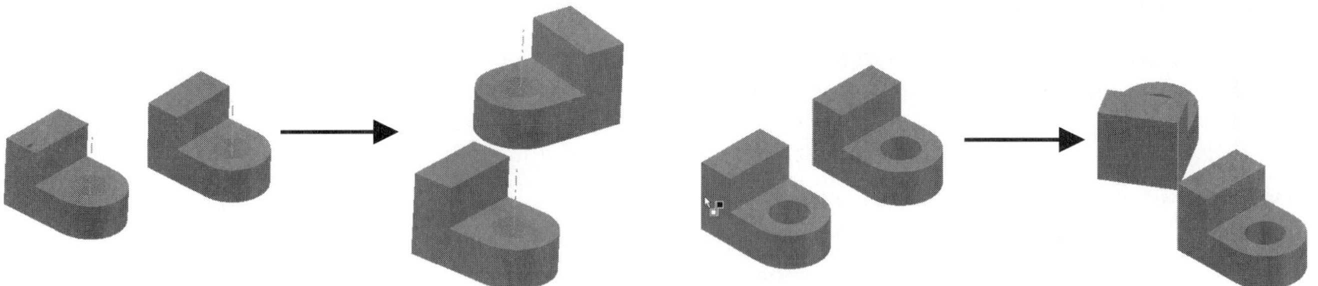

Figure 6.12 - Mate (axis - axis)

Figure 6.13 - Mate (edge - edge)

then selected on each component in turn, after which the constraint is applied. The Place Constraint dialog box is shown in Figure 6.9.

Mate

Mate is the most versatile assembly constraint. It can be used to position two faces so that either the faces mate (i.e., *surface normals* point in opposite directions) or are flush (i.e., surface normals point in the same direction) with one another. See the examples in Figures 6.10 and 6.11. One DOF in translation and two in rotation are removed when two surfaces are mated. An offset distance can also be applied between the faces. The offset is parametric; it has a value (default = 0) and is also assigned a name.

It is also possible to use the mate constraint to align components along an axis or edge. See the examples provided in Figures 6.12 and 6.13. In this case two DOF's, one in translation and one in rotation are removed.

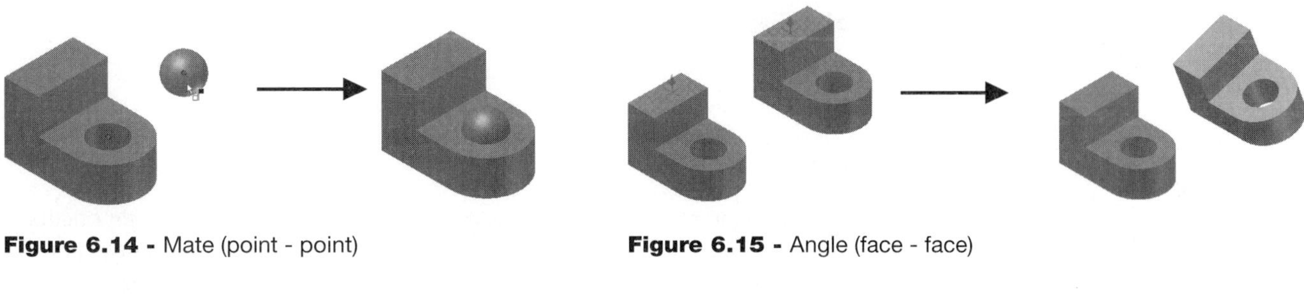

Figure 6.14 - Mate (point - point)

Figure 6.15 - Angle (face - face)

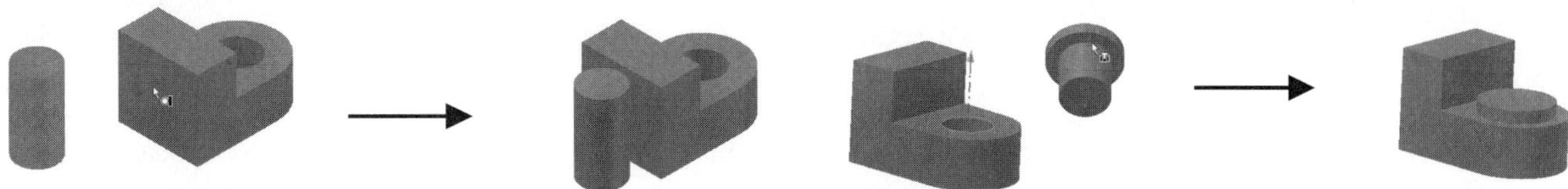

Figure 6.16 - Tangent (surface - face)

Figure 6.17 - Insert

Finally, it is possible to "mate" two points so that they are coincident. An offset distance can also be applied, though the results are somewhat unpredictable. See the example in Figure 6.14.

Angle

The angle constraint is used to position two faces (or edges) at a specific angle to one another. See the example provided in Figure 6.15. In applying an angle constraint, one rotational degree of freedom is removed. The specified angle is parametric.

Tangent

The tangent constraint positions a curved surface so that it is tangent to another surface. The tangent constraint removes one DOF in translation. A parametric offset is created between the two surfaces. See the example shown in Figure 6.16.

Insert

The insert constraint combines a mate (axis - axis) constraint with a mate (face - face) constraint. It is typically used to align a bolt shank in a hole, with the underside of the bolt head mated with a planar face. See Figure 6.17. The insert constraint removes five degrees of freedom; only a single rotational DOF remains. A parametric offset controls the distance between the two surfaces.

As a final note, a component's default reference work features (or other work features) can also be used when adding constraints. This is particularly useful for components that do not have many faces or edges.

Manipulating Components

In the process of applying assembly constraints, it is sometimes helpful to either move or rotate individual components so that component geometry can be more easily selected. The Move Component and Rotate Component tools available on the panel bar can be used for this purpose. Unconstrained, partially constrained, and even fully constrained components can be manipulated in this way. Constrained components that have been temporarily moved and/or rotated snap back to their constrained positions when updated from the standard toolbar.

Assembly Editing

Just as it is possible to edit parts from the part browser, so too can assemblies be edited from the assembly browser.

Constraints

Assembly constraints normally appear in the assembly browser below the components to which they apply, as seen in Figure 6.18. In this example, there is an insert constraint between the handle and the piston, and a mate constraint relating the handle to the cylinder.

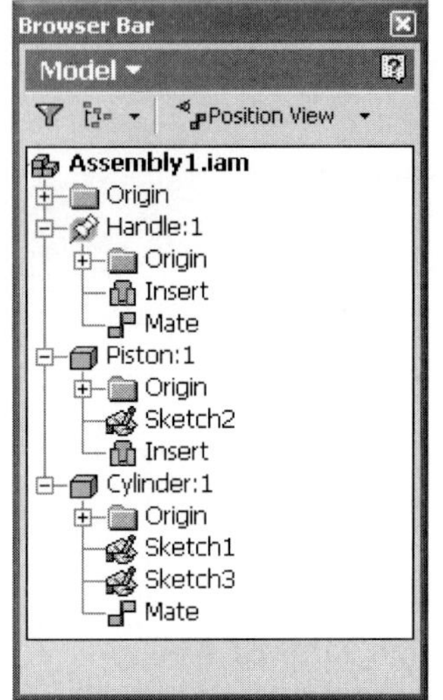

Figure 6.18 - Assembly browser (position view)

Alternatively, by selecting Modeling and Position Views, located at the top of the browser bar, all assembly constraints are organized into a single Constraints folder. See Figure 6.19.

If a constraint is right-clicked in the assembly browser, a context menu appears, as shown in Figure 6.20. Select Edit and the Edit Constraint dialog box appears. This is sometimes done in order to modify the parametric offset (or angle) associated with a constraint.

The name associated with the offset (or angle) parameter can be identified by hovering the cursor over the constraint icon in the browser, as shown in Figure 6.21, or by opening the Parameters f_x table.

Assembly constraints position one part with respect to another part in an assembly. To identify a matching part participating in a constraint, right-click on the constraint and select Other Half, as shown in Figure 6.22 on the left. The "other half" of the constraint associated with the matching part is highlighted, as shown in Figure 6.22 on the right. By selecting Suppress from the same context menu shown in Figure 6.22, the constraint may be suppressed. This means that the constraint is no longer in effect, and parts to which the constraint applies may now be moved.

A constraint can be renamed by clicking on the constraint name, selecting a second time, then entering the new name.

Components/Parts

It is also possible to edit parts (or components) directly in an assembly file. To enter part mode within an assembly file, double-click on the part to be edited, either in the

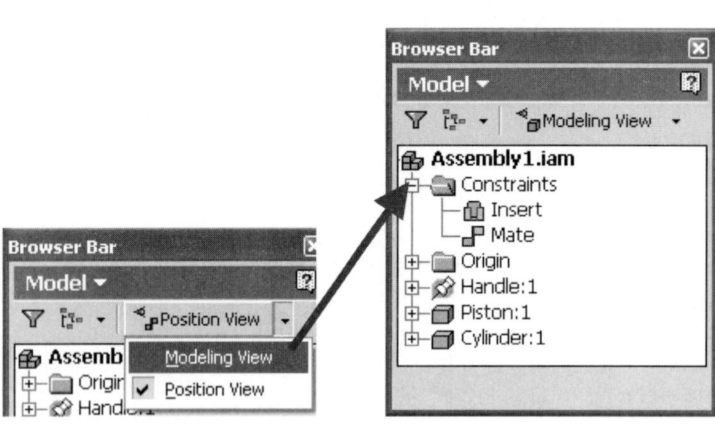

Figure 6.19 - Change from position to modeling view

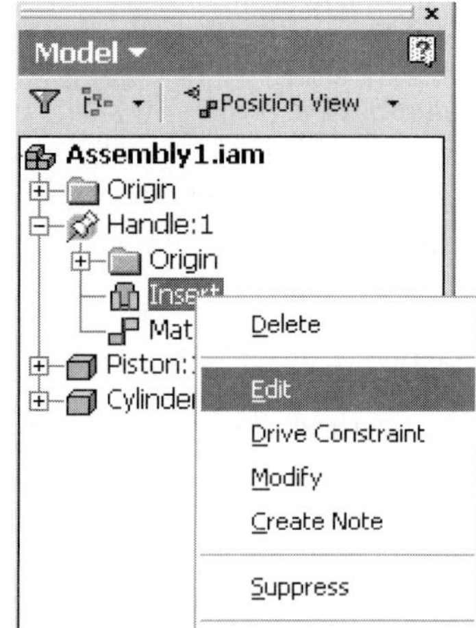

Figure 6.20 - Editing assembly constraints

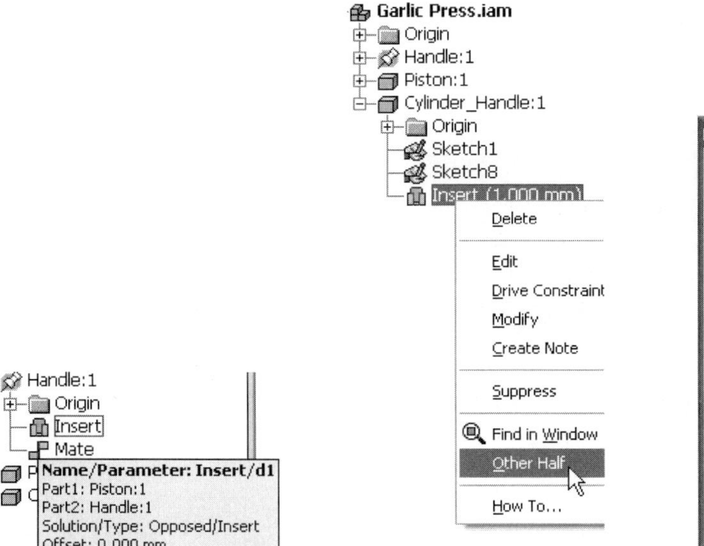

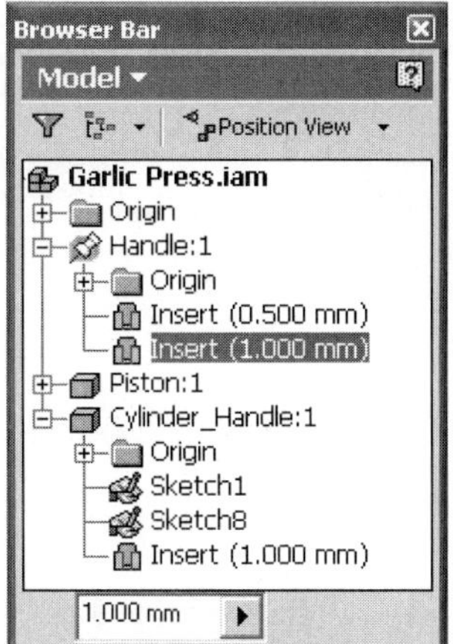

Figure 6.21 - Displaying constraint information

Figure 6.22 - Other Half tool

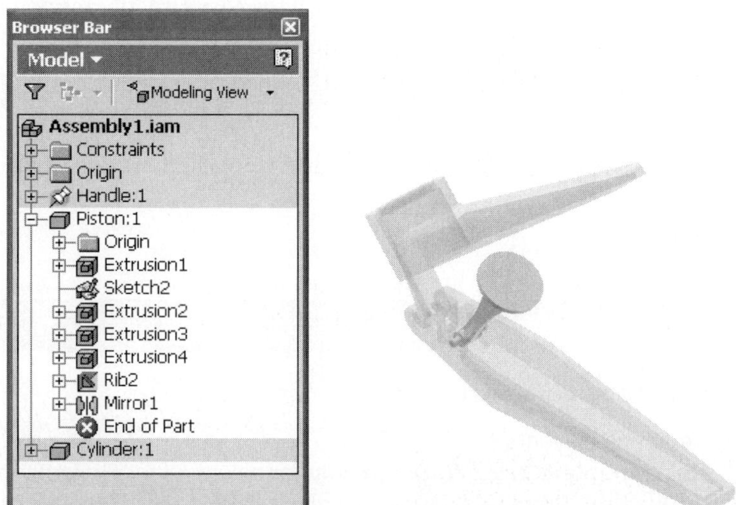

Figure 6.23 - Part editing within the assembly file

graphics area or the browser. In the graphics area all parts other than the one being edited appear in phantom, while in the browser all other components are grayed out. This is shown in Figure 6.23, where the piston part is about to be edited.

At this point the part can be edited in much the same way that it can be edited in a part file. Any changes made will affect both the assembly and the part file. File > Save All should be selected from the menu bar to ensure that the changes are saved.

To return to assembly mode either 1) right-click in the graphics area and select Finish Edit, 2) double-click on the Assembly icon at the top of the browser, or 3) select the Return [Return] button on the standard toolbar.

Figure 6.24 - Ball valve assembly with internal parts visible

Figure 6.25 - Three-quarter and half section views

Assembly Display

The color of a component can be changed within an assembly by highlighting the component (i.e., select in graphics area or in browser), and then selecting a color from the Color control on the standard toolbar. Some of the available colors are translucent, thus allowing assembly views where internal parts are visible, as seen in Figure 6.24.

From the browser it is also possible to control the visibility of components. Components in an assembly also have an *Enabled* setting. If a part is not enabled, it is shown in phantom and it is not selectable.

Finally, the Assembly Section View tool, available from the assembly panel bar, can be used to advantage to display a variety of different shaded section views of an assembly. An example of a three-quarter and a half section of the ball valve is shown in Figure 6.25.

TUTORIAL 17 Garlic Press Assembly

Detailed Subassembly Instructions

1. Change the active projects folder to Garlic Press. From the Open dialog box, select Projects from the What To Do column (or select File > Projects from the menu bar). Once the Projects window is open, click on the Garlic Press project name; then click the Apply button. The Garlic Press Workspace is now the active folder.
2. Start a new file, select the Metric tab, and then choose the metric assembly template Standard (mm).iam.
3. Select Place Component from the assembly panel bar. Select the Handle part from the Garlic Press Workspace folder in the Open dialog box, and then click the Open button. The dialog box closes and one *occurrence* of the Handle part is automatically placed in the assembly. Right-click, and select Done. The Handle appears in the graphics window, as shown in Figure 6.26.
4. Again use the Place Component tool to place the Piston part in the assembly. This time it will be necessary to click within the graphics area to

Figure 6.26 - Handle imported into assembly file

Figure 6.27 - Graphics window and browser

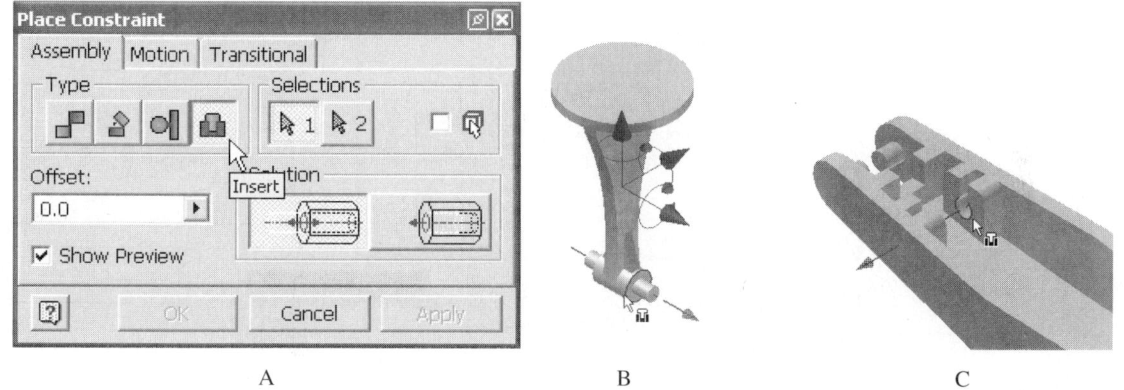

Figure 6.28 - Insert constraint operation

place an occurrence of the part. Select a convenient location and then click. To prevent placement of further occurrences of the piston, right-click and select Done.

5. Turn on the Degrees of Freedom (DOF) symbol. From the menu bar, select View > Degrees of Freedom. The graphics window and assembly browser should now resemble Figure 6.27. In the graphics area the DOF symbol on the piston indicates that this part has six degrees of freedom (three in translation, three in rotation), whereas no DOF symbol appears on the handle, indicating that its position is fixed. Since the handle is the base component, its position is automatically *grounded* (i.e., zero DOF). This is also indicated in the assembly browser, where the handle icon includes a pushpin symbol.
6. Save the assembly file as **Garlic Press.** It is recommended that the file be regularly saved.
7. We will now attach the piston to the handle using an Insert constraint. The insert constraint removes five degrees of freedom, leaving a single degree of freedom in rotation. Select the Place Constraint tool from the panel bar. The Place Constraint dialog box opens. First select the Insert constraint type, as shown in Figure 6.28A. Now move the cursor to the edge of the large diameter cylinder

face on the piston. With the edge highlighted and an arrow (representing the axis of the cylinder) visible (see Figure 6.28B), left-click to select. Now move the cursor to the face on the handle shown in Figure 6.28C. Left-click to select this face and axis. If Show Preview is checked in the Place Constraint dialog box, the piston immediately snaps into position.

In order to center the position of the piston with respect to the handle, it is necessary to change the offset in the dialog box to 0.5, as shown in Figure 6.29. After selecting OK, the graphics window should resemble the image on the right. Notice that the DOF symbol for the piston has changed; a single rotational degree of freedom is now indicated. Try click and dragging on the top part of the piston. You should see that the piston now rotates about the axis specified in the insert constraint operation.

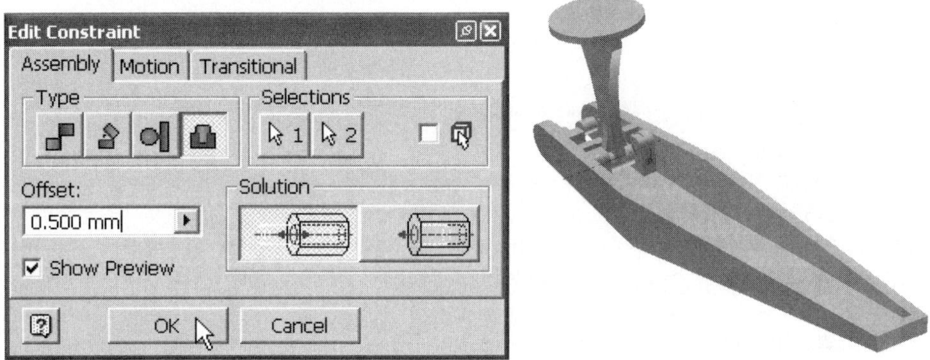

Figure 6.29 - Insert constraint operation (continued)

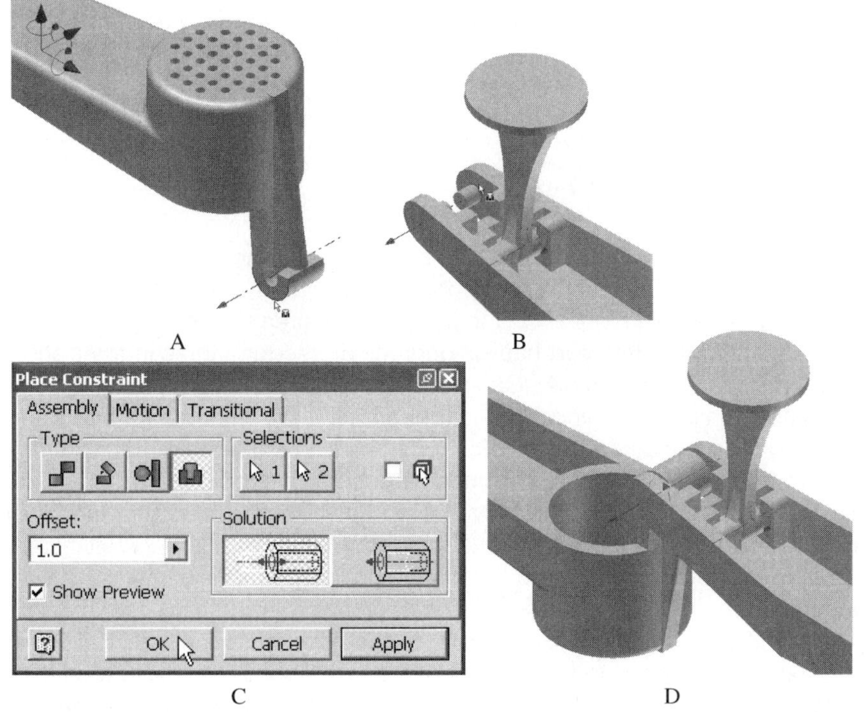

Figure 6.30 - Insert constraint operation

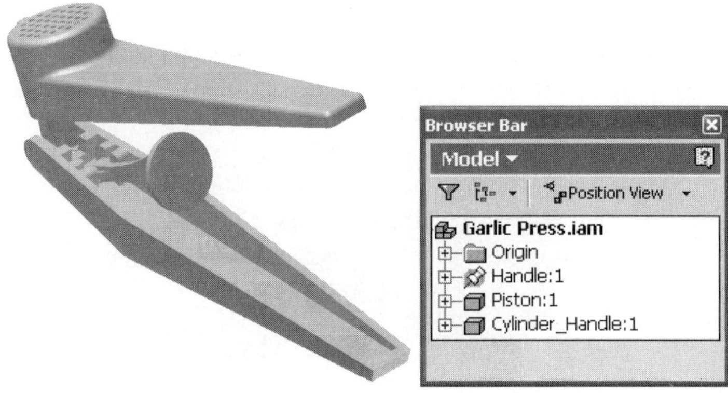

Figure 6.31 - Completed assembly

8. We will now add the third part to the assembly. Select the Place Constraint tool. From the Open dialog box, select Cylinder_Handle and place one instance of the part in the assembly file.
9. Right-click, Isometric View to center the components in the graphics window.
10. Once again, an insert constraint will be used to position the Cylinder_Handle part with respect to the handle. Select Place Constraint tool from the panel bar. After selecting the Insert constraint type, select the edge of the face show in Figure 6.30A. Then select the edge of the face shown in Figure 6.30B. To center the parts, change the offset in the dialog box to 1, as shown in Figure 6.30C, then click OK. The result is shown in Figure 6.30D. Notice that the Cylinder_Handle part now has one degree of freedom in rotation.
11. Isometric View.
12. Turn off the DOF symbols (from the menu bar, select View > Degrees of Freedom).
13. Move the rotating parts so that the garlic press assembly is similar to that shown in Figure 6.31.
14. Save the assembly file as **Garlic Press** in the Garlic Press Workspace folder.

In Tutorials 18, 19, and 20 the Ball Valve parts are assembled. Before starting these tutorials, copy the following files into the Ball Valve Workspace folder; Oring2.ipt, Oring3.ipt, Pin.ipt, Body Cap.ipt. These parts will have been modeled in completing the following problems; P2.1, P2.2, P3.1, and P3.2. Otherwise, complete files for these parts are available on the CD.

TIP: Remember that video versions of all 30 Tutorials are included on the CD accompanying the book. If you experience a problem following a tutorial, consult the corresponding Video Tutorial.

Packing Nut Subassembly — TUTORIAL 18

Detailed Subassembly Instructions

1. Change the active projects folder to Ball Valve. From the Open dialog box, select Projects from the What To Do column. Once the Projects window is open, click on the Ball Valve project name; then click the Apply button. The Ball Valve Workspace is now the active folder.

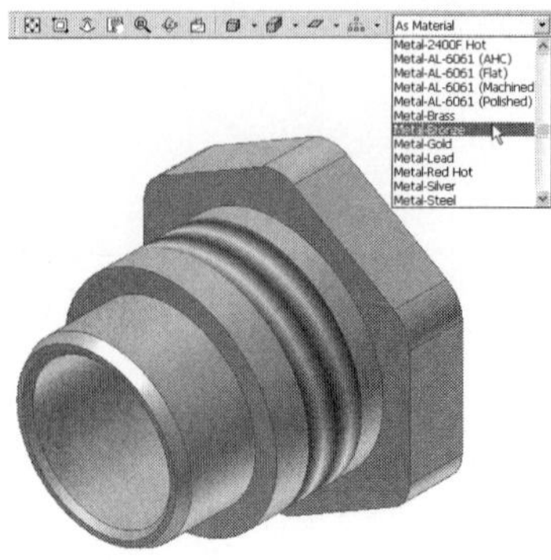

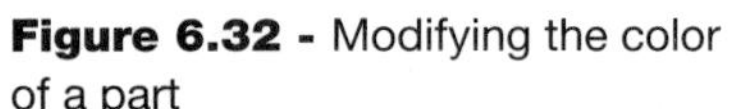

Figure 6.32 - Modifying the color of a part

Figure 6.33 - Current screen and assembly browser view

2. Start a new file, select the Metric tab, and then choose the metric assembly template Standard (mm).iam.
3. Select Place Component from the assembly panel bar, select the Packing Nut part from the Ball Valve Workspace folder, and then click the Open button. The dialog box closes and one *occurrence* of the Packing Nut is automatically placed in the assembly. Right-click, and select Done.

   Since we will be dealing with several different parts in an assembly, it will be easier to distinguish between the different parts if they are assigned different colors.
4. Select the Packing Nut part by clicking on it. From the standard toolbar, select the color control, then scroll down and select a color (Metal-Bronze), as shown in Figure 6.32.
5. Use the Place Component tool to place the Oring3 part in the assembly. Left-click within the graphics area to place an occurrence of the part. To prevent placement of further occurrences of the Oring3 part, right-click and select Done. Your screen and assembly browser should resemble Figure 6.33.
6. Change the color of Oring3 to Rubber (Black).
7. Save the assembly file as **Packing Nut.** It is recommended that the file be regularly saved every five to ten minutes.
8. Select the Place Constraint tool from the panel bar. The Place Constraint dialog box opens. We will first mate the axes of the packing nut and the O-ring. For this constraint, it is not necessary to change settings. Move the mouse over the packing nut until the axis is displayed, as shown in Figure 6.34. With the axis highlighted, left-click. Now move the cursor to the Oring3 part until the axis is highlighted, as shown in Figure 6.35. Once again, left-click.

   If the Show Preview box on the Assembly tab of the Place Constraint dialog box is checked, the O-ring should move into position, as shown in Figure 6.36. Click the OK button to apply the constraint and close the dialog box.
9. Turn on the Degrees of Freedom (DOF) symbol. From the menu bar, select View > Degrees of Freedom. The screen should now resemble Figure 6.37. Note that the placement of the mate constraint has reduced the O-ring's DOF to two, one in rotation and one in translation. Since the Packing Nut is the base component, its position is *grounded* (i.e., zero DOF). This is also indicated in the assembly browser, where the Packing Nut icon includes a pushpin symbol.

   We now need to remove the O-ring's DOF in translation (we will allow it to rotate about its axis). Before we do this, however, we will temporarily move

Figure 6.34 - Selection of packing nut axis

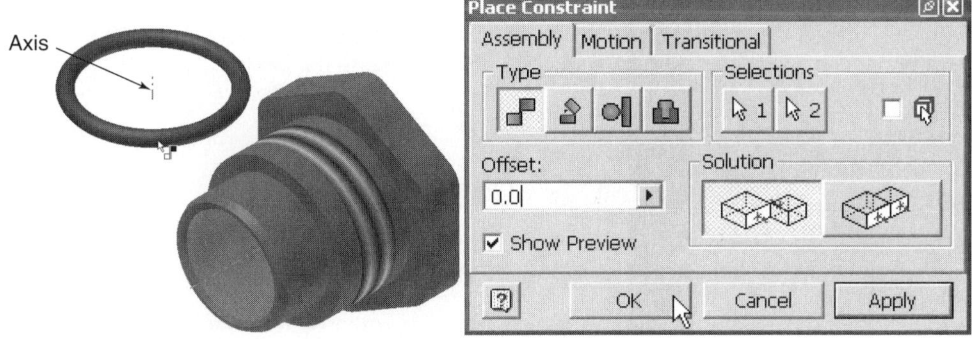

Figure 6.35 - Selection of O-ring axis with place constraint dialog box

Figure 6.36 - Mate (axis - axis) constraint applied

Figure 6.37 - Degrees of freedom symbol displayed

Figure 6.38 - O-ring part temporarily moved

Oring3 away from the packing nut. Note that clicking the Update button on the standard bar will re-position the parts in an assembly, based on previously applied constraints.

10. Select the Move Component tool from the panel bar. Move the cursor over the O-ring until it is highlighted, then click and drag the part away from the base part. See Figure 6.38.
11. The O-ring does not itself have any planar faces that can be used for the purpose of mating it with another part. However, we can use the default reference work features of the part for this purpose. In the assembly browser, expand the Oring3 part, and then expand the part's Origin folder. Turn on the visibility of the XY Plane.
12. Select the Place Constraint tool. Select the packing nut face and the O-ring work plane, change the Offset distance to 1, and then click OK, as shown in Figure 6.39. Note that the cross-sectional diameter of the O-ring is 2.
13. Turn off the visibility of the XY Plane of the Oring3 part. The assembly should be similar to Figure 6.40.
14. Use the Place Component tool to place the Oring2 part in the assembly. Change the color of Oring2 to Rubber (Black).
15. Use the Common View option of the 3D Rotate tool to compose a view of the packing nut similar to that shown in Figure 6.41.
16. Select the Place Constraint tool from the panel bar. Make the selections shown in Figure 6.42, then select OK.

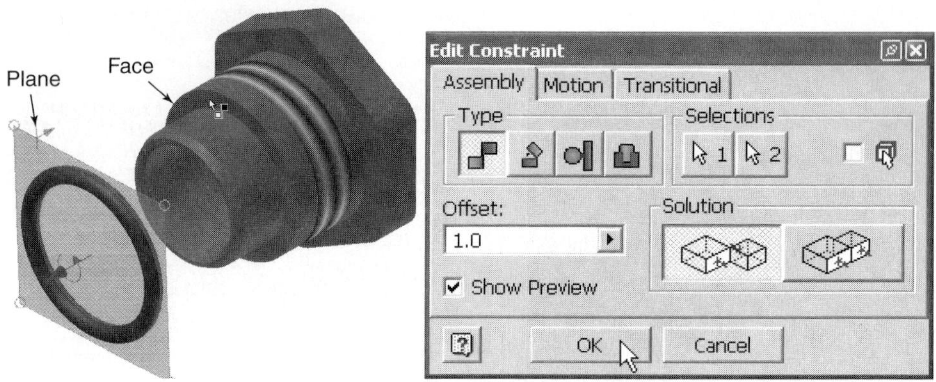

Figure 6.39 - Mate (work plane - face) operation

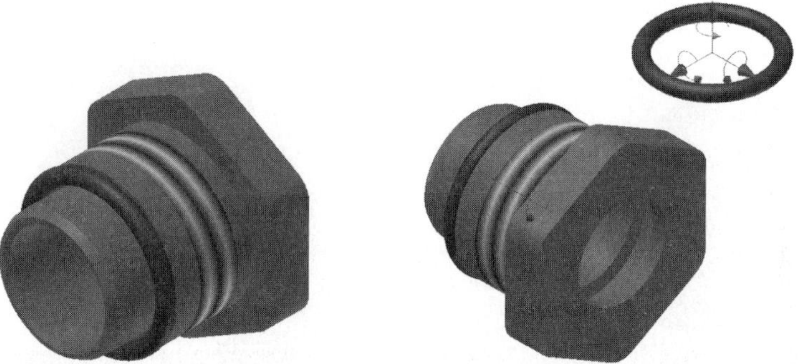

Figure 6.40 - O-ring correctly positioned

Figure 6.41 - Change of view

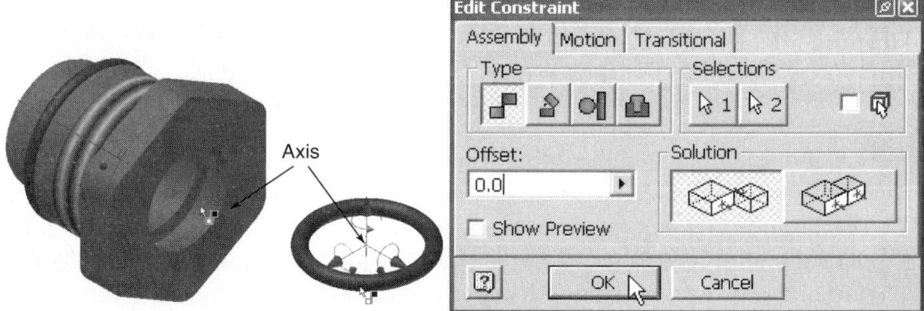

Figure 6.42 - Mate (axis - axis) operation

17. Temporarily turn off the visibility of the packing nut from the assembly browser.
18. Using the Move Component tool, move the Oring2 part.
19. Turn the visibility of the packing nut part back on.
20. Turn on the visibility of the XY Plane of the Oring2 part.
21. Select the Place Constraint tool. Select the packing nut face, the O-ring work plane, change the Offset distance to 1, and then click OK, as shown in Figure 6.43. Note that the cross-sectional diameter of the O-ring is 2.
22. Turn off the visibility of the Oring2 XY Plane.
23. Wireframe display.

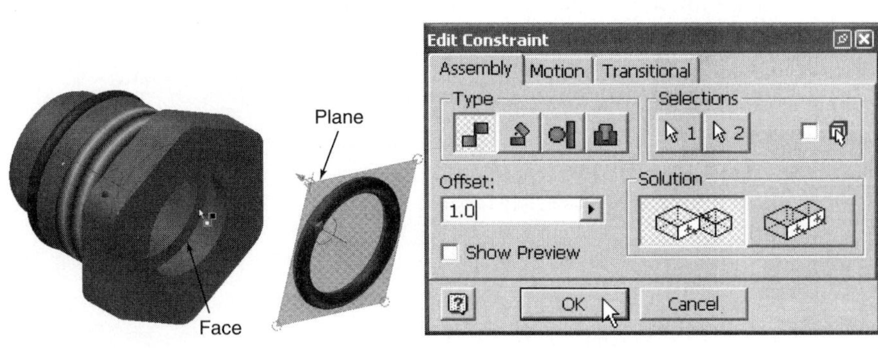

Figure 6.43 - Mate (face - work plane) operation

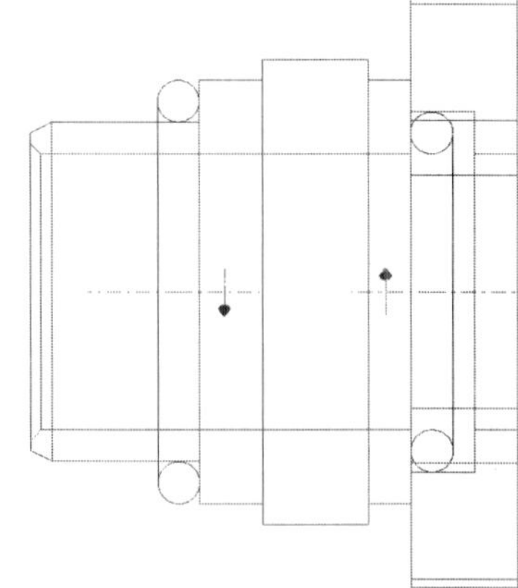

Figure 6.44 - Verify O-ring placement

24. Use the Common View option of the 3D Rotate tool to compose a view of the packing nut similar to that shown in Figure 6.44. From this view it can be seen that the O-rings are correctly positioned with respect to the packing nut.
25. Isometric View.
26. Shaded display.
27. Save the assembly as **Packing Nut** in the Ball Valve folder.

Body Cap Subassembly — TUTORIAL 19

Detailed Subassembly Steps

1. Start a new metric assembly file.
2. Use the Place Component tool to import the Body Cap part file as a base part.
3. In this tutorial, colors will be assigned to parts while in part editing mode. In this way the part color assignments will be attached to the part files, as well as to the assembly file. Double-click on the Body Cap part in either the graphics window, or in the assembly browser. The assembly browser should resemble Figure 6.45.
4. Use the color control on the standard bar to change the color of the Body Cap part to Metal-Bronze.
5. To exit part edit mode, right-click in the graphics area, and select Finish Edit.
6. Use the Place Component tool to import the Oring1 part from the Ball Valve Workspace folder.
7. Double-click on the Oring 1 part to enter part mode.
8. Change the color of the Oring1 part to Rubber (Black).
9. Select Return on the standard toolbar to return to assembly mode.
10. Use the Common View option of the 3D Rotate tool to change the view of the Body Cap part to one similar to that shown in Figure 6.46. Before exiting Common View, right-click and select Redefine Isometric. This will make this view the default isometric view.
11. Save the assembly file as **Body Cap.** Before saving, the Save dialog box shown in Figure 6.47 appears. Select OK to save the assembly file as well as the

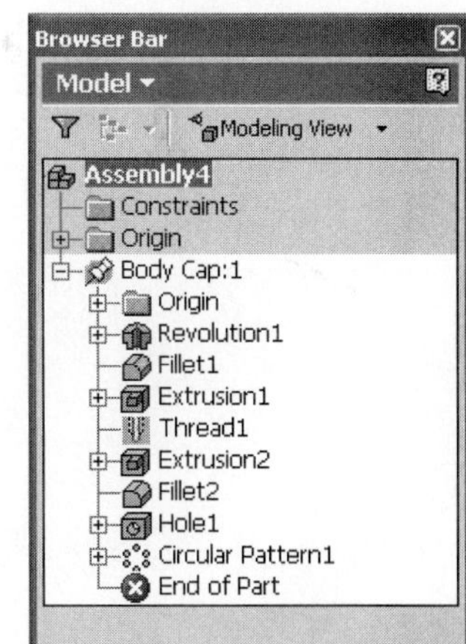

Figure 6.45 - Assembly browser in part edit mode

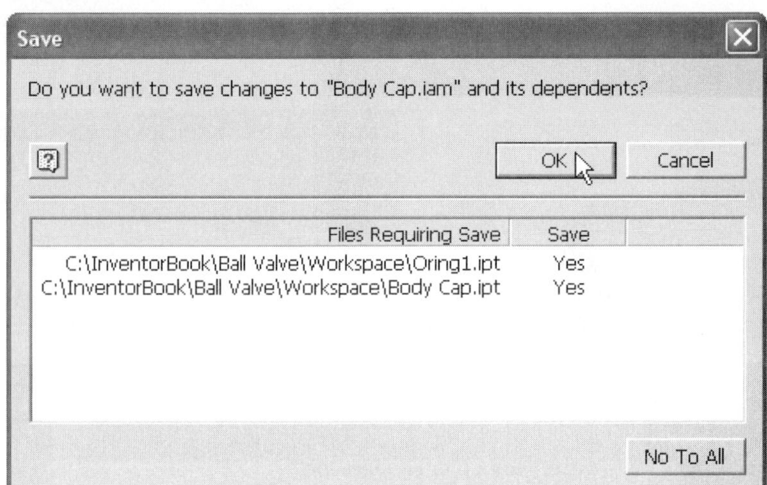

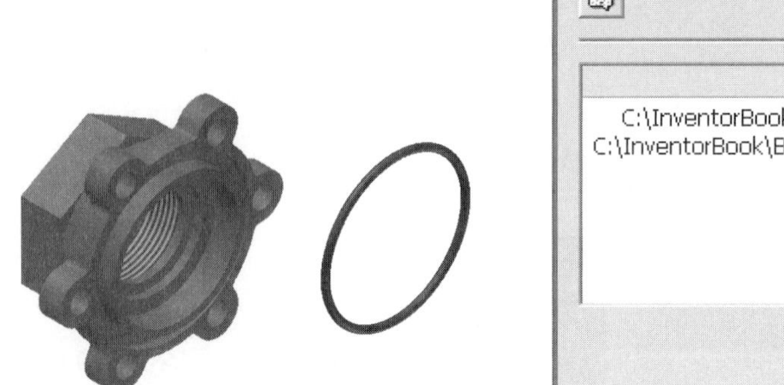

Figure 6.46 - Change view

Figure 6.47 - Saving assembly file and dependents

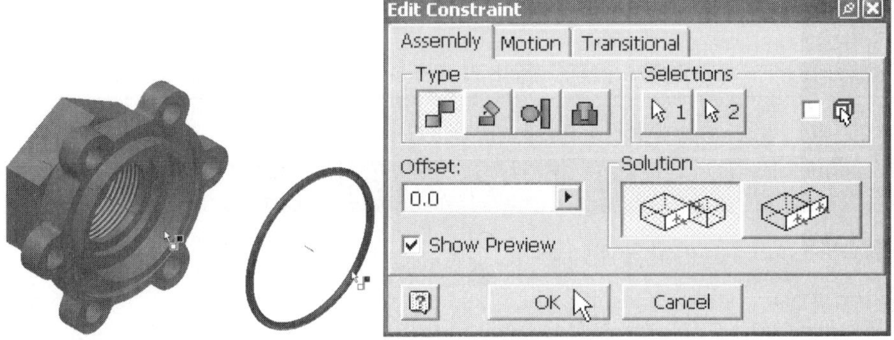

Figure 6.48 - Mate (axis - axis)

changes (i.e., part color) to the dependent part files. It is recommended that the file be regularly saved every five to ten minutes.

12. Use the Place Constraint tool to add the mate (axis - axis) constraint depicted in Figure 6.48.
13. Turn on the Degrees of Freedom (DOF) icon (View > Degrees of Freedom).
14. In the assembly browser, right-click on the Body Cap part, and de-select Enabled. When a part in an assembly is not enabled it is shown in phantom and cannot be selected.
15. Using the Move Component tool, move the Oring1 part away from the base part, so that part geometry on both parts can easily be selected.
16. Enable the Body Cap part.

    Unlike the O-rings in the previous tutorial, the reference work planes for Oring1 do not split the part symmetrically (To verify this, open both the Oring1 and Oring2 parts, and then turn on the visibility of the default reference work planes). This is a direct consequence of not projecting the default work axes onto the sketch plane, and then constraining the sketch to the reference geometry. In the assembly, this forces the need to mate and then offset two

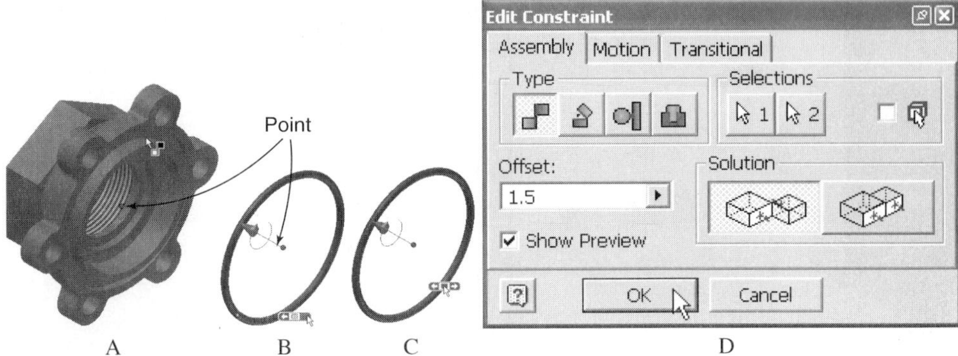

Figure 6.49 - Mate (point - point)

points, a less reliable technique than mating two surfaces (part face, work plane), as was used in the previous tutorial. This is a good example of why it is good practice to use projected reference geometry when creating part models. See Tutorial 2 in Chapter 2.

17. Use the Place Constraint tool to add the mate (point - point) constraint depicted in Figure 6.49. First select the point on the Body Cap shown in Figure 6.49A. In order to select this specific point, the edge shown in this figure must be highlighted (or another concentric edge just outside of the highlighted edge). Next select the point shown in Figures 6.49B and C, on Oring1. It will probably be necessary to use the Select Other tool to do this. Finally, enter an offset of 1.5, as shown in the dialog box in Figure 6.49D (the cross-sectional diameter of the O-ring is 3).
18. Wireframe display.
19. Use the Common View tool to compose the *principal view* shown in Figure 6.50. Note the location of the O-ring. If it is not positioned as shown in the figure, try adjusting the previously applied constraint offset to -1.5.
20. Shaded display.
21. Isometric View.
22. Use the Place Component tool to add the Spring part (in the Ball Valve Workspace folder) to the assembly.
23. Double-click on the Spring part in either the graphics area or the browser to enter part mode.
24. Change the color of the Spring part to Sea Green.
25. Right-click in the graphics area, and select Finish Edit to return to assembly mode.
26. For the Spring part, turn on the visibility of both the XY Plane and the Z Axis.
27. Use the Place Constraint tool to add a mate (axis - axis) constraint, as shown in Figure 6.51. Click OK.
28. After using the Move Component tool to move the spring away from the Body Cap part, use the Place Constraint tool to add a mate/flush (face - work plane) constraint, as shown in Figure 6.52.
29. Turn off the visibility of the XY Plane and the Z Axis of the Spring. The assembly should now resemble Figure 6.53.
30. Use the Place Component tool to add the Seat part.
31. Use the Place Constraint tool to add the mate (axis - axis) constraint shown in Figure 6.54.

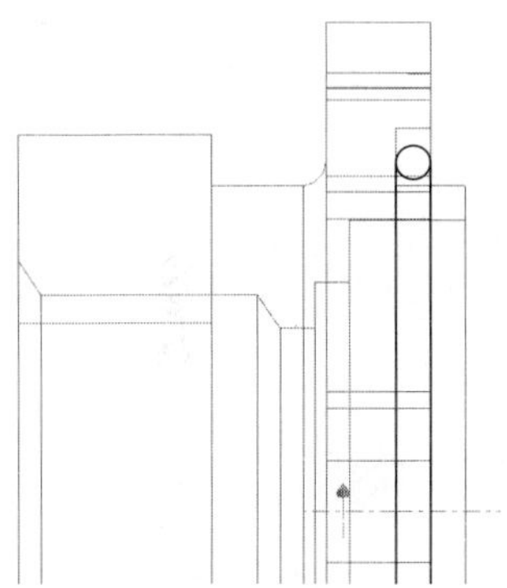

Figure 6.50 - O-ring position with respect to body cap

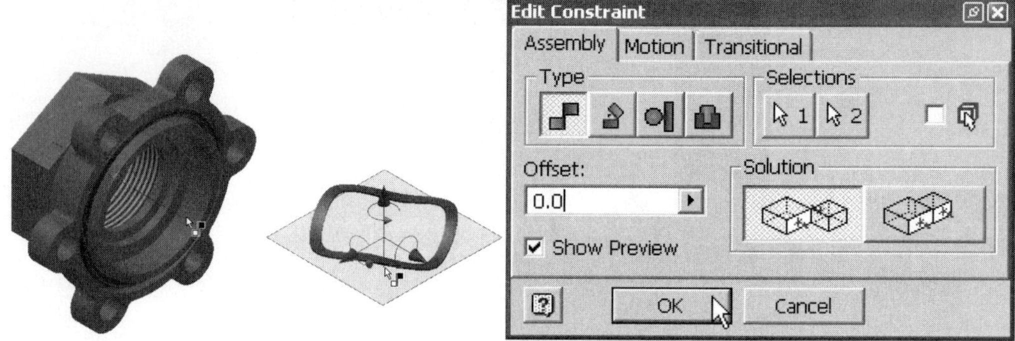

Figure 6.51 - Mate (axis - axis)

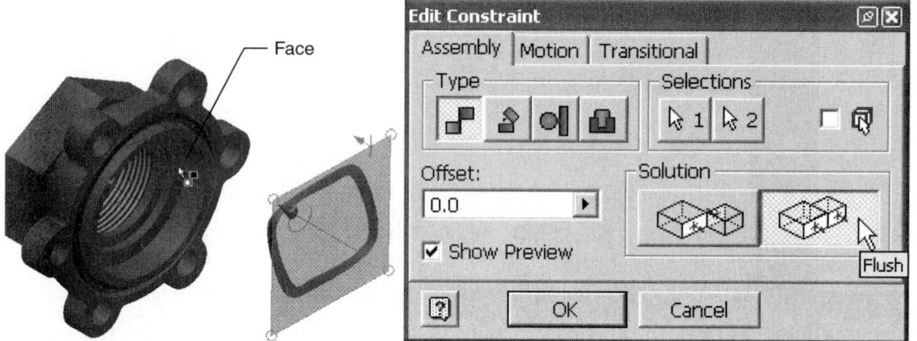

Figure 6.52 - Mate/flush (face - work plane)

Figure 6.53 - Subassembly with constrained parts

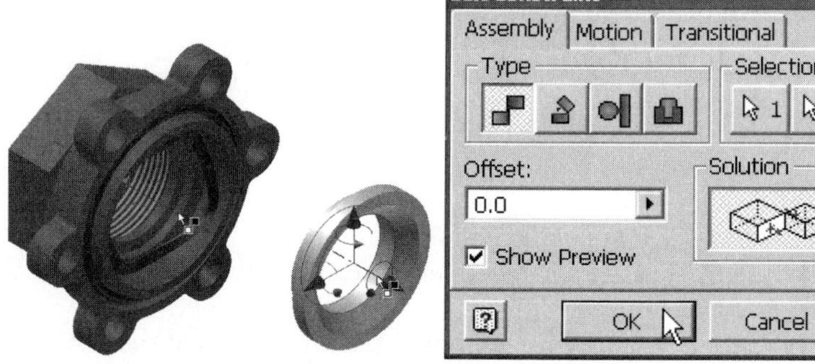

Figure 6.54 - Mate (axis - axis)

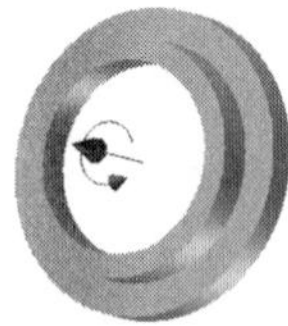

Figure 6.55 - Rotated view

32. Use the Move Component tool to move the Seat away from the assembly.
33. Use the Rotate Component tool to rotate the view of the Seat so that it is similar to Figure 6.55. (Select the tool icon, then the Seat component (or vice versa), then rotate the view of the part.)
34. Use the Place Constraint tool to add the mate (face - face) constraint shown in Figure 6.56.

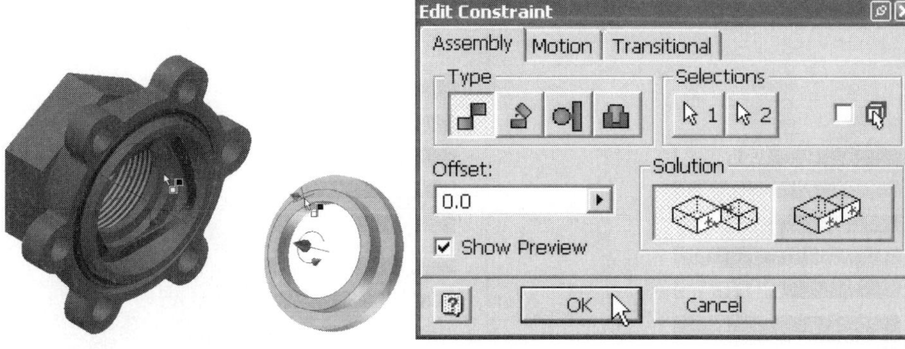

Figure 6.56 - Mate (face - face)

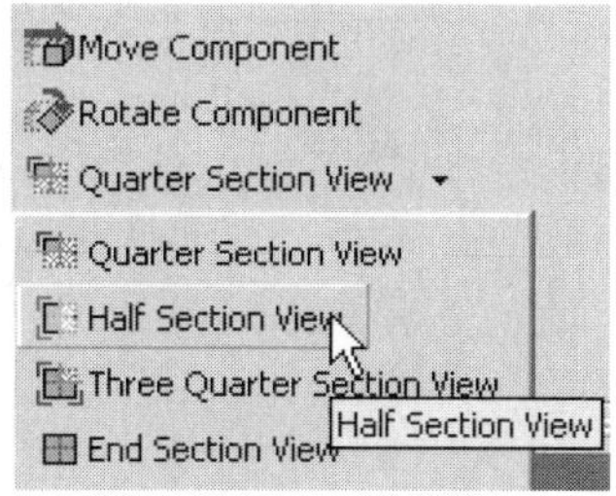

Figure 6.57 - Half section view selection

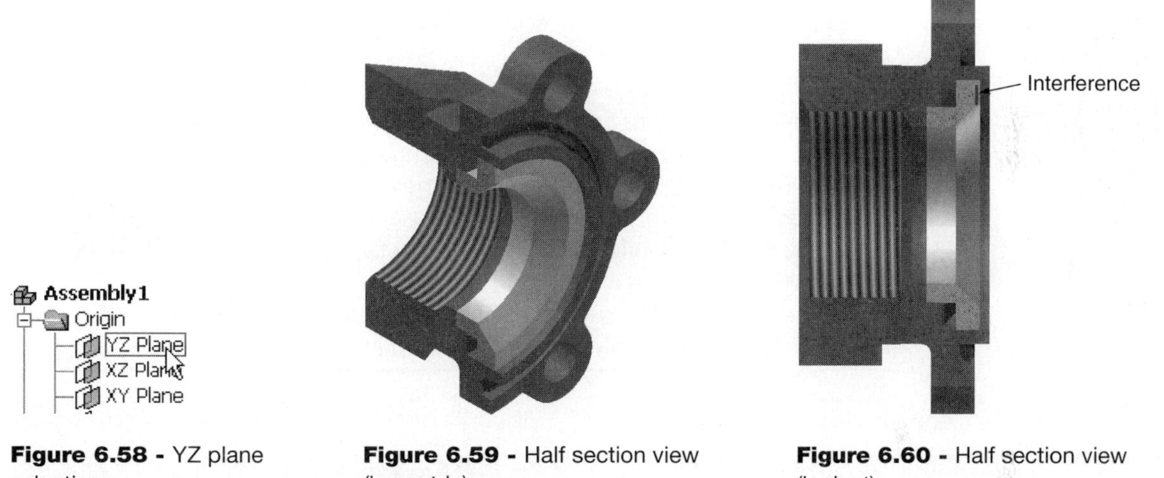

Figure 6.58 - YZ plane selection

Figure 6.59 - Half section view (isometric)

Figure 6.60 - Half section view (look at)

35. Turn off the Degrees of Freedom.
36. Isometric View.
37. Use the Section View tool to create a half section view of the assembly. It will be necessary to click on the down arrow to access the flyout, and then select Half Section View, as shown in Figure 6.57. Expand the Origin folder directly below the Assembly icon in the browser, and then select the YZ Plane (Figure 6.58). This will serve as the cutting plane. In the graphics area right-click and select Done. The screen should be similar to Figure 6.59.
38. Use the Look At tool, selecting the YZ Plane. The screen should resemble Figure 6.60. Note that the Spring and the Seat interfere. For now we will modify the previously applied constraint so that the two parts do not interfere.
39. At the top of the assembly browser, click on the Position View icon and then select Modeling View. This reorders the browser so that, among other things, all the assembly constraints are placed in a single folder.

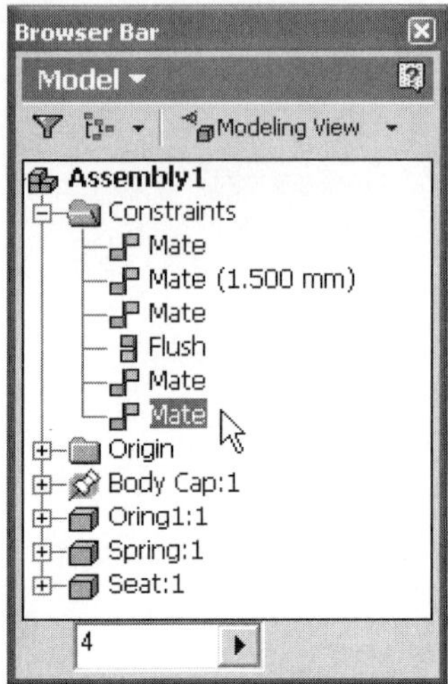

Figure 6.61 - Edit offset constraint

Figure 6.62 - Half section after offset edit

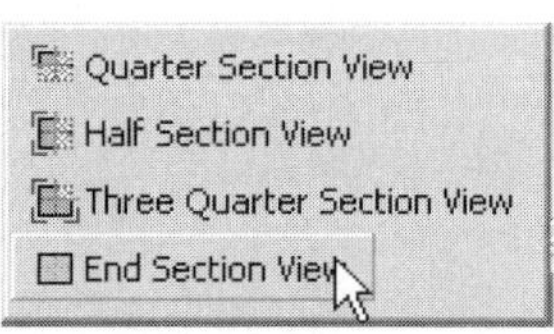

Figure 6.63 - End half section selection

Figure 6.64 - Completed subassembly

40. Expanding the Constraints folder, select the last Mate Constraint. In the Offset box at the bottom of the browser, change the offset to 4, and then hit the Enter key (Figure 6.61). The screen should now look like Figure 6.62. This simulates the Body Cap subassembly with the spring in an uncompressed state.
41. Using the Section View tool, expand the flyout and select End Section View, as shown in Figure 6.63.
42. Isometric View—the subassembly should be similar to Figure 6.64.
43. Save the assembly file as **Body Cap** in the Ball Valve Workspace folder.

Tutorial 20 Ball Valve Assembly

Detailed Assembly Steps

1. Start a new Metric assembly file.
2. Use the Place Component tool to import the Body part file as a base part.
3. Using the color control on the standard toolbar, change the color of the Body part to Metal-Bronze.
4. Use the Common View option of the 3D Rotate tool to change the view of the Body Cap part to one similar to that shown in Figure 6.65. Before exiting Common View, right-click and select Redefine Isometric.
5. Use the Place Component tool to add the Bumper part to the assembly.
6. Use the color control on the standard toolbar to change the color of the bumper to Rubber (Black).
7. Save the assembly file as **Ball Valve.** It is recommended that the file be regularly saved every five to ten minutes.
8. Turn on the Degrees of Freedom (View > Degrees of Freedom, from the menu bar).
9. Use the 3D Rotate tool to change the view slightly so that it is possible to see all the way through the Body part.

Figure 6.65 - Body part, view redefined

10. Use the Rotate Component tool to rotate the view of the Bumper so that the (unchamfered) bottom face of the bumper is visible. Use the Place Constraint tool to add an Insert constraint between the two parts, as shown in Figure 6.66. An insert constraint has the same effect as a mate (axis - axis) constraint coupled with a mate (face - face) constraint. In both cases five degrees of freedom are removed, with only one rotational DOF remaining.
11. It is necessary to first select the Insert constraint type, then to select the insert surfaces.
12. Use the Place Component tool to add the Packing Nut subassembly to the main assembly.
13. You may notice that the colors applied to the Packing Nut subassembly parts in Tutorial 18 have not carried over. This is because part colors, when changed in an assembly file, only apply to the assembly file where the change was made. They do not change the color in the part file, or, as is the case here, when the assembly file is added as a component to another assembly file. In order to avoid this situation, change the part color directly in the part file. The other alternative, used in the preceding tutorial, is to first go into part editing mode in the assembly file, and then change the color of the part.

    Change the color of the O-rings to rubber (black).
14. Use the Rotate Component tool to re-orient the Packing Nut assembly so that it resembles Figure 6.67.
15. Use the Place Constraint tool to add the Insert constraint shown in Figure 6.68. Note the offset of 0.206 mm.

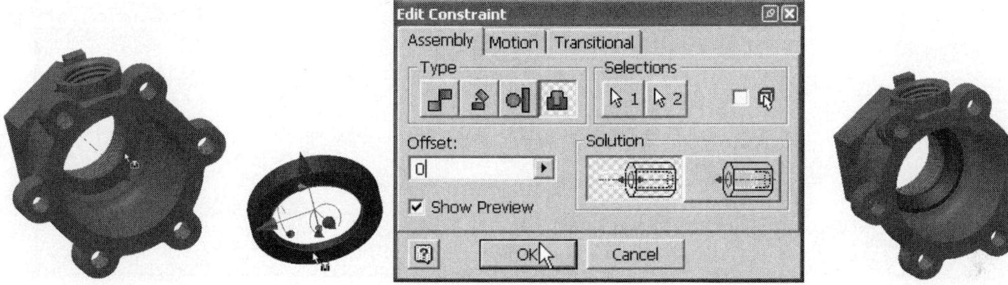

Figure 6.66 - Insert

Figure 6.67 - Rotate view

Figure 6.68 - Insert

16. Use the Place Component tool to add the Stem part to the assembly. Note that because the color of the stem was changed in the part file (see Tutorial 9 in Chapter 4), the beige color carries over in the assembly.
17. Turn off the visibility of both the Body and the Bumper parts.
18. Isometric View.
19. Use the Rotate Component tool to reorient the stem to an orientation similar to that shown in Figure 6.69. This step is necessary to ensure that the Stem is oriented correctly with respect to the Packing Nut subassembly, as shown in Figure 6.70 on the right.
20. Use the Place Constraint tool to add a mate (axis - axis) constraint between the Stem and the Packing Nut, as shown in Figure 6.70.
21. Use the Move Component tool to move the Stem away from the Packing Nut.
22. Zoom in on the parts.
23. Use the Rotate Component tool to rotate the view of the Packing Nut so that the bottom circular edge(s) are visible.
24. Use the Place Constraint tool to add a mate (edge - edge) constraint, as shown in Figure 6.71. Recall that the Packing Nut has both an inside and an outside chamfer. Select an unchamfered edge to be mated. The resulting assembly (visible parts only) should be similar to Figure 6.72. The image on the right is obtained by using Common View to create a principal view of the assembly.

Figure 6.69 - Reoriented stem part

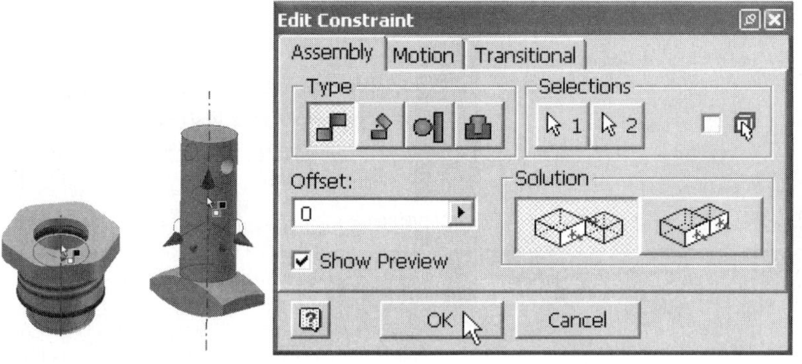

Figure 6.70 - Mate (axis - axis)

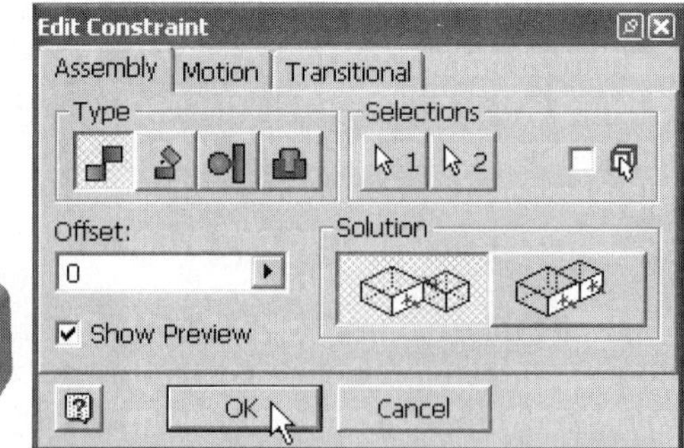

Figure 6.71 - Mate (edge - edge)

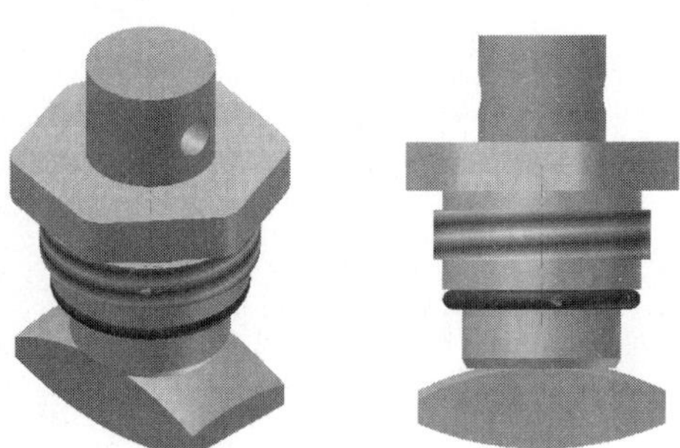

Figure 6.72 - Assemble stem and packing nut (2 views)

25. Turn on the visibility of the Body and Bumper parts.
26. Isometric View.
27. Use the 3D Rotate tool to change the view so that it is possible to see all the way through the Body part.

    NOTE: By clicking on the lower part of the Stem, and then dragging, it is possible to rotate the part about its vertical axis. In fact in the actual ball valve assembly, the stem is a moving part, free to rotate about its vertical axis. In the assembly model we will now temporarily remove this last DOF, in order to simplify the assembly process.

28. Use the Place Constraint tool to add an angle constraint between the two faces shown in Figure 6.73. Note the angle between the faces is 90 degrees.
29. At the top of the assembly browser, click on Position View [Position View], and then select Modeling View.
30. After expanding the Constraints folder, left-click on the text of the Angle constraint. Left-click a second time, and then rename the angle constraint OpenClose, as shown in Figure 6.74.
31. Turn off the visibility of the Body and Bumper parts and the Packing Nut component.
32. Isometric View.
33. Use the Place Component tool to add the Ball part to the assembly.
34. In the browser, locate the Ball part and expand it. Next expand its Origin folder, and turn on the visibility of the Z Axis.
35. Use the Place Constraint tool to add a mate (axis - axis) constraint, as shown in Figure 6.75.
36. Use the Move Component tool to move the Ball away from the Stem. Note that instead of using the Move Component tool, it is possible to simply click on the part, and then drag it.
37. Use the Place Constraint tool to add an angle constraint (0 degrees) between the two faces shown in Figure 6.76.

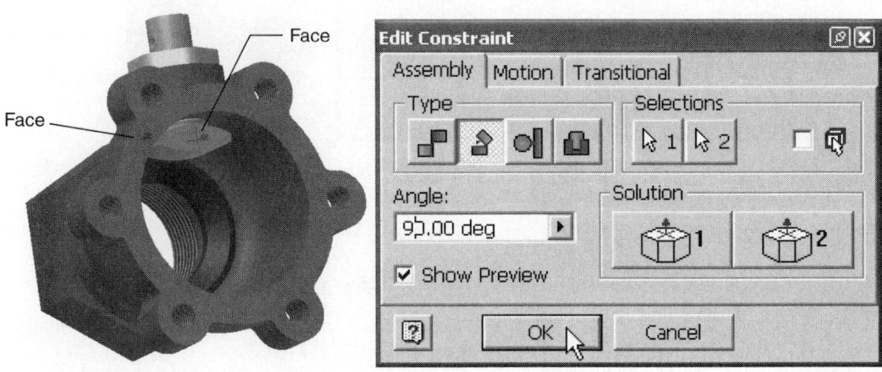

Figure 6.73 - Angle (face - face), 90 degrees

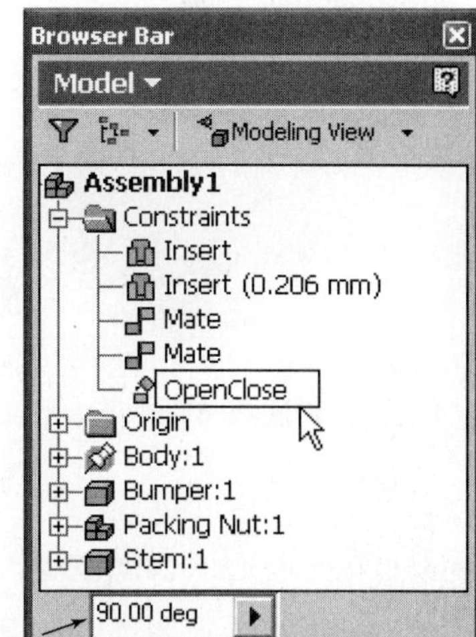

Figure 6.74 - Rename angle constraint

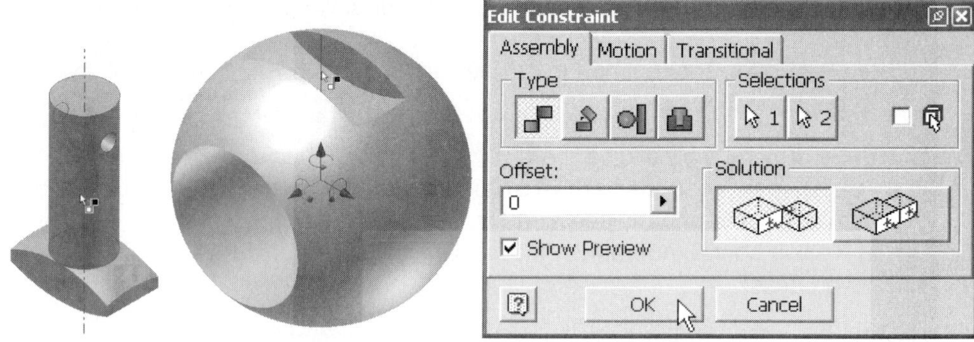

Figure 6.75 - Mate (axis - axis)

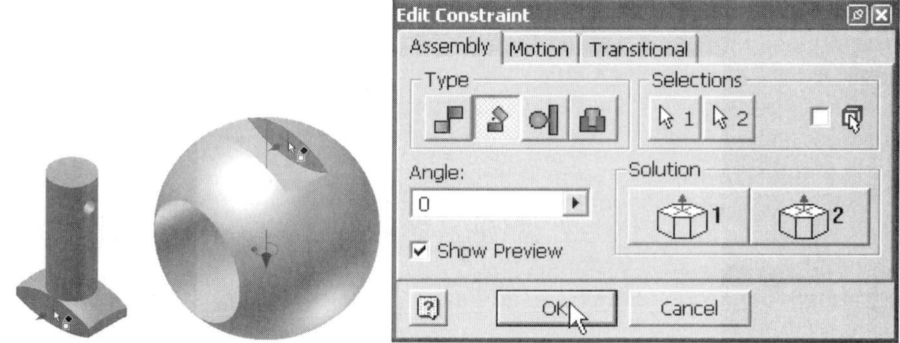

Figure 6.76 - Angle (face - face), 0 degrees

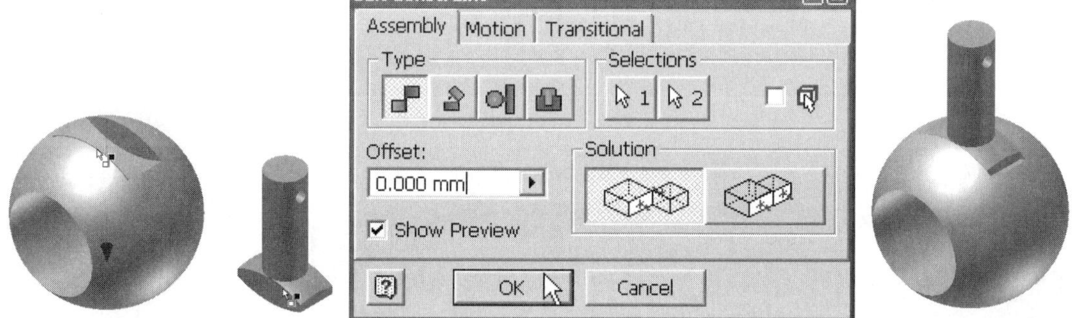

Figure 6.77 - Mate (edge - edge)

38. Use the Place Constraint tool to add a mate (edge - edge) constraint between the two edges shown in Figure 6.77.

39. Turn off the visibility of the Ball part Z axis.

40. In the assembly browser, locate the OpenClose constraint. Right-click on the constraint (icon or text) and select Suppress. In the graphics area it should now be possible to click and drag on either part, and see both parts rotate in unison about a vertical axis.

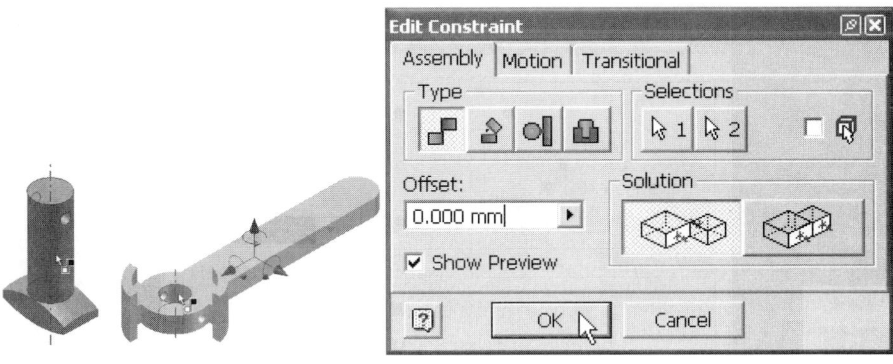

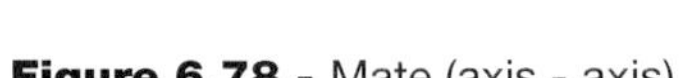
Figure 6.78 - Mate (axis - axis)

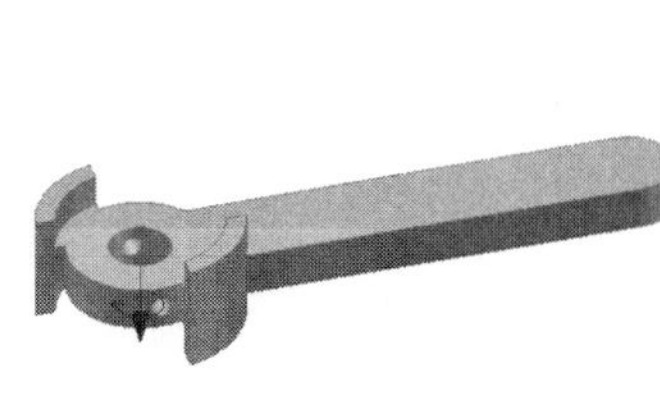
Figure 6.79 - Reorient view

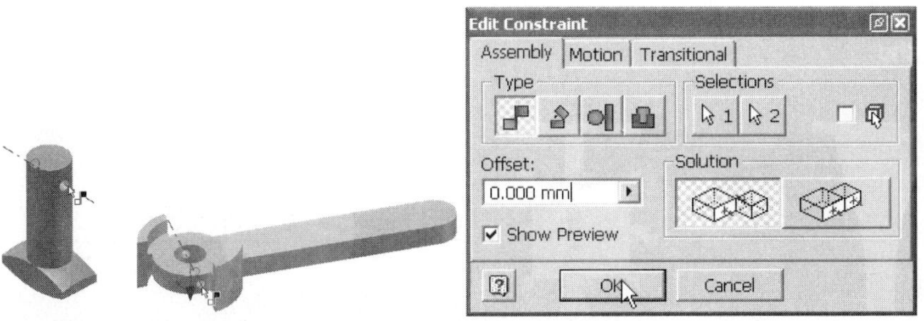

Figure 6.80 - Mate (axis - axis)

Figure 6.81 - Assembly thus far

41. Turn off the visibility of the Ball component.
42. Use the Place Component tool to add the Handle to the assembly.
43. Use the Place Constraint tool to add a mate (axis - axis) constraint between the stem and the handle (Figure 6.78).
44. Move and rotate the handle by clicking and dragging. The handle should be oriented as shown in Figure 6.79.
45. Use the Place Constraint tool to add another mate (axis - axis) constraint between the stem and the handle, this time between the horizontally oriented holes (Figure 6.80).
46. Turn on the visibility of the Body, Bumper, Packing Nut, and Ball components. The assembly model should now resemble Figure 6.81. It should be possible to drag the handle and see the ball rotate. The green cube (visible when the DOF is turned on) indicates that one part moves another part by way of a third part.
47. Use the Place Component tool to add the Pin part to the assembly file.
48. Zoom in on the Handle, Stem and Pin.
49. Rotate the Handle clockwise about 60 degrees.
50. Use the Place Constraint tool to add the mate (axis - axis) constraint, as shown in Figure 6.82.
51. Move the Pin out away from the assembly.
52. Use the Place Constraint tool to add a tangent constraint as shown in Figure 6.83. Note that it will probably be necessary to change the Solution to Inside, and that the offset distance is 1. The Pin should now be positioned as shown in Figure 6.84.

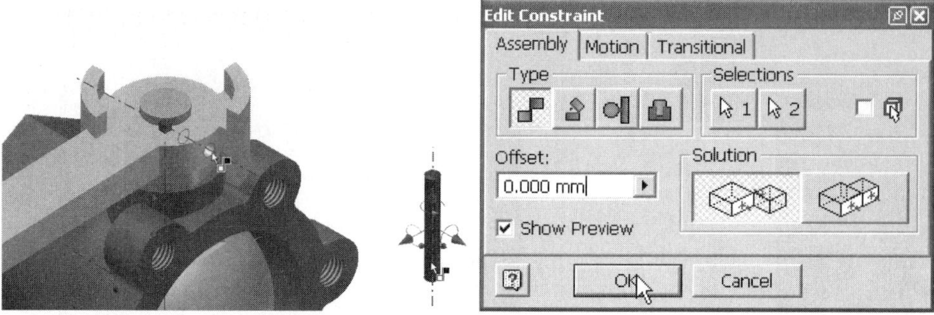

Figure 6.82 - Mate (axis - axis)

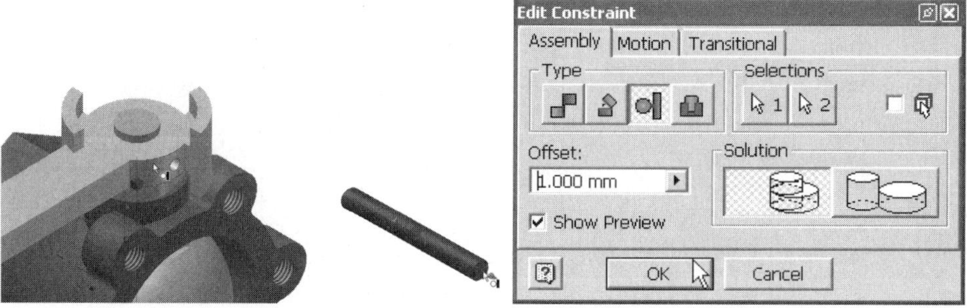

Figure 6.83 - Tangent (surface - face), offset 1

Figure 6.84 - Zoomed in view of pin

Figure 6.85 - Mate (face - face)

53. Isometric View.
54. Use the Place Component tool to add the Body Cap subassembly to the assembly.
55. Use the Place Constraint tool to add a mate (face - face) constraint, as shown in Figure 6.85.
56. Move the Body Cap subassembly away from the main assembly.
57. Use the Place Constraint tool to add a mate (axis - axis) constraint between the Body and the Body Cap, as shown in Figure 6.86. If necessary, turn off the visibility of the Ball part, to ensure that its axis is not selected by mistake. The

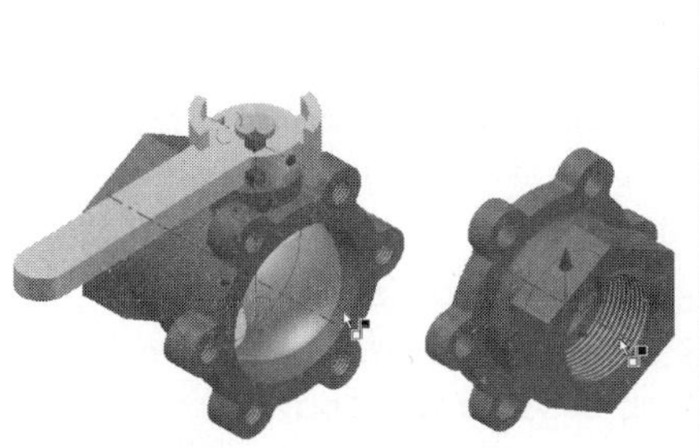

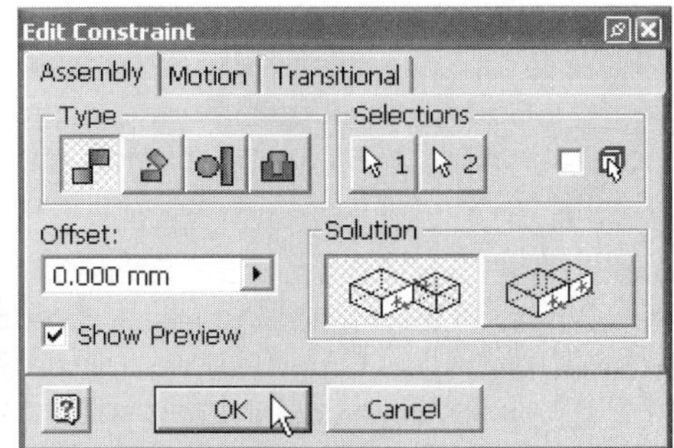

Figure 6.86 - Mate (axis - axis)

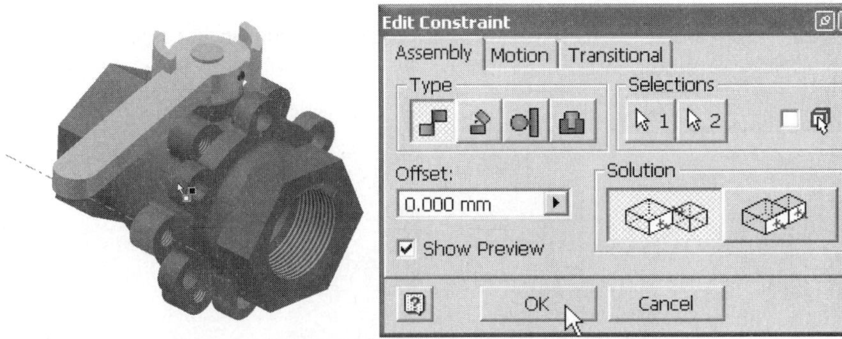

Figure 6.87 - Mate (axis - axis)

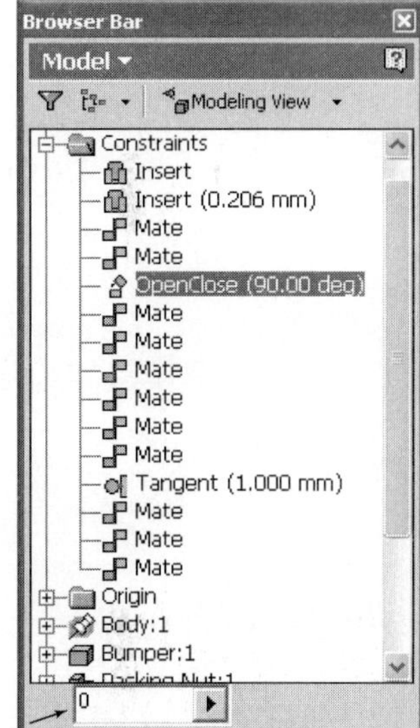

Figure 6.88 - Edit offset

Body Cap subassembly is still free to rotate about the valve axis. Verify this by clicking and dragging the Ball Cap subassembly. Rotate it such that the six bolt holes are not aligned.

58. Turn off the DOF icons (View > Degrees of Freedom).
59. Use the Place Constraint tool to add another mate (axis - axis) constraint, as shown in Figure 6.87.
60. Isometric View.
61. Unsuppress the OpenClose angle constraint (find the constraint in the browser, right-click, highlight and select Suppress to remove the check mark).
62. Change the offset value of the OpenClose angle constraint to 0 degrees, as shown in Figure 6.88.
63. Expand the Body Cap subassembly component in the browser. Double-click the Seat part to edit, and assign it another color (e.g., Yellow (Flat)). Right-click and select Finish Edit to return to assembly mode. It may be necessary to do this more than once.
64. From the browser, right-click on the Body part, and de-select Enabled. Repeat this process for the Body Cap part (only the part, not the entire subassembly).
65. Using the Section View tool, select a Half Section View from the flyout. Now expand the Origin folder from the browser and select the YZ Plane. Right-click in the screen area, and then select Done. The screen should resemble Figure 6.89. Note that if the assembly is not symmetrical about the YZ Plane in the assembly folder, you can use a default reference plane for a part (e.g., Ball) instead. With the assembly shown in half section, interferences between parts will be more easily recognized.

Figure 6.89 - Half section view with body and body cap parts disabled

NOTE: Using the Section View tool can cause significant slowdowns. One possibility is to turn off the visibility of the less relevant parts (i.e., everything except Ball and the Body Cap subassembly). Another alternative is not to use the Section View tool at all. Try using wireframe display, turning the visibility of parts on/off, and changing part colors. Note that some of the available colors are transparent.

66. Enable both the Body and the Body Cap parts.
67. Use Common View to look at the section. See Figure 6.90. All of the parts appear to be correctly positioned, except for the Seat and Spring (Body Cap subassembly). This is because the Spring is still uncompressed. We will now correct for this.
68. Select the Section View tool in the panel bar to expand the flyout, and then choose End Section View.
69. In the browser, double-click on the Spring part. We are now in part mode within the assembly. We can now edit the Spring.
70. Use the Parameters tool from the panel bar to change the Spring height from 7 to 3, as shown in Figure 6.91.

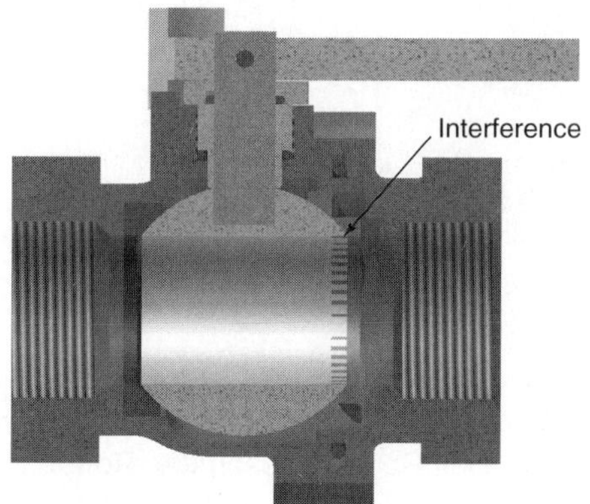

Figure 6.90 - Half section view with uncompressed spring

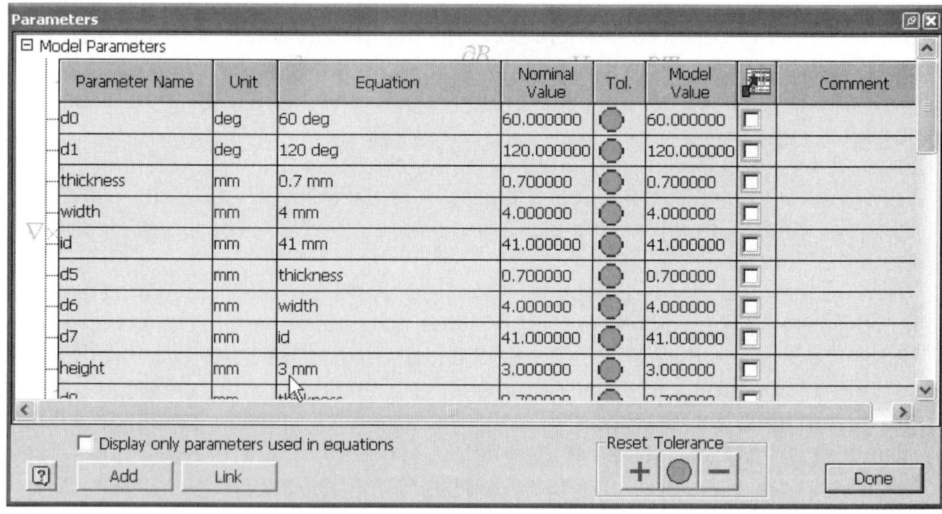

Figure 6.91 - Edit spring height parameter

71. Click the Update icon on the standard toolbar in order to have the Spring height change take effect in the assembly.
72. To return to assembly mode either double-click on the Assembly icon at the top of the browser, or right-click in the graphics area and select Finish Edit.
73. It is still necessary to re-position the Seat to account for the Spring compression. Double-click on the Body Cap subassembly component (not the part) icon in the browser to activate the subassembly. Locate a mate constraint with a 4 mm offset (between the Seat and Body Cap parts). Upon selecting it, an offset edit box appears. Change the offset to 0, as shown in Figure 6.92.
74. Right-click in the graphics area, and select Finish Edit.
75. Using the Section View tool, compose a half section view again. The screen should resemble Figure 6.93. The Spring and Seat parts are now correctly positioned.
76. Use the Section View tool to change to an End Section View.
77. Isometric View.
78. We will now use the Autodesk Inventor Content Library to add a hex head bolt to the Ball Valve assembly. In the title browser bar, click the arrow next to Model, and then select Library, as shown in Figure 6.94.
79. The content library browser opens. Double-click on the Standard Parts icon, as shown in Figure 6.95.
80. Move through the library tree structure, double-clicking on each folder in succession:
    - ISO
      - Screws and Threaded Bolts
        - Hex Head Types
          - ISO 4017 (Regular Thread)
81. On the ISO 4017 (regular thread) part page, change the nominal diameter to 8, and the nominal length to 25, as shown in Figure 6.96.
82. Upon moving the cursor over the image bit map associated with the part, the cursor changes to an empty eyedropper. Click and hold down the left mouse

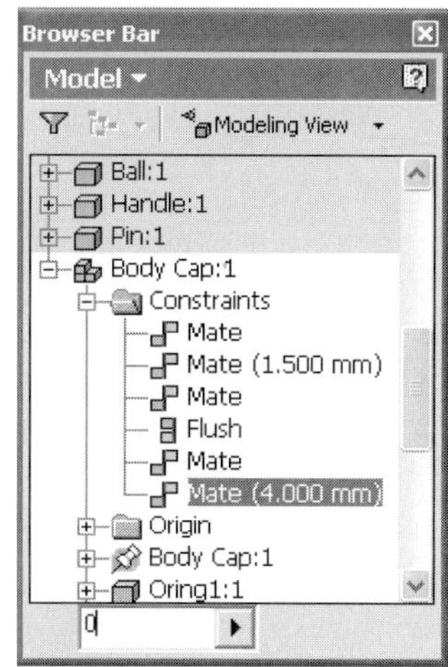

Figure 6.92 - Edit offset parameter

Figure 6.93 - Half section with compressed spring

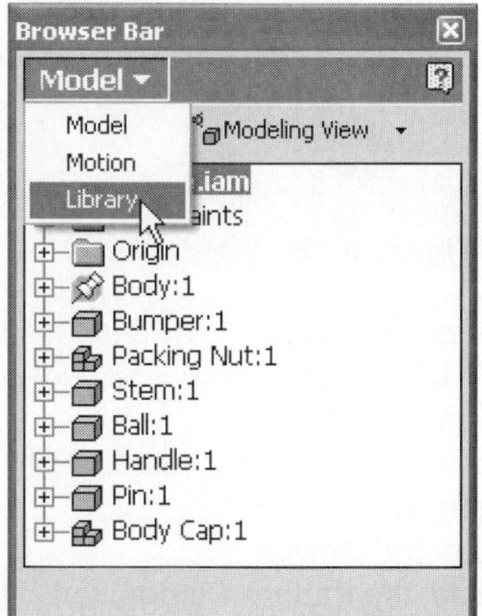

Figure 6.94 - Library browser selection

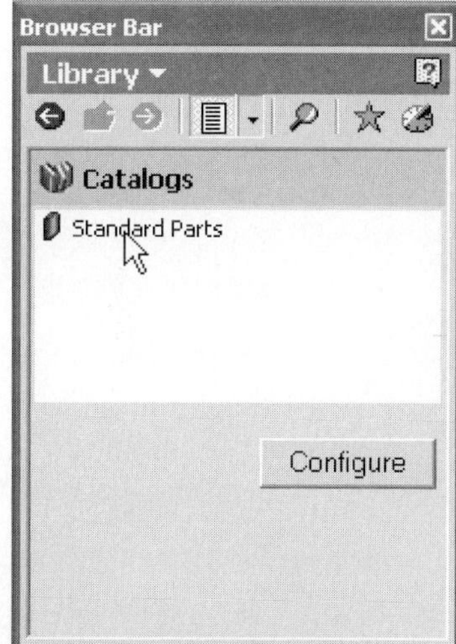

Figure 6.95 - Content library browser

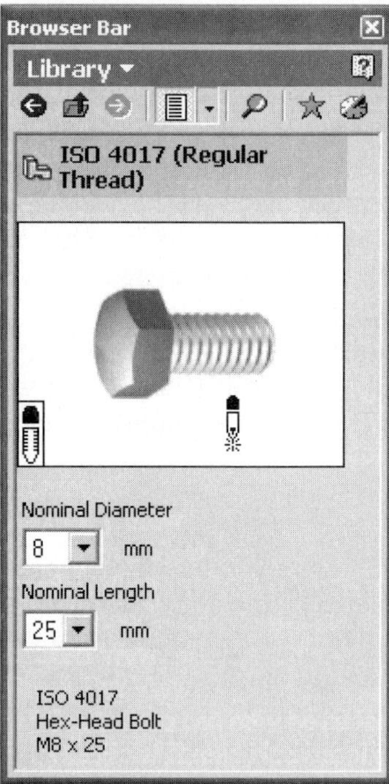

Figure 6.96 - ISO 4017 part page with parameters set

Figure 6.97 - Standard hex head bolt added to assembly

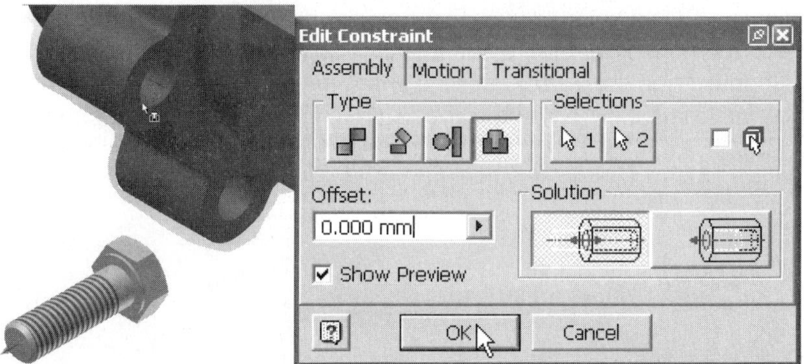

Figure 6.98 - Insert

button. The cursor now changes to a full eyedropper. Now drag the mouse into the graphics area, and release (Figure 6.97).

83. Zoom in.

84. Use the Place Constraint tool to add an insert constraint, as shown in Figure 6.98.

85. Isometric View.

86. Select the Pattern Component tool from the panel bar. Click on the Circular tab, and select the bolt. Select the Axis Direction arrow, and then select a feature that highlights the correct rotational axis, as shown in Figure 6.99. Now enter the parameters in the dialog box as shown, and click OK.

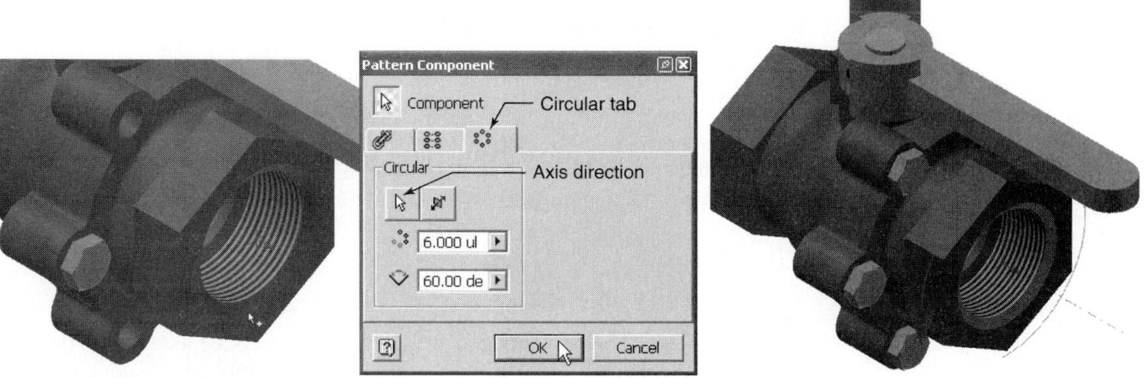

Figure 6.99 - Circular pattern operation

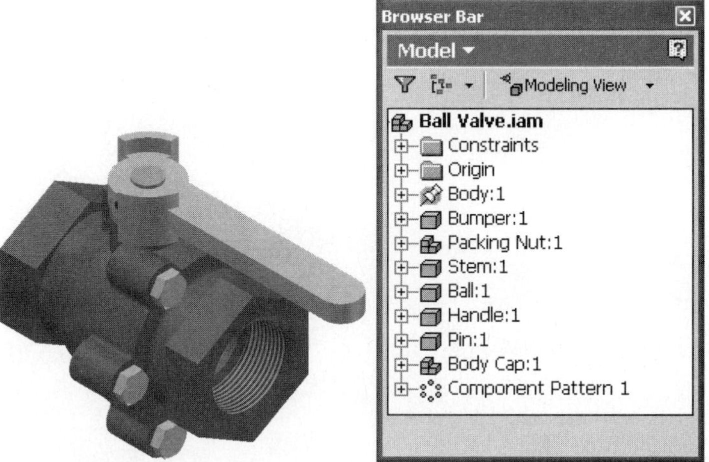

Figure 6.100 - Completed ball valve assembly with part tree

87. Change back to the assembly browser (in the title area, click the arrow next to Library, and then select Model. The completed assembly and component hierarchy are shown in Figure 6.100.

88. Save the assembly file in the Ball Valve workspace folder.

QUESTIONS

1. T F Unlike the part environment, the assembly environment does not provide a hierarchical tree structure.
2. Which of the following is not a method for adding a component to an assembly file?
   - **a.** Place component
   - **b.** Create component
   - **c.** Mirror component
   - **d.** Content library
   - **e.** Pattern component

3. T F Only one assembly constraint can be applied to a part in the assembly environment.
4. Give the number of degrees of freedom for each:
   - **a.**
   - **b.**
   - **c.**
5. Match the assembly constraint icons with the corresponding assembly constraint names.
   - Insert
   - Angle
   - Mate
   - Tangent
6. T F Once the components have been inserted into the assembly environment and the constraints have been applied, it is not possible to manipulate the components.
7. T F It is not possible to edit parts directly from the assembly file.

PROBLEM

1. Use the parts in the Butterfly Valve Parts folder on the CD to create the assembly shown in the figure below.

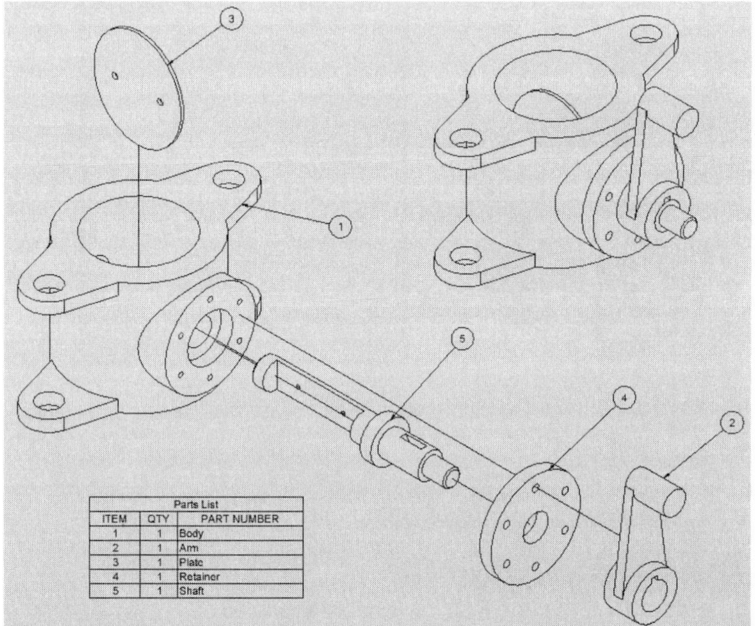

Parts List		
ITEM	QTY	PART NUMBER
1	1	Body
2	1	Arm
3	1	Plate
4	1	Retainer
5	1	Shaft

CHAPTER 7

Part Documentation

LEARNING OBJECTIVES

- Select the sheet size, orientation, and placement of title block and border in the drawing environment
- Describe the arrangement of the principal drawing views when using third angle projection
- Use the Dimension Styles dialog box to modify the dimension style of a drawing
- Use the Text Styles dialog box to modify the text style of a drawing
- Use the Base View tool to add a base view to a drawing
- Use the Project View tool to project views from another view
- Use the Auxiliary View tool to create an auxiliary view showing the true size and shape of an inclined surface
- Use the Section View tool to create full, half, and offset section views
- Use the Detail View tool to create detail views from an existing view
- Use the Break Out View tool to display both the internal and external detail of an existing view
- Use the Edit View dialog box to modify display labels, view scale, and view style
- Use the Automated Centerlines tool to add centerlines, center marks, and centered patterns
- Use the Center Line, Center Mark, Center Line Bisector, and Centered Pattern commands to manually add centerlines
- Use the Hole/Thread Notes tool to display hole and thread notes
- Use the Get Model Dimensions command to display the parametric dimensions of a part in a drawing
- Use the General Dimension tool to add nonparametric reference dimensions to geometry
- Use the Text tool to add text or symbols to drawing files
- Modify the hatch style in a section view
- Use the Properties dialog box to add entities to the title block
- Use the Custom View window to change the view orientation of an object
- Edit part dimensions in the drawing environment

Introduction

One of the common features of parametric solid modelers is the ability to derive traditional two-dimensional engineering drawings from part and assembly models. Like part and assembly files, 2D drawing data in Autodesk Inventor is maintained in a separate file format. Inventor drawing files have an IDW file extension.

Inventor part and drawing files are fully *associative*. This means that changes made in a part file are reflected in the related drawing file. Further, if parametric dimensions are modified in a drawing file, the part file will also change.

In this chapter we will start by looking at the steps required to set up a drawing in Inventor. We will then look at the tools available for adding (as well as editing) views of a part to a drawing. Next we look at drawing annotations, specifically how to add centerlines, as well as hole and thread notes. Dimensioning is then discussed. The chapter concludes with tutorials describing the creation of a few part drawings.

Drawing Setup

Drawing setup is fairly straightforward in Inventor. The principal setup tasks include the selection of the sheet (or paper) size, as well as its orientation, the placement of a title block, and the configuration and placement of a border. In addition, drafting standards may need to be modified, as well as dimension and text styles. It is also possible to create custom title blocks, borders, drafting standards, and dimension and text styles.

Sheets, title blocks and borders are all specified from the browser bar in an Inventor drawing file. When a new drawing file (i.e., IDW file extension) is started, the browser bar and the corresponding drawing sheet will be similar to that shown in Figure 7.1. Note that by default a border (i.e., Default Border) and title block (i.e., ANSI—Large) are attached to the drawing sheet (i.e., Sheet:1). Either of these can be deleted by right-clicking on the icon or associated text in the browser, and selecting Delete from the context menu.

To specify the sheet size and orientation, it is probably easier to edit the default sheet, rather than to create a new one. To do this, right-click on the sheet icon in the browser, and select Edit Sheet . . . from the context menu, as shown in Figure 7.2.

Figure 7.1 - Drawing file browser and graphics window at start-up

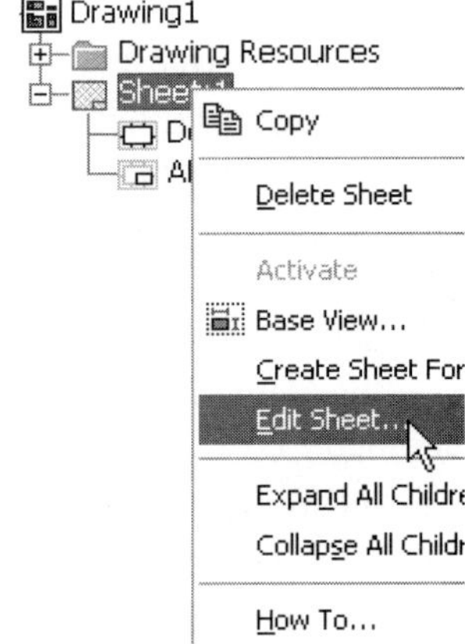

Figure 7.2 - Edit sheet access

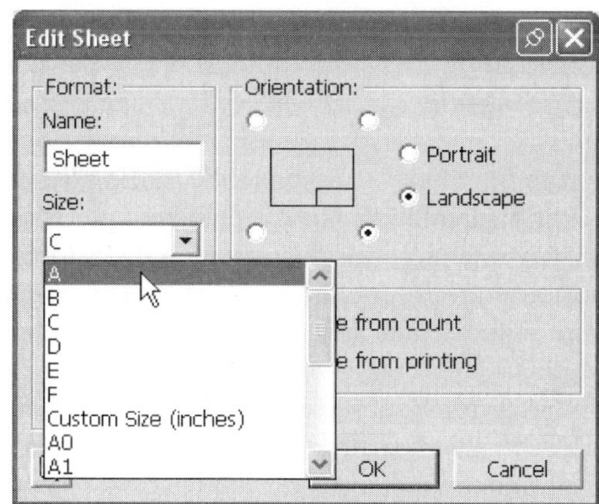

Figure 7.3 - Edit sheet dialog box

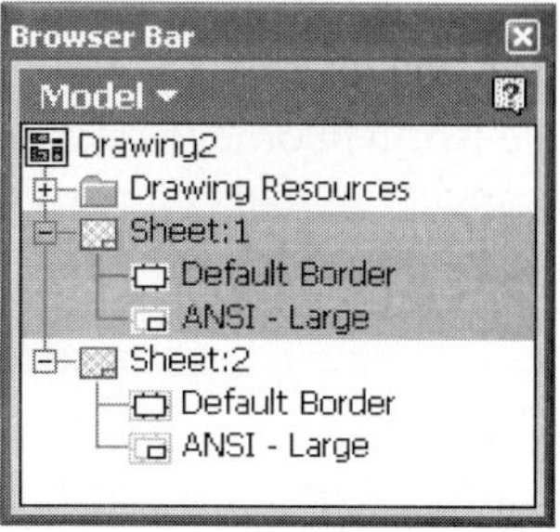

Figure 7.4 - Adding a new sheet to a drawing

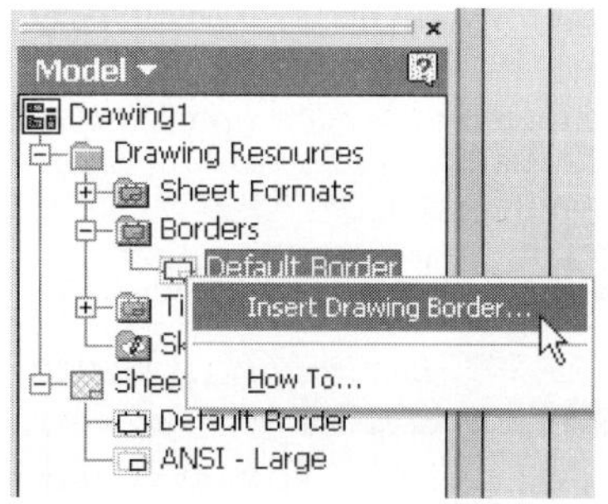

Figure 7.5 - Insert drawing border access

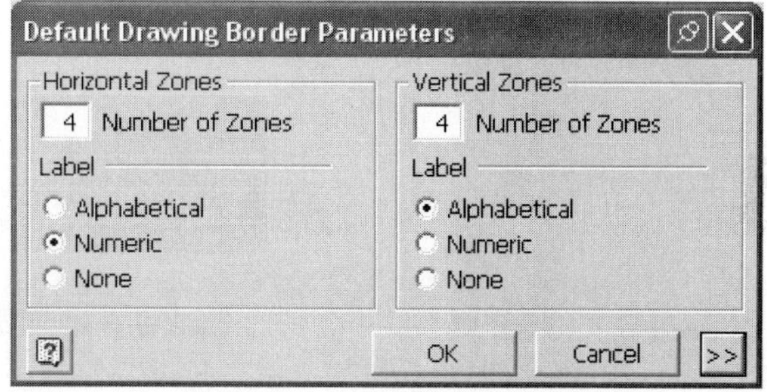

Figure 7.6 - Border parameters dialog box

The Edit Sheet dialog box opens, as shown in Figure 7.3. Here a number of standard metric and English sheet sizes can be selected from, as well as a custom size and standard sheet sizes with predetermined view configurations. The tutorials in this book use A size (8½ × 11) sheets. Note that the paper orientation (i.e., Portrait or Landscape) is also specified here.

Alternatively a new sheet can be created from the menu bar, by selecting Insert > Sheet . . . This opens the New Sheet dialog box, which is identical to the Edit Sheet dialog box shown in Figure 7.3. A title block and border must be added manually to a sheet created in this way.

More than one sheet can be included in a drawing file. Right-clicking in the browser and then selecting New Sheet can be used to add new sheets. A new sheet created in this way will inherit the border and title block of an existing sheet. In Figure 7.4, a new sheet has been added in this way. The new sheet (i.e., Sheet:2) is currently active. Double-clicking on the Sheet:1 icon will activate it.

The Drawing Resources folder is included with all drawing files. It provides access to borders and title blocks. To add a border to a sheet, expand both the Drawing Resources and the Borders folder. Right-click on the Default Border icon, and select Insert Drawing Border . . . from the context menu (see Figure 7.5).

This opens the Default Drawing Border Parameters dialog box, seen in Figure 7.6. Note that in addition to specifying zones (that serve to partition the drawing sheet), there is a More >> button, where additional border parameters can be specified.

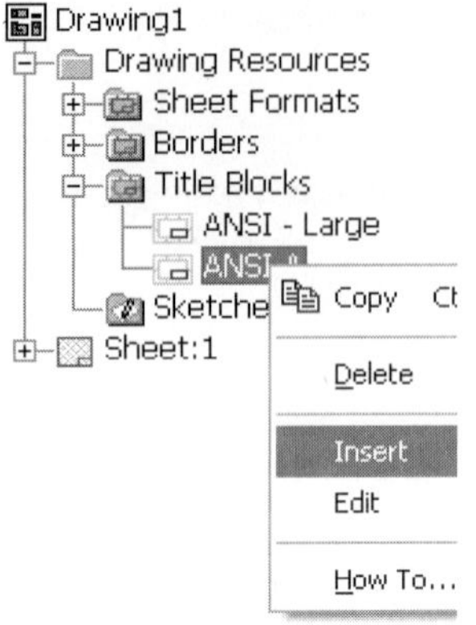

Figure 7.7 - Insert or edit title block access

To add a title block to a sheet, expand both the Drawing Resources and the Title Blocks folders in the browser. Right-clicking on a title block icon opens the context menu seen in Figure 7.7. Choose Insert to add the title block to the current sheet, or Edit to modify an existing title block.

Choosing Format > Standards . . . from the menu bar opens the Drafting Standards dialog box, shown in Figure 7.8. Many drafting standards have been developed throughout the world in order to standardize the way in which engineering drawing information is presented. In this country the two most commonly used drafting standards are ANSI (from the American National Standards Institute), and ISO (from the International Organization for Standardization).

In addition to specifying standards governing such drawing elements as dimensions, tolerances, linetype usage, surface finish, etc., drafting standards also specify how different drawing views are to be arranged with respect to one another. There are two different techniques for determining this arrangement: *first angle projection* and *third angle projection*. Third angle projection is typically employed in the United States. This projection technique results in a top view of an object being positioned above (and vertically in line with) a front view, and a right view being positioned to the right of (and horizontally in line with) a front view. For more information on first and third angle projection, see the PowerPoint slides maintained on the accompanying website.

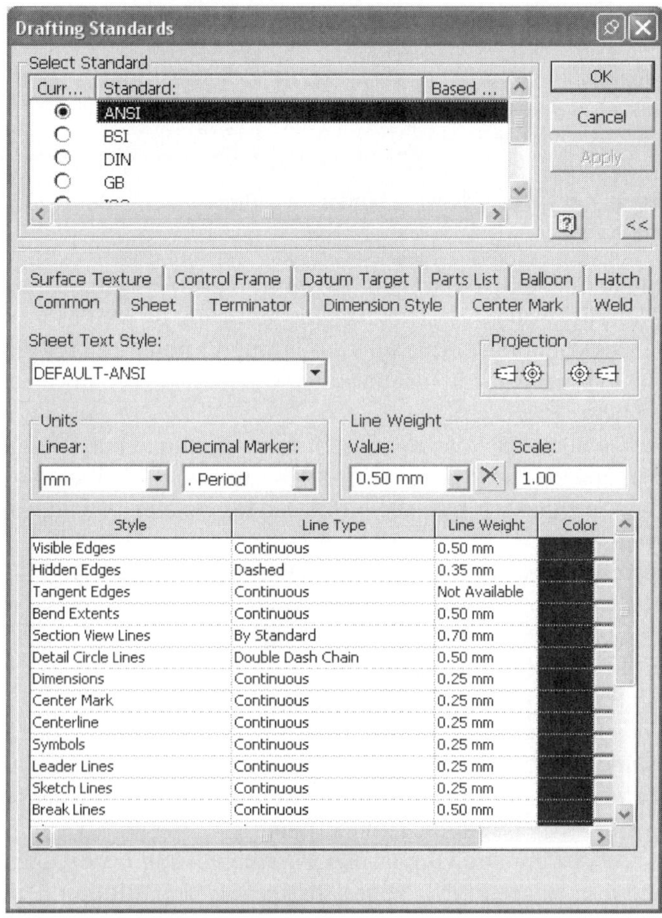

Figure 7.8 - Drafting standards dialog box

Whereas the ANSI standard employs *third angle projection* as the default, the ISO standard uses *first angle projection* as the default. The tutorials in this book use the ANSI standard. Shown in Figure 7.9 is a section of the Common tab on the drafting standards dialog box where the type of projection is specified. Note that in addition to providing access to several different drafting standards, a new standard can also be created here, based on an existing standard.

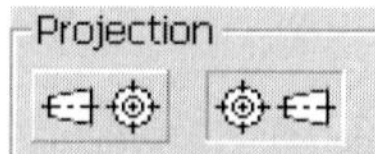

Figure 7.9 - First and third angle projection specification

The dimension style of a drawing can be modified by selecting Format > Dimension Styles . . . from the menu bar. This opens the Dimension Styles dialog box, shown in Figure 7.10. Finally, drawing text styles can be modified by selecting Format > Text Styles . . . from the menu bar. This opens the Text styles dialog box, shown in Figure 7.11.

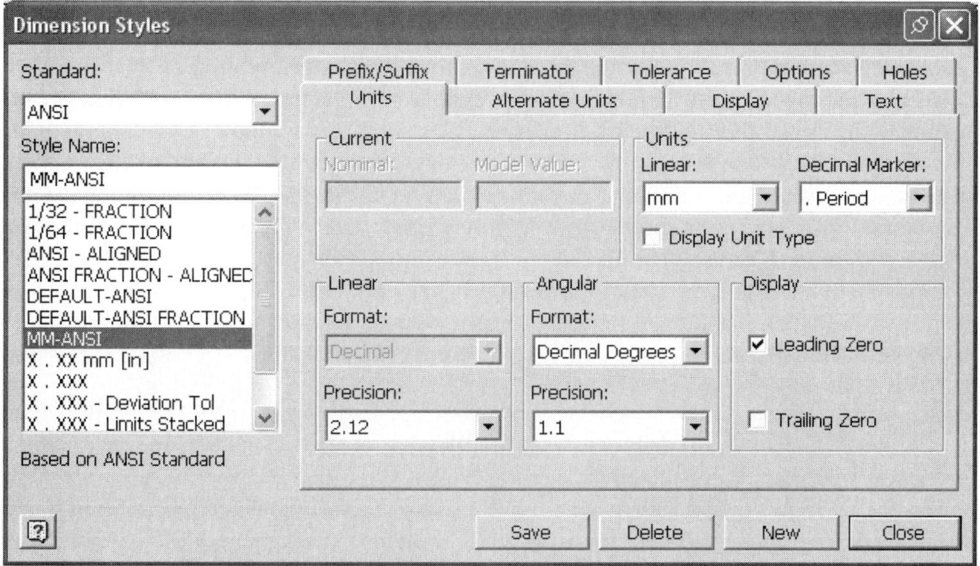

Figure 7.10 - Dimension styles dialog box

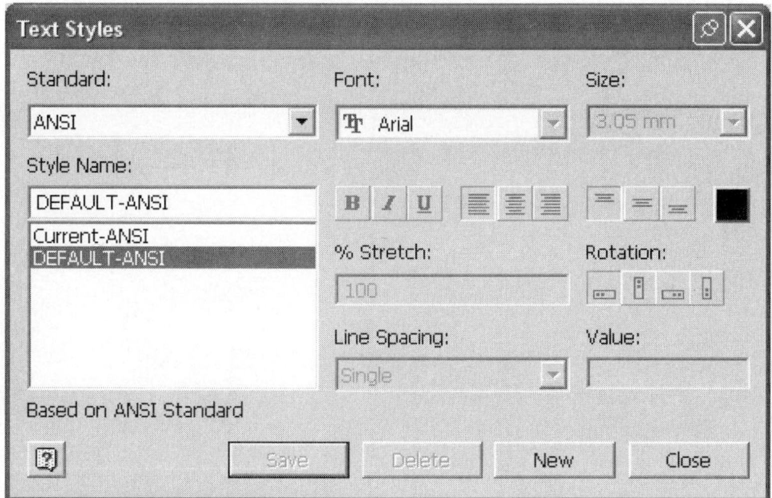

Figure 7.11 - Text styles dialog box

Drawing Views

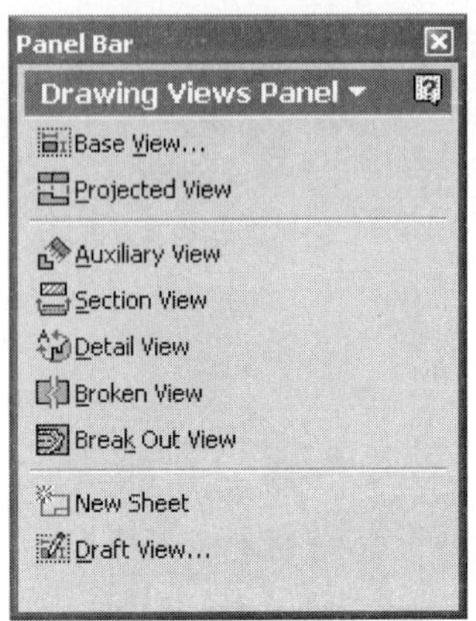

Figure 7.12 - Drawing views tools

Views of Inventor part, assembly and presentation files are added to a drawing by using the Drawing Views panel from the panel bar. These tools are shown in Figure 7.12.

Base View

Base View . . . is used to add an initial, base view of a part to a drawing. Although a drawing can have more than one base view, it is generally better to add a single base view to a sheet, and then to project other views from the base view. In this way proper alignment between adjacent and related views is maintained.

Upon selecting the Base View tool, you will see the Drawing View dialog box open (Figure 7.13).

In the File section on the Component tab, any open files will be listed in the drop-down list box. Otherwise click on the Explore button to browse for a file. Once a file has been selected, use the Orientation section to select the orientation of the base view. Either select from one of the standard principal or isometric views, or click on the Change view orientation icon to open the Custom View window. An example using the Custom View window is provided in one of the tutorials at the end of the chapter. The scale between the view and the model can be set in the Scale section, using the drop-down list box. The type of view display is specified in the Style area. Choices include with hidden lines, with hidden lines removed and shaded. Other view options are available from the Options tab.

Project View

Once a base view has been placed on a drawing sheet, other views can be projected from this view using the Projected View tool. In Figure 7.14, the base view is in the center. All of the other views were created at the same time using the Projected View tool. To use the tool, first select the view to be projected from. As the cursor is moved, a profile image of the view to be projected is visible. Left-clicking establishes that a view is to be projected. Once all of the projections have been selected, right-click and select Create from the context menu.

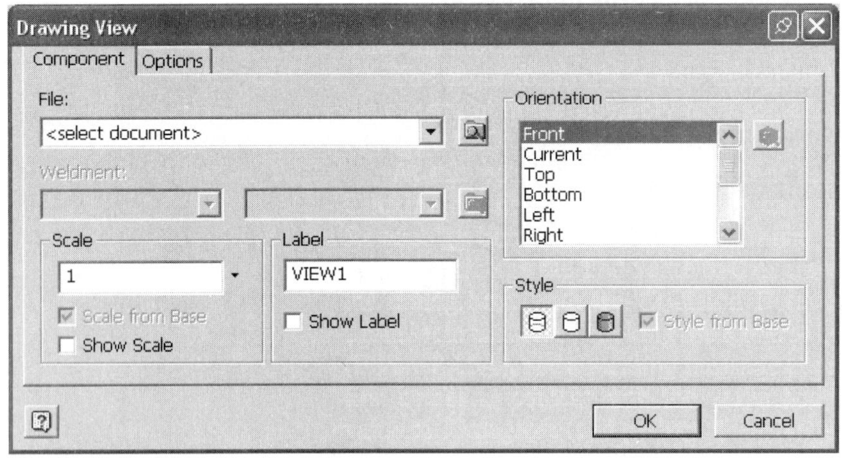

Figure 7.13 - Drawing view dialog box

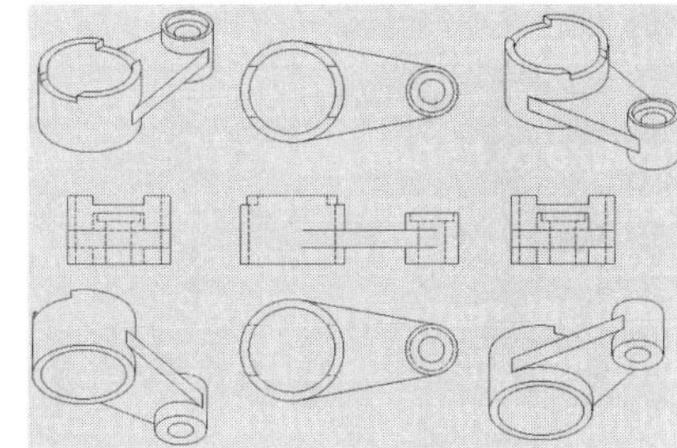

Figure 7.14 - Projected views

Figure 7.15 - Part with an inclined surface

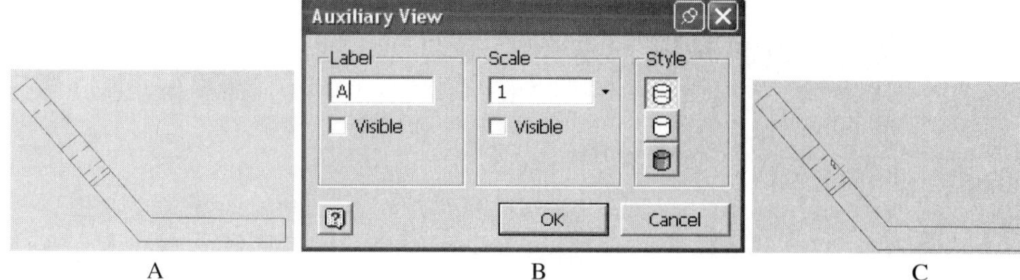

Figure 7.16 - Auxiliary view creation

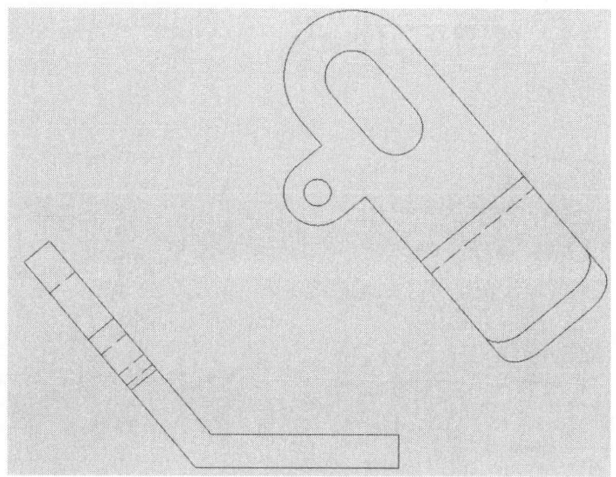

Figure 7.17 - Auxiliary view

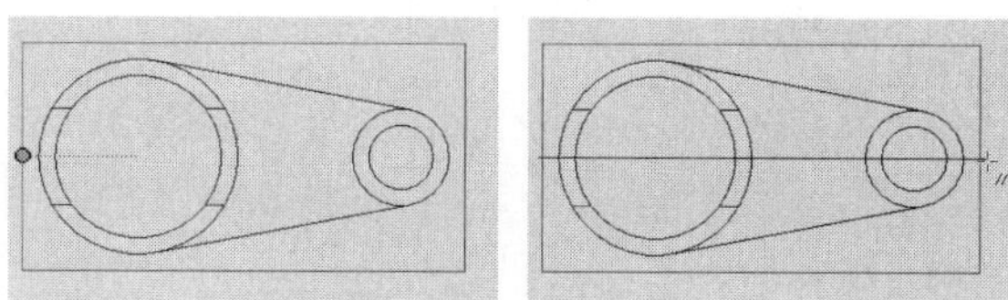

Figure 7.18 - Cutting plane path definition

Auxiliary View

Auxiliary views are created with the Auxiliary View tool. Auxiliary views are typically used to show the true size and shape of an inclined surface. Figure 7.15 shows a part with an inclined surface. After selecting the Auxiliary View tool, select an existing drawing view where the inclined surface is seen on edge (Figure 7.16A). The Auxiliary View dialog box opens (Figure 7.16B). Now select the inclined surface (Figure 7.16C). Move the cursor to position the auxiliary view. Left-click when ready and the auxiliary view is created (Figure 7.17).

Section View

Full, half, and offset section views can all be created with the Section View tool. To create a section view, first select the view where the cutting plane will appear on edge, and then define the path of the cutting plane in the selected view, as shown in Figure 7.18. Handles are available as the cutting plane is being defined, ensuring that the path will lie along lines of symmetry. Be sure that the path extends beyond the part geometry.

Once the path is defined, right-click and select Continue. The Section View dialog box opens (Figure 7.19A), and the cutting plane is visible in the graphics area, as is a

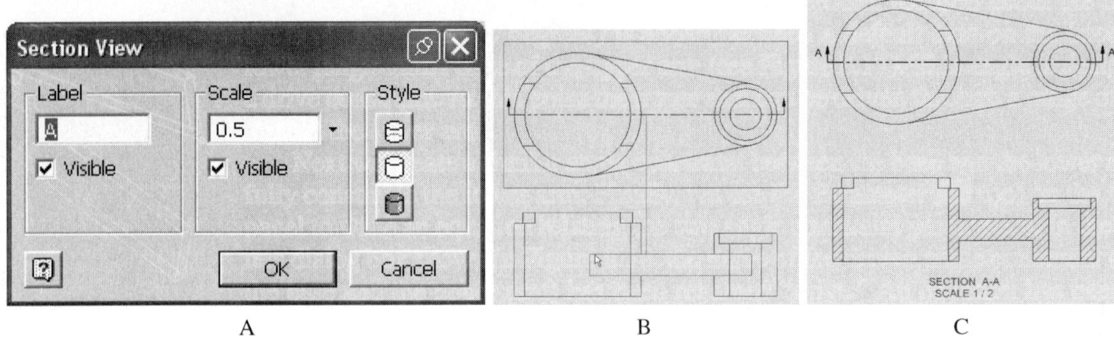

A B C

Figure 7.19 - Section view creation

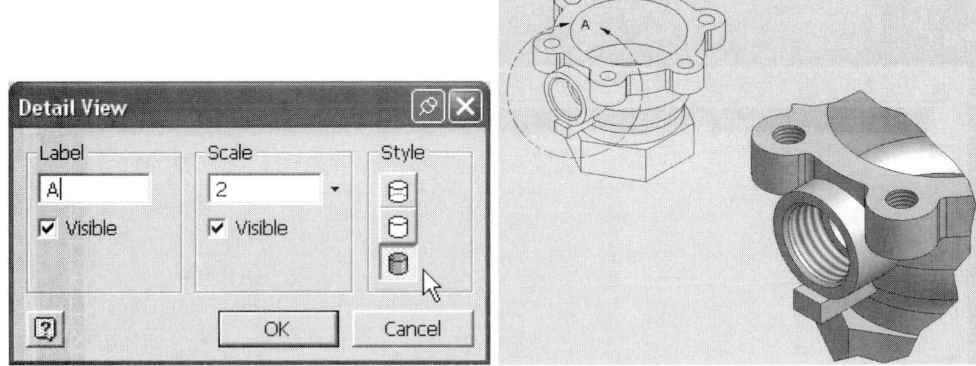

Figure 7.20 - Detail view

profile image of the section view (Figure 7.19B). If necessary, make changes in the dialog box. The placement of the section is determined by left-clicking in the graphics area. Figure 7.19C shows the resulting section view.

Detail View

In certain situations it is useful to create a detail view from an existing view at an enlarged scale. Use the Detail View tool to create detail views. An example of a detail view is shown in Figure 7.20. After selecting the Detail View tool, select the view that the detailed view will be derived from. The Detail View dialog box opens, as shown in Figure 7.20. Note that the style of the view has been changed to Shaded. Left-click to specify the center of the detail view. A circle appears. Everything inside the circle will appear in the detail view. When satisfied with the area of the detail view, left-click. Now move the cursor to an empty drawing area. Left-click to place the detail view.

Break Out View

A Break Out View tool has been added in Release 6 of Inventor. Break out views permit the display of both internal and external detail in the same view, as shown in Figure 7.21. In order to create a break out view, it is necessary to create a closed profile sketch that is associated with an existing view. Figure 7.22 shows the sketch used to

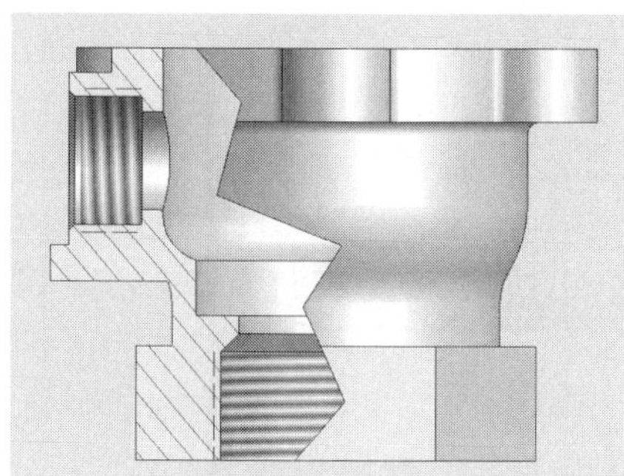

Figure 7.21 - Break out view

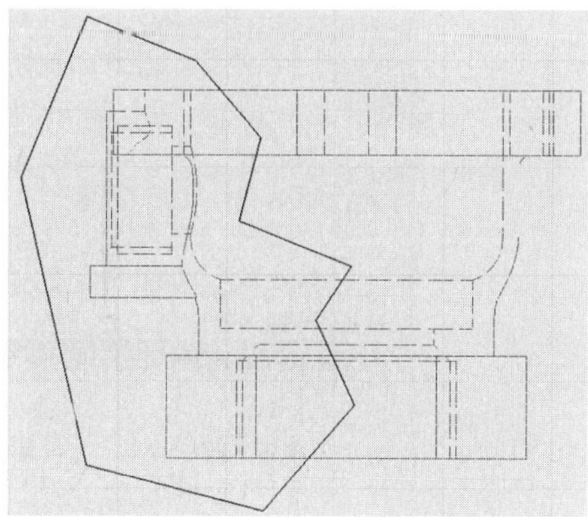

Figure 7.22 - Break out view sketch

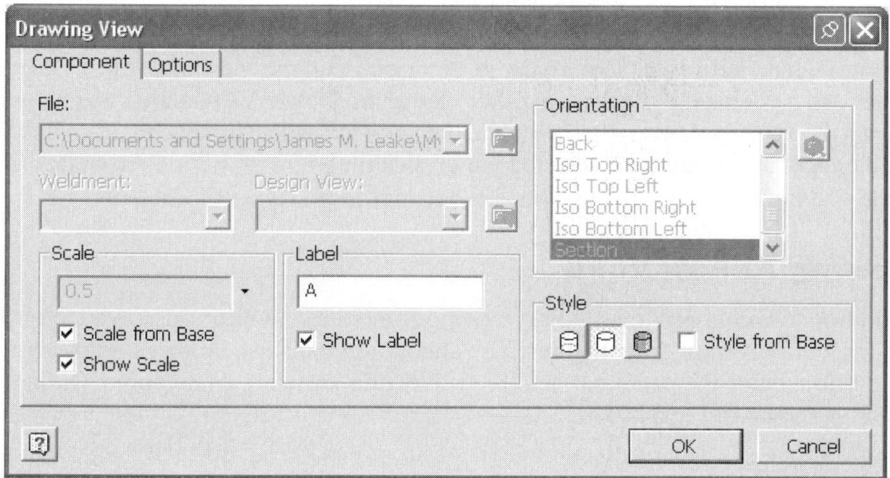

Figure 7.23 - View editing

create the view shown in Figure 7.21. This sketch must be associated with the view. Tutorial 22 provides an example where a break out view is created.

View Editing

Views can be edited by right-clicking on the view, and then selecting Edit View . . . from the context menu. The Edit View dialog box appears, as shown in Figure 7.23. Note that among other things, it is possible to display view labels, change the view scale, and change the view style from this dialog box.

Views can be moved by simply clicking and dragging. If a base view is moved, and other principal views were projected from this parent view, then all the views move so that alignment is maintained. Conversely, a projected view can only be moved in the direction of alignment with the parent view.

Alignment between views can be added or removed. This is accomplished by right-clicking on a view, selecting Alignment, and then selecting one of the alternatives shown in Figure 7.24.

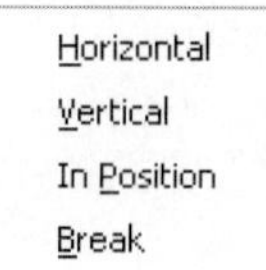

Figure 7.24 - View alignment options

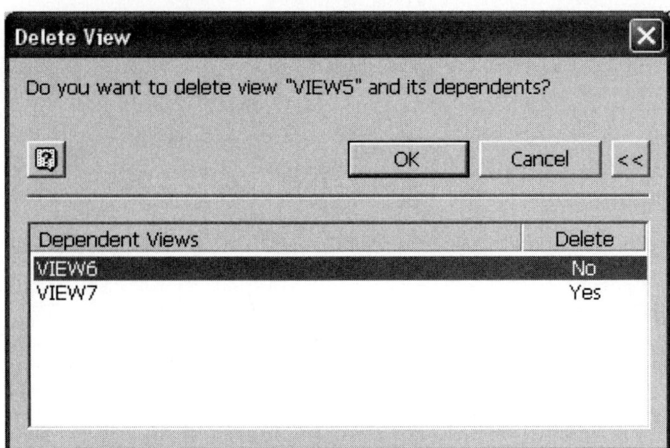

Figure 7.25 - Delete view dialog box

Views can also be deleted. To do so, right-click on the view and select Delete. If the view to be deleted is a parent to other views, the Delete View dialog box shown in Figure 7.25 appears. If you want to delete the view as well as its dependents, click OK. If you only want to delete the selected view, click the More >> button, then click in the Delete column to change Yes to No.

Drawing Annotations

In addition to Drawing Views tools, Drawing Annotation tools are also available when working with Inventor drawing files. The annotation tools can be accessed by clicking in the title area of the panel bar, and then selecting Drawing Annotation. These tools appear as shown in Figure 7.26.

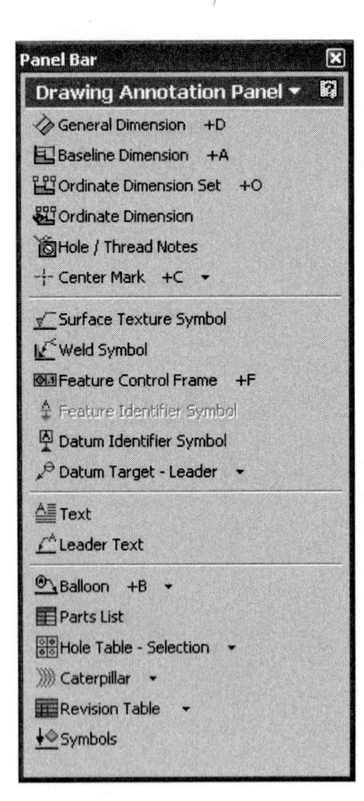

Figure 7.26 - Drawing annotation tools

Centerlines

New with Release 6 is the Automated Centerlines tool. To add centerlines to a view, right-click on the view and select Automated Centerlines . . . The Centerline Settings dialog box, as shown in Figure 7.27, appears. The Apply To area is used to select the features to which automated centerlines are applied. The Projection section determines the type of projection for which automated centerlines are applied; axis normal creates center marks when the circular edge is normal to view plane and axis parallel creates centerlines when the circular edge is parallel to view plane.

If automated centerlines are not used, centerlines and center marks can be added manually to drawing views using one of four different tools: Center Mark, Center Line, Center line bisector, and Centered Pattern. All of these tools can be accessed from the panel bar. If the tool of interest is not displayed, expand the centerline tools by clicking the down arrow, as shown in Figure 7.28.

To see Show Me animations of the different centerline tools, select the Visual Syllabus icon from the standard toolbar. Now open the Drawings palette from the drop-down list, and select the Centerlines icon, as shown on the left in Figure 7.29. A menu of different centerline options (Figure 7.29, on the right) appears.

Centerlines can be extended once they are placed. Move the cursor over the centerline so that the green circle handles appear. Click and drag on a handle to extend the length. See Figure 7.30.

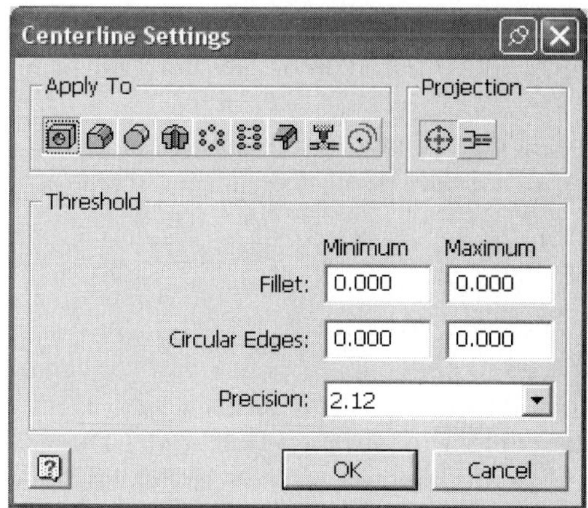

Figure 7.27 - Automated centerline settings

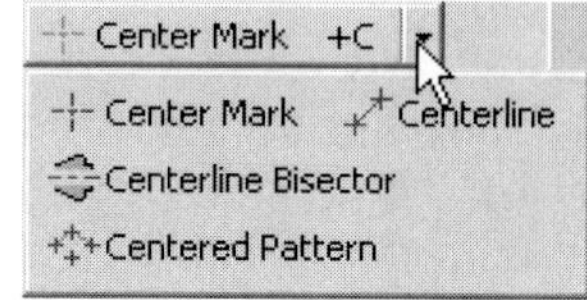

Figure 7.28 - Centerlines and center marking tools

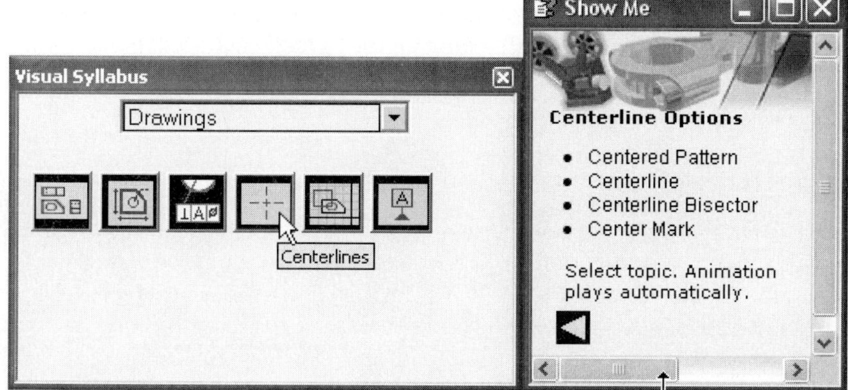

Figure 7.29 - Centerline Show Me animations

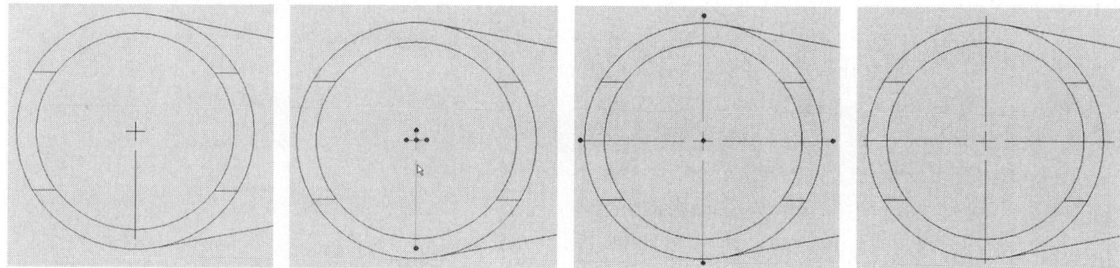

Figure 7.30 - Editing centerlines

Hole and Thread Notes

If the part being viewed in a drawing has features created with the Hole or the Thread tools, use the Hole/Thread Notes tool to display callout notes in a drawing view. To add hole notes, after selecting the tool, move the cursor over the hole feature in its circular view, and then left-click. Move the cursor to position the note, and click again. Some examples of hole notes placed in a view are shown in Figure 7.31.

An example of a drawing view with a thread note is shown in Figure 7.32. Note that in this case it was not necessary to place the thread note in the threads' circular view.

Dimensions

The parametric (or model) dimensions used to define features in a part file can be displayed in a drawing file. To do this, move the cursor over a drawing view, right-click, select Get Model Annotations from the context menu, then Get Model Dimensions. Any parametric dimensions associated with this view should now be displayed. An example can be seen in Figure 7.33.

Once the parametric dimensions are displayed, they will need to be moved to improve the clarity of the drawing. Move the mouse over a dimension until the dimension handles (e.g., green circles) are displayed. The dimension can now be modified by clicking and dragging. The dimensions have been cleaned up and centerlines added in Figure 7.34.

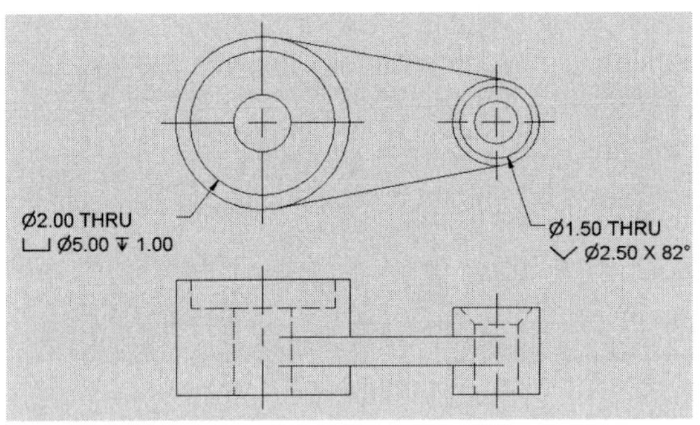

Figure 7.31 - Hole note examples

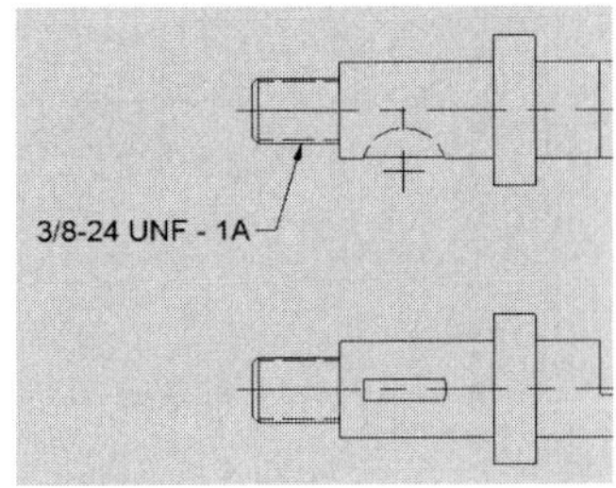

Figure 7.32 - Thread note example

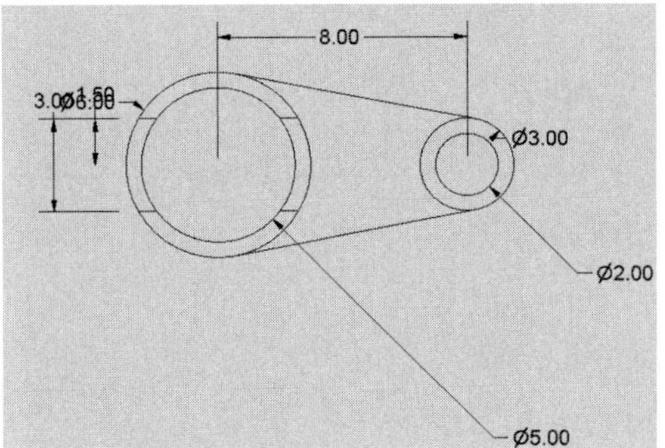

Figure 7.33 - Get model dimensions

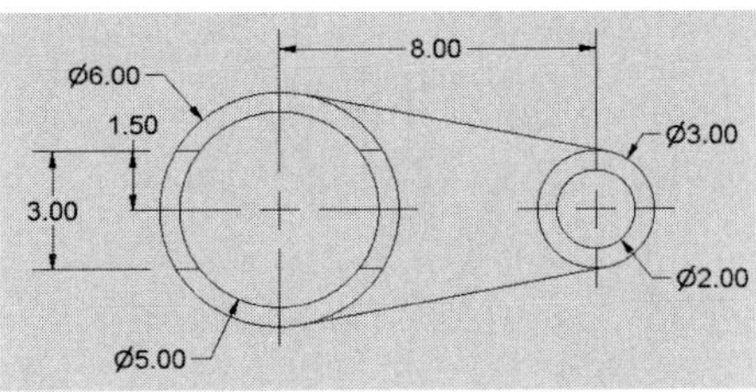

Figure 7.34 - Clean up dimensions

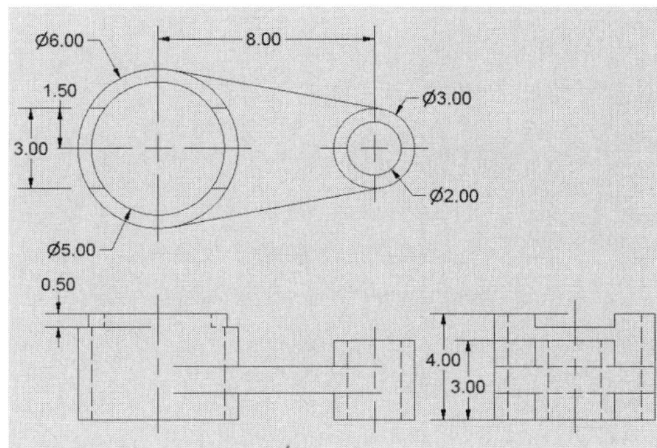

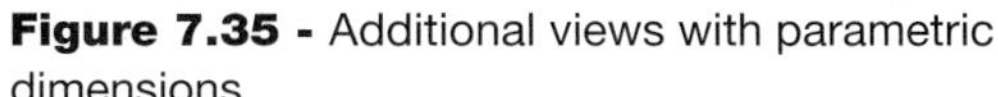

Figure 7.35 - Additional views with parametric dimensions

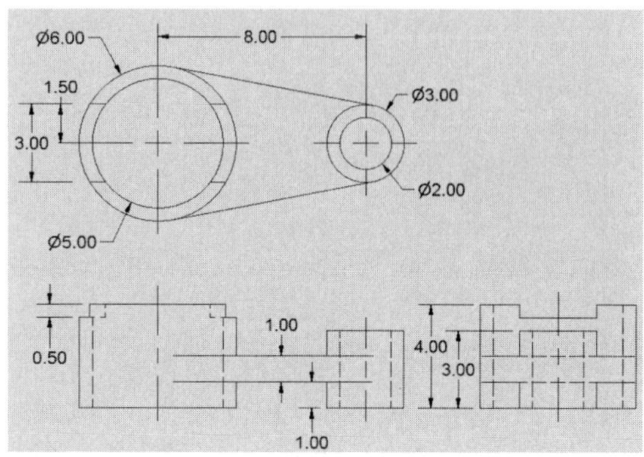

Figure 7.36 - Fully dimensioned drawing

Additional principal views of the same part are displayed in Figure 7.35, along with the parametric dimensions associated with the views. Although these three height dimensions (0.5, 3.0, and 4.0) have been re-positioned, their appearance is still not entirely satisfactory. There should be a gap between the extension lines of a dimension and the object lines of the part. Since they are parametric dimensions, however, they cannot be decoupled from the model geometry that they define.

To address this situation we can add drawing (or reference) dimensions. Drawing dimensions are created with the General Dimension tool. Dimensions created in this way are non-parametric. They do not drive the part geometry.

When a model dimension is deleted (by right-clicking, then selecting Delete), it is only deleted from a drawing. The parametric dimension still exists in the part file; it is simply not displayed in the drawing.

In this case it is not necessary to delete the height dimensions, though. It is easier to right-click on the view, select Annotation Visibility, and then remove the check mark next to Model Dimensions. The General Dimension tool was then used to add these dimensions (as well as two other dimensions) to the now fully dimensioned drawing shown in Figure 7.36.

It will be demonstrated in a tutorial at the end of this chapter that Inventor part and drawing files are *bidirectionally associative*. This means that if the value of a parameter is modified in a part file, then the associated drawing file will likewise be modified, and vice versa. In the drawing file, this is accomplished by right-clicking on a model dimension, and then selecting Edit Model Dimension . . .

Text

Text is added to drawing files with the Text tool. When the Text tool is selected, the cursor changes to a yellow circle and cross-hairs. Select the text insertion point, and the Format Text dialog box appears, as show in Figure 7.37. Enter the text in the edit box, and then click the OK button in order to display the text in the drawing. From the Format Text dialog box, the text justification, rotation angle, color, font, and text size can all be changed. Common drafting symbols can also be accessed and inserted into a drawing (see Figure 7.38).

It is also possible to insert parameter values (associated with a component on which the drawing is based) into the text stream. Parameters associated with the

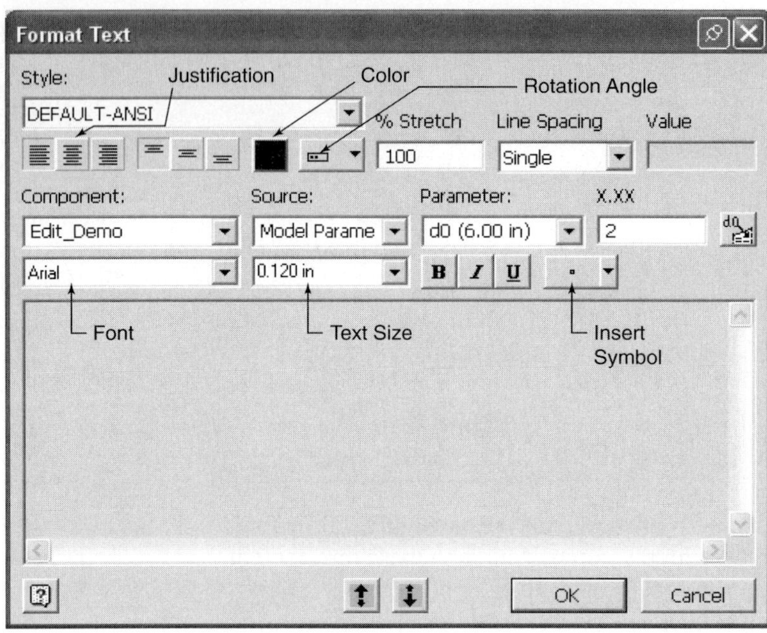

Figure 7.37 - Format text dialog box

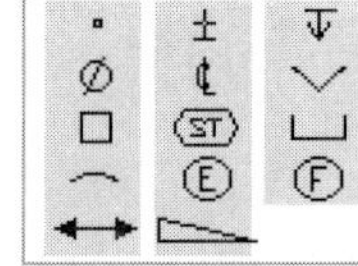

Figure 7.38 - Common drafting symbols

component are accessed from the Parameter drop-down list. After selecting a parameter, click the Add Parameter button to place the parameter value into the text window.

Existing text can be modified by right-clicking on the text and selecting Edit Text . . . The Format Text dialog box opens. Edit the text, and click OK. Note that if you wish to change the format of existing text, for example by increasing the font size, you will first need to highlight the text in the edit box, and then change the formatting.

TUTORIAL 21 Ball Valve Seat Drawing

Detailed Drawing Creation Steps

1. The Ball Valve project folder should already be active. (If not, select File > Projects . . . from the menu bar, select Ball Valve, and click Apply). Start a new file, select the Metric tab, and then choose the ANSI metric drawing template ANSI (in).idw file. The Inventor drawing environment is shown in Figure 7.39.
2. In the browser, right-click on Sheet1, and then select Edit Sheet. . . . Change the sheet size from C to A, as shown in Figure 7.40.
3. In the browser, right-click on the ANSI—Large title block icon, and then select Delete.
4. In the browser, expand the Drawing Resources folder, and then expand the Title Blocks folder. Right-click on the ANSI A title block icon, then select Insert (or just double-click).

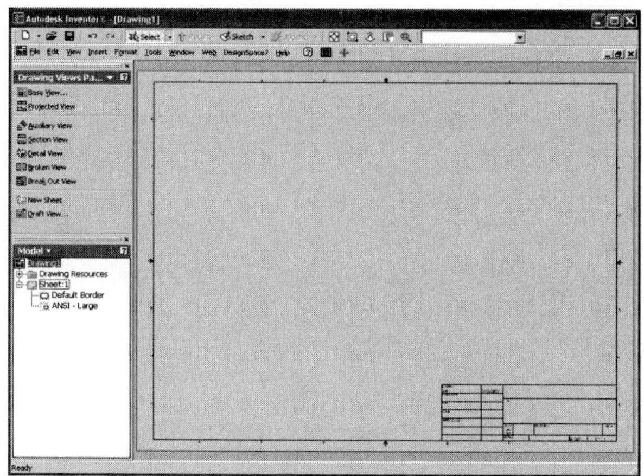

Figure 7.39 - Inventor drawing environment

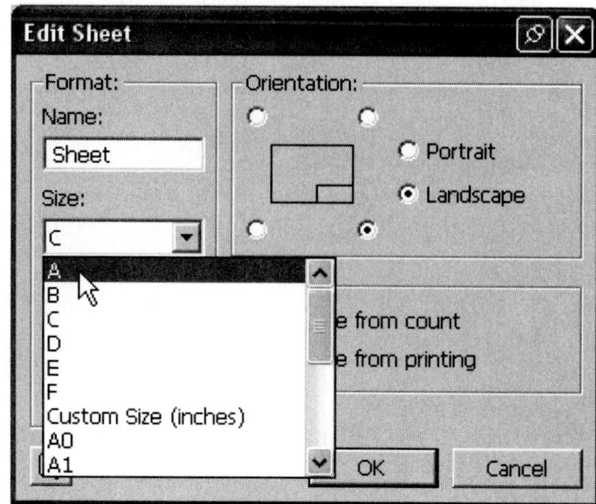

Figure 7.40 - Change sheet size

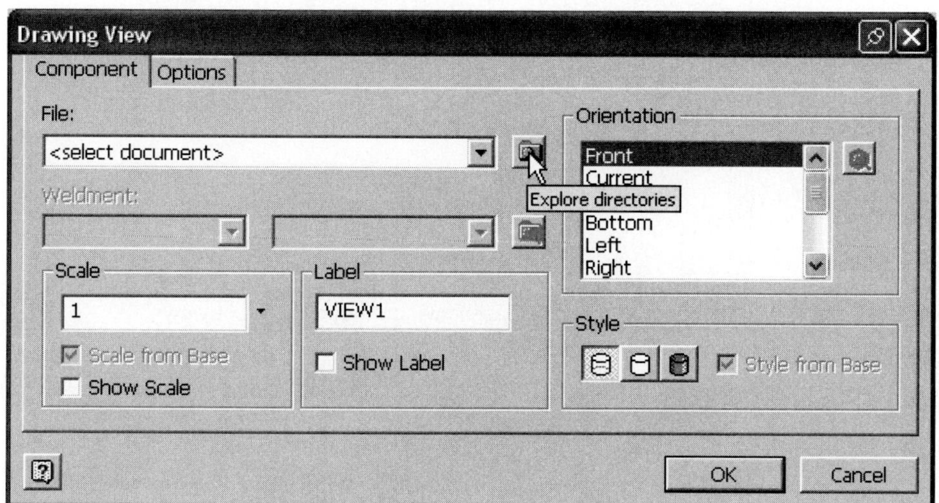

Figure 7.41 - Drawing view dialog box

5. Select the Base View . . . tool from the Drawing Views panel. In the Drawing View dialog box, click on the Explore directories button, as shown in Figure 7.41.
6. In the Open dialog box that appears, select the Seat part in the Ball Valve Workspace folder, and then select Open.
7. The Drawing View dialog box reappears. In addition, a view of the seat part appears in the graphics window. Position the view on the left side of the drawing area; left-click to place the view. The dialog box closes, and a front view of the seat is visible in the drawing area, as shown in Figure 7.42.

   NOTE: If the view on your screen is not the same as that shown in the Figure 7.42, select different views from the Orientation column in Drawing View dialog box until one matching the figure appears.

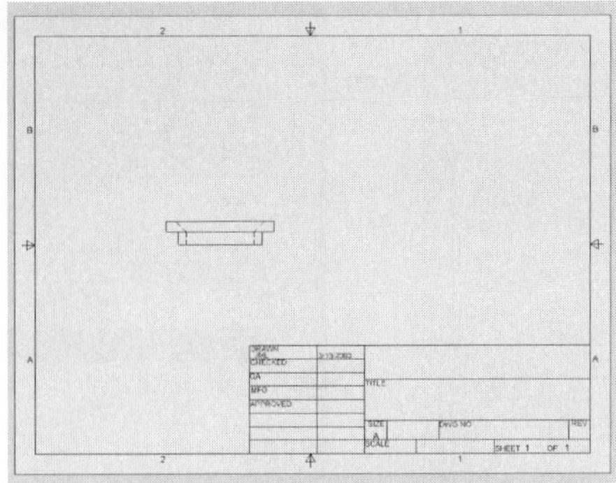

Figure 7.42 - Base view placed

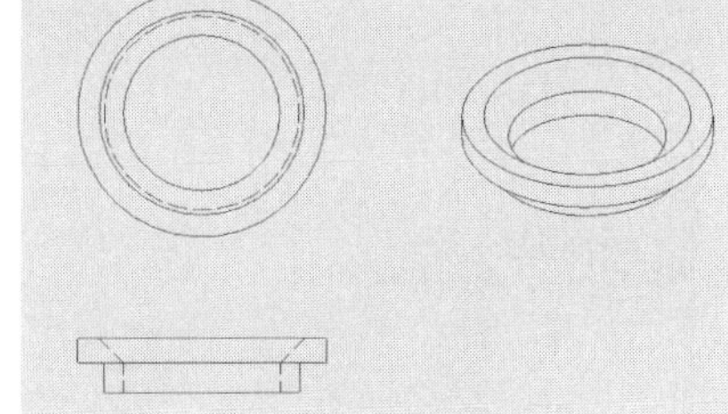

Figure 7.43 - Projected views

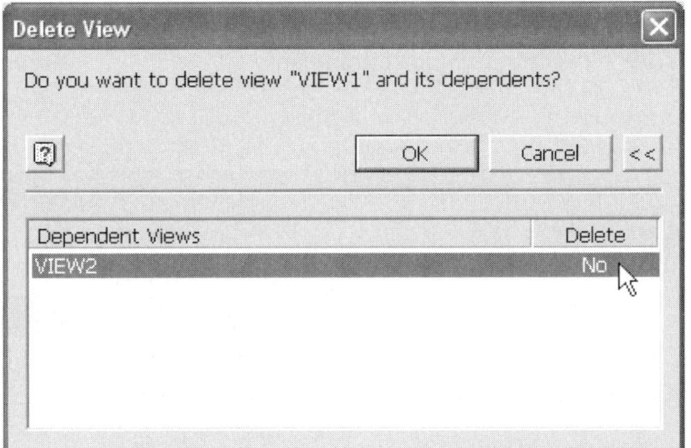

Figure 7.44 - Shaded view style

Figure 7.45 - Delete base view (but not dependent view)

8. Select the Projected View tool. Move the cursor over the base view so that the outline is highlighted and then left-click. Now move the mouse above the base view. A projected top view appears. When you are satisfied with the position of the projected view, left-click to place the view. Next move the cursor above and to the right of the base view, until an isometric view of the part appears. Once again, left-click to place the view. Now right-click and select Create from the context menu. Your screen should now resemble Figure 7.43.
9. Move the cursor over the isometric view in the upper right corner until the bounding area of the view is highlighted. Right-click and then select Edit View . . . The Edit View dialog box opens. In the Style area of the dialog box, select Shaded, as shown in Figure 7.44, and then select OK. The isometric view in the drawing is now shaded.
10. In looking at the three views placed thus far, it might be preferable to replace the base front view with a section view before continuing. To do this, move the cursor close to the base view so that the bounding area appears, and then right-click and select Delete from the context menu. The Delete View dialog box appears. We would like to delete VIEW1, the front view, but not its dependent view (i.e., top view). Select the More >> button, then click on the VIEW2 Delete cell to change the response from Yes to No (see Figure 7.45), and then select OK. At this point only the top and the isometric views remain.

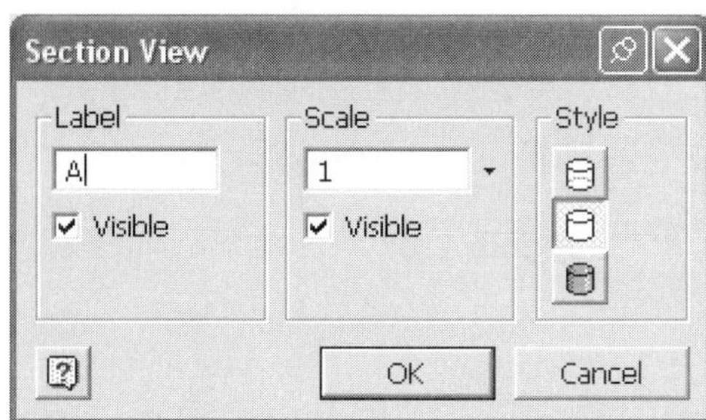

Figure 7.46 - Section view dialog box

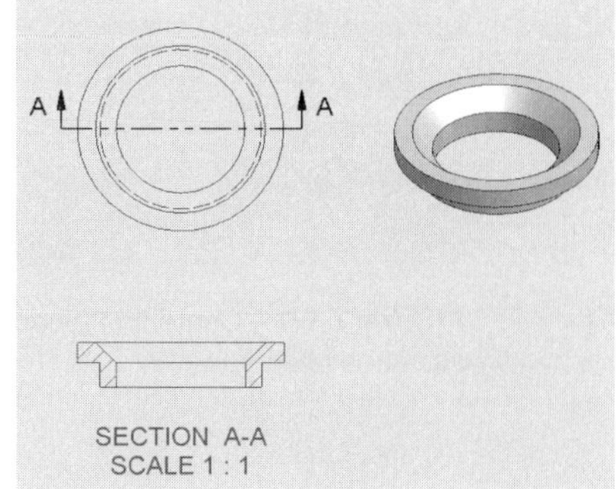

Figure 7.47 - Section view

11. Select the Section View tool from the panel bar, and then select the top view. Move the cursor to the left quadrant of any one of the concentric circles until a green circle appears. Now move the cursor further to the left, outside the view. The green circle turns yellow and a horizontal line appears. Left-click to accept the first endpoint of the cutting plane on edge. Now move the cursor horizontally to the right, outside of the largest diameter circle. You should see a horizontal line. Left-click to select the other cutting plane endpoint. Now right-click and then select Continue from the context menu. The Section View dialog box opens, as shown in Figure 7.46. Move the cursor down, and then left-click to place the section view. The screen should now be similar to Figure 7.47.
12. Zoom in on the section view. We will now modify the hatch pattern so that the hatch pattern angle is different from the chamfer angle. Move the cursor over one of the hatching lines until the entire hatch pattern is highlighted, right-click and select Modify Hatch . . . from the context menu. The Modify Hatch Pattern dialog box opens. Change the Scale to 1:2 (0.50), and the Angle to 35 degrees, and then select OK, as shown in Figure 7.48.
13. Zoom out so that all views are visible.
14. Save the file as **Seat** in the Ball Valve Workspace. Save the file every 5 to 10 minutes. Note that because part and drawing files have different extensions (i.e., .ipt, .idw) the same file name can be re-used.

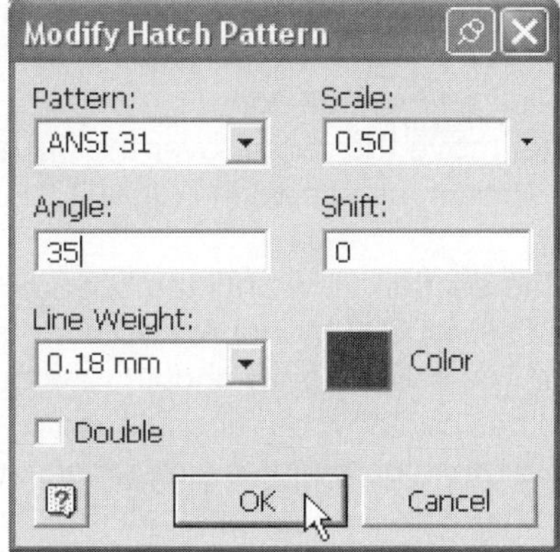

Figure 7.48 - Modify hatch pattern

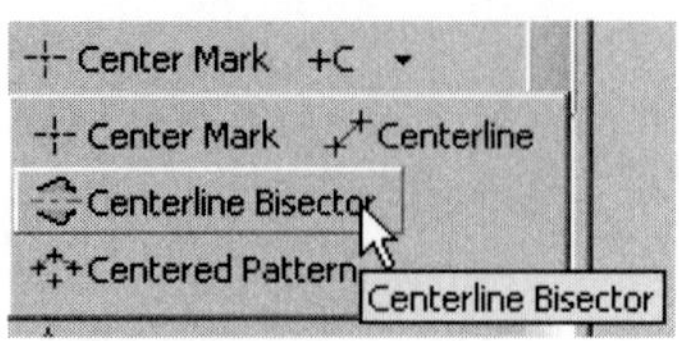

Figure 7.49 - Centerline bisector tool access

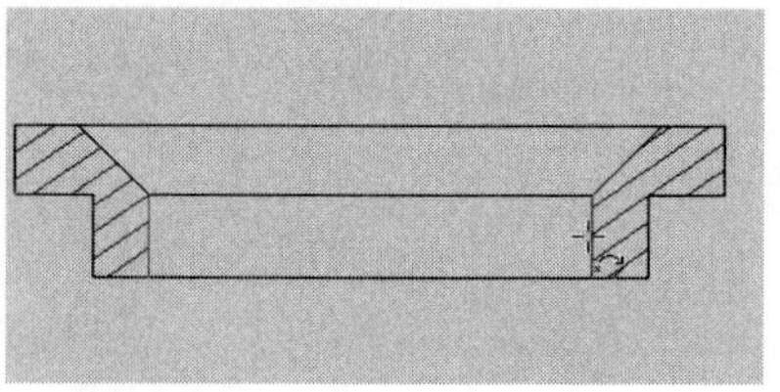

Figure 7.50 - Centerline bisector selection lines

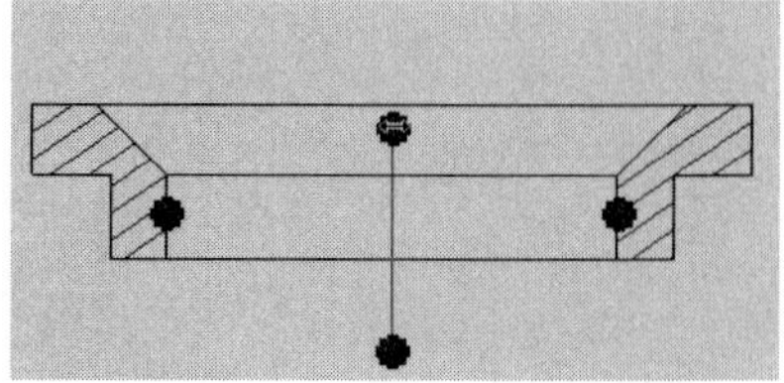

Figure 7.51 - Centerline extension

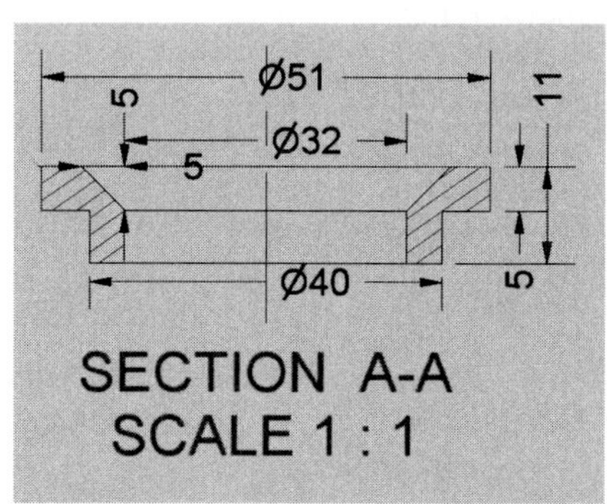

Figure 7.52 - Model dimensions displayed

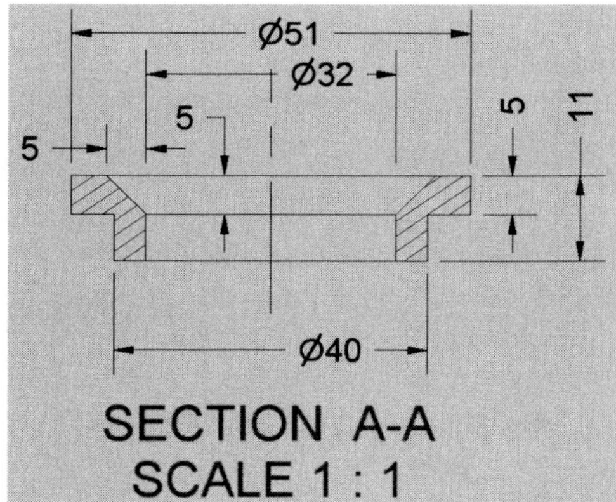

Figure 7.53 - Re-positioned dimensions

15. On the panel bar, switch from the Drawing Views panel to the Drawing Annotation panel by left-clicking in the panel bar title area, and selecting Drawing Annotation Panel.
16. Click on the Center Mark tool, and then left-click on the outermost concentric circle in the top view. Right-click and choose Done (or hit the Escape key) to exit Center Mark mode.
17. Zoom in on the section view.
18. Access the Centerline Bisector tool by expanding the flyout, as shown in Figure 7.49. Select the two symmetrical vertical lines shown in Figure 7.50. A centerline is created between the two selected lines. To exit Centerline Bisector mode, right-click and select Done.
19. To extend the centerline, move the cursor over the upper part of the centerline until four green circles appear, as shown in Figure 7.51. Left-click and drag to extend the centerline.
20. Move the cursor over the section view until the bounding area of the view appears, right-click and select Get Model Annotations > Get Model Dimensions from the context menu. The section view should now be similar to Figure 7.52.
21. The parametric dimensions can now be repositioned by moving the cursor over the dimension until the dimension handles (i.e., green circles) appear. At this point dragging on the dimension text will relocate the dimension text, extension lines, and arrows. Adjust the dimensions so that the view is similar to Figure 7.53.
22. Zoom All .

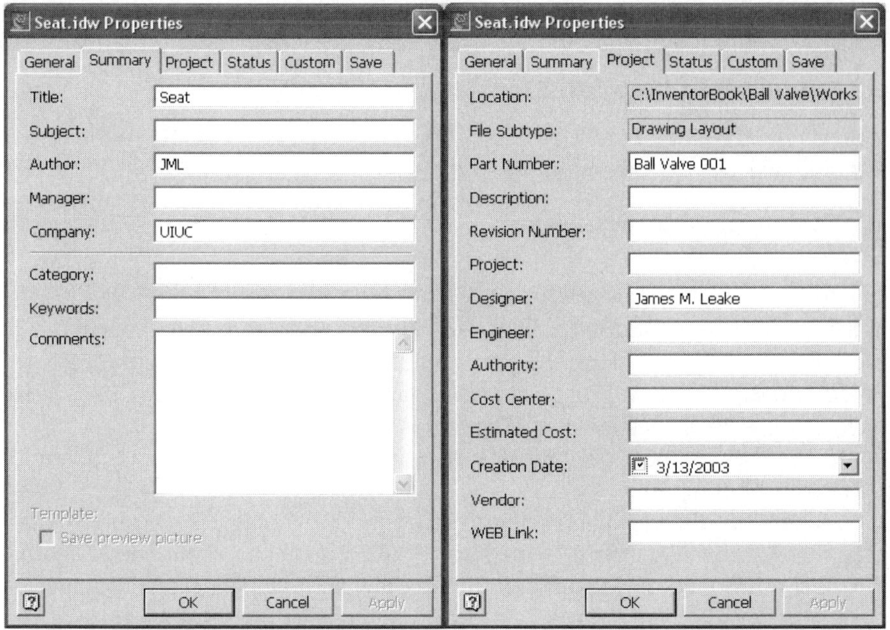

Figure 7.54 - Drawing file properties

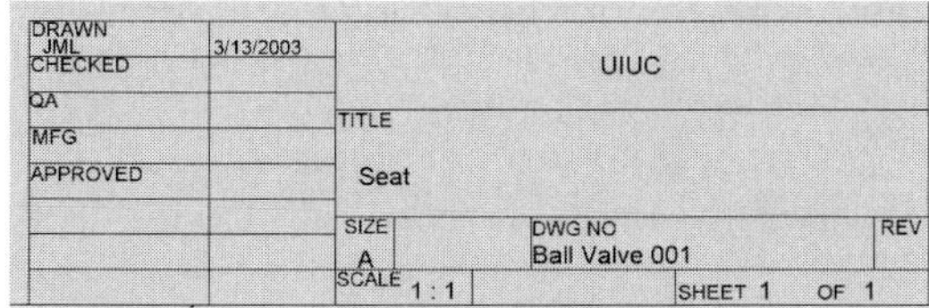

Figure 7.55 - Completed title block

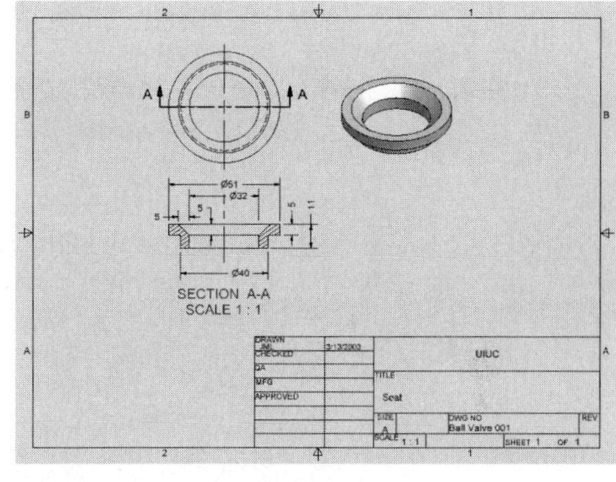

Figure 7.56 - Completed drawing

23. Use Zoom Window to zoom in on the title block.
24. Select File > iProperties . . . from the menu bar (or right-click in the browser and select Properties). In the Properties dialog box, enter the values shown in Figure 7.54 on both the Summary and the Project tabs, click the Apply button, and then OK.
25. After using the Text tool to add the scale, the title block should now resemble Figure 7.55.
26. Zoom All.
27. Save the drawing file. The completed drawing should be similar to Figure 7.56.
28. Choose File > Print . . . from the menu bar. The Print Drawing dialog box opens. Click on the Preview . . . button to verify that the drawing and printer are properly configured, and then print.

TUTORIAL 22 Ball Valve Body Drawing

Detailed Drawing Creation Steps

1. Start a new file, select the Metric tab, and then choose the ANSI metric drawing template file.
2. In the browser, right-click on Sheet1, and then select Edit Sheet. . . . Change the sheet size from C to A.
3. In the browser, right-click on the ANSI—Large title block icon, and then select Delete.
4. In the browser, expand the Drawing Resources folder, and then expand the Title Blocks folder. Right-click on the ANSI A title block icon, then select Insert.
5. Select the Base View tool from the Drawing Views panel. In the Base View dialog box, click on the Explore Directories button.
6. In the Open dialog box, select the Body part in the Ball Valve Workspace folder, and then select Open.
7. The Base View dialog box reappears, as does a view of the Body part in the graphics window. Change the Scale to 1:2 (0.5), as shown in Figure 7.57. With the view positioned on the left side of the drawing area, left-click. The dialog box closes, and a front view of the valve body is visible in the drawing area, as shown in Figure 7.58.
8. Use the Projected View tool to project the views shown in Figure 7.59.
9. Save the file as **Body** in the Ball Valve workspace. Save regularly every 5 to 10 minutes.
10. We will now change the right side view to a break out view. Select the Right view. Now select the Sketch tool from the standard toolbar. Use the Line tool to create a closed profile sketch similar to that shown in Figure 7.60.
11. Right-click, Finish Sketch.
12. Select the Break Out View tool, and then select the Right view. The Break Out View dialog box should open, and both the view and sketch should be highlighted, as shown in Figure 7.61.

    NOTE: If a message appears indicating that the view does not contain a sketch with a closed profile, then the sketch is not associated with the right view. Redo step 10.

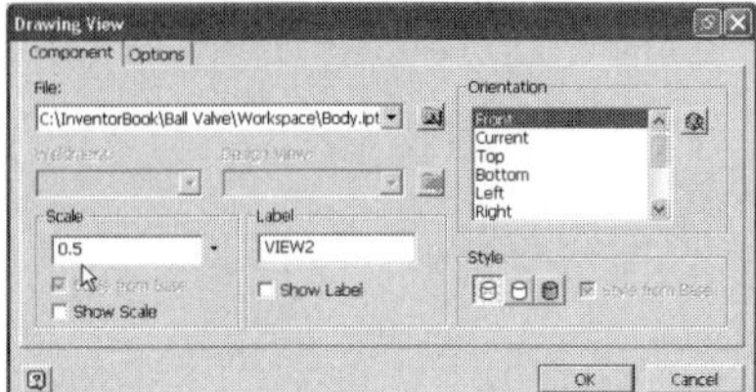

Figure 7.57 - Create view dialog box

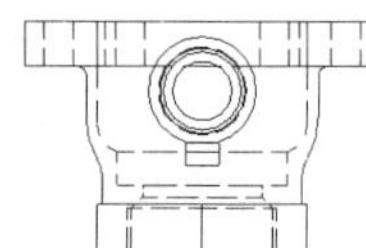

Figure 7.58 - Base view

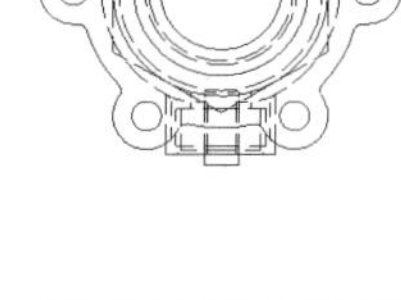

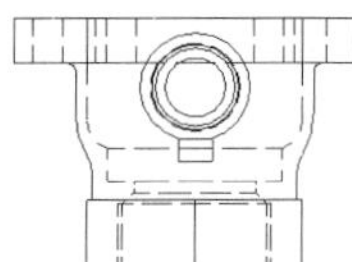

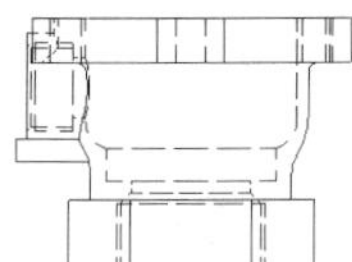

Figure 7.59 - Project views

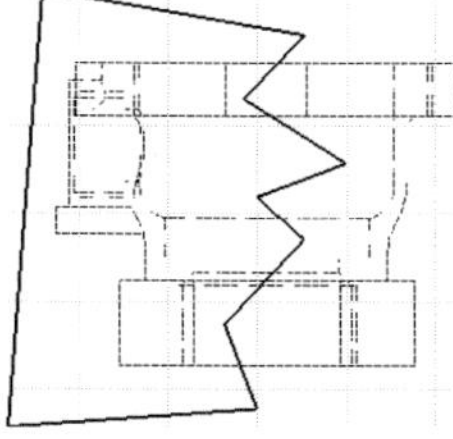

Figure 7.60 - Closed profile sketch

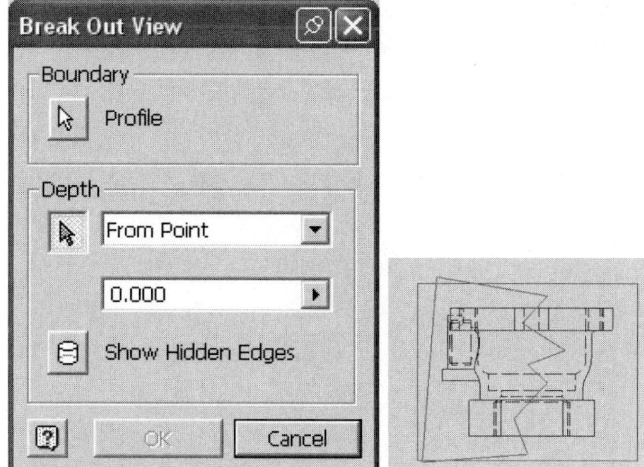

Figure 7.61 - Break out view operation

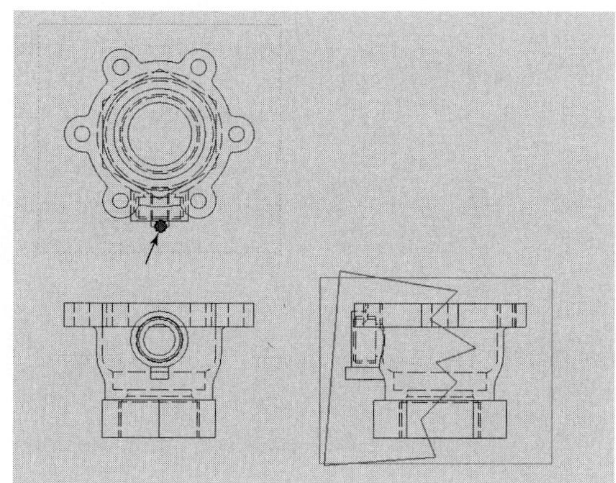

Figure 7.62 - Specification of break out depth

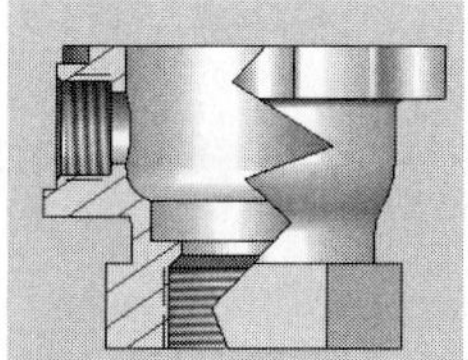

Figure 7.63 - Shaded break out view

13. Using the default From Point depth type, we will use another view to specify the break out depth. Move the cursor to the Top view. Position the cursor so that the green circle appears, as shown in Figure 7.62. With the green circle visible, left-click to specify the depth. Click OK in the dialog box to create the break out view.
14. We will now change the style of the break out view to shaded. Highlight the right side view, right-click, and select Edit View . . . The Drawing View dialog box opens. In the Style area, remove the check next to Style From Base, select Shaded, and click OK. The right side view should now resemble Figure 7.63.
15. Right-click on the Top view, and select Automated Centerlines . . . Under Apply To, select Hole, Revolve, and Circular Pattern Features. Under Projection, select

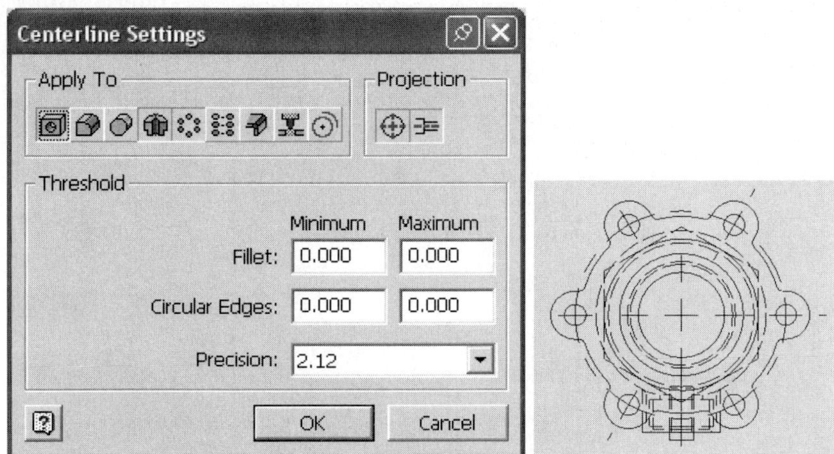

Figure 7.64 - Automated centerlines

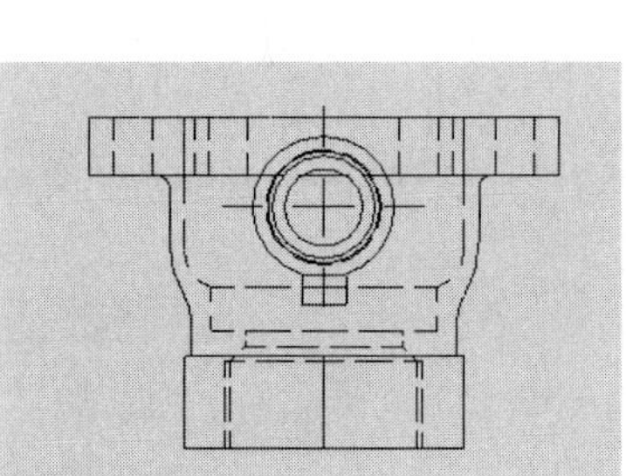

Figure 7.65 - Automated centerlines (front view)

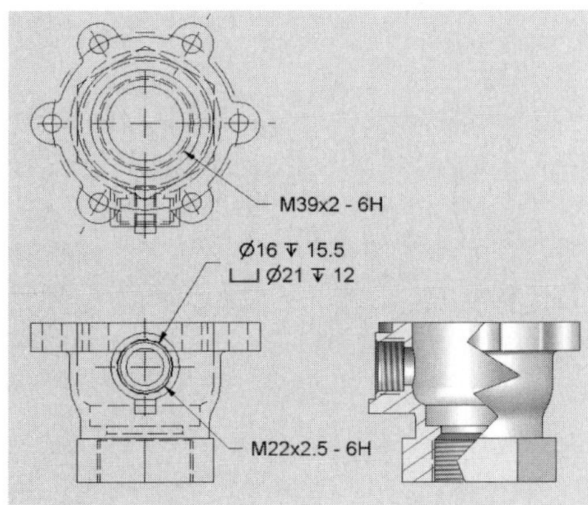

Figure 7.66 - Hole and thread notes

both Axis Normal and Axis Parallel (See Figure 7.64 left). Click OK to place the centerlines. The top view should now resemble Figure 7.64 right.

16. Right-click on the Front view, and select Automated Centerlines . . . Under Apply To, select Cylindrical Features, under Projection, select Axis Normal, and then click OK. See Figure 7.65.
17. On the panel bar, switch from the Drawing Views panel to the Drawing Annotation panel by left-clicking in the panel bar title area, and selecting Drawing Annotation Panel. There are many parametric dimensions associated with this part. Cleaning up these dimensions involves a significant amount of time. We will consequently leave this part un-dimensioned, except for hole and thread notes, which will be added as follows.
18. Use the Hole/Thread Notes tool to add callouts for the counterbore hole and the two sets of internal threads contained on this part. With the tool selected, move the cursor over the hole/thread note in its circular view until the correct geometry is highlighted. Left-click, and the note appears. Once satisfied with the position of the note, left-click again to place the note. See Figure 7.66 for the hole and thread notes to be added to the drawing.

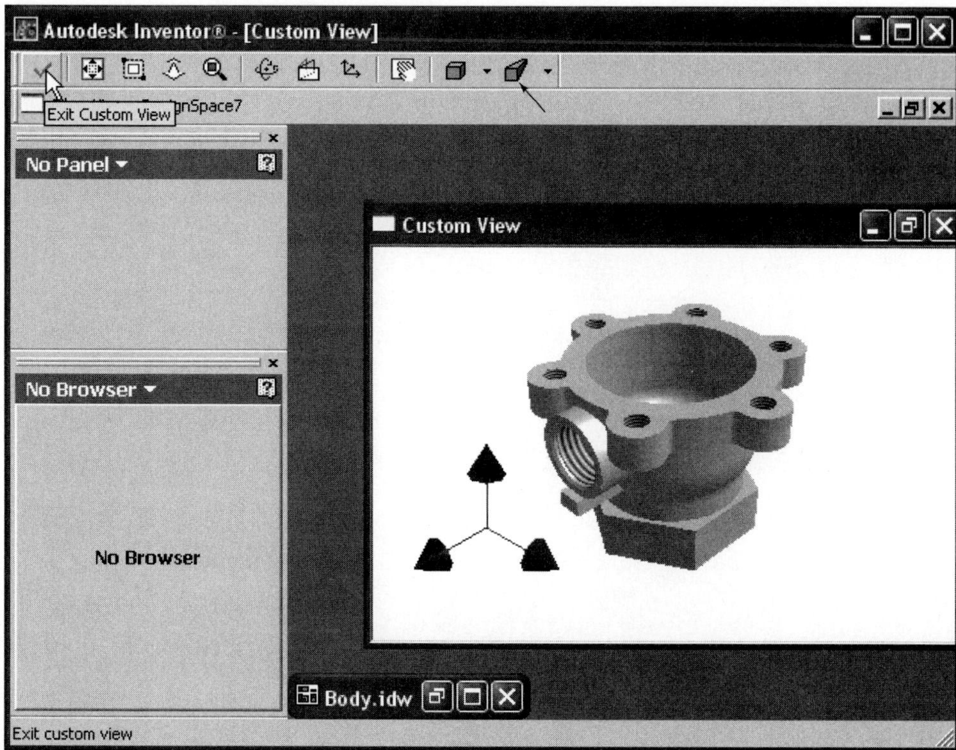

Figure 7.67 - Custom view

19. We will now add a shaded perspective view to the drawing. Perspective views must be added as a base view. In the panel bar, switch back to the Drawing View tools. Select the Base View tool. With the Drawing View panel open, use Explore directories to locate the Body part. In the Orientation section of the Drawing View dialog box, select the Change view orientation button. The Custom View window opens. Right-click and select Isometric View. On the tool bar docked at the top of the window, change to Perspective Camera mode, then click on Exit Custom View. See Figure 7.67.

    The Drawing Views dialog box reappears. Change the style to shaded. In the graphics window select a position in the upper right corner to place the perspective view. The graphics area should now resemble Figure 7.68.
20. Zoom All.
21. Use Zoom Window to zoom in on the title block.
22. Use the Text tool to add the scale (1:2) to the title block.
23. Use the Drawing Properties dialog box (File > iProperties . . .) to modify the title block so that it resembles Figure 7.69.
24. Zoom All.
25. Save the drawing file. The completed drawing should be similar to Figure 7.70.
26. Choose File > Print . . . from the menu bar. The Print Drawing dialog box opens. Click on the Preview . . . button to verify that the drawing and printer are properly configured, and then print.

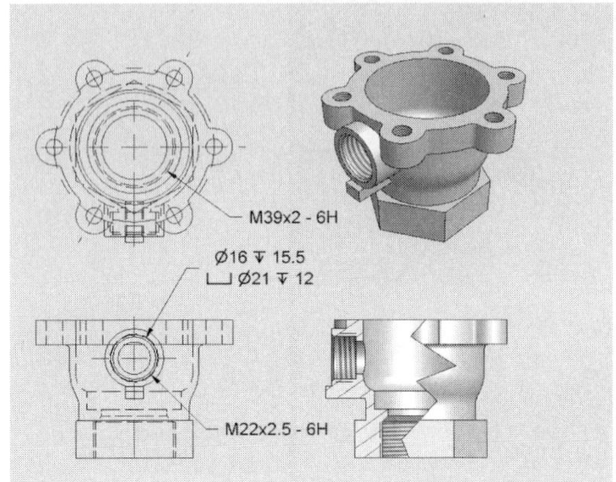

Figure 7.68 - Drawing with perspective view

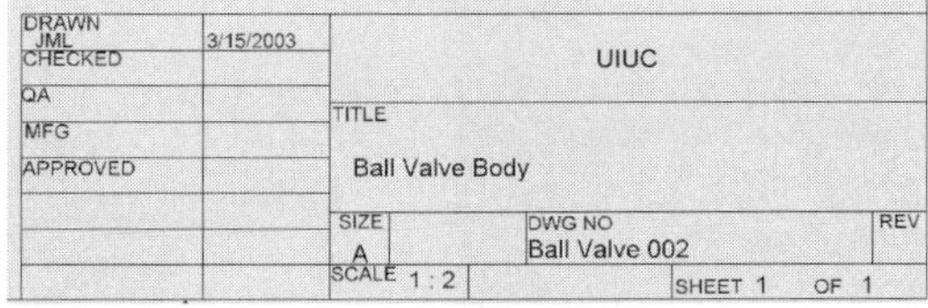

Figure 7.69 - Title block

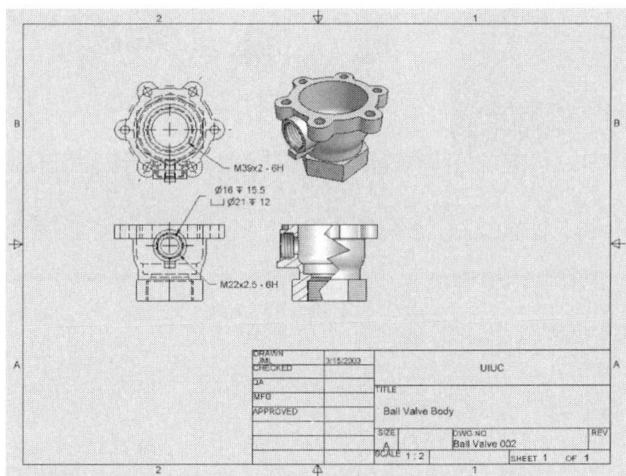

Figure 7.70 - Completed drawing

TUTORIAL 23 Garlic Press Piston Drawing

Detailed Drawing Creation Steps

1. Select File > Projects . . . from the menu bar, then make the Garlic Press project active.
2. Start a new file, select the Metric tab, and then choose the ANSI metric drawing template file.
3. In the browser, right-click on Sheet1, and then select Edit Sheet. . . . Change the sheet size from C to A. Also change the paper orientation to Portrait, as shown in Figure 7.71.
4. In the browser, right-click on the ANSI—Large title block icon, and then select Delete.
5. In the browser, expand the Drawing Resources folder, and then expand the Title Blocks folder. Right-click on the ANSI A title block icon, then select Insert.
6. Select Base View from the Drawing Views panel. The Drawing View dialog box opens. Select the Explore directories button.

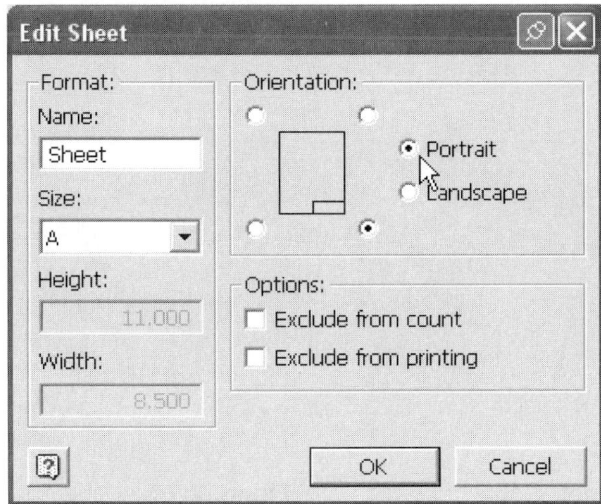

Figure 7.71 - Edit sheet

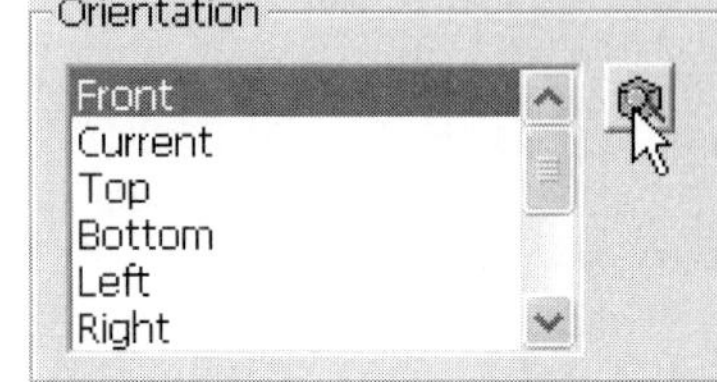

Figure 7.72 - Change view orientation

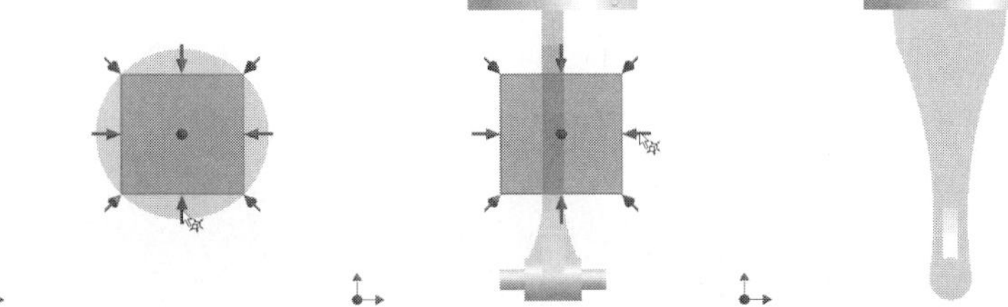

Figure 7.73 - Custom view selection

7. In the Open dialog box that appears, select the Piston part in the Garlic Press Workspace folder, and then select Open.
8. Rather than accept one of the principal (i.e., Front, Top, etc.) views available from the Orientation area on the dialog box, choose the Change view orientation button, as shown in Figure 7.72.

   The Custom View window opens, showing the Piston part file. Using the 3D Rotate tool (available from the docked toolbar at the top of the window), Common View mode, change the view to that shown on the right in Figure 7.73. After exiting the 3D Rotate tool, select the Exit Custom View button from the toolbar.

   The drawing sheet reappears, as does the Drawing View dialog box. Change the scale to 2:1 (2.00), and then place the custom view in the lower left of the drawing sheet by left-clicking. The screen should resemble Figure 7.74.
9. Save the file as **Piston** in the Garlic Press workspace. Save the file every 5 to 10 minutes.
10. Use the Projected View tool to add the views shown in Figure 7.75.
11. Add centerlines to the principal drawing views, as shown in Figure 7.76. Use either Automated Centerlines, or the manual centerline tools (e.g., Center Mark, Centerline Bisector) available from the Drawing Annotations panel. Note that undesired centerlines can always be deleted (from the right mouse button context menu).

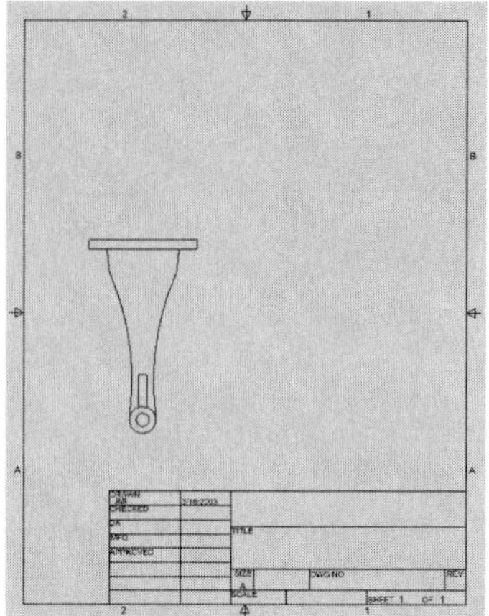

Figure 7.74 - Base view

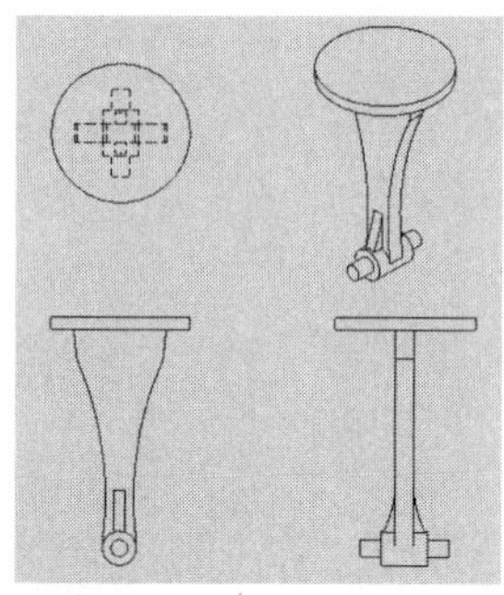

Figure 7.75 - Projected views

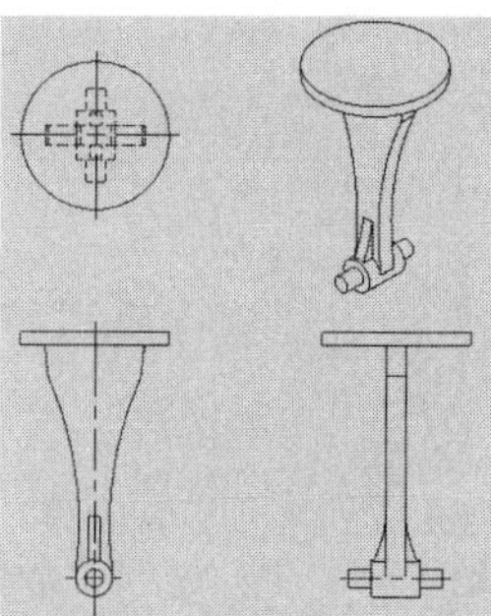

Figure 7.76 - Centerlines added

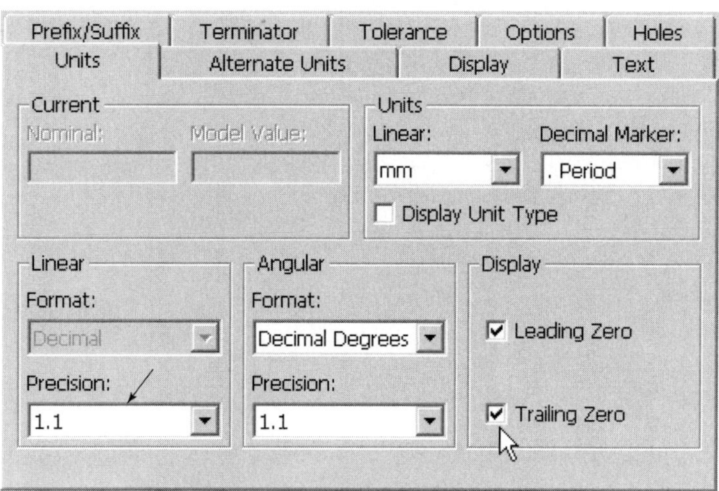

Figure 7.77 - Edit dimension styles

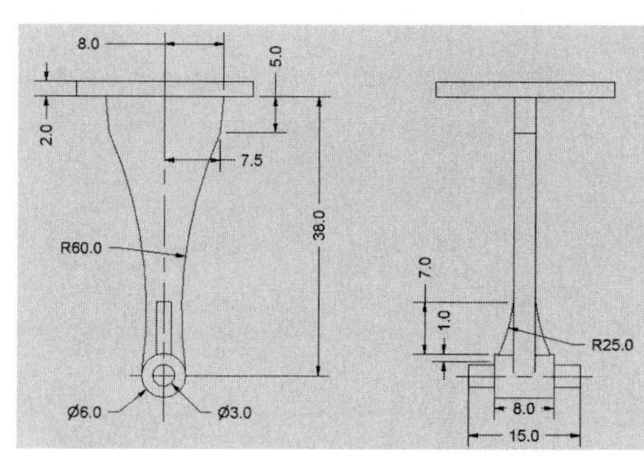

Figure 7.78 - Front and right view model dimensions

12. Move the cursor over the Front view so that the view's bounding area is highlighted, right-click and select Get Model Annotations > Get Model Dimensions. Repeat this process for both the Top and the Right Side views.
13. Select Format > Dimension Styles . . . from the menu bar. On the Units tab of the Dimension Styles dialog box, change the Precision to one decimal place and, under Display, check the Trailing Zero box (Figure 7.77). Click the Save button, then Close.
14. Zoom Window on the lower two views.
15. Delete one of the R60.0 dimensions by first highlighting it, right-clicking and then selecting Delete. Note that deleting a model dimension only deletes the dimension in the drawing file, not in the associated part file.
16. Reposition the other model (i.e., parametric) dimensions as necessary. The repositioned front and right view model dimensions are shown in Figure 7.78.

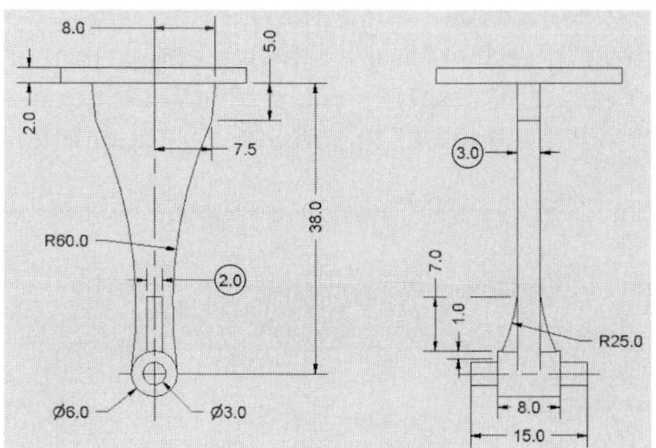

Figure 7.79 - Added drawing dimensions (encircled)

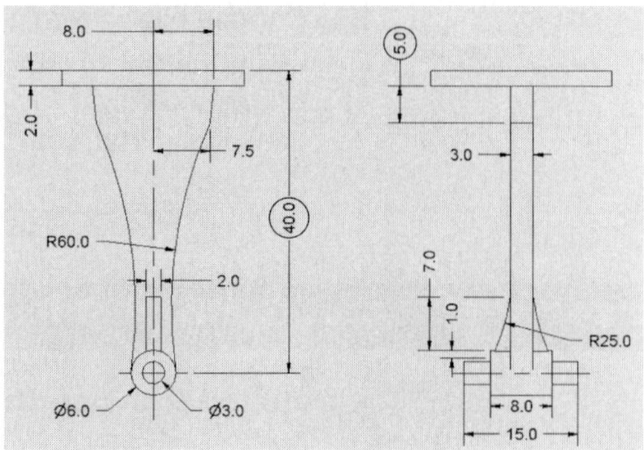

Figure 7.80 - Model dimensions replaced with drawing dimensions (encircled)

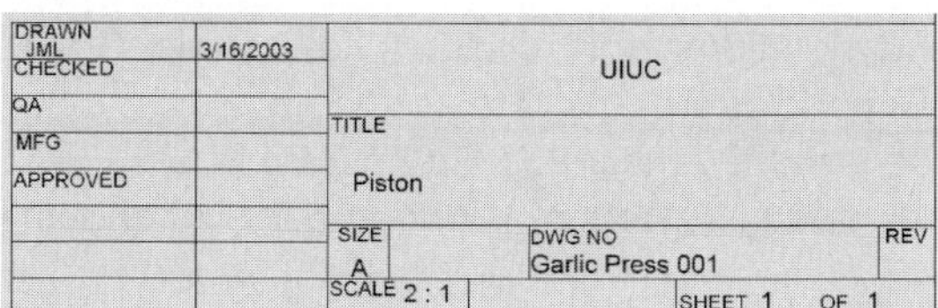

Figure 7.81 - Title block

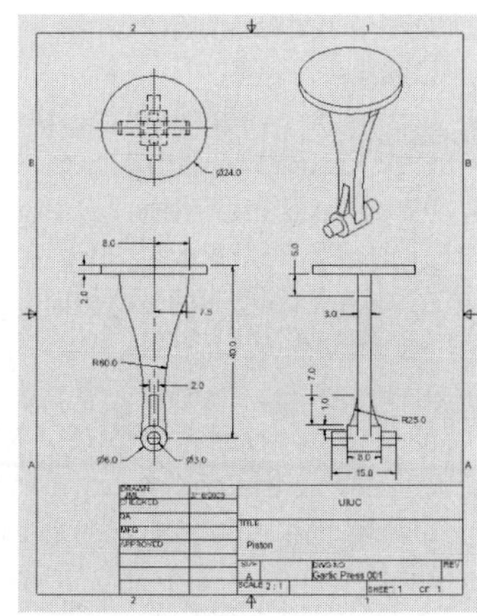

Figure 7.82 - Completed Piston drawing

17. A few dimensions are still missing. Use the General Dimension tool to add drawing dimensions (Figure 7.79). Note that drawing dimensions are not associated with the model geometry.
18. It is sometimes desirable to replace model dimensions with drawing dimensions that more closely adhere to good dimensioning practice. Delete the 5.0 and 38.0 height dimensions in the front view. Use the General Dimension tool to replace them with the dimensions shown in Figure 7.80.
19. Zoom All.
20. Use Zoom Window to zoom in on the title block.
21. Use the Text tool to add the scale (2:1) to the title block.
22. Use the Drawing Properties dialog box (File > iProperties . . .) to modify the title block, as shown in Figure 7.81.
23. Zoom All.
24. Save the drawing file. The completed drawing should be similar to Figure 7.82.

25. Choose File > Print . . . from the menu bar. The Print Drawing dialog box opens. Click on the Preview . . . button to verify that the drawing and printer are properly configured, and then print. Note that it will probably be necessary to use File > Print Setup . . . from the menu bar to change the paper orientation to Portrait.

TUTORIAL 24 Associativity of Part and Drawing Files

Detailed Drawing Creation Steps

1. Open the file ch7_editdemo.ipt located on the CD.
2. Start a new drawing file, select the English tab, and then choose the ANSI (in) drawing template ANSI (in).idw.
3. In the browser, right-click on Sheet1, and then select Edit Sheet. . . . Change the sheet size from C to A.
4. In the browser, right-click on the ANSI—Large title block icon, and then select Delete.
5. In the browser, expand the Drawing Resources folder, and then expand the Title Blocks folder. Right-click on the ANSI A title block icon, then select Insert.
6. Use the Base View and Projected View tools to create the views shown in Figure 7.83. Note that because the model file is already open, it is not necessary to use Explore directories to locate the file. Use a drawing scale of 1:4 (i.e., 0.25) for all views.
7. Turn on the visibility of the model dimensions (right-click on view, select Get Model annotations > Get Model Dimensions) in the three principal views.
8. Clean up the model dimensions so that they resemble Figure 7.84.
9. Using the Window menu on the menu bar, select ch7_editdemo.ipt, as shown in Figure 7.85.
10. In the part browser, right-click on the Base Cylinder feature, and then select Edit Feature. After changing the extruded depth from 4 to 6, as shown in Figure 7.86, click OK.
11. Using the Window menu on the menu bar, select ch7_editdemo. The drawing file should now resemble Figure 7.87. Note that the height of the base cylinder is now 6 inches.
12. Move the cursor over the model dimension of 8.00, as shown in Figure 7.88. Right-click and then select Edit Model Dimension . . . The Edit Model Dimension

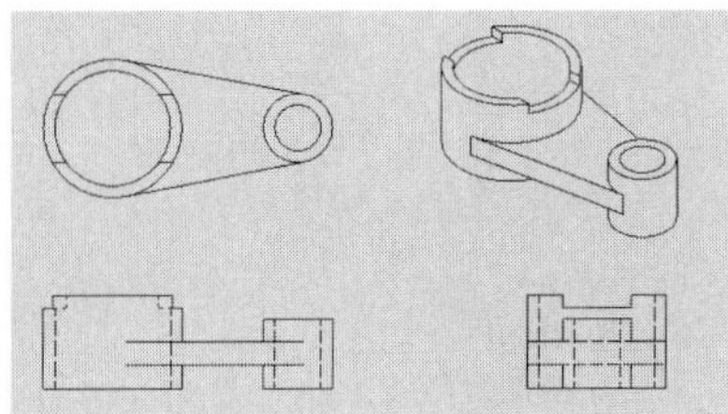

Figure 7.83 - Base and projected views

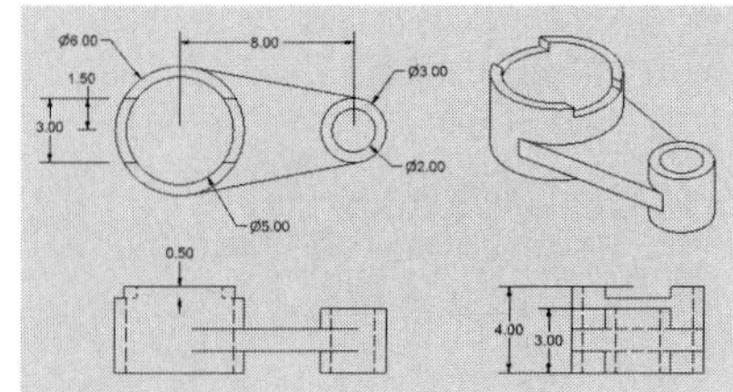

Figure 7.84 - Repositioned model dimensions

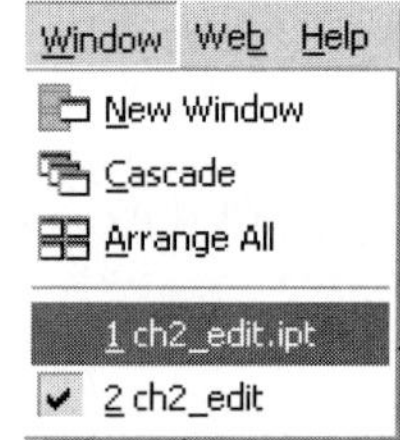

Figure 7.85 - Change active file

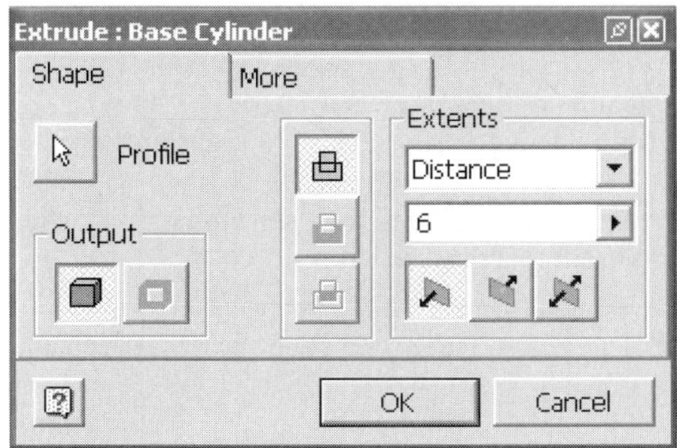

Figure 7.86 - Edit part feature

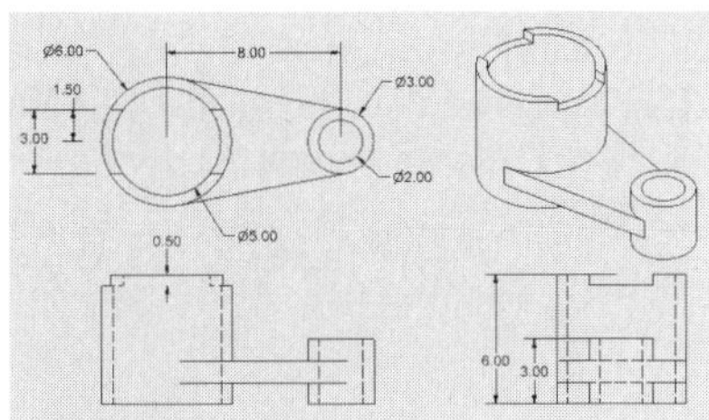

Figure 7.87 - Part - drawing associativity

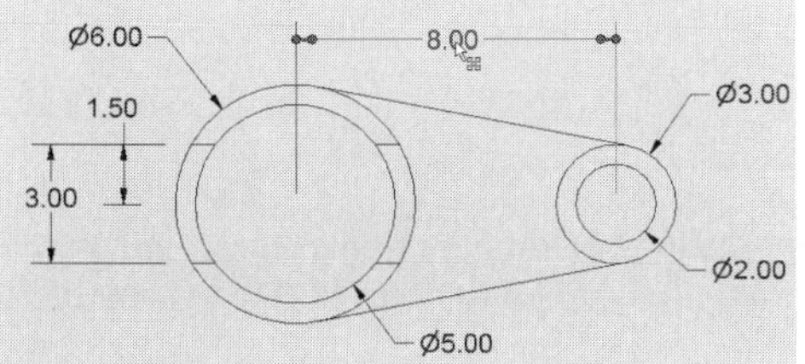

Figure 7.88 - Select model dimension

Figure 7.89 - Edit model dimension

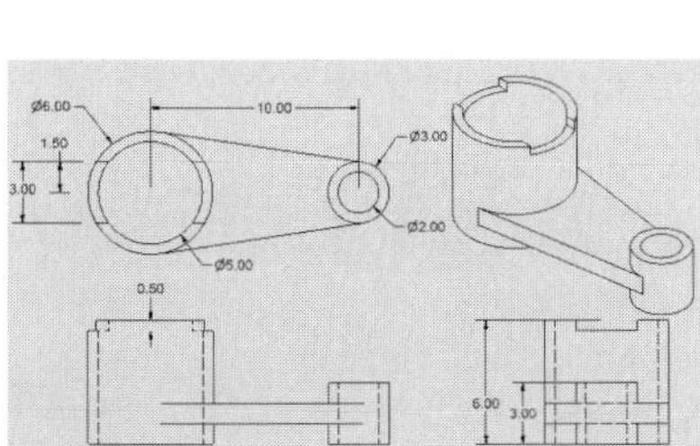

Figure 7.90 - Modified drawing views

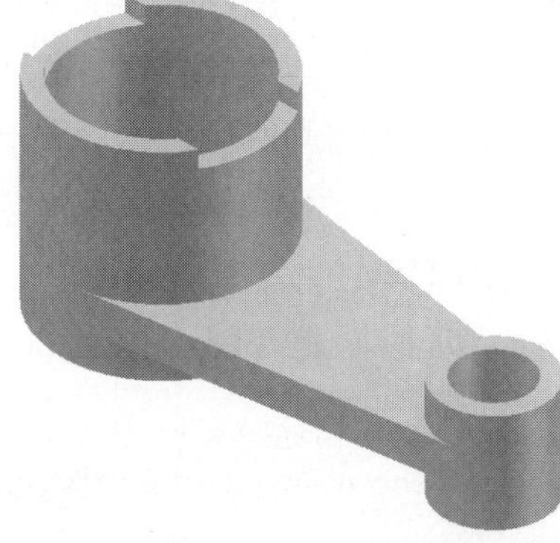

Figure 7.91 - Modified part file

box appears. Enter 10, and then click Apply (green check mark), as shown in Figure 7.89. The drawing file should now be similar to Figure 7.90.

13. Again using the Window menu on the menu bar, select ch7_editdemo.ipt. The part should now resemble Figure 7.91. It is not necessary to save either file. Note that upon installation of Inventor, if the bi-direction associativity option from drawing to part files was not selected, then the drawing file edit made in step 12 would not affect the model.

QUESTIONS

1. T F The third angle projection technique results in the top view of an object being positioned below the front view and the left view being positioned to the right of the front view.
2. T F The units precision of a dimension cannot be modified in the drawing environment
3. The Text tool cannot be used to add which symbol to a drawing file?
   - **a.** ±
   - **b.** ∅
   - **c.** ≠
   - **d.** °
4. The base view can be created using the ________ tool, while the ________ tool is used to create other views from the primary view.
5. Which of the following is not a supplementary view that can be created in the drawing environment?
   - **a.** Auxiliary View
   - **b.** Descriptive View
   - **c.** Section View
   - **d.** Detail View
   - **e.** Broken View
   - **f.** Break Out View
6. T F Centerlines can only be added by manually using the Centerline command.
7. T F The Hole/Thread Notes tool can only be applied to holes or threads that have been created using the Hole or Thread tool.
8. Which cannot be modified from the Edit View dialog box?
   - **a.** Display labels
   - **b.** Model dimensions
   - **c.** View scale
   - **d.** View style
9. T F Only parametric (model) dimensions of a part can be displayed in the drawing environment (i.e. reference dimensions cannot be added).
10. T F The hatch pattern angle can be modified.
11. T F Model dimensions can be edited from the drawing environment.

PROBLEMS

1. Create a drawing of the **Ball** part, similar to that shown in the figure below.

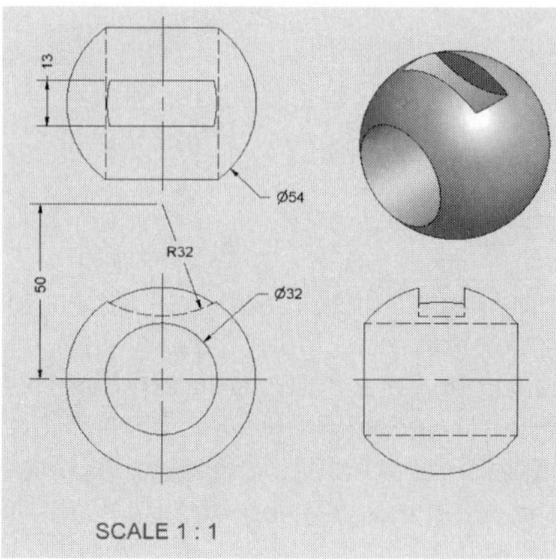

2. Create a drawing of the **Handle** part, similar to that shown in the figure below.

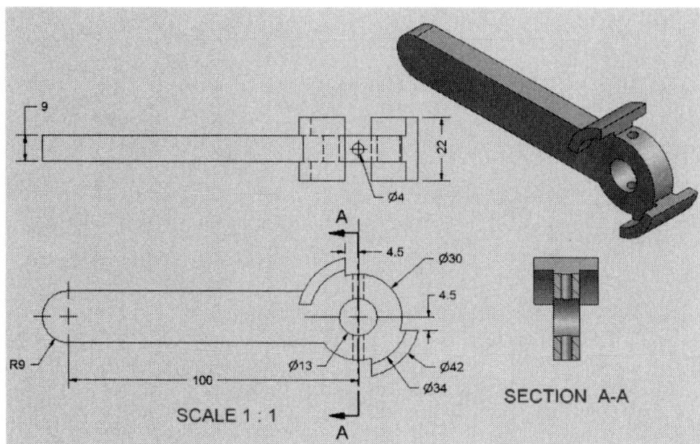

3. Create a drawing of the **Packing Nut** part, similar to that shown in the figure below.

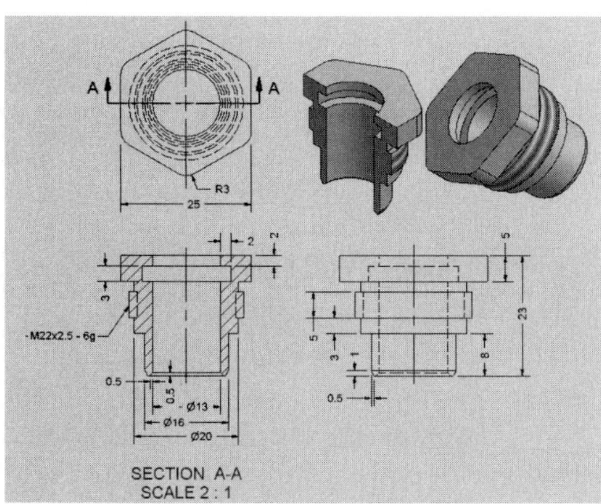

8 CHAPTER

Advanced Assembly Modeling

LEARNING OBJECTIVES

- Use either the Properties or the Materials dialog box to assign physical properties to a part
- Use the Center of Gravity tool to display the coordinates of the center of gravity of an assembly
- Use the Bill of Materials tool to organize, display, and export part properties
- Use the Interference tool to detect physical interferences between parts
- Use the Drive Constraint tool to drive a single assembly constraint through a range of motion
- Use the record option from the Drive Constraint tool to create an AVI animation file showing the motion of an assembly with moving parts

Introduction

In this chapter we will look more closely at assembly files. Within the assembly environment we will assign materials and calculate the mass properties of Inventor parts. We will then create a bill of materials for the assembly, as well as display and determine the center of gravity of the assembly. Next we will see how to detect interferences between solid parts within an assembly.

Solid modeling software is gradually being transformed into what might be called product development or virtual prototyping software. Inventor's ability to simulate the motion of moving parts within an assembly is one indication of this trend. In the last part of this chapter we will explore Inventor's ability to determine the range of motion of moving parts within an assembly. We will also learn how to make animation files that show this motion.

Physical Properties

Inventor can be used to automatically calculate the physical properties (e.g., volume, surface area, center of gravity) of modeled parts. If a material is assigned to the part, then the mass of the part is also determined. These mass and material properties can then be included in a bill of materials or used for weight estimating.

The physical properties of Inventor parts can be accessed either from part or assembly files. From a part file, select File > iProperties . . . from the menu bar (or right-click on the part icon in the browser, and select Properties from the context menu). From an assembly file, double-click on the part in either the graphics window or the browser to move to part mode within the assembly. Now select File > iProperties . . . from the menu bar (or right-click on the part icon in the browser, and select Properties from the context menu). In either case, the Properties dialog box opens. Select the Physical tab.

A default material is automatically assigned to Inventor parts. Default materials have the same density as fresh water (1000 kg/m^3 or 62.4 lb/ft^3). Figure 8.1 shows the Physical properties tab for a part that has been assigned the properties of mild steel. This was done by selecting a material from the Material drop-down list, and then clicking either the OK or the Apply button. From the information provided in Figure 8.1, we know that the density of mild steel is 0.284 lbm/in^3, and that the weight of the part is 0.217 pounds. The center of gravity of a part is referenced from the center point of the part's reference work planes (accessed from the Origin folder in the browser).

The physical properties of Inventor parts can also be specified by using material formats. To access the material formats, select Format > Material from the menu bar. The Materials dialog box shown in Figure 8.2 opens.

The Material List shows the materials available in the file. Select a material from the drop down list to display the properties of the material. Note that a Rendering Style

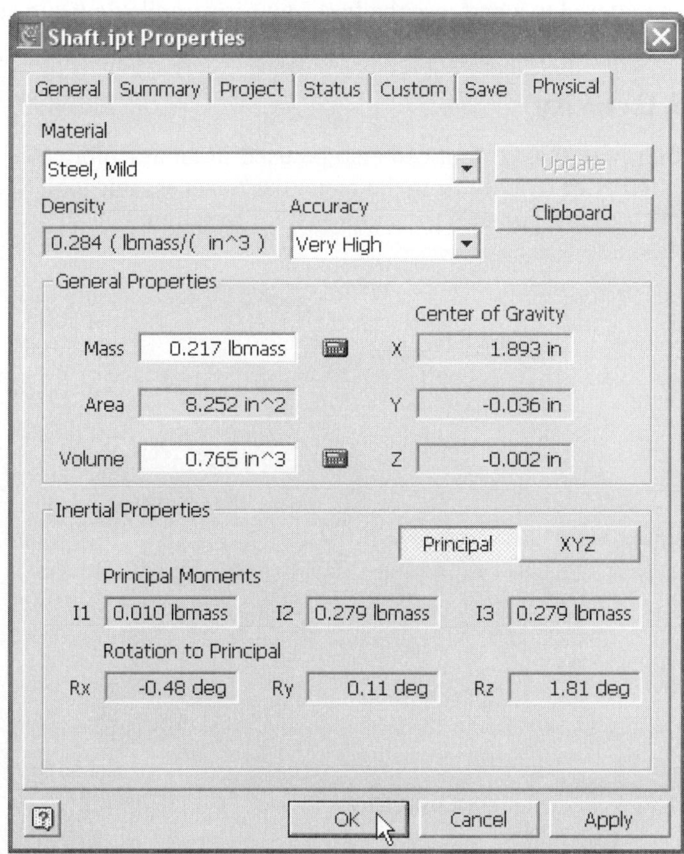

Figure 8.1 - Physical properties

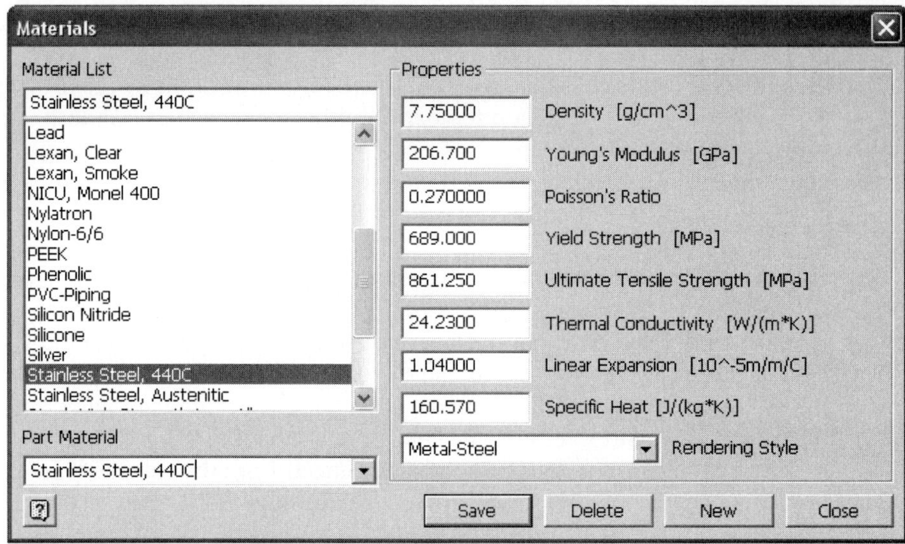

Figure 8.2 - Material formats

is associated with a given material. Although the Rendering Style controls the color of the part, this color can be changed, if desired.

To assign a material to a part, use the Part Material dropdown list to select a material. Select Close to close the dialog box.

Center of Gravity

The Center of Gravity (COG) tool can also be used in an assembly (see Figure 8.3). Select View > Center of Gravity from the menu bar to access this tool. See the discussion of the COG tool in Chapter 4 for additional information.

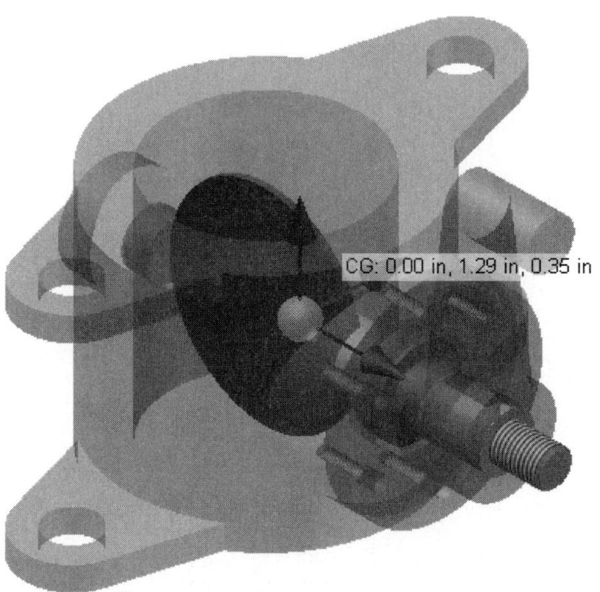

Figure 8.3 - Butterfly valve assembly with COG displayed

Bill of Materials

A Bill of Materials tool can be accessed from within an assembly by selecting Tools > Bill of Materials . . . from the menu bar. Figure 8.4 shows the Bill of Materials dialog box for a butterfly valve.

In the upper left corner of the dialog box, two buttons, Column Chooser and Export, are particularly useful. Selection of the Column Chooser button opens the Bill of Material Column Chooser dialog box, shown in Figure 8.5.

In Figure 8.5 the mass property is about to be added to the bill of materials (highlight the mass property in the list on the left, then click the Add button). Properties listed on the right will be displayed in the bill of materials. To remove a property, highlight it on the right, and then click Remove. A modified BOM is shown in Figure 8.6. To export the bill of materials as an Excel spreadsheet, select Export.

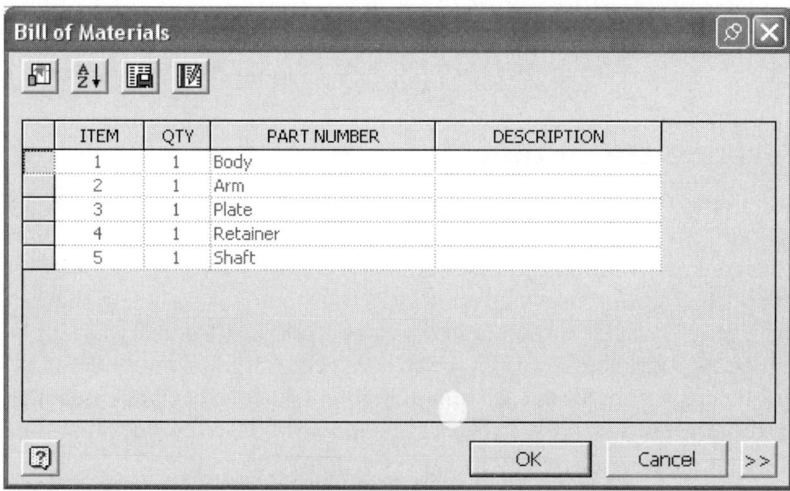

Figure 8.4 - Butterfly valve BOM

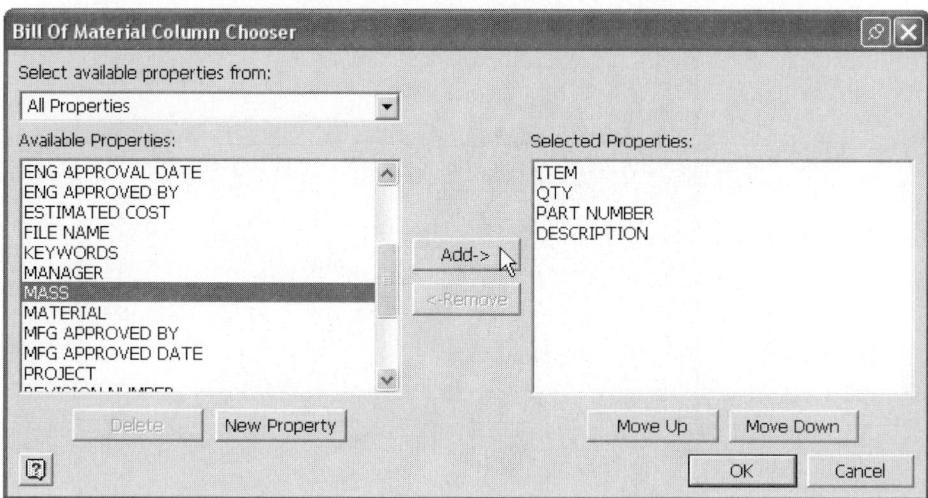

Figure 8.5 - Column chooser

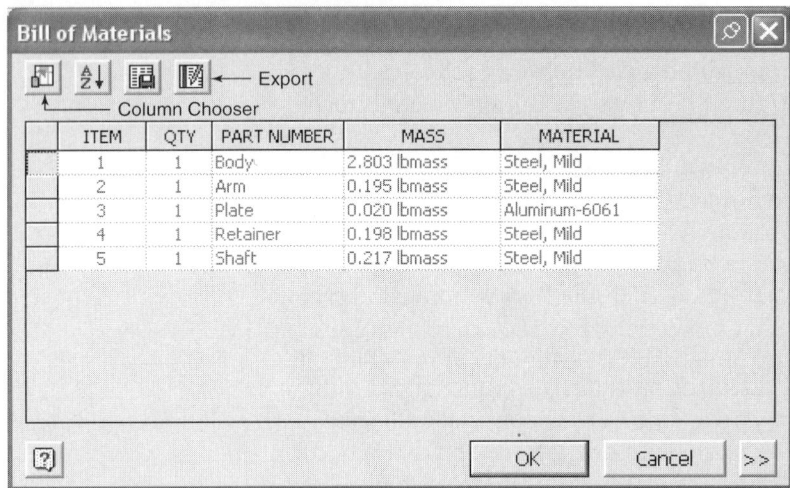

Bill of Materials

ITEM	QTY	PART NUMBER	MASS	MATERIAL
1	1	Body	2.803 lbmass	Steel, Mild
2	1	Arm	0.195 lbmass	Steel, Mild
3	1	Plate	0.020 lbmass	Aluminum-6061
4	1	Retainer	0.198 lbmass	Steel, Mild
5	1	Shaft	0.217 lbmass	Steel, Mild

Figure 8.6 - Modified BOM

Interference Analysis
Define Set # 1
Define Set #2
OK
Cancel

Figure 8.7 - Interference analysis tool

Interference Detection

In addition to the ability to compute physical properties, another important assembly modeling tool provides the ability to detect physical interferences between parts. The Interference tool in Inventor is accessed from the menu bar by selecting Tools > Analyze Interference. The Interference Analysis dialog box opens, as shown in Figure 8.7.

Either one or two sets of components can be selected for analysis. If two sets are selected, then interferences between the two sets are found. If only one set is selected, then interferences within the set are found. Figure 8.8 shows a clamp assembly that has been checked for interferences. In this case a total of 5 interferences have been found.

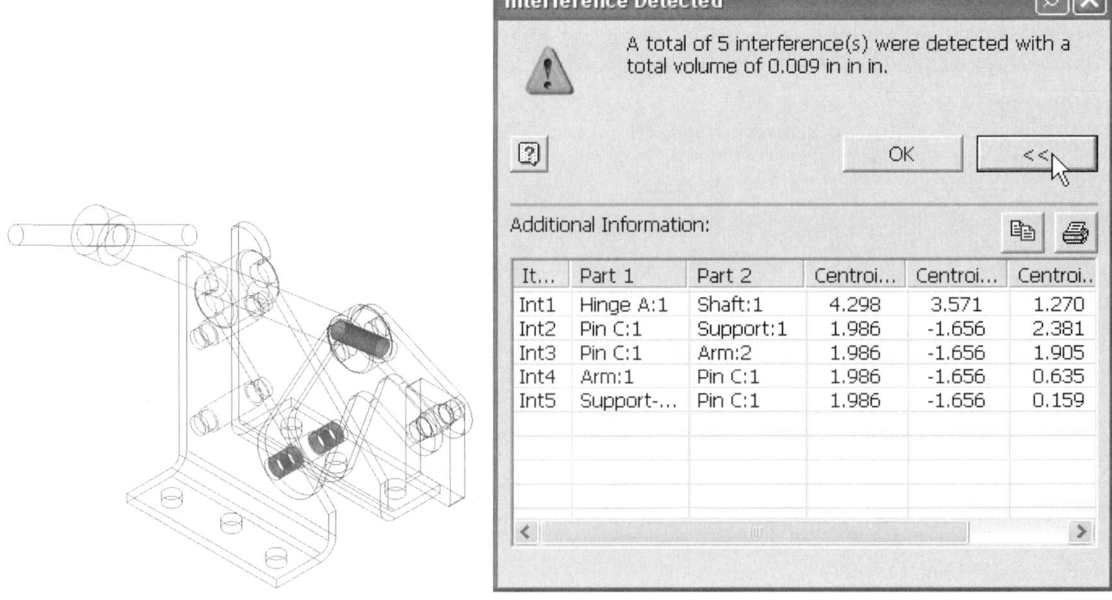

Interference Detected

A total of 5 interference(s) were detected with a total volume of 0.009 in in in.

Additional Information:

It...	Part 1	Part 2	Centroi...	Centroi...	Centroi..
Int1	Hinge A:1	Shaft:1	4.298	3.571	1.270
Int2	Pin C:1	Support:1	1.986	-1.656	2.381
Int3	Pin C:1	Arm:2	1.986	-1.656	1.905
Int4	Arm:1	Pin C:1	1.986	-1.656	0.635
Int5	Support-...	Pin C:1	1.986	-1.656	0.159

Figure 8.8 - Analyze interference

On the assembly model interference solids are temporarily displayed to show the locations where parts interfere. Clicking the More >> button on the message box reveals a table of the interferences identifying the offending parts. With this information, it is fairly easy to edit the parts directly within the assembly in order to eliminate the interferences.

Motion Simulation

Driving an Assembly Constraint

In Chapter 6 we saw that by using assembly constraints to position parts relative to one another in an assembly, it is possible to simulate the motion of any moving parts within the assembly. This is because the moving parts still have unconstrained degrees of freedom. The motion is accomplished by clicking and dragging on the partially constrained parts. In the case of both the ball valve and the garlic press assemblies however, this motion can result in interferences between solids that are not possible in the real world. Examples of this kind of interference are shown in Figure 8.9.

Inventor's Drive Constraint tool can be used to drive a single assembly constraint through a range of motion. This motion can then be saved in an animation file. As we will see, the drive constraint tool can also be used to restrict this range of motion so that part collisions such as those depicted in Figure 8.9 are avoided.

In order to use the drive constraint tool, part motion within an assembly must first be restricted by adding another constraint. Having added this constraint, the moving parts should now be fully constrained within the assembly; they cannot move. The newly added constraint is then designated as a *drive constraint,* and a range of motion can be specified.

Figure 8.10 shows a butterfly valve assembly. Nonmoving parts are shown in phantom. At this point within the assembly file it is possible to click and drag on the handle so that the handle, shaft, and elliptical plate all rotate about the shaft axis for a full 360 degrees.

The problem with the previously described situation is that in reality the elliptical plate closes the valve passage well before it reaches a horizontal orientation. Figure 8.11 shows a wireframe display of the assembly with the elliptical plate oriented horizontally. The Interference tool has been used to identify a physical interference between the plate and the valve body. Clearly this is not physically possible.

Figure 8.9 - Part interferences

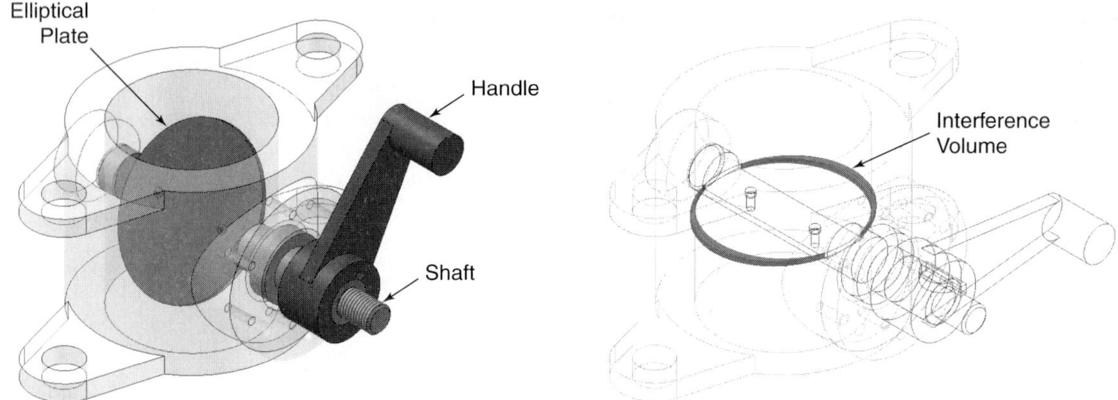

Figure 8.10 - Butterfly valve

Figure 8.11 - Interference detected

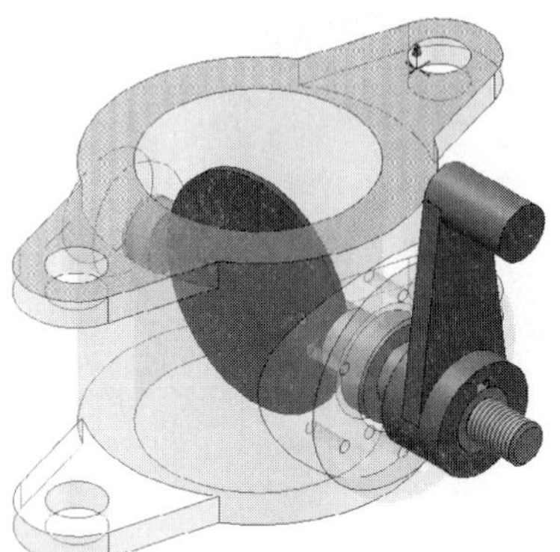

Figure 8.12 - Angle constraint (90 degrees) applied

Figure 8.13 - Drive constraint designation

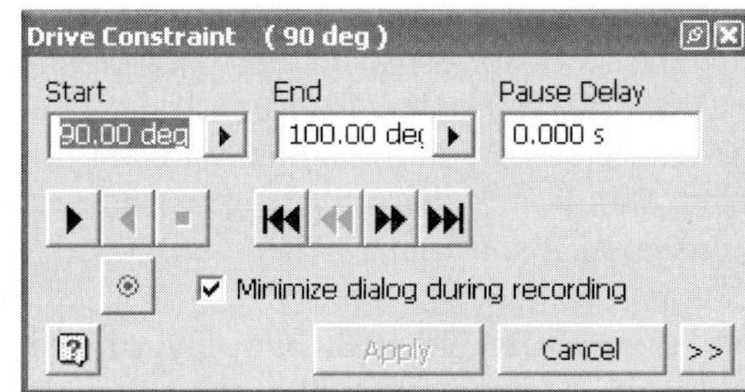

Figure 8.14 - Drive constraint dialog box

Figure 8.12 shows an angle constraint of 90 degrees being added between the plate and the top face of the valve body. Once this constraint is applied, none of the valve parts can move.

The angle constraint is now designated as a drive constraint by right-clicking on the constraint in the assembly browser, and then selecting Drive Constraint, as shown in Figure 8.13.

At this point the Drive Constraint dialog box opens, as shown in Figure 8.14. Clicking on the Forward button causes the mechanism to advance; clicking the Reverse button returns the mechanism to the start (i.e., vertical) position.

Now change the End value to 0 degrees and select the More button to expand the dialog box. With the Collision Detection box checked, the assemblage is once again set in motion by clicking the Reverse button. This time, motion stops at 32 degrees (Figure 8.15A), and a warning message appears (Figure 8.15B). This is where the plate first comes in contact with the valve body (Figure 8.15C). After adjusting the End value to 33 degrees (and perhaps the Start value to −33), the Record button can be used to create an AVI animation file showing the complete range of motion of the assembly.

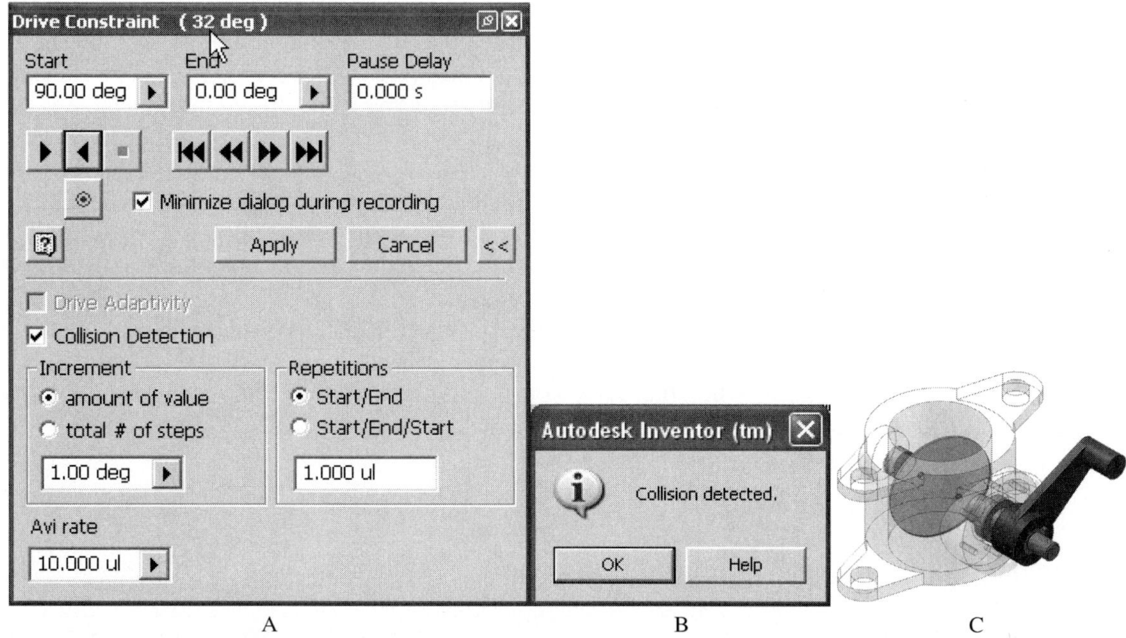

Figure 8.15 - Drive constraint with collision detection

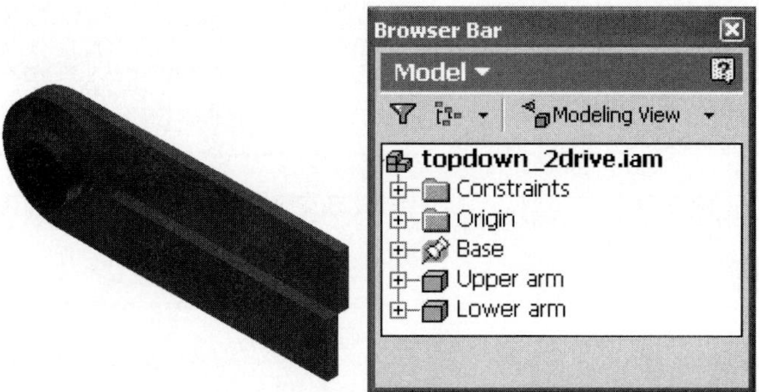

Figure 8.16 - Scissor assembly

The Increment area of the expanded Drive constraint dialog box is used to specify the smoothness of the animation. As currently set in Figure 8.15A, a snapshot will be taken in one degree increments, starting at 90 degrees. Increasing the increment angle will cause the animation to be faster but choppier.

To capture the motion from beginning to end and back again, from the Repetitions area choose Start/End/Start and change the number of repetitions to 2.

Only one constraint can be designated as driven at any one time. However, it is possible to use parameters and equations to simultaneously drive more than one constraint at a time. The simple scissor assembly shown in Figure 8.16 consists of 3 parts; the fixed base and two moving parts, the upper and lower arms. Their motion is such that when the upper arm rotates counterclockwise about the axis of the base, the lower arm rotates clockwise.

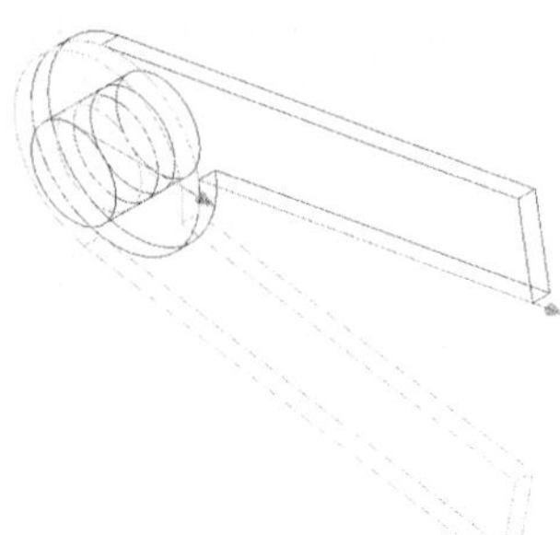

Figure 8.17 - Angle constraint

Initially both arms have a single degree of freedom, rotation about the base axis. An angle constraint (0 degrees) is added between a default axis on the base and the lower longitudinal edge of the upper arm (see Figure 8.17). This angle is designated as a drive constraint. At this point the upper arm can be driven through a range of motion, but not the lower arm.

Next we create a similar angle constraint between the base and the lower arm. This fully constrains the lower arm. Finally, we use the Parameter f_x tool on the assembly panel to write an equation relating the two angle constraint parameters. In the Parameters dialog box shown in Figure 8.18, the name of the driven angle constraint (between the base and the upper arm) has been renamed "drive." An equation has been written for the value of the second angle constraint, d13. Since the motion of the lower arm is equal and opposite to that of the upper arm, the equation is d13 = −drive.

Finally, when the Drive constraint tool is applied, the two arms both move, but in opposite directions (see Figure 8.19).

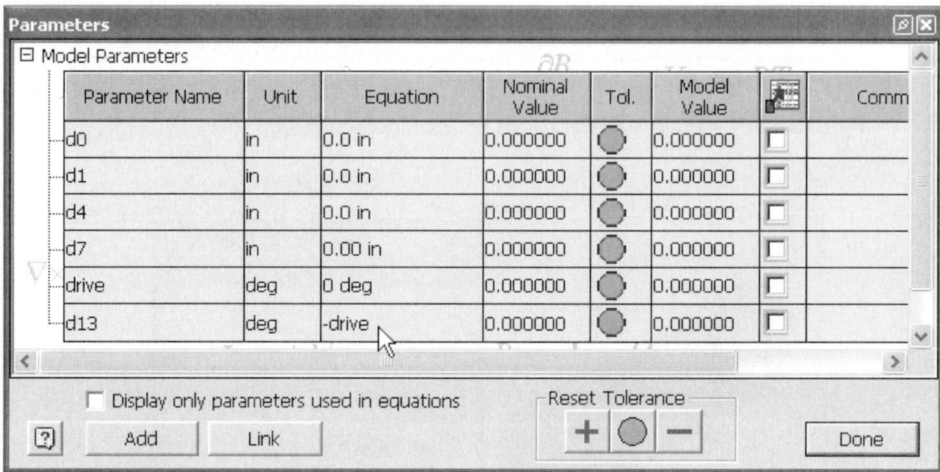

Figure 8.18 - Parameters dialog box

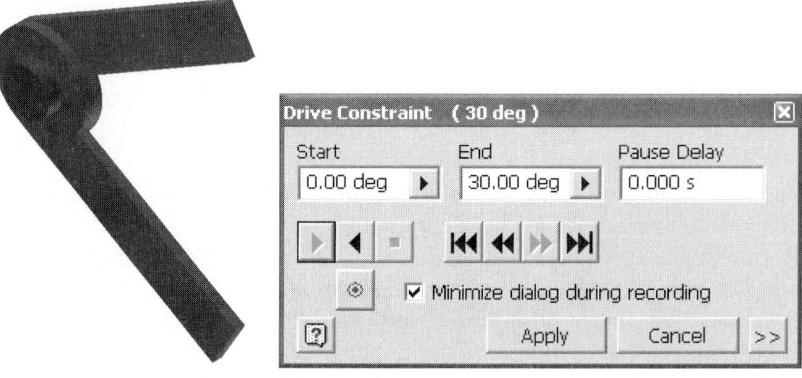

Figure 8.19 - Drive two related motions

Detailed Steps

1. Select File > Projects . . . from the menu bar, then make the Ball Valve project active.
2. Using the information provided in Table 8.1, open each of the Ball Valve part files and assign the material listed in the table. Note that in some cases it is also necessary to change the color of the part. Use either the Properties dialog box (File > iProperties . . . > Physical) or the Materials dialog box (Format > Material) to assign materials. Once the material (and, if necessary, the color) has been assigned, Save and Close the file.
3. Open the **Ball Valve** assembly file. In the graphics window the assembly should resemble Figure 8.20.
4. We will now check for part interferences. Select Tools > Analyze Interference from the menu bar. The Interference Analysis dialog box opens, as shown in Figure 8.21. Select all the assembly components from the browser (only one occurrence of the hex head bolt need be selected), then click OK.

   A message box appears indicating that an interference has been found. Selecting the More >> button reveals that the interference is between the Spring

Table 8.1 Ball Valve Parts

Ball Valve Part	Material	Color (Override)
Ball	Stainless, Steel, 440C	
Body	Bronze. Soft Tin	
Body Cap	Bronze. Soft Tin	
Bumper	Silicone	
Handle	Steel, Mild	
Oring1	Silicone	
Oring2	Silicone	
Oring3	Silicone	
Packing Nut	Bronze. Soft Tin	
Pin	Steel, Mild	Red (Flat)
Seat	Steel, Mild	Yellow (Flat)
Spring	Steel, Mild	Sea Green
Stem	Bronze. Soft Tin	Beige (Light)

Figure 8.20 - Ball valve assembly

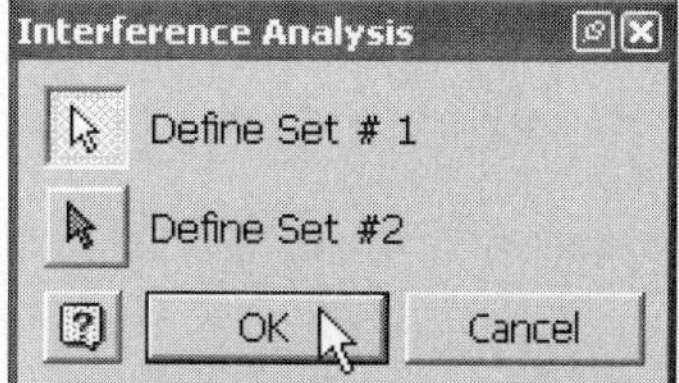

Figure 8.21 - Interference analysis dialog box

and the Seat parts, as shown in Figure 8.22. We will now eliminate this interference.

5. Both the Spring and Seat parts are components of the Body Cap subassembly. Double-click on the Body Cap subassembly in the browser to activate it, and then turn off the visibility of the other parts in the subassembly (e.g., Body Cap and Oring1). The browser and screen should now resemble Figure 8.23.

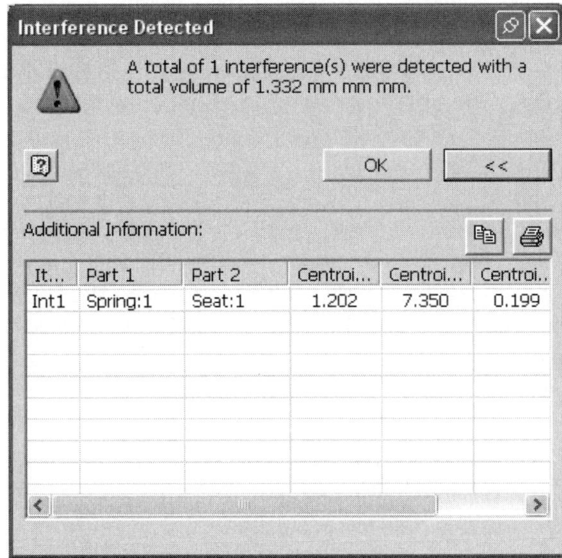

Figure 8.22 - Interference found

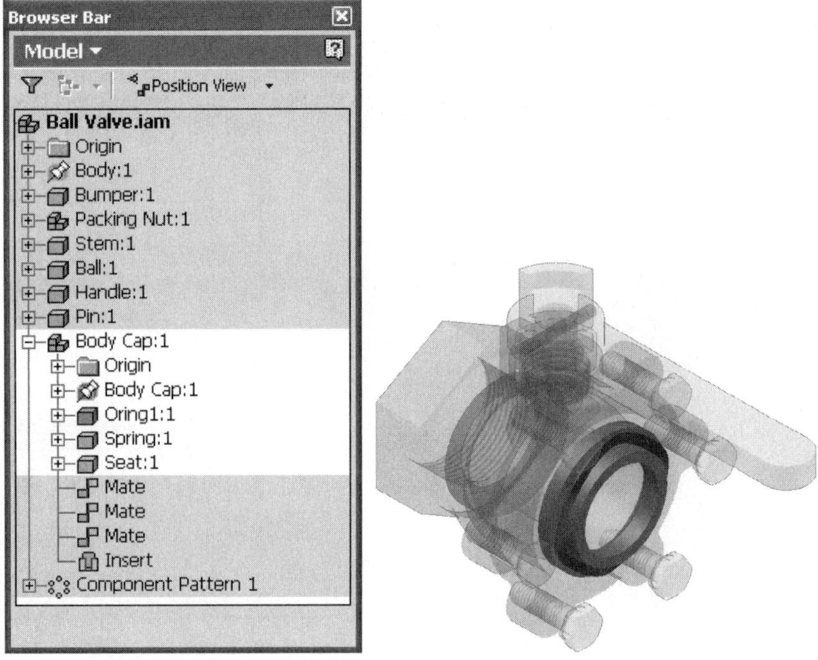

Figure 8.23 - Preparation to correct interference

6. Choose Tools > Analyze Interference from the menu bar. Select the Spring and Seat parts, and then click OK. The same interference is reported. The interference between the two parts is shown in Figure 8.24.
7. To eliminate the interference we will modify the Spring part. After clicking OK to close the Interference message box, double-click the Spring (in either the graphics window or the browser) to activate the part.
8. Recall that in modeling the Spring several parameters were named. We will now change the value of the "id" parameter. In the panel bar select the Parameters f_x tool. The Parameters table for the Spring opens. Locate the "id" parameter, and change its value from 41 to 42, as shown in Figure 8.25. Click Done to close the Parameters table.

Figure 8.24 - Interference volume

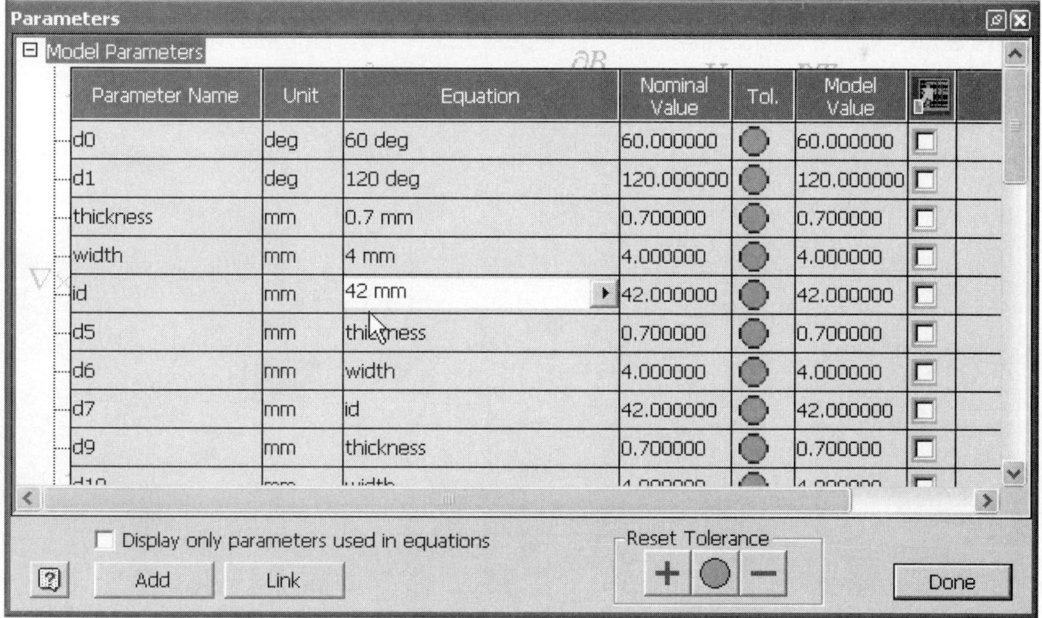

Figure 8.25 - Change "id" parameter value

9. Right-click and select Finish Edit; we are now at the level of the Body Cap subassembly.
10. Select Tools > Analyze Interference. After once again selecting the Spring and Seat parts, and clicking OK, you should see the message shown in Figure 8.26.
11. Turn on the visibility of the Body Cap and Oring1 parts.
12. Right-click and select Finish Edit; we are now back at the assembly level.
13. Isometric View.
14. Once again check for interferences. If other interferences exist, use the approach described in the previous steps to eliminate them.
15. We will now use the Center of Gravity tool. Choose View > Center of Gravity from the menu bar. The message box shown in Figure 8.27 appears. Click OK. The center of gravity symbol is now displayed. See Figure 8.28.
16. To display the coordinates of the center of gravity on screen, first change the selection mode to Select Features, as shown in Figure 8.29. The Select tool is available on the menu bar.
17. Now move the cursor over the COG symbol in the graphics area. Use the Select Other tool to cycle through the selectable features until you see the coordinates displayed (Figure 8.30).
18. Save the **Ball Valve** assembly file. This completes the tutorial.

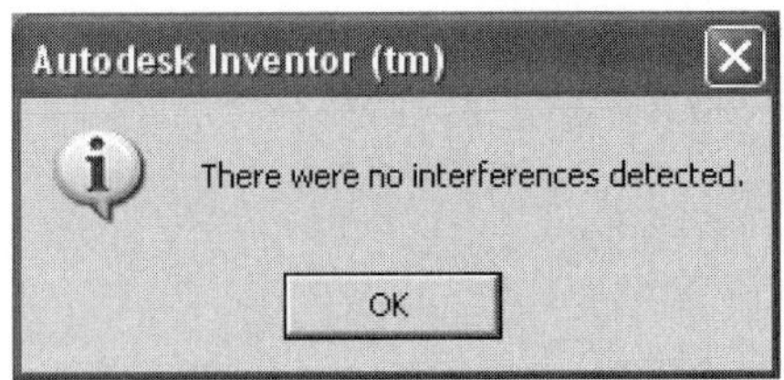

Figure 8.26 - No interferences message

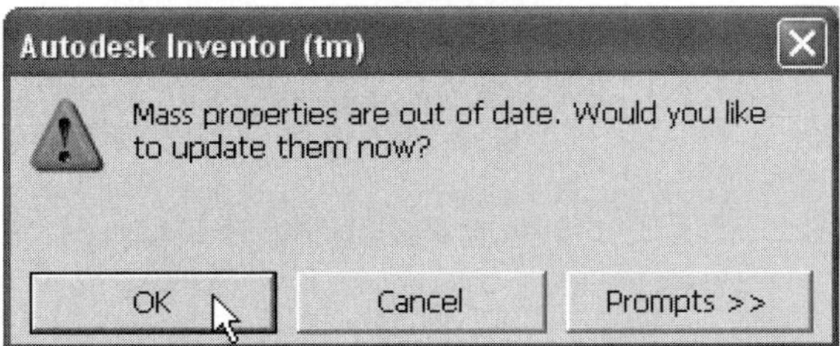

Figure 8.27 - Mass properties out of date message

Figure 8.28 - Ball valve with COG symbol

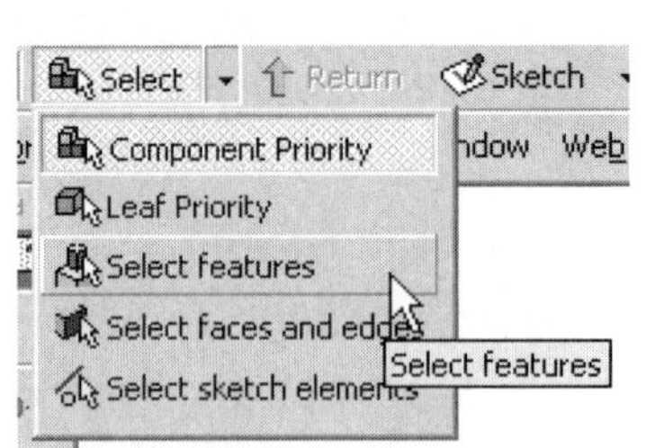

Figure 8.29 - Change selection mode

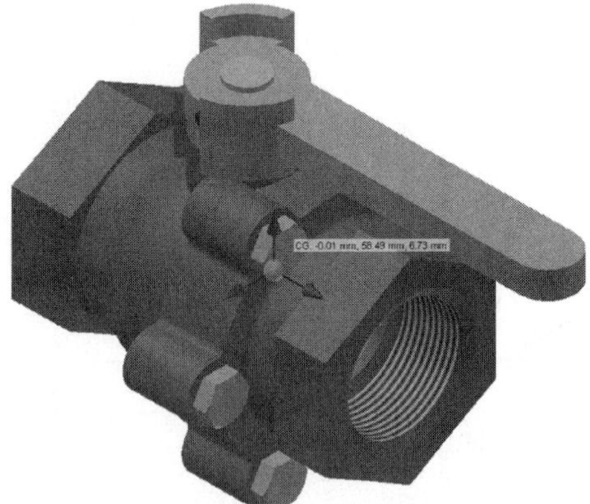

Figure 8.30 - Ball valve with CG coordinates (–0.01 mm, 58.49 mm, 6.73 mm) displayed

Ball Valve Drive Constraint — TUTORIAL 26

■ Detailed Steps

1. Open the **Ball Valve** assembly file.
2. An angle constraint of 90 degrees was placed between faces on the Body and Stem parts in the Ball Valve assembly tutorial in Chapter 6 (step 28, Tutorial 20). This constraint was then named OpenClose (step 30). We will now suppress this constraint, replacing it with a constraint that performs the same function, but makes more sense from a physical standpoint. At the top of the assembly browser, change to Modeling View so that all of the assembly constraints are contained in a single constraints folder in the assembly browser.
3. After expanding the Constraints folder, locate the OpenClose angle constraint, right-click and then select Suppress.
4. Use the Common View tool to change the isometric view to one similar to that shown in Figure 8.31.
5. Zoom in.
6. Use the Place Constraint tool to add the angle constraint shown in Figure 8.32.
7. In the browser, rename this angle constraint; call it Drive.
8. In the browser right-click on the Drive constraint and select Drive Constraint, as shown in Figure 8.33. The Drive Constraint dialog box opens.
9. In the Drive Constraint dialog box, click on the Forward button. The handle should rotate 10 degrees in a clockwise direction away from the stop on the valve body. Now click the Reverse button to return the handle to the closed position.
10. Change the value in the End box to 90 degrees, and then click on the More >> button. After making the settings changes shown in Figure 8.34, click the Forward button. On your screen you should have seen the valve handle first rotating clockwise, and then counterclockwise 90 degrees, simulating the opening and closing of the valve.
11. Zoom out, and then use the 3D Rotate tool to adjust the view of the Ball Valve assembly so that it resembles Figure 8.35.

Figure 8.31 - View re-oriented

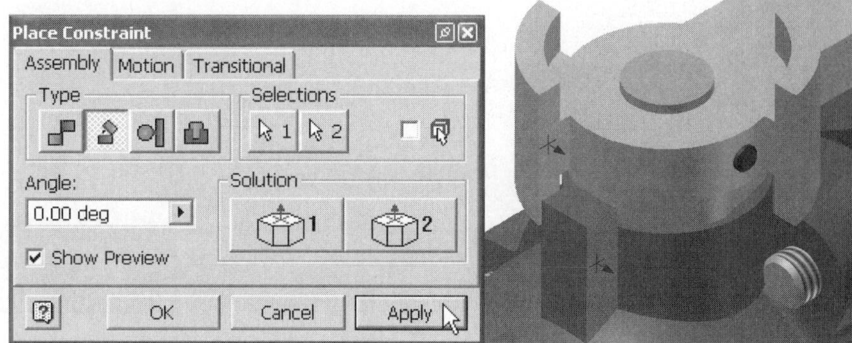

Figure 8.32 - Angle constraint

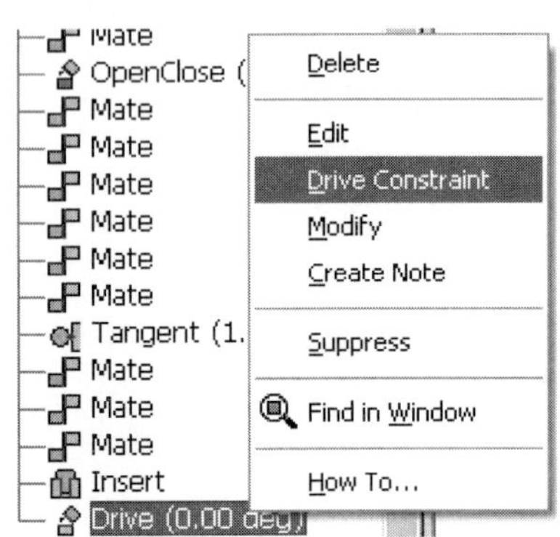

Figure 8.33 - Drive constraint access

Figure 8.34 - Drive constraint dialog box

Figure 8.35 - Re-oriented view

12. Change to a Perspective Camera view, using the Camera tool on the standard toolbar, as shown in Figure 8.36. We are about to create a movie showing the range of movement of the Ball Valve. In order to reduce the size of the movie file, it is a good idea to first reduce the size of the Inventor window.
13. If the Inventor window is maximized, left-click on the middle button in the upper right corner of the Inventor window to reduce the size of the window. The size of the window can now be further adjusted by moving the cursor to a corner of the window and dragging.
14. If the Drive Constraint window is closed, reopen it by right-clicking on the Drive constraint in the browser, and then selecting Drive Constraint. We are now ready to record. Click the Record button.
15. Before the movie is recorded, it must be named and placed in a folder. In the Open dialog box, navigate to the Ball Valve project workspace, if necessary. Give the file a suitable name, and then click Open, as shown in Figure 8.37.

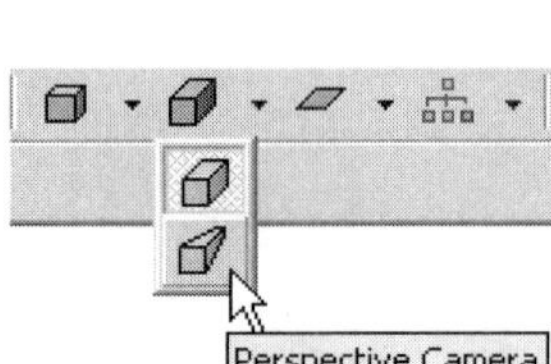

Figure 8.36 - Perspective camera view

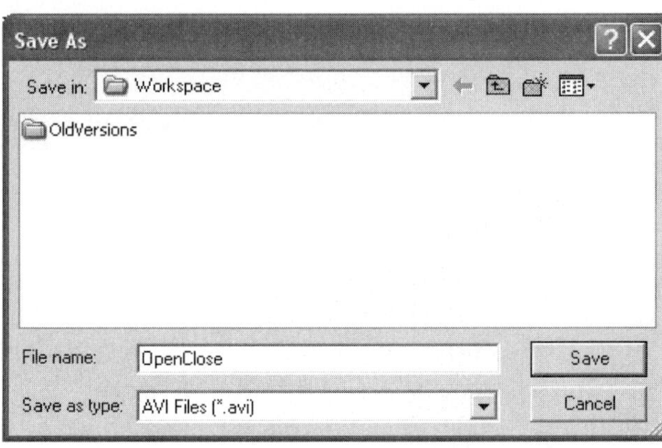

Figure 8.37 - Name AVI file

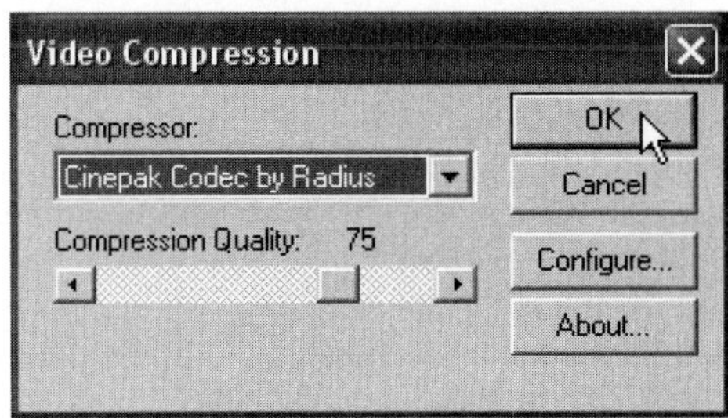

Figure 8.38 - Video compression selection

16. A Video Compression window opens. Accept the default, and click OK as shown in Figure 8.38.

17. Any objects overlapping the graphics window will be recorded; make sure that the graphics window is completely visible. Click on the Forward button on the Drive Constraint dialog box to start recording. Once the motion is complete, click on the Record button to stop the recording. In order to view the Movie, open the AVI file in a video player (e.g., Windows Media Player).

18. Save the **Ball Valve** assembly file. This completes the drive constraint tutorial. ■

QUESTIONS

1. T F The default material density of a part in Inventor is that of steel (0.284 lb/in^3).
2. T F The Bill of Materials tool can be used to export part properties to Microsoft® Excel, as well as other application software programs.
3. T F While the Interference tool will highlight the interferences on the assembly model, it does not give a list of the parts that interfere.
4. T F Only an angle constraint can be driven using the Drive Constraint tool.
5. Which button is used to record an AVI animation showing the motion of an assembly?

   a.

   b.

   c.

   d. >>

PROBLEM

1. For the clamp shown in the figure below:

- Assign material properties to the clamp parts. Assume that all of the parts are made of mild steel.
- Check for interferences, eliminating any that may be found.
- Create a bill of materials for the clamp. The bill of materials should have the following columns: Item, Quantity, Part Number, Mass, Material.
- If available, export the bill of materials to Microsoft Excel. How much does the clamp weigh?
- Use the drive constraint tool to create animation showing the range of motion of the clamp. Note that the Shaft component on the clamp is currently free to translate through the hole in the Hinge B component.

NOTE: The clamp parts and the assembly file are located in the QAclamp Problem folder on the CD.

CHAPTER 9

Presentation Files and Working Drawings

LEARNING OBJECTIVES

- Use the Create View tool to bring an assembly file into the presentation environment
- Use the Tweak Components tool to move components in a presentation file
- Use the Filter button on the browser to change the display of the presentation file data
- Group sequences so that tweaks occur simultaneously
- Change the camera view for an individual sequence
- Play the tweaks or sequences in an animation
- Use the Animation tool to record animations of the tweaks or sequences
- Use the Animation tool and the Edit Sequences and Tasks dialog box to review and edit animations
- Create a working drawing of an assembly, including both assembled and exploded views
- Use the Parts List tool to create a parts list for an exploded view
- Use the Column Chooser button in the Parts List environment to modify the properties given in the parts list
- Use the Balloon tool to add balloons relating the exploded view parts to the parts list
- Add additional sheets to a drawing set-up

Introduction

Thus far we have seen that with a parametric solid modeler like Autodesk Inventor parts can be modeled, 2D drawing information can be extracted from the parts, and parts can be combined to form assemblies. We have also seen that all of this data, stored either in part, drawing, or assembly files, is interlinked and easily modified.

In this chapter we will look at documenting assemblies. Presentation files are the basis for creating an exploded view of an assembly. In conjunction with this, an animation file showing the assembly of the product is generated. Once the exploded view has been created in the presentation file, working drawings can be created that include exploded and assembly section views, along with a parts list (i.e., bill of materials).

Presentation Files

In addition to part, drawing, and assembly files, Inventor also includes the Presentation file format, which has an .ipn file extension. The primary function of the presentation mode is to create an *exploded view* of an assembly that can later be used in a working drawing. An exploded view of the butterfly valve is shown in Figure 9.1. Tools are also available in the presentation environment that allow the exploded view creation process to be recorded and played back, either in Inventor or in an outside application.

The Presentation panel bar is shown in Figure 9.2. The Create View tool is used to bring an assembly file into the presentation environment. Although it is possible to apply a uniform explosion factor to the assembly as it is inserted (see the Select Assembly dialog box in Figure 9.3), it is generally better to move or tweak the parts manually, using the Tweak Components tool. A *tweak* is a transformation (translation or rotation) made to a component in an exploded view.

Figure 9.1 - Exploded view of butterfly valve

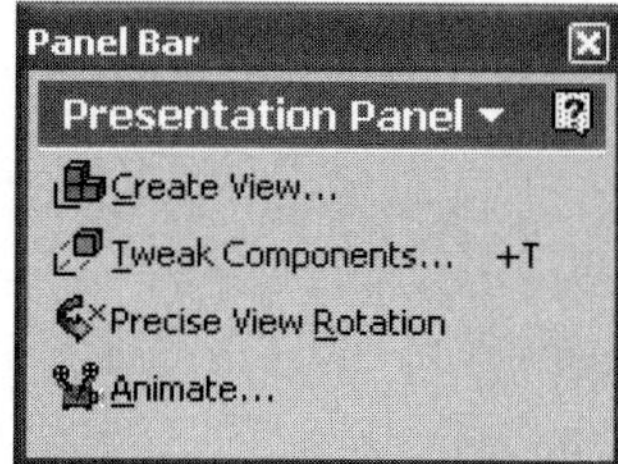

Figure 9.2 - Presentation panel tools

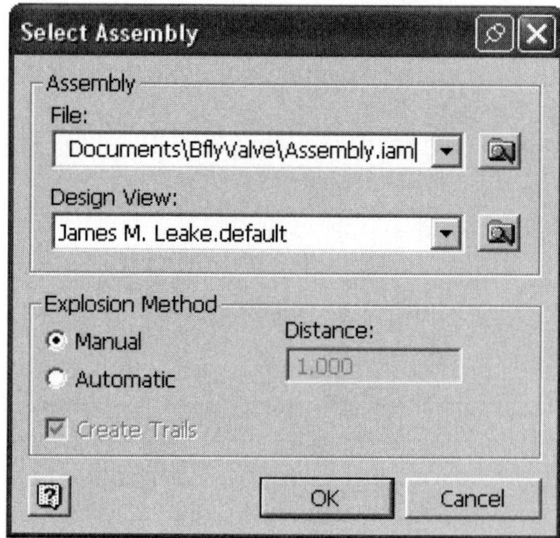

Figure 9.3 - Select assembly to place in presentation file

The Tweak Component dialog box is shown in Figure 9.4. Here it is necessary to specify components to be tweaked, as well as a tweak direction. The tweak direction refers to the transformation to be applied to the selected component(s). The transformation can either be a translation along or a rotation about an axis.

Direction triads appearing on the model geometry (when the Direction button is selected) are used to specify the tweak direction. Tweak direction and components can be selected in any order. Although the direction triads are associated with the geometry of a specific component, this direction can be used to tweak any component(s) selected. A transformation or tweak magnitude (distance or angular rotation) must also be specified.

Once the components and the transformation are set in the dialog box, clicking the green check mark applies the tweak. The Clear button clears the current dialog box settings in preparation for specifying the next tweak.

As each tweak is created, it is also listed in the presentation browser. Using the Filter icon at the top of the browser, presentation file data can be displayed in three different ways, as shown in Figure 9.5. The butterfly valve exploded view hierarchy when set to *Sequence View* is shown in Figure 9.6.

Explosion1 was created when the Create View tool was used to place the assembly into the presentation file. A *task* is an animation showing the assembly (or disassembly) of an assembly. A task consists of one or more sequences. A *sequence* is a segment of time within an animation consisting of at least one tweak. Within a sequence components can be hidden, and the camera view angle can be changed. It is also possible to group tweaks within a sequence so that they occur simultaneously within the animation. This is shown in Figure 9.7. In Figure 9.7A, the plate will initially be rotated −90 degrees (Sequence2), and then it will be translated 4 inches (Sequence3). After grouping the two sequences (see Figure 9.7B), the shaft will translate and rotate simultaneously (see Figure 9.7C).

Tweaks (and sequences) are played in an animation in the reverse order in which they were initially created. This means that we start with an assembled view; we then apply tweaks in order to disassemble (i.e., explode) the view. Since the animation is played in reverse order, however, it shows the components being assembled. This ordering is reflected in the browser. In Figure 9.7A, Sequence5 was created first, then Sequence4, and so on. When the animation is played, Sequence1 is seen first, then Sequence2, etc. Sequences can be reordered in the browser by clicking and dragging them up or down the hierarchy.

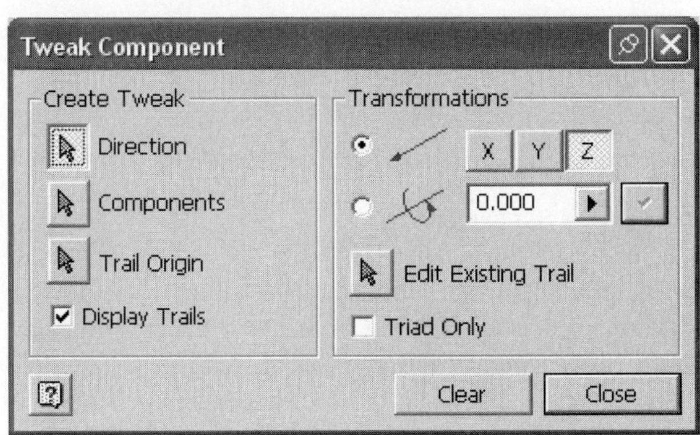

Figure 9.4 - Tweak component dialog box

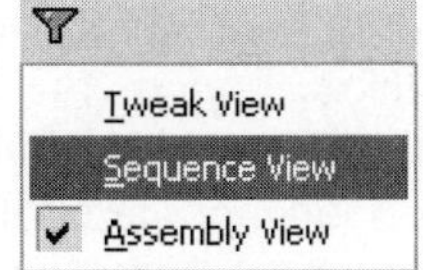

Figure 9.5 - Presentation browser view selection

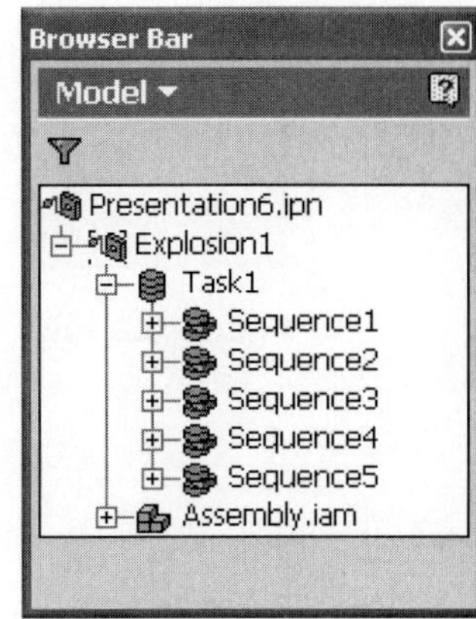

Figure 9.6 - Butterfly valve sequence view

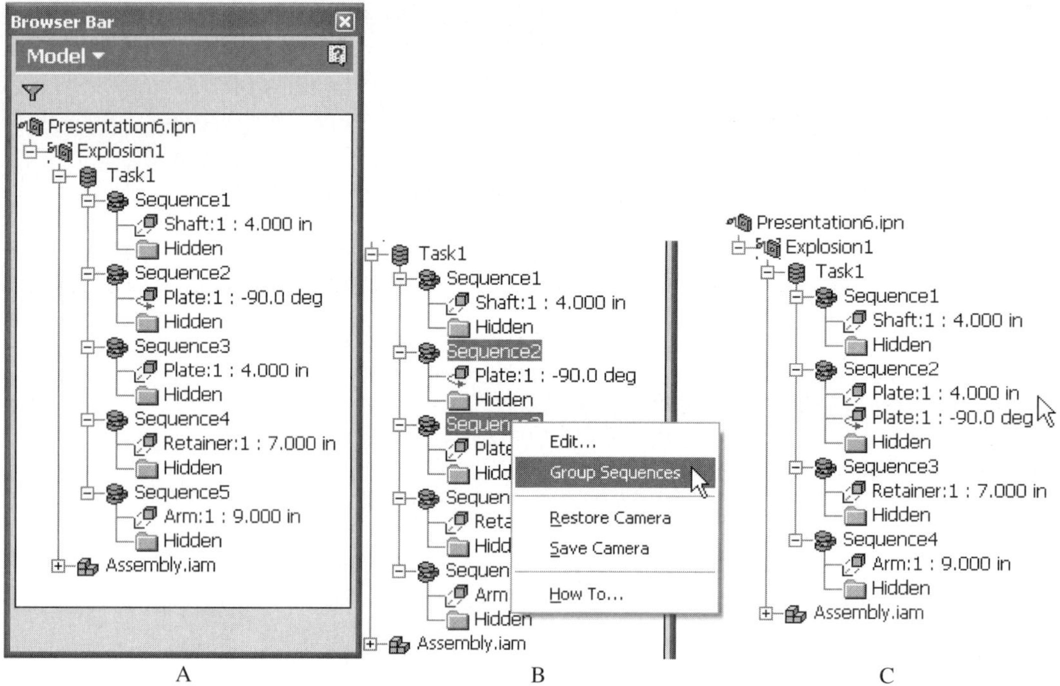

Figure 9.7 - Group sequence

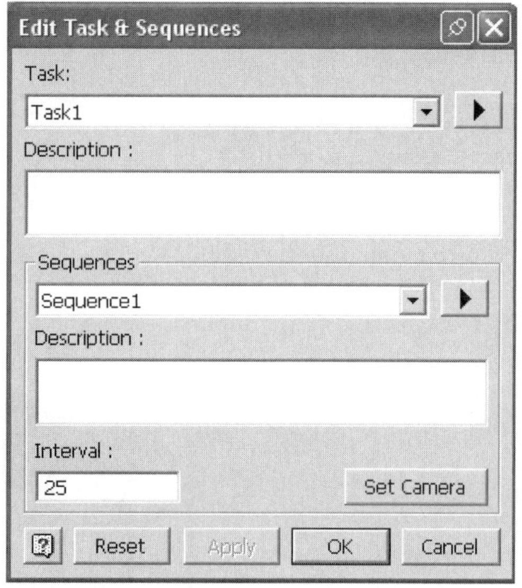

Figure 9.8 - Edit tasks and sequences

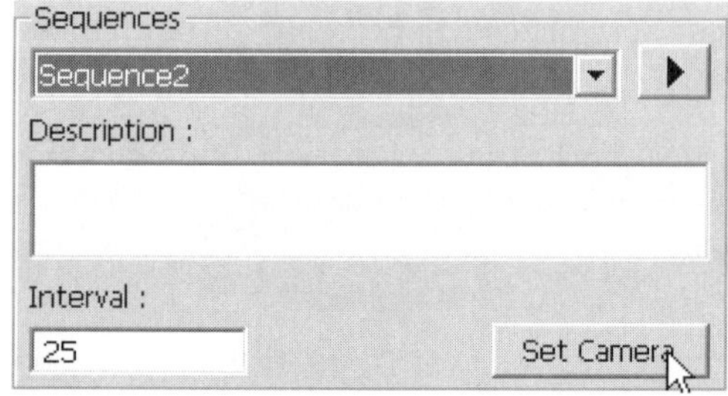

Figure 9.9 - Set camera

Right-clicking on a Task icon in the browser, and then selecting Edit . . . (or double-clicking) opens the Edit Task & Sequences dialog box, shown in Figure 9.8. Clicking the Forward button in the Task area plays the entire animation.

In the Sequences area of the dialog box (see Figure 9.9), sequences can be selected from a drop-down list. Click the Forward button and the sequence plays. The Set Camera button sets the current view (vector and zoom) as the camera angle for the selected sequence. Any of the viewing tools on the standard toolbar can be used to

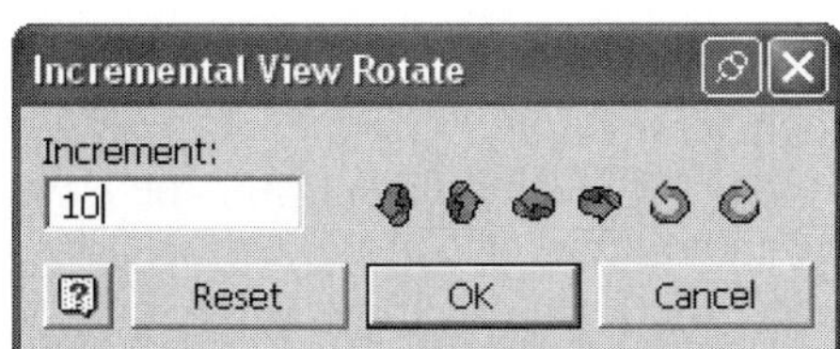

Figure 9.10 - Incremental view rotate dialog box

Figure 9.11 - Animation dialog box

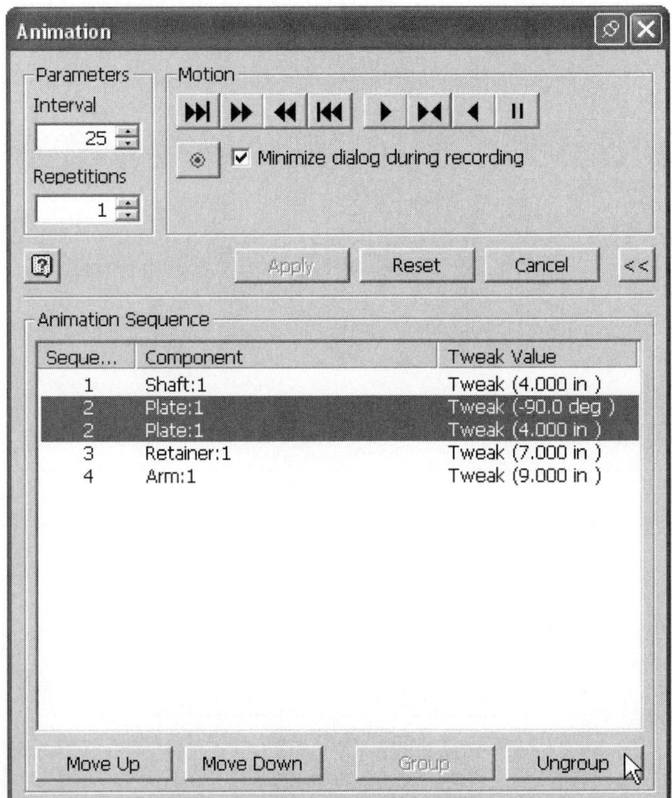

Figure 9.12 - Expanded animation dialog box

specify the view. Each sequence can have its own camera angle. If no specific camera angle is assigned to a sequence, then the current view is used during playback. The Reset button resets the view to its exploded state.

Precision View Rotation is used to tweak the view. The Incremental View Rotate dialog box is shown in Figure 9.10.

The Animate tool on the panel bar opens the Animation dialog box, shown in Figure 9.11. The main function of this dialog box is to Record animations so that they can be saved and played outside Inventor.

Like the Edit Sequences and Tasks dialog box, the Animation tool can also be used to review and edit animations. An expanded Animation dialog box is shown in Figure 9.12. Note that the previously created group sequence (see Figure 9.7) is about to be ungrouped. It is also possible to Group and reorder sequences (Move Up, Move Down) in the Animation dialog box.

Working Drawings

After the presentation file containing the exploded view has been created, a *working drawing* of the assembly can be made using a drawing file template. Using the Base View tool, specify a presentation (.ipn) file format in the Files of type: drop-down list in order to insert an exploded view (see Figure 9.13). An assembled view can be added to a drawing in the same way, this time by specifying an assembly (i.e., .iam) file in the Files of type: drop-down list.

Once an exploded view has been added to the drawing sheet, a parts list can also be created using the Parts List tool, available on the Drawing Annotation panel. After the parts list has been inserted, it can be edited. The Edit Parts List dialog box is shown in Figure 9.14. See the discussion of the Column Chooser tool in the section

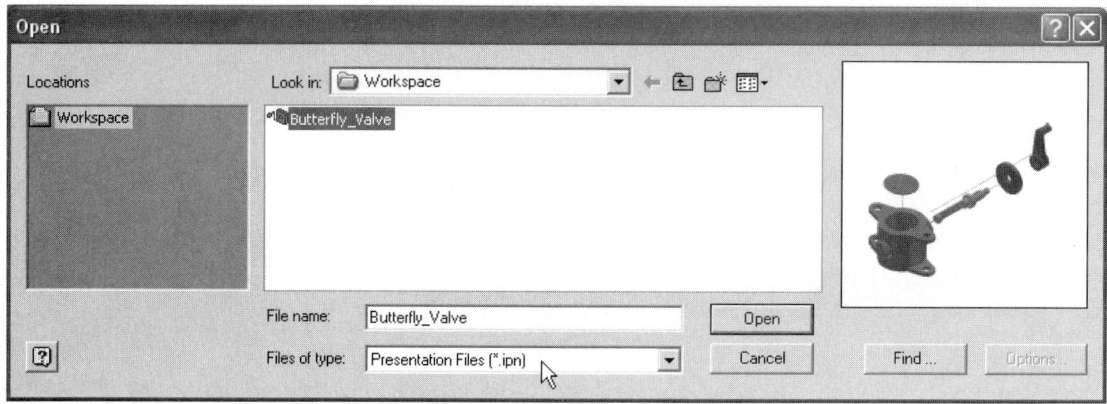

Figure 9.13 - Insert exploded view

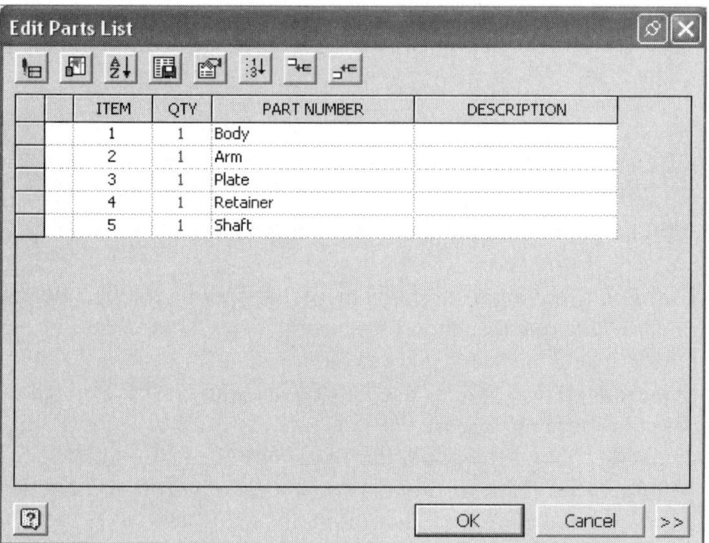

Edit Parts List

ITEM	QTY	PART NUMBER	DESCRIPTION
1	1	Body	
2	1	Arm	
3	1	Plate	
4	1	Retainer	
5	1	Shaft	

OK Cancel >>

Figure 9.14 - Edit parts list dialog box

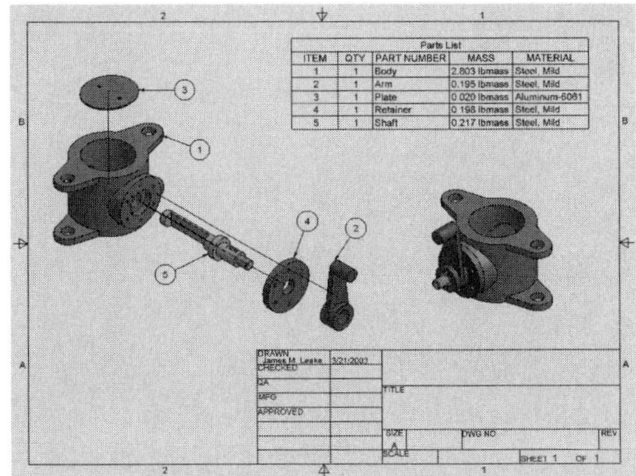

Figure 9.15 - Working drawing

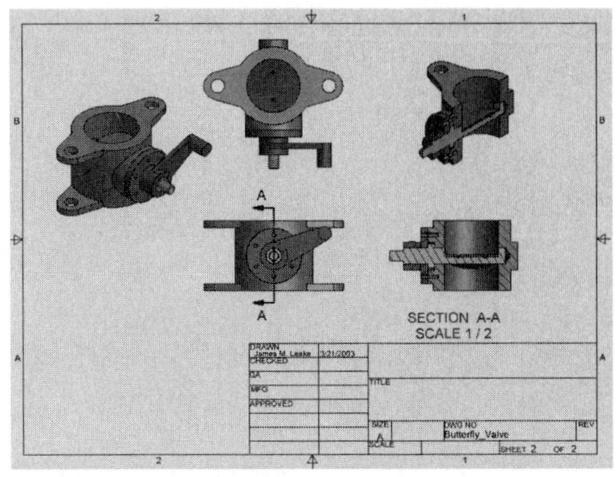

Figure 9.16 - Working drawing (sheet 2)

on the Bill of Materials tool in Chapter 8. Clicking the Help button in the lower left corner of this dialog box is a good way to learn about parts list customization.

Balloons relating the exploded view parts to the parts list can be added using the Balloon tool, also available on the Drawing Annotation panel. A drawing sheet showing an exploded view of the butterfly valve is shown in Figure 9.15. Note that the drawing includes a parts list and balloons, as well as an assembled isometric view of the valve.

Additional drawing sheets can be added to a drawing by right-clicking in the browser, and then selecting New Sheet. Seen in Figure 9.16 is another drawing sheet of the butterfly valve assembly.

Ball Valve Presentation File — TUTORIAL 27

1. Select File > Projects . . . from the menu bar, then make the Ball Valve project active.
2. Start a new metric presentation file. From the menu bar, select File > New. From the Open dialog box, select the Metric tab, and then select the presentation template file Standard.ipn.
3. From the panel bar, select the Create View tool.
4. From the Select Assembly dialog box, click on the File Explore button. After locating the Ball Valve.iam assembly file, click Open. Now click OK, as shown in Figure 9.17.
5. Select the Tweak Components tool. Move the cursor over the assembly until a coordinate system similar to the one shown in Figure 9.18 appears, then left-click.
6. Expand the browser until the Ball Valve components are visible. Select all of the hex head bolts (use the Shift key), as shown in Figure 9.19. Now that the Direction (Z axis) and Components (6 bolts) have been selected, enter 500 for the Transformation distance in the Tweak component dialog box, then left-click on the green check mark, as shown in Figure 9.20. Now click the Clear Clear button. This serves to de-select the direction, components, and distance.
7. Zoom All . The screen should resemble Figure 9.21.

Figure 9.17 - Select assembly dialog box

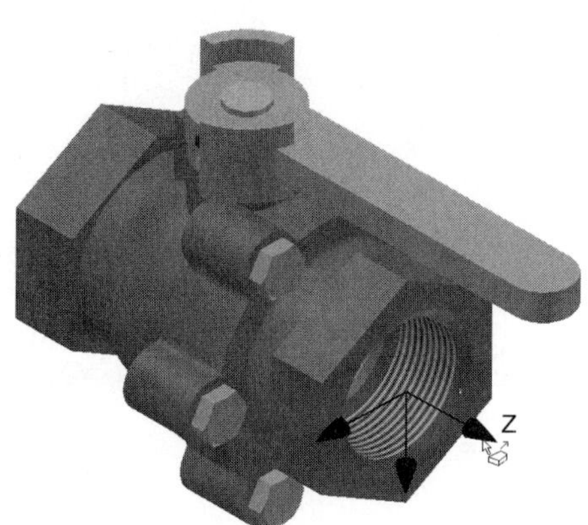

Figure 9.18 - Tweak direction selection

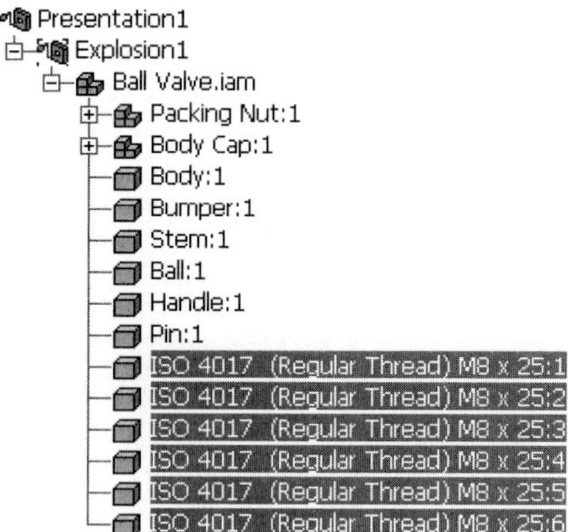

Figure 9.19 - Tweak component selection in browser

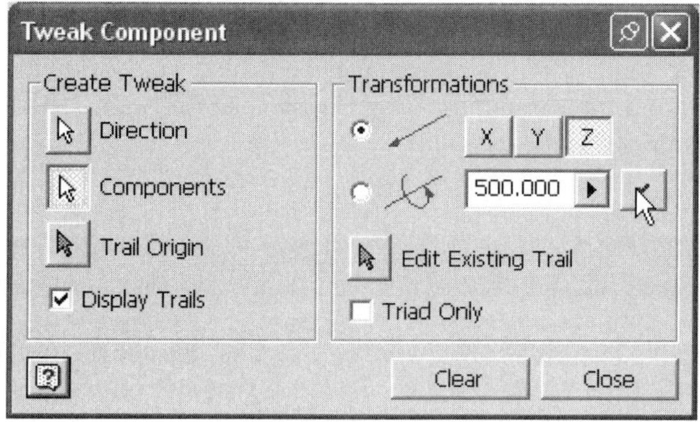

Figure 9.20 - Tweak component dialog box

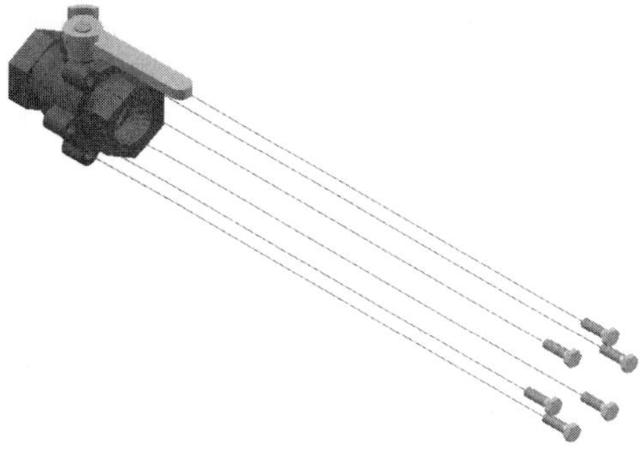

Figure 9.21 - Tweak translation (6 bolts)

8. Save the file as **Ball Valve.** Remember to save the file every 5 to 10 minutes.
9. Use the Tweak Components tool to open the Tweak Component dialog box. For the next tweak, use the same Z axis direction used to tweak the bolts. In the browser, select the Body Cap *subassembly*. After entering 350 as the tweak distance, click on the green checkbox as shown in Figure 9.22. After left-clicking the Close button in the dialog box, the screen should resemble Figure 9.23.
10. We will now look at an animation of the tweaks created thus far. At the top of the browser, left-click on the Filter icon. Select Sequence View.
11. There is now a Task1 item in the browser. Double-click on Task1. The Edit Tasks and Sequences dialog box opens. Click on the Task Forward button, as shown in Figure 9.24. After the animation has played, left-click on the OK button.
12. Expand the Body Cap subassembly in the browser.

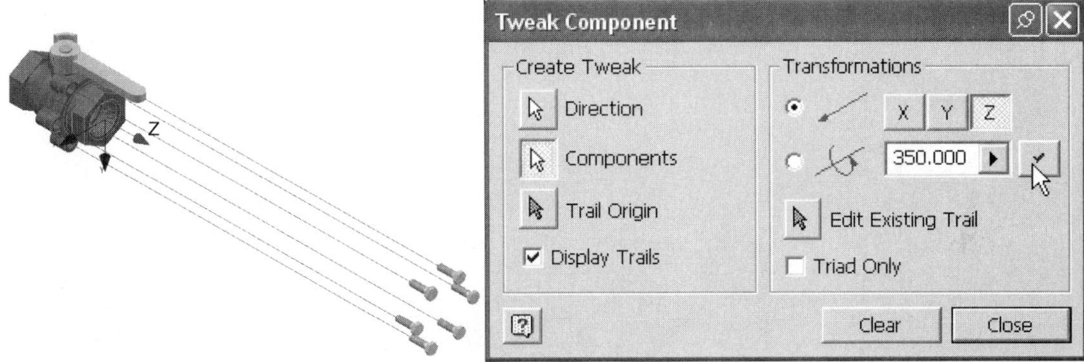

Figure 9.22 - Translation tweak (body cap subassembly)

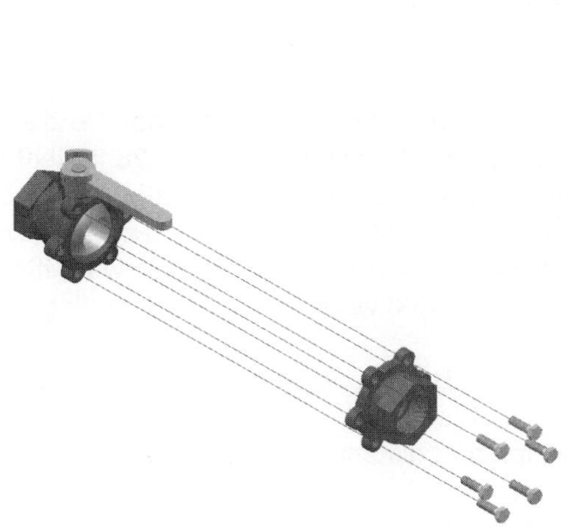

Figure 9.23 - Body cap subassembly tweak

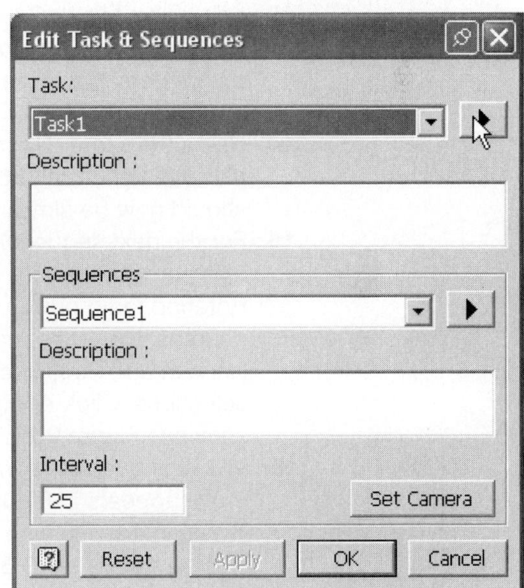

Figure 9.24 - Edit Tasks & Sequences dialog box

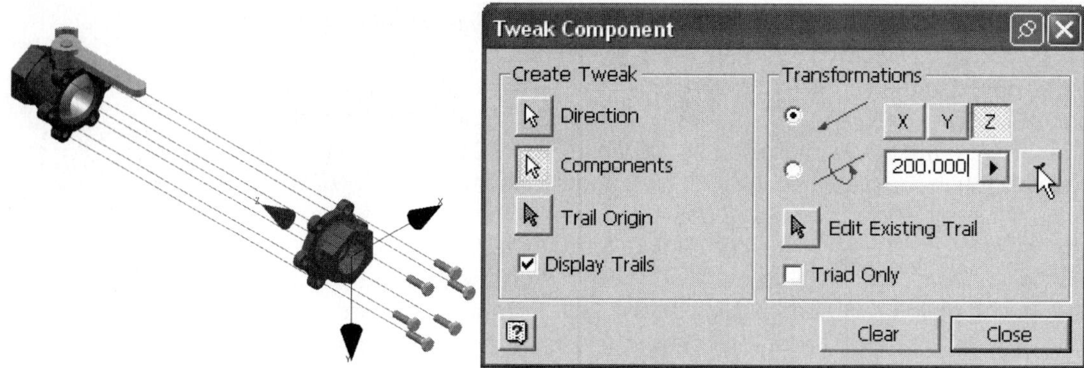

Figure 9.25 - Translation tweak (Oring1)

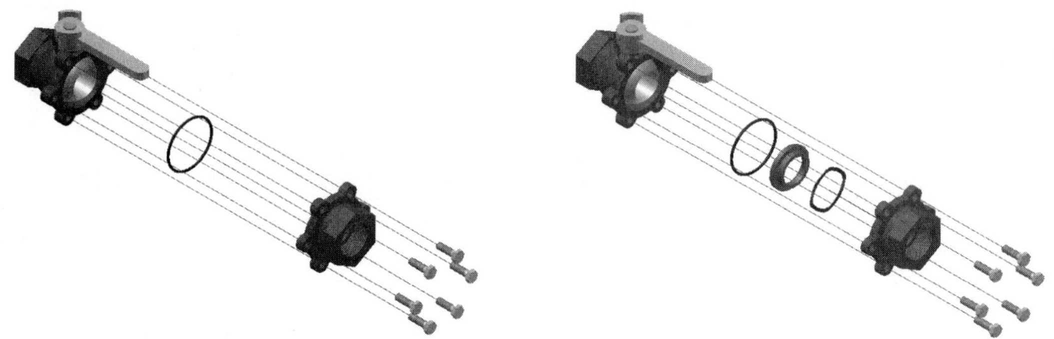

Figure 9.26 - Oring1 tweak

Figure 9.27 - Seat and spring tweaks

13. Select the Tweak Components tool. See Figure 9.25 for the Direction, as well as the distance to be moved. From the browser, select the Oring1 part as the component. Click the green check mark to apply the move. Click Clear to de-select. The screen should resemble Figure 9.26.
14. Using the same direction as was applied to the Oring1 part, first move the Seat part a distance of 150, and then the Spring a distance of 100. The graphics area should now be similar to Figure 9.27.
15. For the next sequence, select the coordinate system such that the Z axis is colinear with the axis of the Stem part, as shown in Figure 9.28. Select the Rotation radio button in the dialog box, and change the rotation angle to 90 degrees. Now select the following components in the browser: Ball, Stem, Handle, and Pin. Hold down the Ctrl key while selecting to allow multiple selections. Click Apply (green check mark) when ready. The resulting tweak is shown in Figure 9.29. Click Clear.
16. The next tweak is a translation of the Ball shown in Figure 9.30. The resulting move appears in Figure 9.31. Click Clear.
17. Translate the Pin with the direction and distance shown in Figure 9.32. The result is shown in Figure 9.33. Click Clear.
18. Next translate the Handle and Pin vertically a distance of 200, as shown in Figure 9.34. After zooming out, the screen should be similar Figure 9.35. Clear.

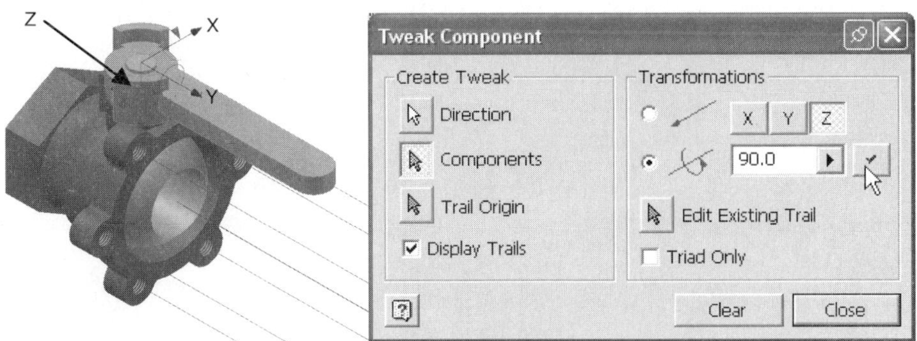

Figure 9.28 - Rotation tweak (ball, stem, handle, pin)

Figure 9.29 - Ball, stem, pin, handle tweak

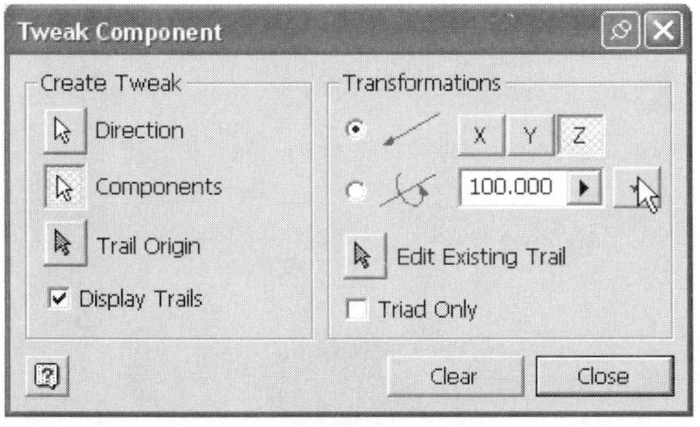

Figure 9.30 - Translation tweak (ball)

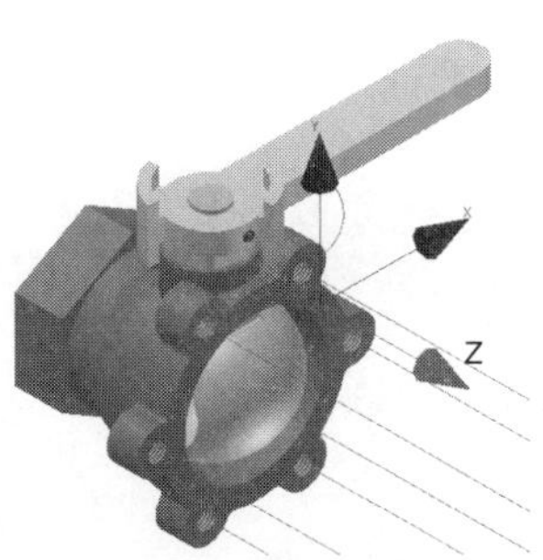

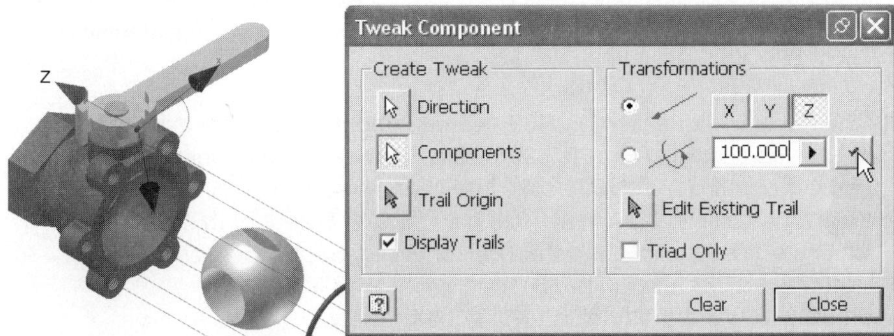

Figure 9.31 - Ball translated

Figure 9.32 - Translation tweak (pin)

Figure 9.33 - Pin tweak

19. The next three sequences involve translations of the Stem. Using the coordinate system shown in Figure 9.36, translate the stem 40 units in the negative Z direction, then 50 units in the negative X direction, and finally 100 units in the negative Z direction. Clear when finished.

20. Use the coordinate system shown in Figure 9.37 to first translate the Bumper 80 units in the Z axis direction, then 100 units in the X axis direction.

21. Translate the Packing Nut *subassembly* 100 units in the vertical axis direction, as shown in Figure 9.38.

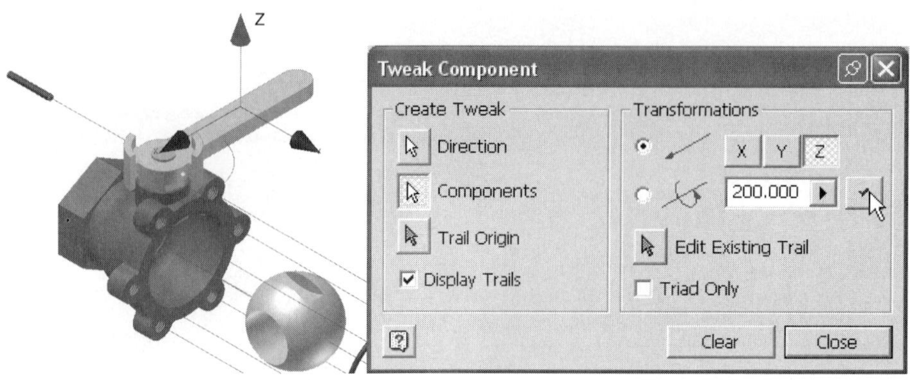

Figure 9.34 - Vertical translation tweak (handle and pin)

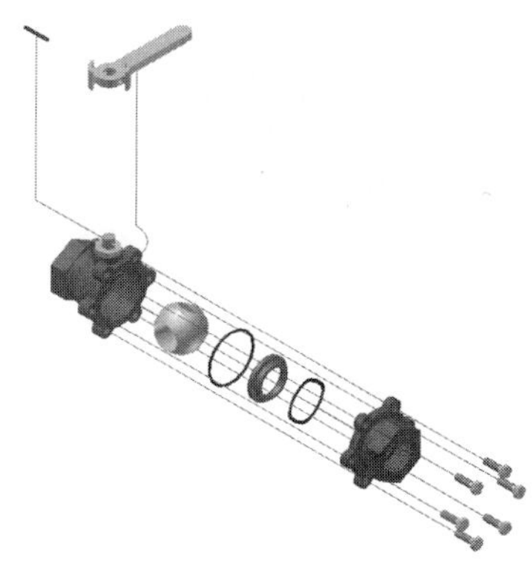

Figure 9.35 - Handle and pin tweak

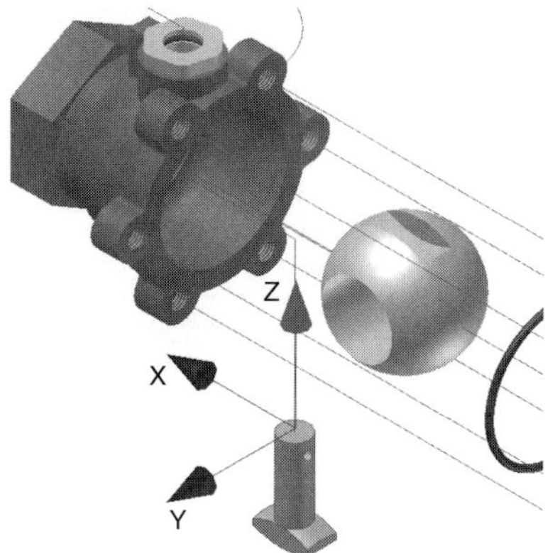

Figure 9.36 - Multiple translation tweaks (stem)

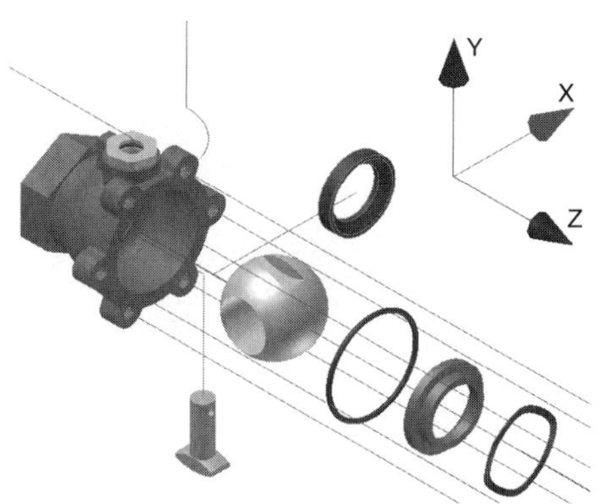

Figure 9.37 - Multiple translation tweaks (bumper)

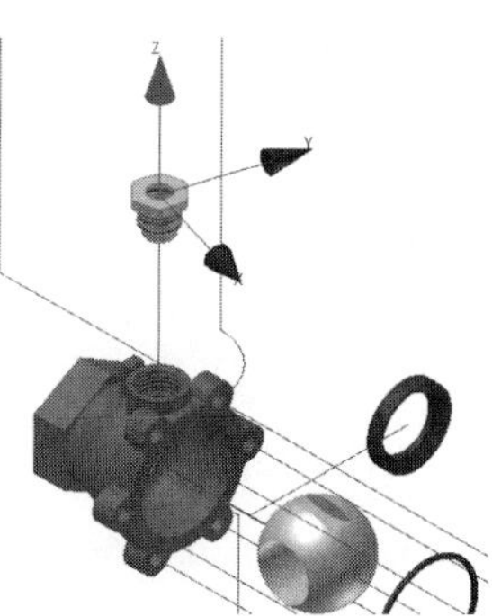

Figure 9.38 - Vertical translation tweak (packing nut assembly)

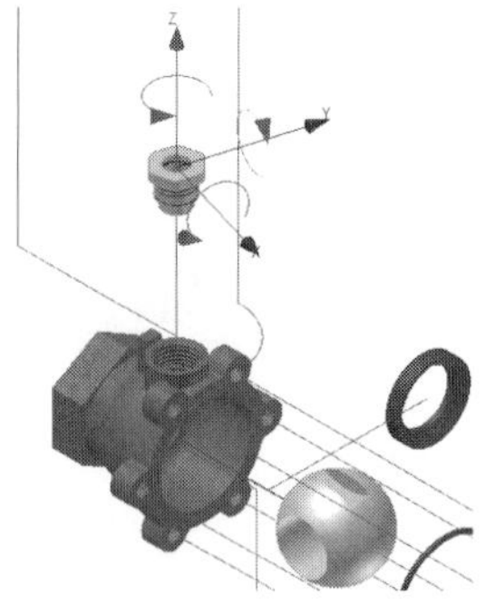

Figure 9.39 - Rotation tweak (packing nut subassembly)

22. Rotate the Packing Nut assembly 360 degrees about the vertical axis (of the Packing Nut), as shown in Figure 9.39. Close the Tweak Component dialog box.
23. We will now group the last two tweaks into a single sequence so that they occur simultaneously. In the browser, expand Task1. Select both Sequence1 and Sequence2. Right-click, and then select Group Sequences, as shown in Figure 9.40.
24. In the browser, expand the Packing Nut subassembly.
25. Select the Tweak Components tool. Translate Oring2 up 50 units along a vertical axis, and Oring3 down 50 units along a vertical axis, as shown in Figure 9.41. Close the Tweak Component dialog box.
26. In the browser select Sequence1 and Sequence2, right-click and select Group Sequences.
27. To review the animation sequence, double-click on the Task1 icon in the browser, and then click on the Play button in the Task area. It is also possible to review one sequence at a time in the Sequences area of the Edit Task & Sequences dialog box. Select a sequence from the sequence list, and then click Play. Click the Reset button at the bottom of the dialog box before selecting another sequence to play.

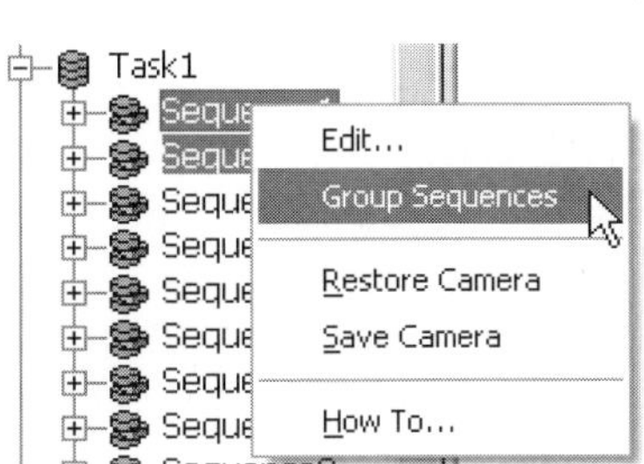

Figure 9.40 - Group sequences

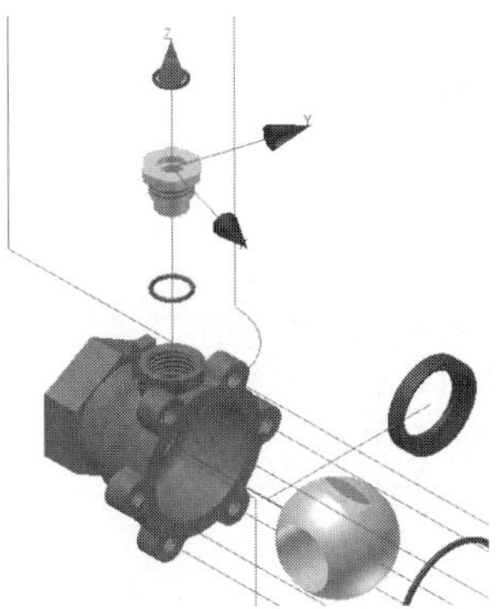

Figure 9.41 - Multiple translation tweaks (Oring2 and Oring3)

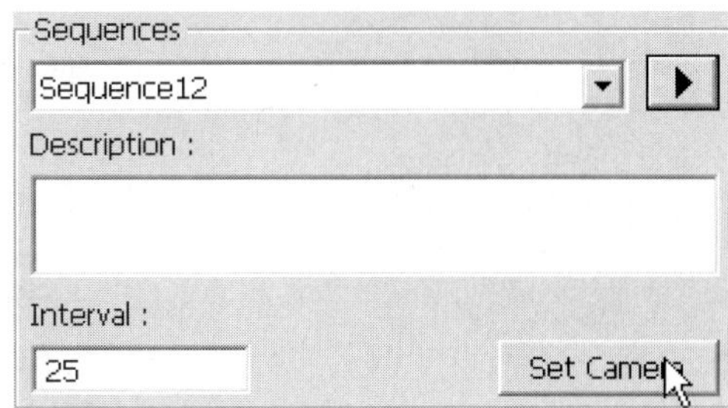

Figure 9.42 - Set camera for sequence

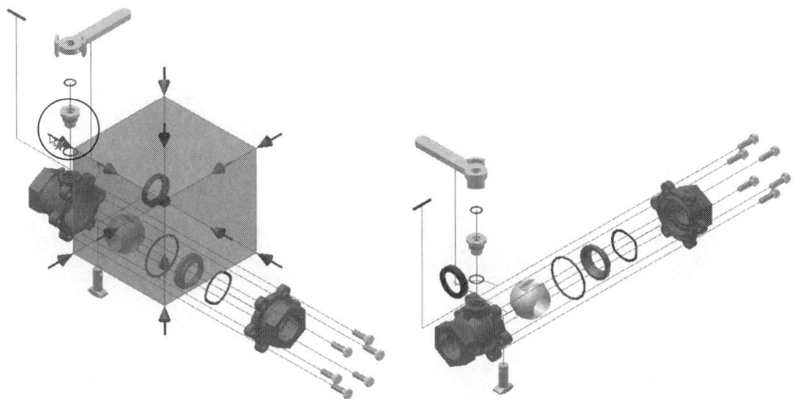

Figure 9.43 - Change camera view

28. It is possible to change the camera angle for different sequences. Select Sequence12 from the Sequence list. Play this sequence to verify that this sequence moves the Spring to the Body Cap. Reset, then left-click on the Set Camera button, as shown in Figure 9.42. Now use the Common View tool to change the camera view, as shown in Figure 9.43. After setting the camera view, click the Apply button at the bottom of the dialog box.
29. Now select Sequence15 from the sequence list. This should be the sequence where the Body Cap subassembly translates to the main assembly. Click the Set Camera button.
30. Now use Common View to change the view back to the default isometric view, as shown in Figure 9.44. After changing the camera view, click the Apply button.
31. While the Edit Tasks & Sequences dialog box is still open, click on Play in the Task area to view the completed assembly movie. When satisfied, click OK.
32. On the panel bar, click the Animate tool. The Animation dialog box opens. Before making the animation, you may want to reduce the size of the Inventor graphics area in order to keep the movie size small. Click the Record button,

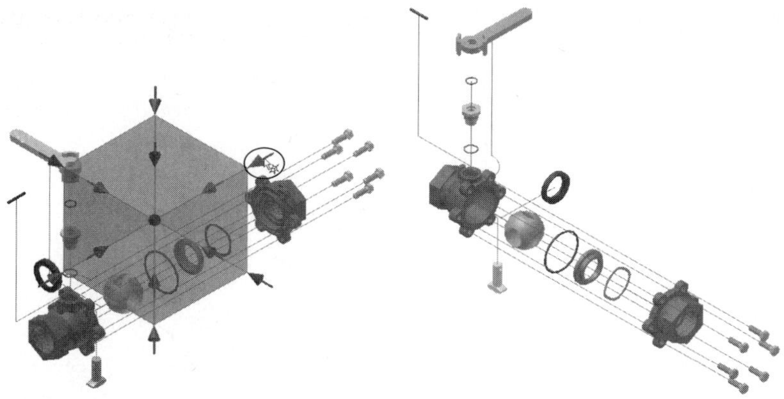

Figure 9.44 - Change camera view

Figure 9.45 - Animation dialog box

as shown in Figure 9.45. Save the file as Explosion in the Ball Valve workspace. Accept the default Video Compression. To start recording, click the Play Forward button on the Animation dialog box. When the animation is complete, uncheck the Record button. You should now be able to view the movie using a video player like the Windows Media Player.

33. Save the presentation file in the Ball Valve workspace folder.

TUTORIAL 28 Ball Valve Exploded View Assembly Drawing

Detailed Drawing Creation Steps

1. Start a new file, select the Metric tab, and then choose the ANSI metric drawing template ANSI (mm).idw file.
2. In the browser, right-click on Sheet1, and then select Edit Sheet. . . . Change the sheet size from C to A.
3. In the browser, right-click on the ANSI - Large title block icon, and then select Delete.
4. In the browser, expand the Drawing Resources folder, and then expand the Title Blocks folder. Right-click on the ANSI A title block icon, then select Insert.

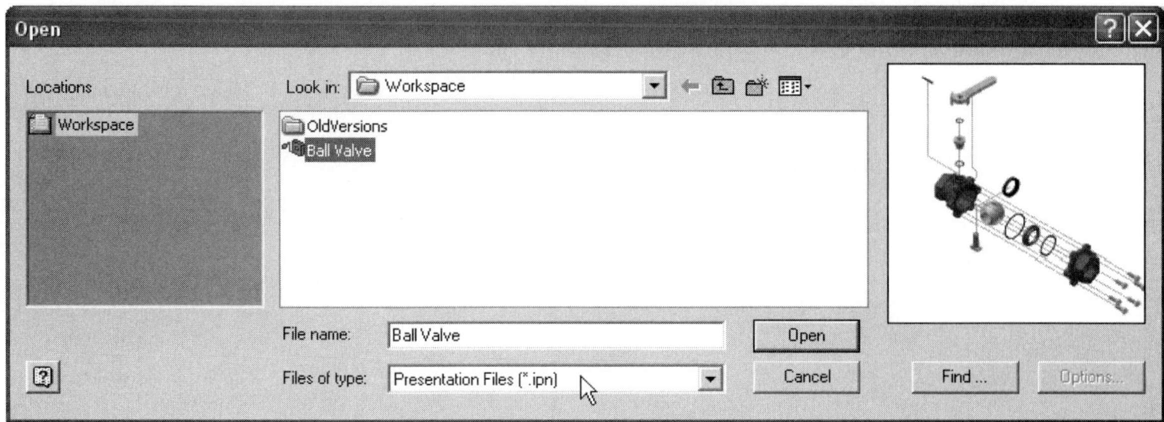

Figure 9.46 - Open presentation file

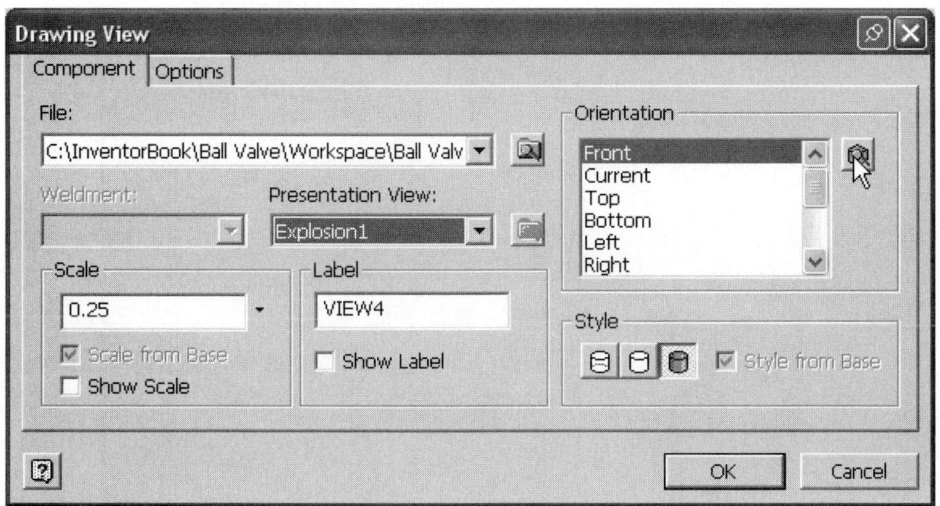

Figure 9.47 - Drawing view dialog box

5. Select Base View from the Drawing Views panel. In the Base View dialog box, click on the Explore Directories button.
6. In the Open dialog box, in the Files of type: drop-down list, select Presentation Files (*.ipn), as shown in Figure 9.46. Highlight the Ball Valve.ipn file in the Ball Valve Workspace, and then select Open.
7. In the Drawing View dialog box, set the Scale to 1:4 (0.25), and the display to Shaded. Select the Change View Orientation button, as shown in Figure 9.47. In the Custom View window use the viewing tools on the standard toolbar to change the view to that shown in Figure 9.48. When the view is correct, click Apply (green check mark). Now place the exploded view in the drawing by clicking, as shown in Figure 9.49.
8. Again using the Base View tool, add a shaded isometric view of the Ball Valve assembly with a 0.5 scale to the drawing. When finished, the screen should resemble Figure 9.50.

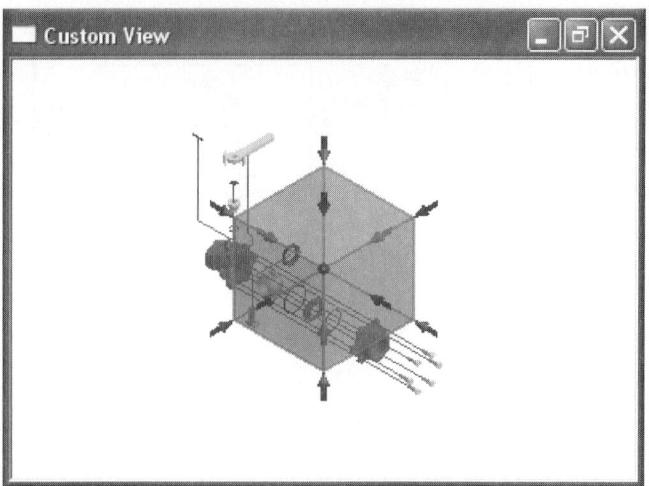

Figure 9.48 - Custom view window

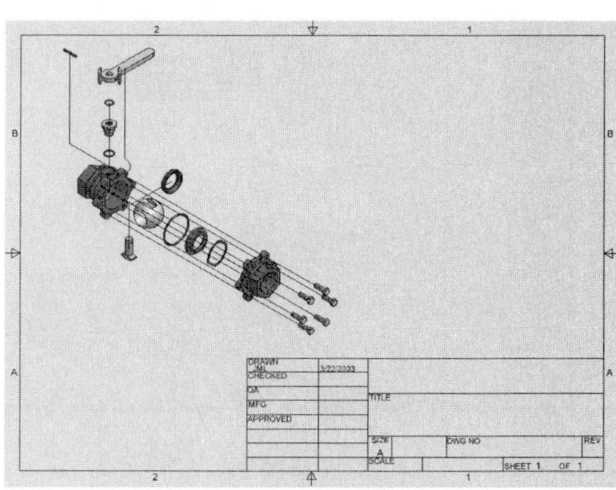

Figure 9.49 - Exploded view placed

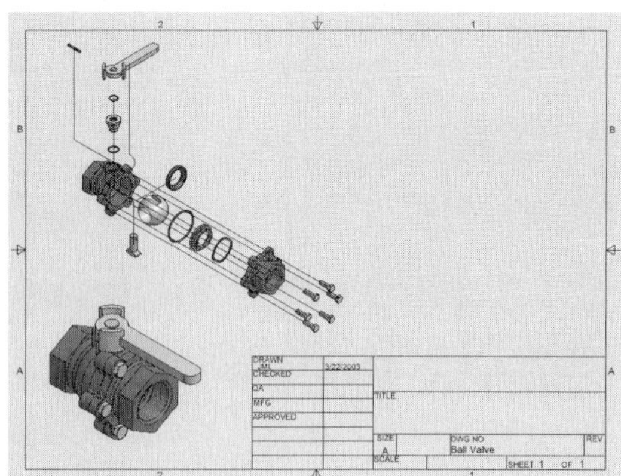

Figure 9.50 - Assembled view placed

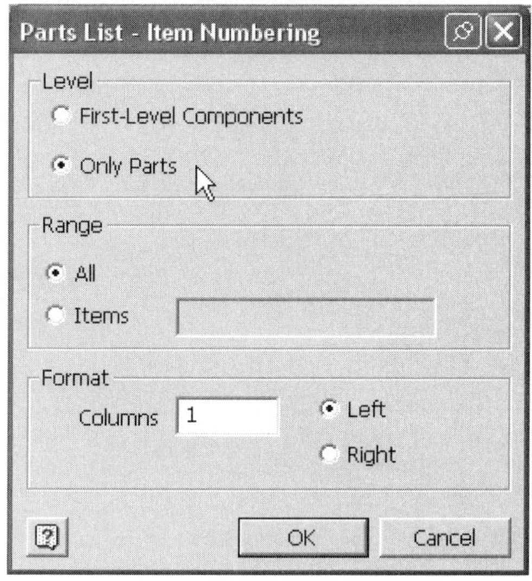

Figure 9.51 - Parts list - item numbering

9. Save the file as **Ball Valve Assembly** in the Ball Valve workspace folder. Be sure to save every 5 to 10 minutes.
10. In the panel bar switch to the Drawing Annotation tools.
11. Select the Parts List tool. Left-click on the exploded view. The Parts List - Item Numbering dialog box appears. Make the selections shown in Figure 9.51, then select OK.
12. Click in the upper right corner of the drawing to place the parts list.
13. Zoom in on the parts list. It should resemble Figure 9.52.
14. When highlighted, right-click on the parts list, and then select Edit Parts List . . . The Edit Parts List dialog box appears. See Figure 9.53.
15. Click on the Column Chooser button. The Parts List Column Chooser dialog box appears. Add the Mass and Material columns (highlight in Available

Parts List			
ITEM	QTY	PART NUMBER	DESCRIPTION
1	1	Body	
2	1	Bumper	
3	1	Packing Nut	
4	1	Oring3_3	
5	1	Oring2_3	
6	1	Stem	
7	1	Ball	
8	1	Handle	
9	1	Pin	
10	1	Body Cap	
11	1	Oring1	
12	1	Spring	
13	1	Seat	
14	6	ISO 4017 - M8 x 25	Hex-Head Bolt

Figure 9.52 - Parts list (unmodified)

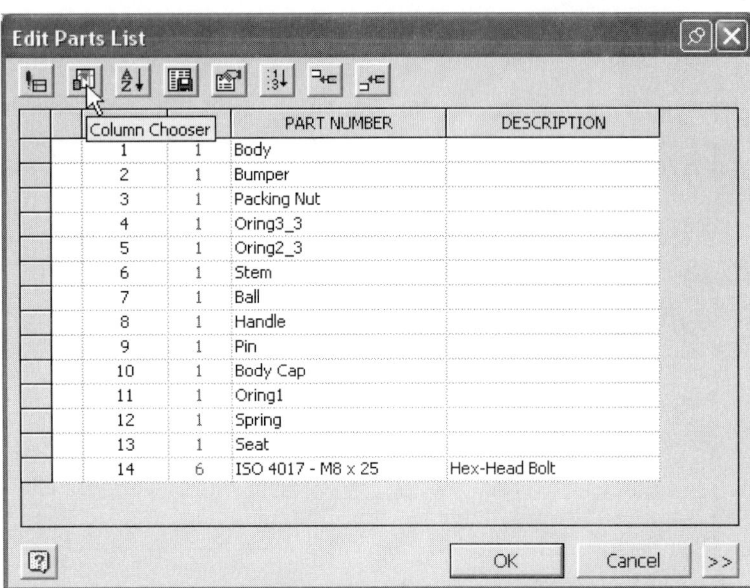

Figure 9.53 - Edit parts list dialog box

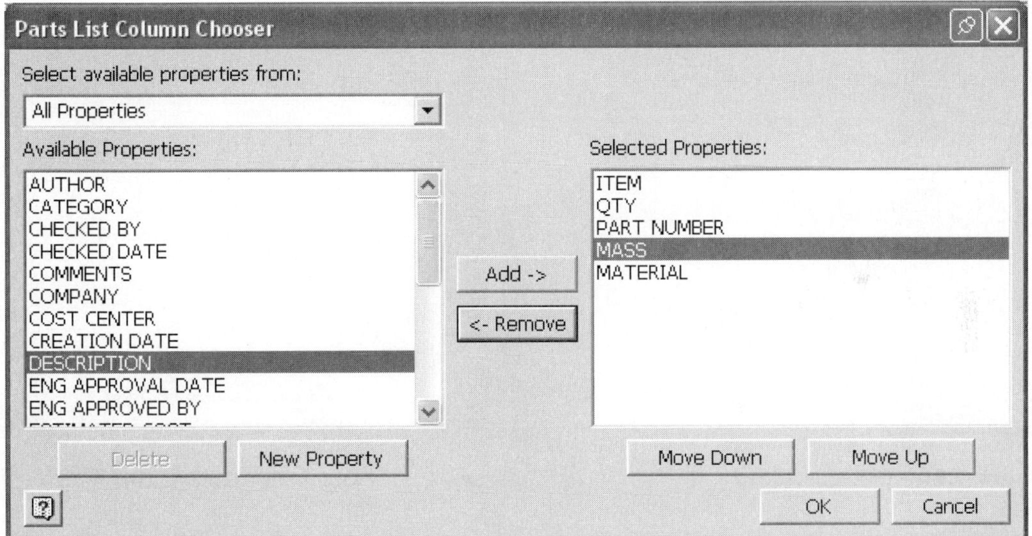

Figure 9.54 - Parts list column chooser

Properties list, then click Add) and remove the Description column (highlight on right, click Remove). See Figure 9.54. Click OK when done.

16. The Edit Parts List dialog box re-appears. Reduce the column widths by placing the cursor along a column edge in a title cell, as shown in Figure 9.55, and then click OK.
17. Zoom All .
18. If necessary, re-position the parts list by moving the cursor over a handle (green circle) in one of the corners of the parts list.

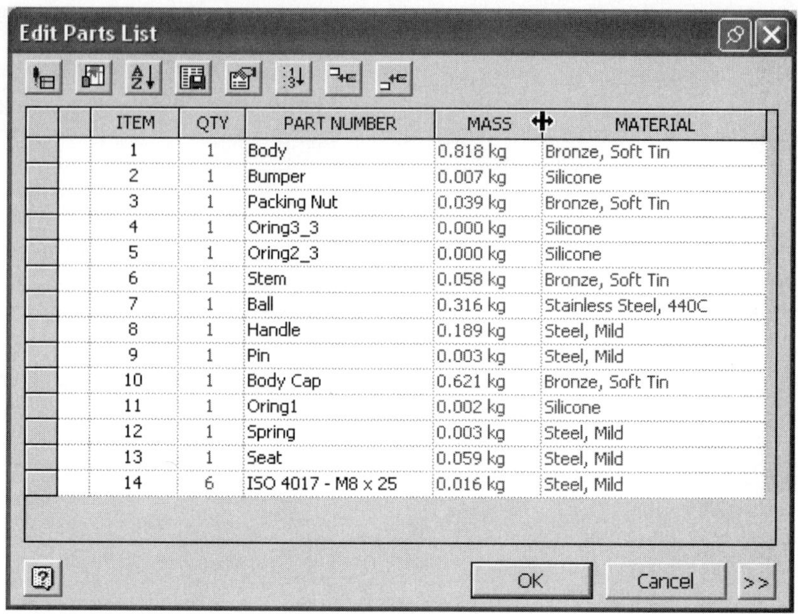

ITEM	QTY	PART NUMBER	MASS	MATERIAL
1	1	Body	0.818 kg	Bronze, Soft Tin
2	1	Bumper	0.007 kg	Silicone
3	1	Packing Nut	0.039 kg	Bronze, Soft Tin
4	1	Oring3_3	0.000 kg	Silicone
5	1	Oring2_3	0.000 kg	Silicone
6	1	Stem	0.058 kg	Bronze, Soft Tin
7	1	Ball	0.316 kg	Stainless Steel, 440C
8	1	Handle	0.189 kg	Steel, Mild
9	1	Pin	0.003 kg	Steel, Mild
10	1	Body Cap	0.621 kg	Bronze, Soft Tin
11	1	Oring1	0.002 kg	Silicone
12	1	Spring	0.003 kg	Steel, Mild
13	1	Seat	0.059 kg	Steel, Mild
14	6	ISO 4017 - M8 x 25	0.016 kg	Steel, Mild

Figure 9.55 - Edit parts list (modified)

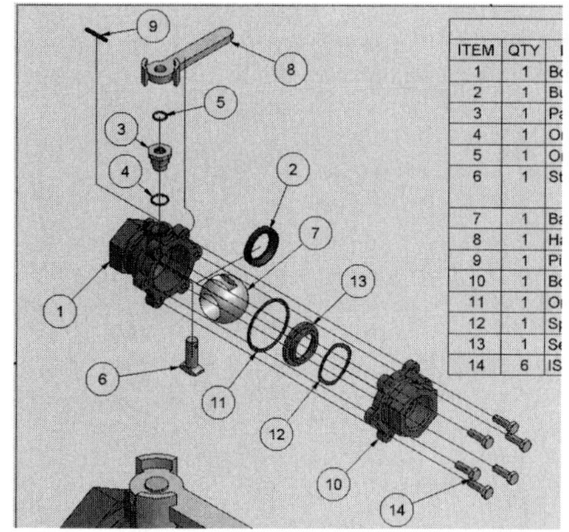

Figure 9.56 - Balloons added

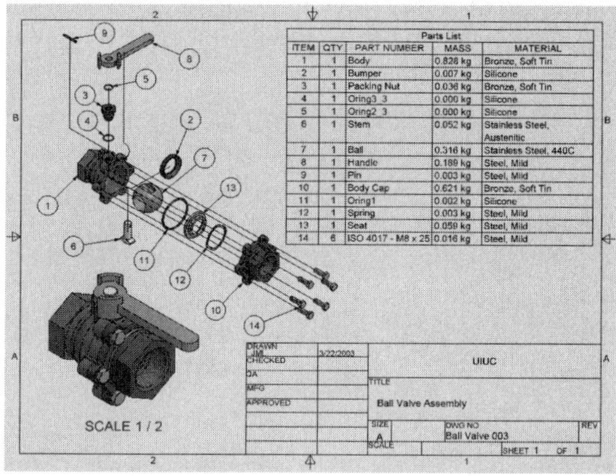

Figure 9.57 - Completed working drawing

19. Zoom in on the exploded view.
20. Select the Balloon tool from the Drawing Annotations Panel. Move the cursor to an edge of a part until a (coincident) constraint appears. Left-click to accept. Now move the cursor a short distance to an empty area away from the part. Left-click again. The balloon appears. Now right-click and select Continue.
21. Continue adding balloons in this way for all of the parts. When finished, the exploded view should resemble Figure 9.56.
22. Zoom All.
23. After completing the title block information and displaying the drawing scale on the assembled view, the drawing should be similar to Figure 9.57.
24. Save the drawing file.

QUESTIONS

1. T F The Tweak Components tool can only move components in translation.
2. The icon [filter icon] is designated as the
   a. Filler
   b. Filter
   c. Funnel
   d. Flexural
3. T F Tweaks or sequences can be grouped such that they occur simultaneously.
4. T F Tweaks and sequences cannot be reordered in the browser bar.
5. T F The Set Camera button on the Edit Task & Sequence dialog box allows for the modification of the camera view in an animation.
6. T F The numbering sequence for the parts list and balloons must be added manually for a part.

PROBLEMS

1. Create an AVI animation file showing the assembly of the clamp used in Problem 1 at the end of Chapter 8.
2. Create a working drawing of the clamp assembly, similar to that shown in the Figure below.

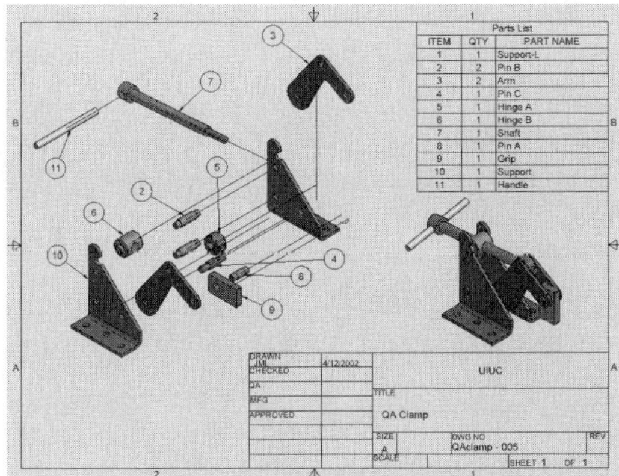

3. Add a second sheet to the Ball Valve assembly drawing created in Tutorial 28. The second sheet should resemble the figure below.

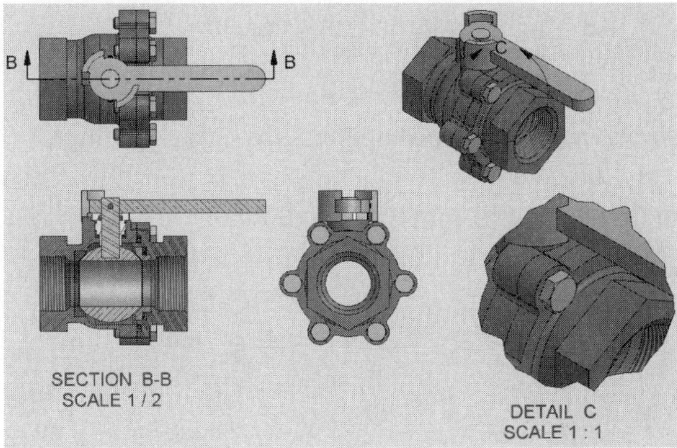

10 CHAPTER

Hybrid Modeling

LEARNING OBJECTIVES

- Control the shape of a spline using the following spline controls:
  - Spline Tension
  - Display Curvature Comb
  - Close Spline
  - Insert Point
  - Fit Method: Smooth, Sweet, AutoCAD
  - Bowtie Method options: Handle, Curvature, Flat
- Use the common sketched feature controls (i.e. extrude, revolve, sweep, and loft) to create a surface feature
- Use the Thicken/Offset tool to:
  - Add or remove thickness on an existing solid
  - Offset the face of an existing solid, resulting in an offset surface
  - Add thickness to a surface, converting the surface to a solid
  - Offset a surface
- Use the Delete Face tool to delete one or more faces from a solid part, converting the solid to a surface
- Use the Replace Face tool to replace an existing face with another face
- Use the Stitch Surface tool to combine surfaces that share a common edge
- Use rail curves to control the shape of a lofted feature
- Use the Emboss tool to create a raised or recessed feature on an existing face
- Use the Insert Image tool to add a bitmap image to a sketch
- Use the Decal tool to convert a sketch plane image to a decal feature placed on a part face
- Use work points and grounded work points to help create complex shapes
- Create work planes normal to a point on a sketch curve
- Use the 3D Intersection Tool to produce a 3D curve from the intersection of two surfaces

Introduction

Probably the most significant improvement in Autodesk Inventor R6 is the addition of a truly hybrid part modeling environment. Users can now seamlessly combine surfaces with parametric solids to create complex sculpted parts. Several new part tools have been added, while many of the previously existing sketch and part creation tools have been improved.

Solid modelers were first developed in such a way that they could only create valid solids, i.e., shapes having a closed volume. Such a shape is called a *manifold* model. This is in many ways a useful restriction, because it guarantees that only real world, manufacturable parts can be created.

From the standpoint of the designer, though, it is often helpful to be able to work abstractly, creating two-dimensional and even one-dimensional geometry. Surface modeling programs, for example, have the ability to create extremely complex, free-form surfaces. Even though these surfaces have no thickness, they are still useful design tools.

Splines

Splines are freeform curves defined by a series of points. Although there are many different kinds of splines, all splines created in Inventor pass through or *interpolate* their defining points. In the 2D sketch environment the Spline tool is available from the panel bar by clicking the down arrow beside the Line/Spline tool, as shown in Figure 10.1. The spline is interactively created by left-clicking to locate spline points in the graphics window. To terminate the spline, double-click at the endpoint, or right-click, and select Continue. Using the latter method, the spline tool remains active, and can be used to add other splines. Figure 10.2 shows a spline with four spline points.

Spline points can be constrained using either dimensions or geometric constraints. Splines with unconstrained points are easily modified by clicking and then dragging the spline points. In Figure 10.3 the spline shown in the previous figure has been modified by moving the second point on the left.

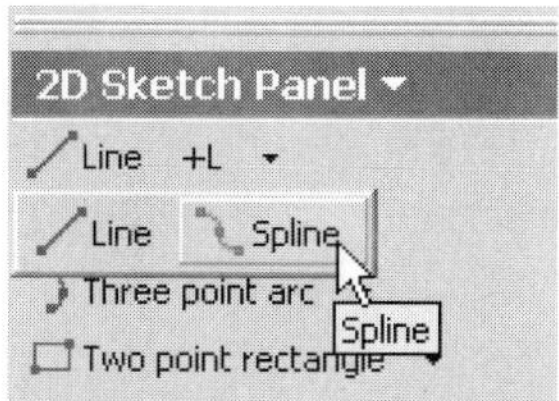

Figure 10.1 - Spline tool access

Spline Controls

There are many different ways to control the shape of a spline, several of which are used in the tutorials at the end of this chapter. To access these controls, right-click on the spline in the graphics window. Included among the context menu options are the spline controls shown in Figure 10.4.

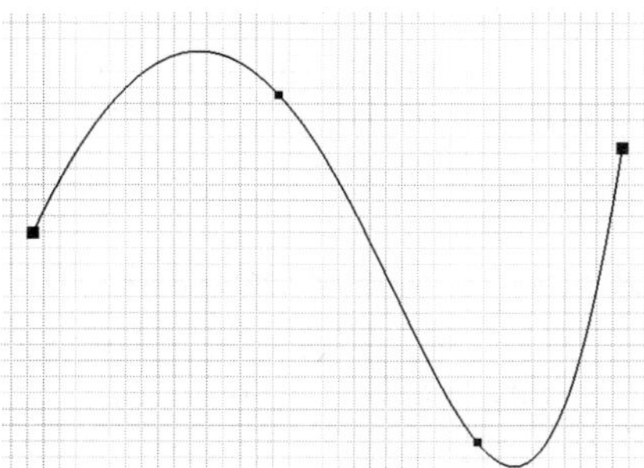

Figure 10.2 - Spline curve

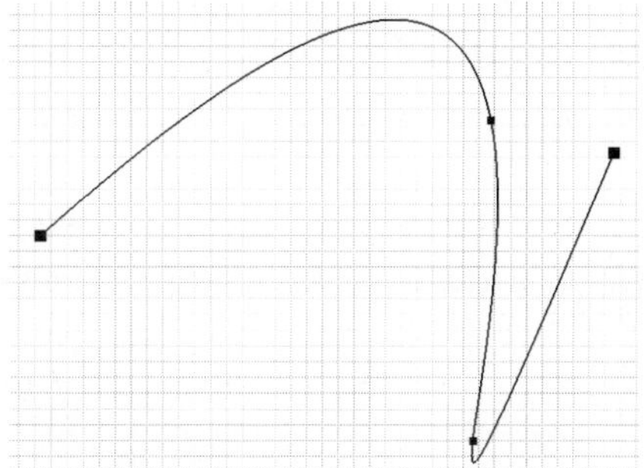

Figure 10.3 - Modified spline

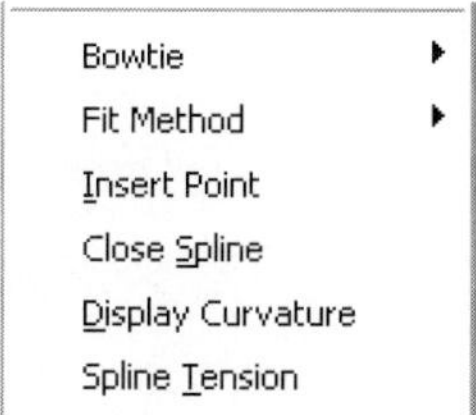

Figure 10.4 - Spline controls

Spline Tension

Selecting spline tension opens the slide bar shown in Figure 10.5 on the left. A spline's tension is initially set to zero; increasing the tension modifies the spline, as shown on the right in Figure 10.5.

Display Curvature

Selecting Display Curvature reveals the spline *curvature comb,* as shown in Figure 10.6. The curvature comb displays a comb of spines along the spline. Each spine represents the spline curvature at that point, adjusted for view scale. The longer the spine, the greater the curvature. The curve on the left, with its two *inflection points,* has fairly dramatic shifts in curvature. The curve on the right has less curvature, and the transition in curvature along the spline is gentler. Once the curvature comb is turned on, it remain active until it is turned off (right-click, de-select Display Curvature).

Close Spline

The Close Spline option, as the name implies, closes the spline so that it forms a closed loop. In Figure 10.7 the spline shown in Figure 10.2 has been closed.

Insert Point

The insert point option allows the user to add internal points to the spline. In Figure 10.8, an additional point has been added to the spline originally shown in Figure 10.2.

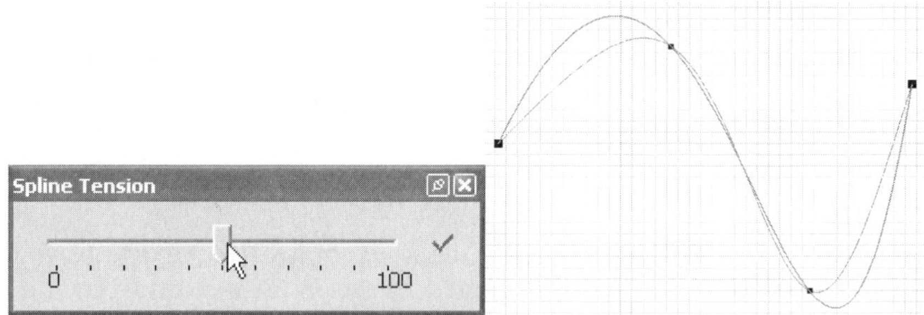

Figure 10.5 - Spline tension control

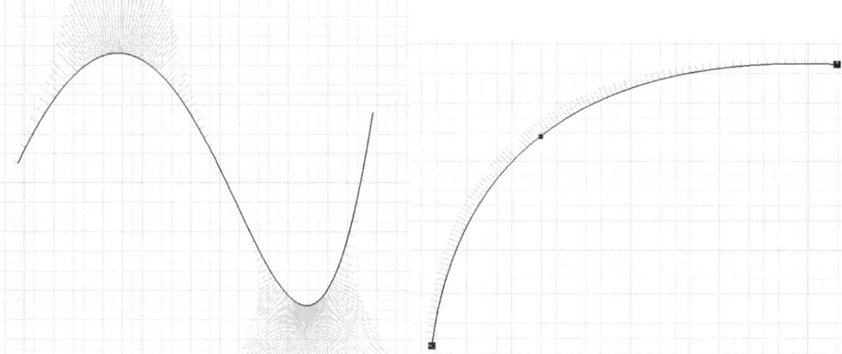

Figure 10.6 - Display curvature

Fit Method

In Figure 10.9 the three possible fit methods, Smooth, Sweet, and AutoCAD are respectively displayed from left to right.

Bowtie Method

There are three different bowtie options: Handle, Curvature, and Flat. They are all used to adjust the shape of the spline at a specific spline point. Use Handle to control the slope of the curve at a spline point. After right-clicking on the spline and then selecting Bowtie > Handle, a handlebar with grips appears at the nearest spline point, as shown on the left in Figure 10.10. The grips can then be dragged to adjust the tangency

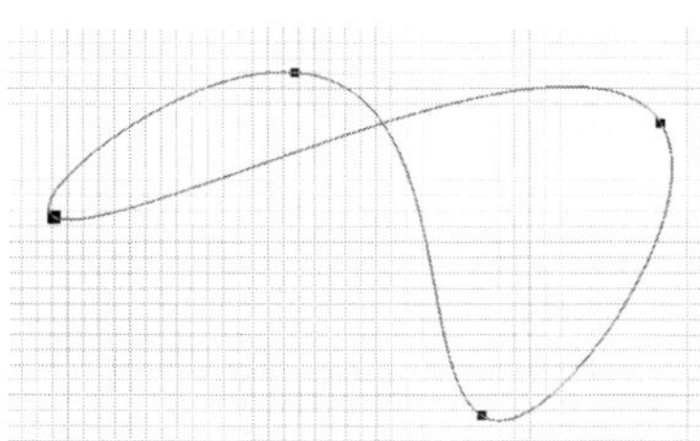

Figure 10.7 - Close spline

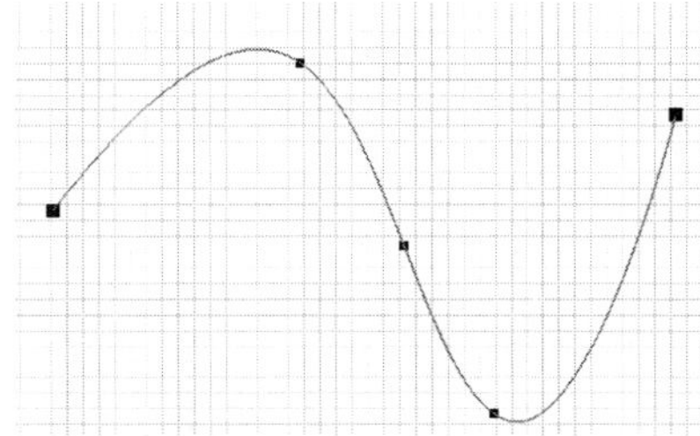

Figure 10.8 - Insert point

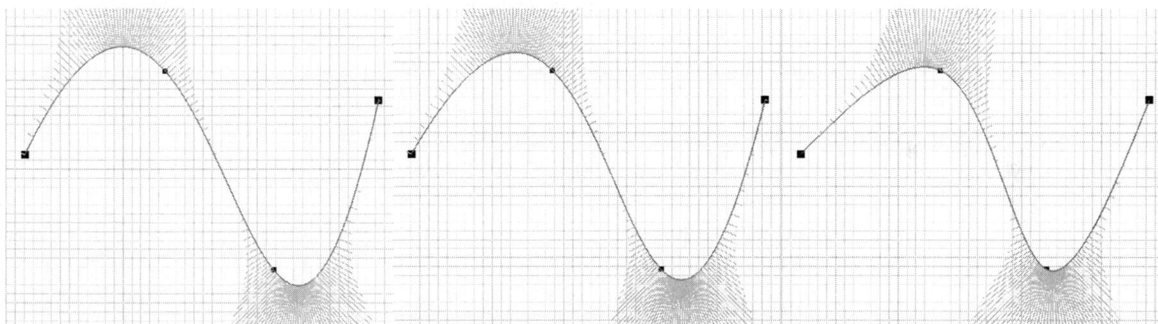

Figure 10.9 - Fit methods: smooth, sweet, AutoCAD

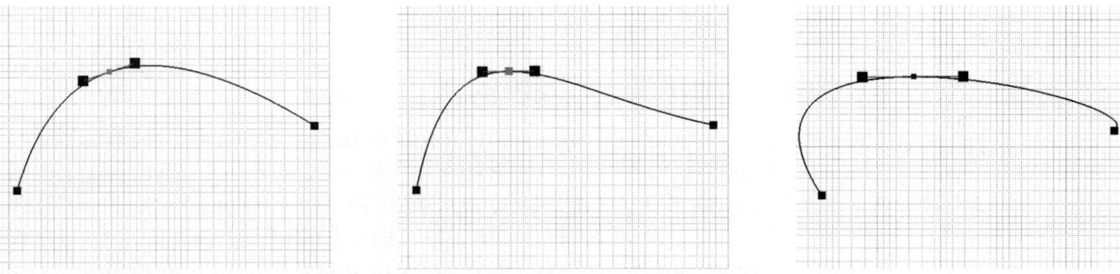

Figure 10.10 - Bowtie handle control

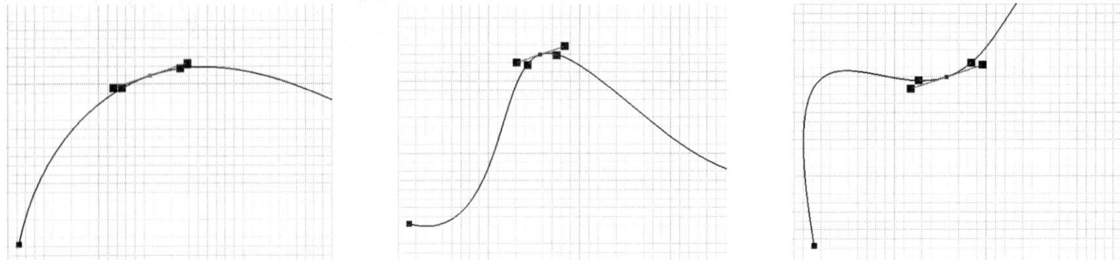

Figure 10.11 - Bowtie curvature control

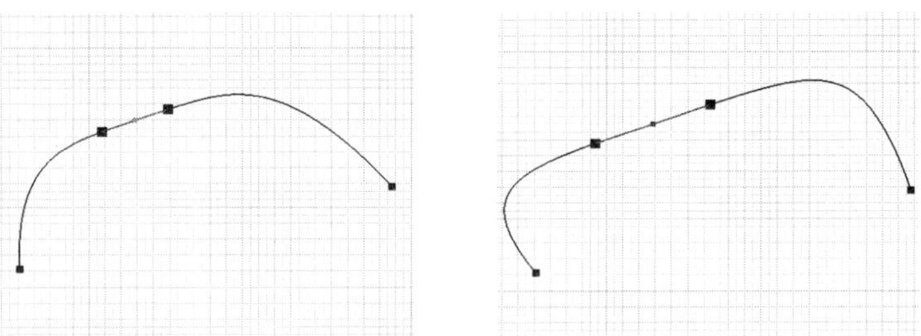

Figure 10.12 - Bowtie flat control

of the curve at the spline point, as shown in the middle figure. In this case the handlebar is approximately horizontal, indicating a zero slope (and a local maximum) at this point. In the rightmost figure, the grip has been dragged outward and the handlebar is lengthened. The slope of the curve is still zero at this spline point, but the curve is flattened in the vicinity of the spline point. In other words, the effect of the tangency at this spline point is distributed over a greater portion of the curve.

The bowtie curvature option allows the user to control the curvature of a spline at a spline point. After right-clicking on the spline in the vicinity of the spline point of interest, and then selecting Bowtie > Curvature, a curvature bar (and a handlebar) with grips appear, as shown on the left in Figure 10.11. By dragging on the grips of the curvature bar, the curvature at the spline point can be increased (middle) or reversed (right).

The Bowtie Flat option modifies the spline at a point by creating a straight, constant slope segment in the vicinity of the spline point. See Figure 10.12 on the left. If the handlebar is lengthened by dragging on the grips, then the length of the straight segment increases, as shown on the right.

Surface Creation Tools

The same tools used to create solid features in Inventor are also used to create surfaces. These sketched feature tools are Extrude, Revolve, Sweep, and Loft. All of these tools require 2D sketch profiles in order to create the feature. In the case of a solid feature, these profiles must be closed. Surface features, on the other hand, can be created from either closed or open profiles. The Output setting on the feature dialog box, either solid or surface, determines whether a solid or a surface is created. In the part browser surface features are assigned indexed names like ExtrusionSrf1 or RevolutionSrf2. Once a surface (or solid) feature has been created with any one of these tools, it cannot be edited and changed to a solid (surface). The display of a surface can either be set to translucent or opaque. By default surfaces are displayed translucently.

New Hybrid Modeling Tools

In addition to these tools, new hybrid modeling tools included in Release 6 include Thicken/Offset, Delete Face, Replace Face, and Stitch Surface. All of these tools work with existing features. The first two tools provide a means for converting between a solid and a surface.

Thicken/Offset

Thicken/Offset is a versatile tool that can be used to: 1) add or remove thickness to an existing solid, 2) offset the face of an existing solid, resulting in an offset surface, 3) add thickness to a surface and, in so doing, convert the surface to a solid, and 4) offset a surface. The Thicken/Offset dialog box is shown Figure 10.13.

In Figure 10.14 the Thicken tool is used to alternately add material to a face (middle), as well as to remove material from a face (right). For both of these thicken operations the Output in the Thicken/Offset dialog box was set to Solid.

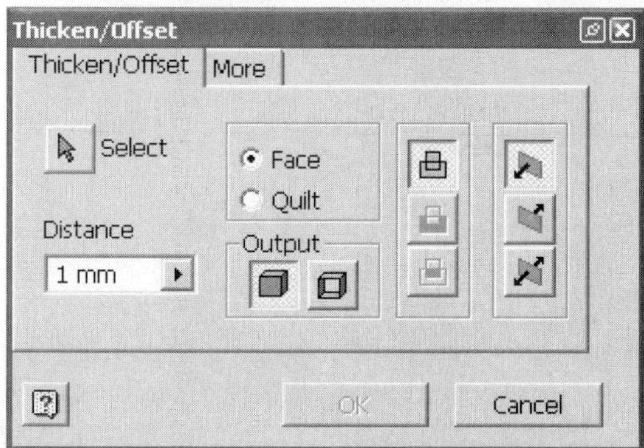

Figure 10.13 - Thicken/Offset dialog box

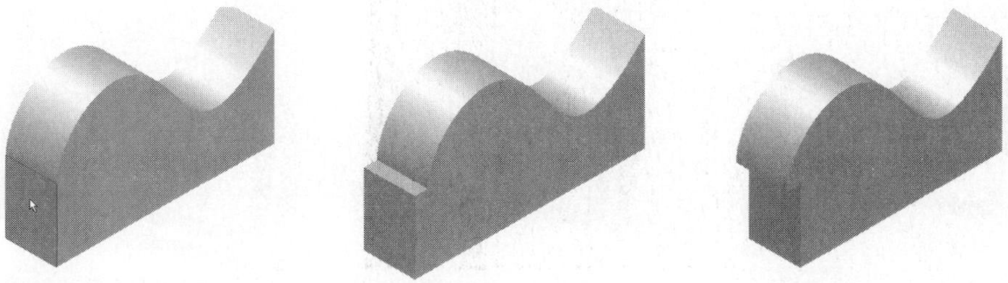

Figure 10.14 - Use Thicken tool to add (or remove) thickness to a solid

In Figure 10.15, the Thicken/Offset tool is used to offset a face. For this operation, the Output was set to Surface . The result is a new offset surface. If the Thicken/Offset tool is used on a surface, with the Output set to Solid, a solid feature results. In Figure 10.16B the select mode was set to Face, whereas in Figure 10.16C the select mode was set to *Quilt.*

Finally, if the Thicken/Offset tool is used on a surface and the Output is set to surface, an offset surface results, as shown in Figure 10.17. In the middle the select mode is set to Face, on the right, to Quilt.

Delete Face

The Delete Face tool can be used to delete one or more faces from a solid part. In so doing the solid part is converted to a surface. In the browser a delete face feature is placed in the browser hierarchy. The resulting surface can later be turned back to a solid with the Thicken tool.

In Figure 10.18 the Delete Face tool has been used to delete the top face of the guitar body. Note that a Delete Face1 icon has been added to the browser, and that the part icon has changed from solid to surface .

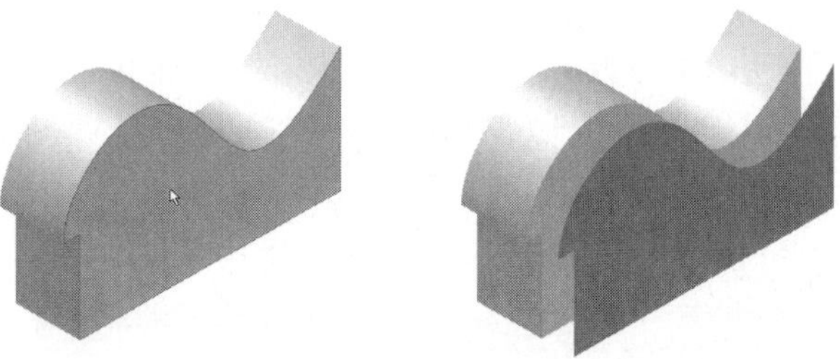

Figure 10.15 - Use Thicken tool to offset a face, convert to surface

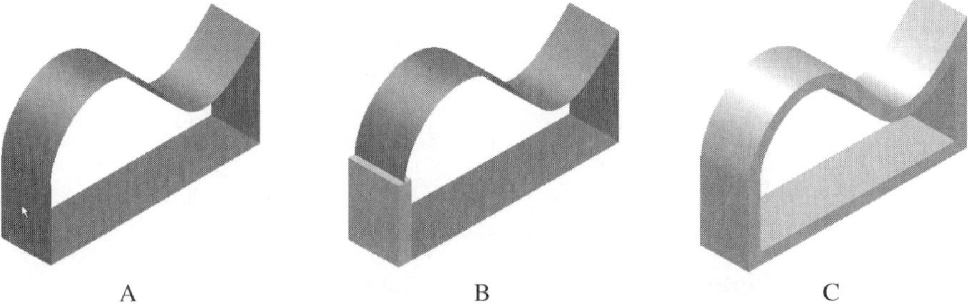

Figure 10.16 - Use Thicken tool to add thickness to a face, convert to solid

Figure 10.17 - Use Thicken tool to offset a surface

Replace Face

The Replace Face tool can be used to replace an existing part face with another face. The part must completely intersect with the replacement face. The Replace Face tool is demonstrated in the Figure 10.19. After selecting the top face of the cylinder (Figure 10.19A), the New Faces button is selected (Figure 10.19B) and the extruded surface is chosen. After selecting OK, the planar face is replaced. In order to clearly see the modified part (Figure 10.19C), it is necessary to turn off the visibility of the undulating surface.

Stitch Surface

The Stitch Surface tool can be used to combine surfaces that share a common edge. The resulting single joined surface is called a *quilt*. Figure 10.20 shows two surfaces that share a common edge. In the left foreground is a lofted surface, while in the right background there is an extruded surface. A close inspection at the seam between the two surfaces reveals slight imperfections; pin pricks of light at the boundary indicate that the seam is not closed.

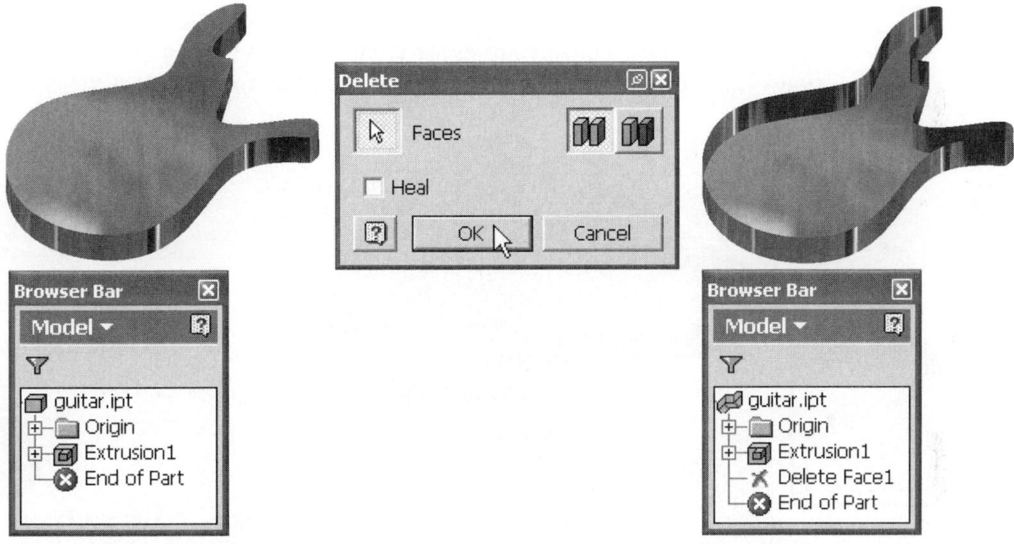

Figure 10.18 - Delete Face operation

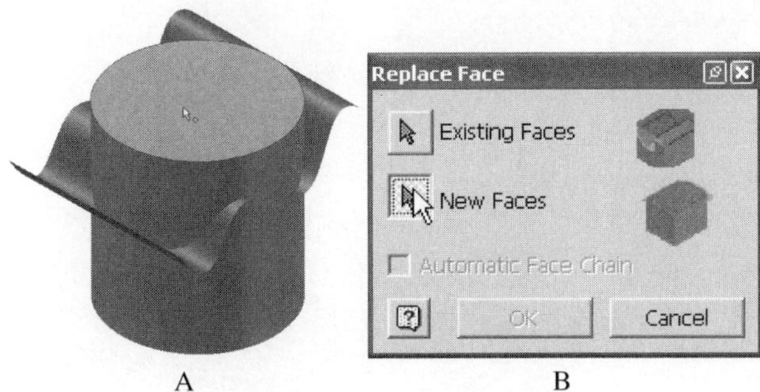

Figure 10.19 - Replace Face operation

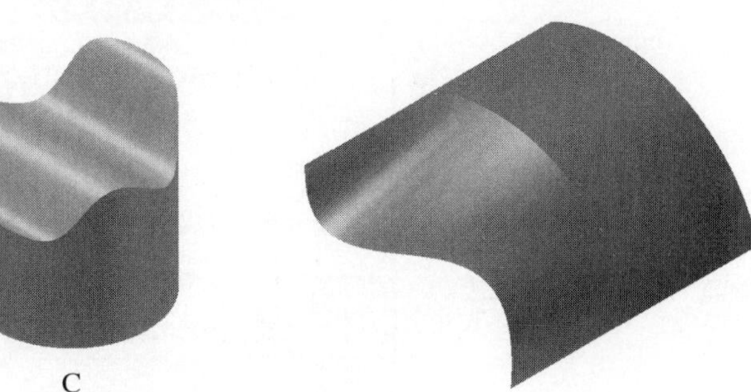

Figure 10.20 - Two surfaces with a common edge

After using the Stitch Surface tool to combine the two surfaces, as shown in Figure 10.21, the two surfaces are combined into a single quilt. The pin prick imperfections are no longer visible. Note that the Stitch Surface tool does not consume the other (loft, extrude) surfaces. In order to arrive at the stitched surface shown on the right, it was necessary to turn off the visibility of the loft and extrude surfaces. The translucency of the stitched surface was also turned off.

Loft

With Inventor Release 6 it is now possible to define rail curves in order to better control the shape of lofted features. Prior to Release 6 closed section curves, like those shown in Figure 10.22 on the left, determined the shape of a feature created with the Loft tool (Figure 10.22 on the right).

Figure 10.23 shows an example of a lofted feature defined with rail curves and the same sections as those used to create the feature shown in the previous figure. After selecting the section curves, the user now has the option of also selecting rail curves.

Rail curves must intersect the section profile curves. To ensure that this happens, Work Points are used to mark the intersections of the section curves with the work planes that are used as rail sketch planes. These work points are then projected onto the rail sketch planes. The rail curves are created so that they snap to the work point projections. Examples of this procedure are provided in Tutorial 30 at the end of the chapter.

Figure 10.21 - Stitch Surface tool

Figure 10.22 - Loft without rails

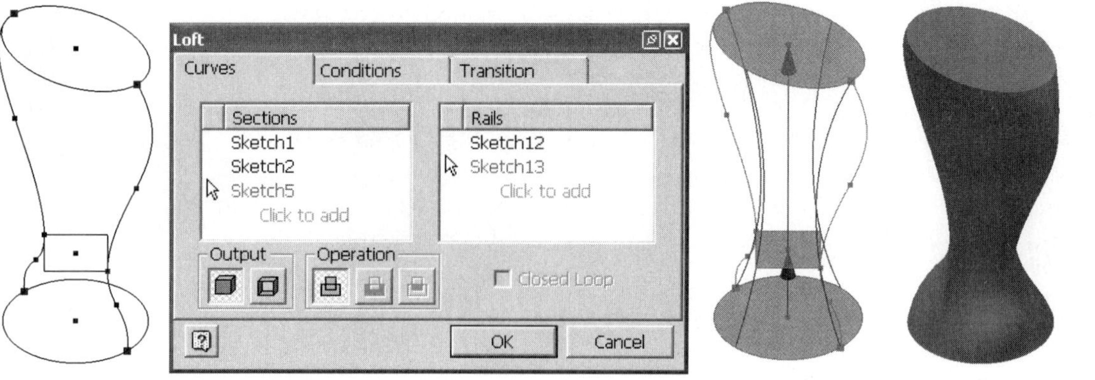

Figure 10.23 - Loft with rails

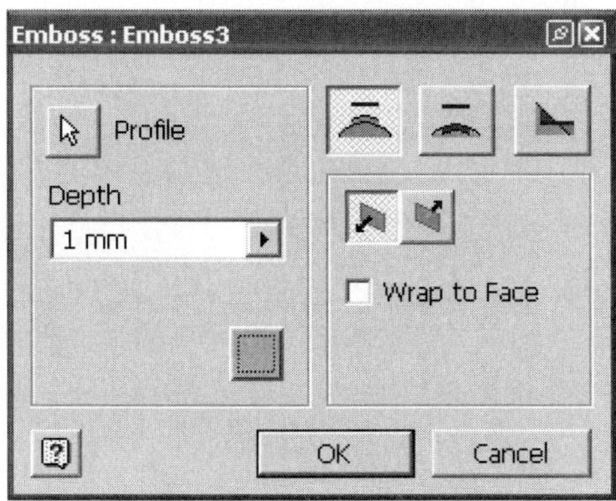

Figure 10.24 - Emboss tool dialog box

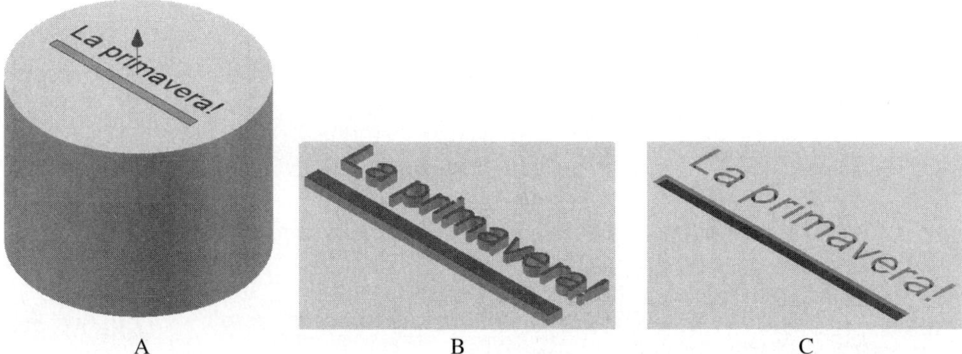

Figure 10.25 - Emboss / engrave feature on planar face

Emboss and Decal

Emboss

The Emboss tool uses a sketch profile to create a raised (i.e., embossed) or recessed (i.e., engraved) feature on an existing solid face. The sketch profile is most commonly text created with the Text tool; however any closed profile can be embossed (engraved) on a solid face. Embossed (engraved) features are used to represent the logos, symbols, instructions, etc. that commonly appear on manufactured parts.

If the emboss (engrave) feature is created on a planar face, the sketch can be created either directly on the face or on a workplane offset from the face.[1] Figure 10.24 shows the Emboss dialog box.

In Figure 10.25A, a sketch plane has been created on the top planar face of a cylinder. After adding text and a rectangle to the sketch, the Emboss tool is selected. Both the text and the closed profile (i.e., the rectangle) are selected (use the Ctrl key). Figure 10.25B shows the result using the following options: Emboss ,

[1]In the event that the planar face is used as the sketch plane, you may have trouble selecting the sketch geometry. In this case turn off the default automatic projection options. To do this, from the menu bar select Tools > Application Options . . . > Sketch tab. Clear the check marks on Autoproject edges during curve creation and Automatic reference edges for new sketch.

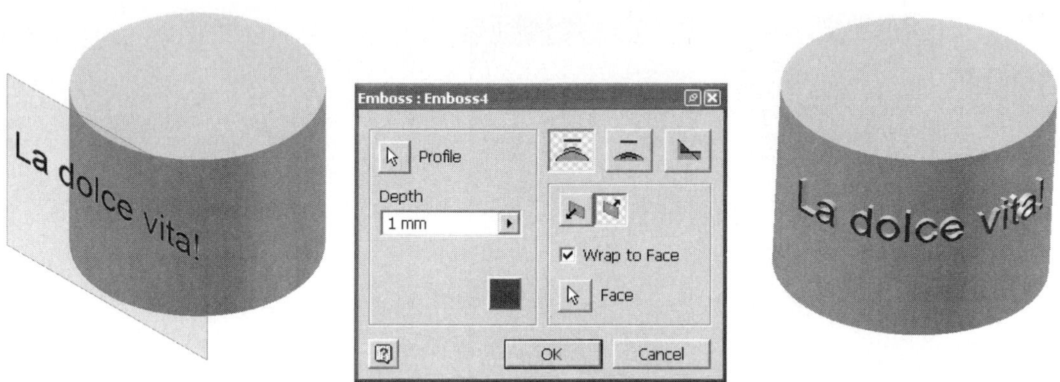

Figure 10.26 - Emboss feature wrapped to cylindrical face

Figure 10.27 - Emboss feature on curved face

Depth = 1 mm, Grape top face color. Figure 10.25C shows the result with Engrave, Depth = 0.5 mm, Green top face color.

To create an embossed feature on a curved solid face, a work plane must be used as the sketch plane. Figure 10.26 shows an embossing operation on the curved face of a cylinder. Note that the Wrap to Face option is checked. With this option checked an embossed feature can wrap a full 360 degrees around a cylindrical or conical face. Unchecked, the profile is projected directly onto the face. Note that it is also possible to create embossed features on most arbitrarily curved surfaces, as shown in Figure 10.27.

Like other Inventor features, embossed features can be edited. To edit embossed feature text, first edit the sketch from the browser, and then use the right mouse button context menu to edit the text.

Decal

The Decal tool is used to convert an image placed on a sketch to a decal feature placed on a part face. The decal feature can be used to represent labels, brand name art, logos, warranty seals, etc. A decal feature can be placed on surfaces or solid faces.

Figure 10.28A shows a workplane and a thickened solid created from an extruded spline surface. A sketch is created on this workplane. The Insert Image tool, available from the 2D Sketch panel, is then used to place a bitmap (.bmp) file on the sketch. Note that the bitmap images can be resized by dragging on an edge of the image; however, the aspect ratio of the image cannot be changed. Figure 10.28B shows the inserted image after exiting sketch mode. Figure 10.28C is the result of projecting the image onto the thickened solid using the Decal tool. Here the workplane visibility has been turned off.

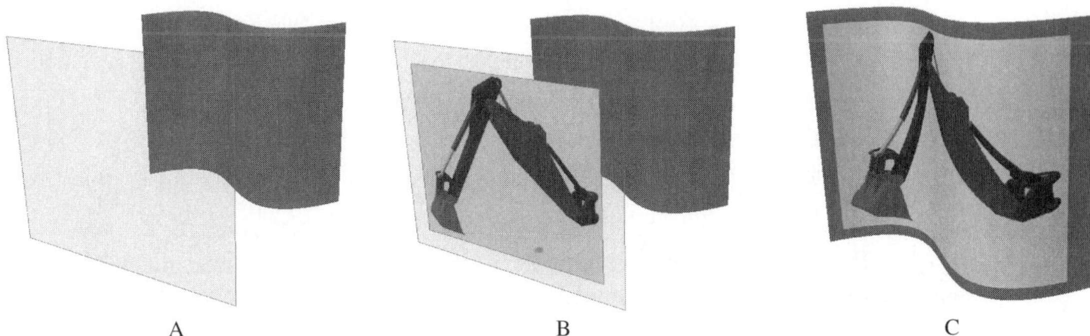

Figure 10.28 - Decal feature creation

New Work Geometry Construction Methods

Work construction geometry is particularly useful when creating complex shapes with hybrid modeling tools. Consequently, a number of new work geometry construction methods have been added in Inventor Release 6.

Work Points

Work points are especially useful in finding the intersection point(s) between two geometric entities (e.g., a curve and a plane). In this way work points can be used to tie different sketch geometries (appearing on two orthogonal work planes) together. A particularly good example is a lofted feature with rails. Since the rail curves must intersect the section curves, work points are used to tie the two sets of curves together. The general procedure is to mark the key point(s) on one sketch with a work point and then to project the work point onto an orthogonal sketch plane so that it can be used in another sketch.

Using the Work Point tool from the panel bar, parametric work points can be added to a vertex or to the midpoint of an edge. Figure 10.29 shows work points that have been placed at a vertex and at the midpoint of an edge, for both a solid and a surface. Figure 10.30 shows several examples where the Work Point tool has been used

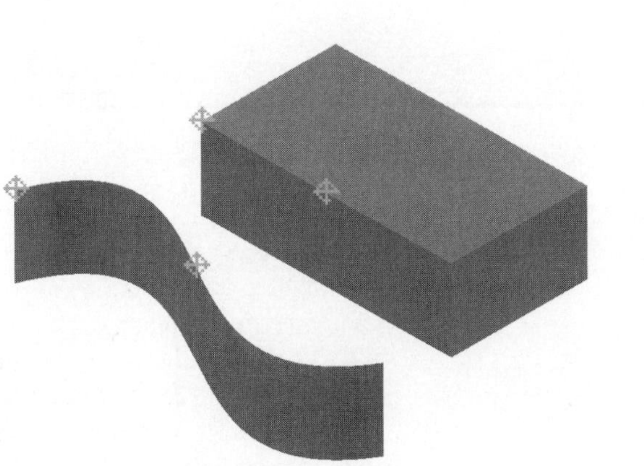

Figure 10.29 - Work points (single selection)

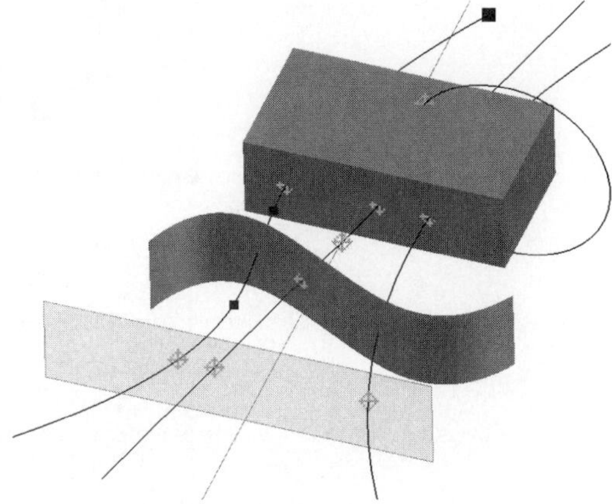

Figure 10.30 - Work points (two selections)

to create work points at the intersection between two selected geometric entities. The possibilities include:

- Planar face (or work plane) and a line (or an edge or axis)
- Surface and a line (or an edge or axis)
- Planar face (or work plane) and a curve (arc, circle, ellipse, spline, curved edge)
- Two lines (sketched line, edge, or axis)

Work points created with the Work Point tool are parametric; this means that they will move if the underlying geometry is modified.

Grounded Work Points

The Grounded Work Point tool is new to Release 6. Grounded work points are non-parametric. Once a work point has been created with the Grounded Work Point tool, its position is fixed in space. Note that this is not the case with ungrounded work points created with the Work Point tool; the location of an ungrounded work point can be changed by adding constraints or dimensions.

The 3D Move/Rotate tool is used to locate and place grounded work points. After selecting the Grounded Work Point tool, a starting point must then be selected, using either the graphics window or browser. Once the start point is selected, the 3D Move/Rotate triad appears at that point; the 3D Move/Rotate dialog box also opens. Both are shown in Figure 10.31.

The triad must now be maneuvered until its position locates the desired position of the grounded work point. If the triad sphere is selected, all three coordinate boxes on the dialog box are activated. The triad moves as relative coordinates are entered. Now click OK to place the work point. Selection of any of the three triad work planes constrains the translation to that plane. By selecting a triad arrow (red–x, green–y, blue–z), the translation is constrained so that only movement along the selected axis is permitted.

To reorient, or rotate the coordinate system triad, select the axis to be rotated about. The dialog box changes to something like that shown in Figure 10.32. By

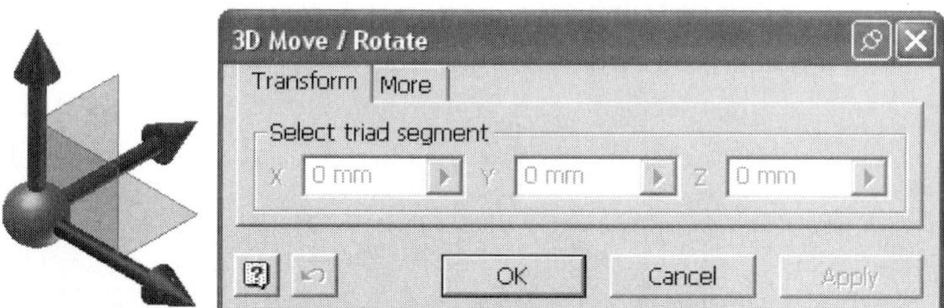

Figure 10.31 - 3D Move/Rotate: triad and dialog box

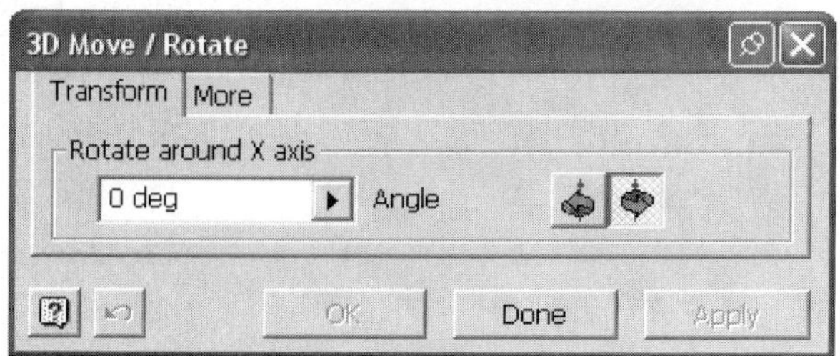

Figure 10.32 - 3D Move/Rotate dialog box; rotation transformation

entering an angle, and then clicking Apply, the coordinate system triad is rotated. Further transformations can now be made. Once the triad is in the position where the grounded work point is to be placed, click OK to create the work point.

Work Plane Normal to a Point on a Sketch Curve

Work planes normal to a spline curve can now be created at the curve's spline points. See Figure 10.33. This capability is useful for creating both swept and lofted features. After selecting the Work Plane tool, select the spline point and the spline in any order to create the work plane.

3D Intersections

Although it is still not possible to trim surfaces in Autodesk Inventor Release 6, the new 3D Intersection tool, available from the 3D Sketch tool panel, is helpful in situations where it is necessary to find the intersection curve between two complex surfaces.

The 3D Intersection tool creates a 3D curve from the intersection of surfaces, work planes, or parts. Figure 10.34A shows two intersecting surfaces. To find the 3D intersection curve between these surfaces, first create a new 3D Sketch. Upon selecting the 3D Intersection tool, the 3D Intersection Curve dialog box opens, as shown in Figure 10.34B. Select the two surfaces, and then click OK. Figure 10.34C shows the intersection curve. The 3D intersection curve can now be used to create other features. Figure 10.35, for example, depicts a lofted surface operation that makes use of the 3D intersection curve.

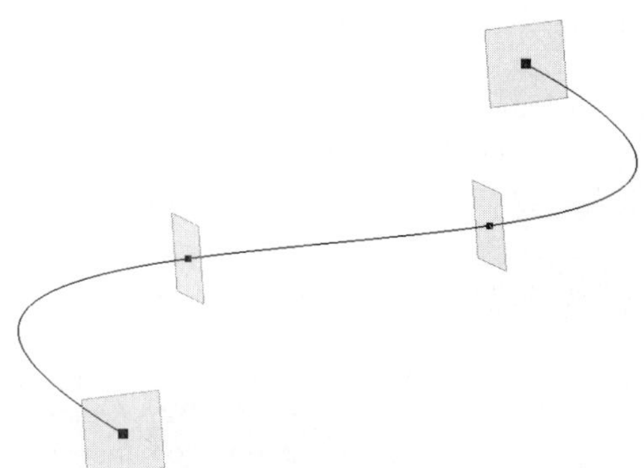

Figure 10.33 - Work planes normal to spline curve at spline points

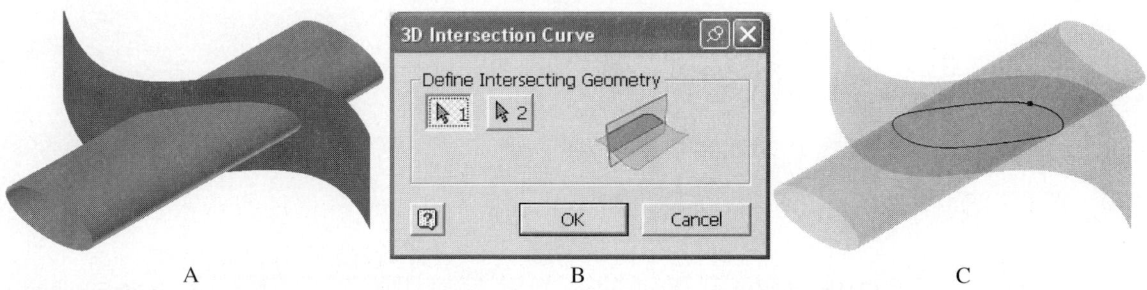

Figure 10.34 - 3D intersection curve operation

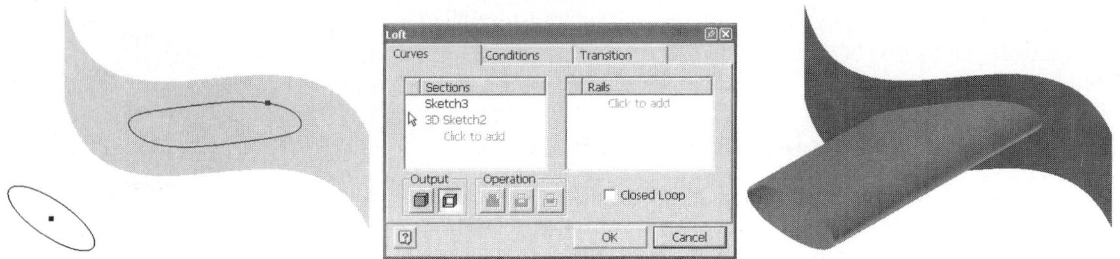

Figure 10.35 - Loft operation with 3D intersection curve

TUTORIAL 29 Camera Body

Build Strategy

1. Base feature (Figure 10.36).
2. Sketch splines on top and bottom faces (Figure 10.37).
3. Lofted surface (Figure 10.38).
4. Replace face (Figure 10.39).

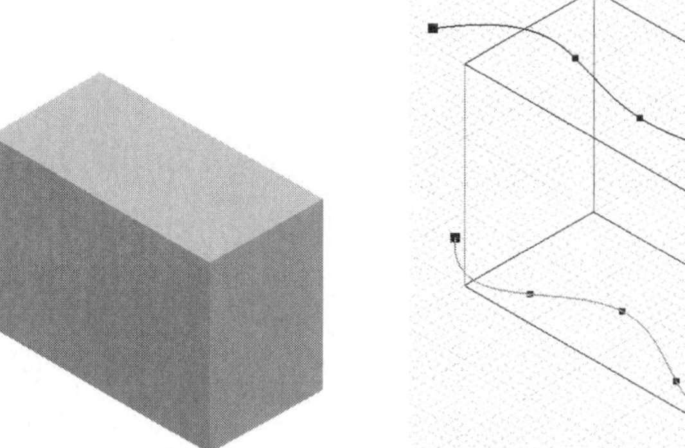

Figure 10.36 - Base feature

Figure 10.37 - Spline sketches (top and bottom faces)

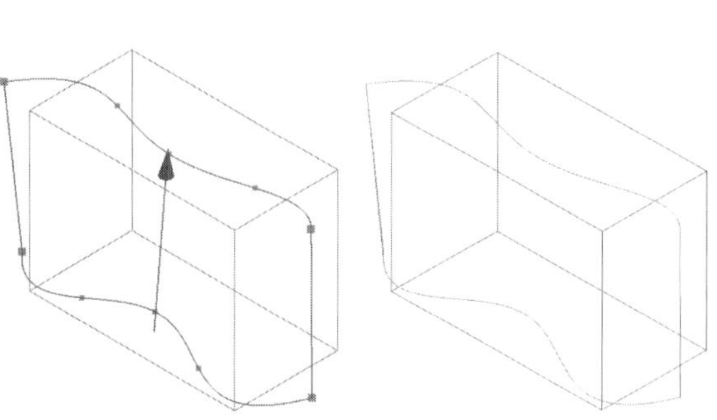

Figure 10.38 - Lofted surface

Figure 10.39 - Replace face

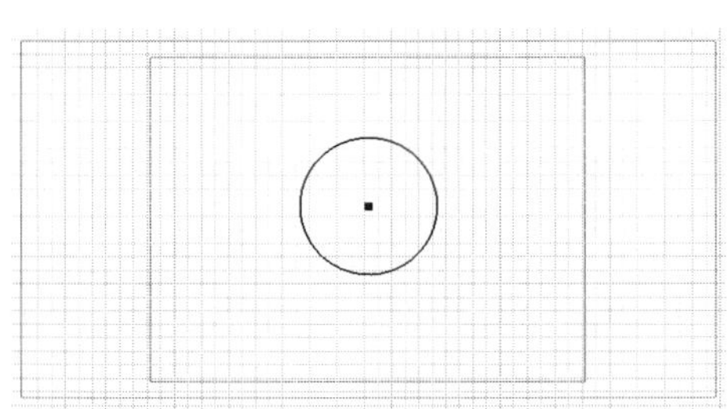

Figure 10.40 - Sketch on offset work plane

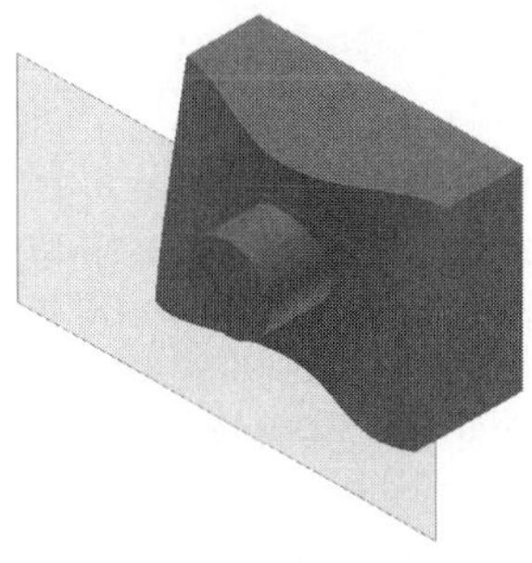

Figure 10.41 - Extrude (join)

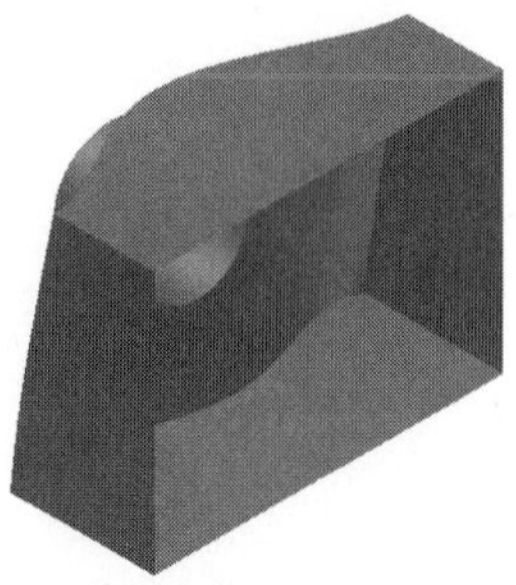

Figure 10.42 - Delete (back) face

Figure 10.43 - Add miscellaneous fillets

Figure 10.44 - Thicken

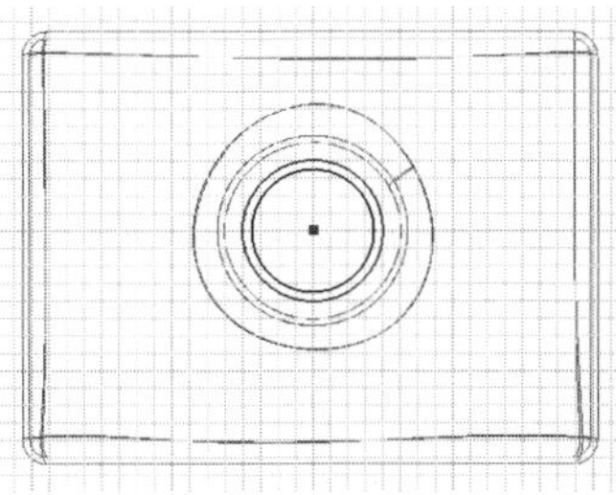

Figure 10.45 - Sketch on face

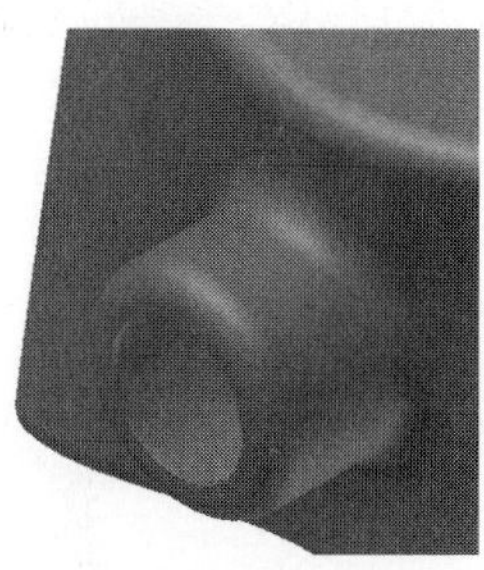

Figure 10.46 - Extrude (cut)

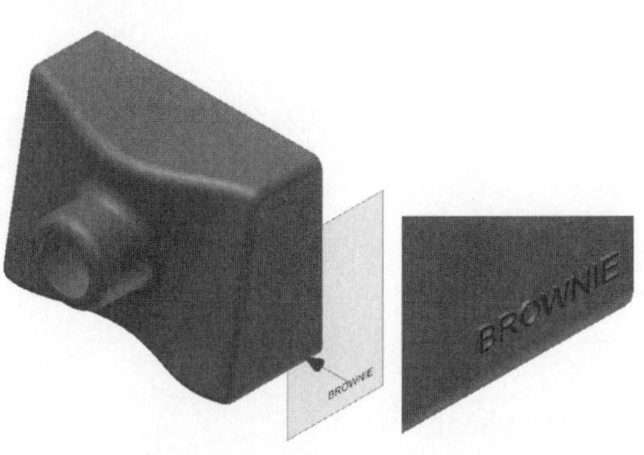

Figure 10.47 - Emboss

Figure 10.48 - Decal

Figure 10.49 - Completed camera body

5. Sketch on offset work plane (Figure 10.40).
6. Extrude (Figure 10.41).
7. Delete Face (Figure 10.42).
8. Fillets (Figure 10.43).
9. Thicken (Figure 10.44).
10. Sketch on face (Figure 10.45).
11. Extrude (cut) (Figure 10.46).
12. Emboss (Figure 10.47).
13. Decal (Figure 10.48).
14. Completed camera body (Figure 10.49).

■ Detailed Modeling Steps

1. Select File > Projects . . . from the menu bar. In the Open dialog box, set the Project name to Default (select Default and then click Apply, or double-click on Default). When the project name is set to Default, the Open and Save As dialog boxes will default to the most recently used folder.
2. Start a new English part file.
3. Use the rectangle tool to add a rectangle, and then dimension it as shown in Figure 10.50.
4. Finish Sketch.
5. Isometric View.
6. Extrude the sketch as shown in Figure 10.51.
7. Save the file as **Camera Body.** Place the file in a folder of your choice. It is recommended that the file be regularly saved every five to ten minutes.
8. Start a new sketch on the top face of the box.
9. Look At the sketch.
10. Use the Spline tool to sketch a spline similar to that shown in Figure 10.52. To complete the spline, right-click and select Continue, then right-click and select Done.
11. Finish Sketch.
12. Isometric View.
13. Wireframe Display.
14. Start a new sketch on the bottom face of the box. It will be necessary to use the Select Other tool to select this new sketch plane.

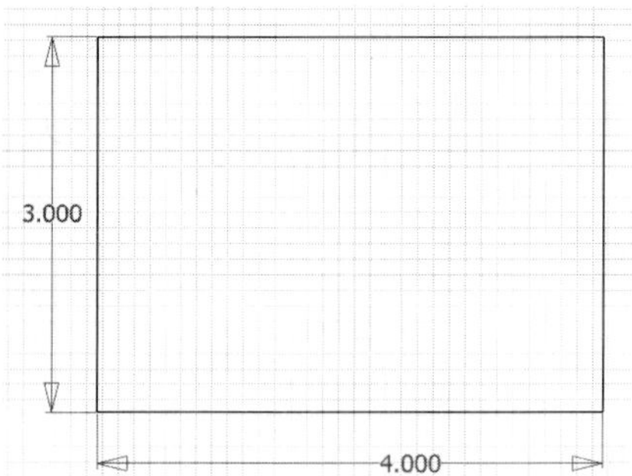

Figure 10.50 - Base feature sketch

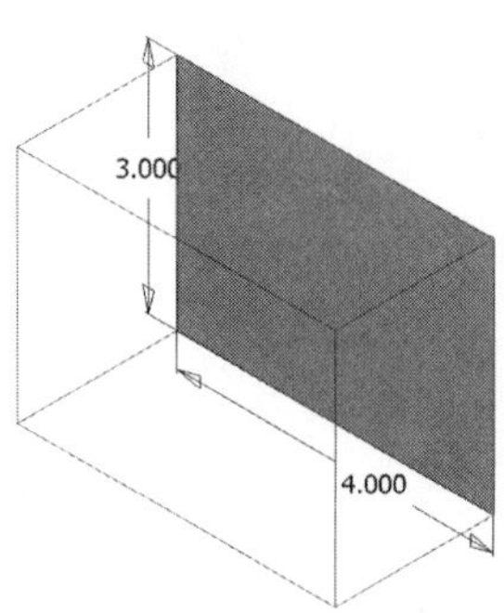

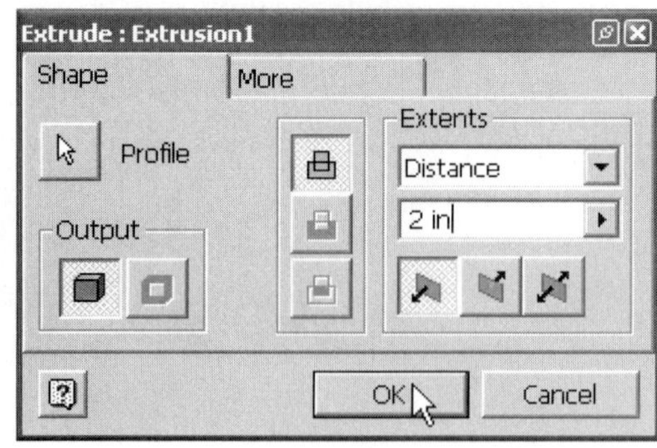

Figure 10.51 - Extrude base feature

15. Again use the Spline tool to sketch another spline similar to the one shown in Figure 10.53.
16. Finish Sketch.
17. Use the Loft tool to create a lofted surface between the two splines. In the loft dialog box, first select Surface Output. It is then necessary to click in the Sections area of the dialog box on the "Click to Add" text. You will now be prompted to select a sketch. Select one of the splines, and then select the same spline again. This ensures that only the spline sketch (and not the rectangle) is selected. Now select the other spline twice. The screen should now resemble the image on the left of Figure 10.54. Click OK to create the surface.
18. Shaded Display.
19. Use the Replace Face tool to substitute the undulating surface for the frontal face of the solid. In the Replace Face dialog box, with the Existing Faces button selected, select the frontal face of the box. Next select the New Faces button, and select the surface. Click OK when ready (Figure 10.55).
20. In the browser turn off the visibility of the lofted surface (LoftSrf1).
21. Use the color control on the standard toolbar to change the color of the part.
22. At this point you may wish to modify the splines. To do so, expand the loft surface in the assembly browser to reveal the two spline sketches. Right-click on either sketch and then select Edit Sketch. At this point it is possible to select

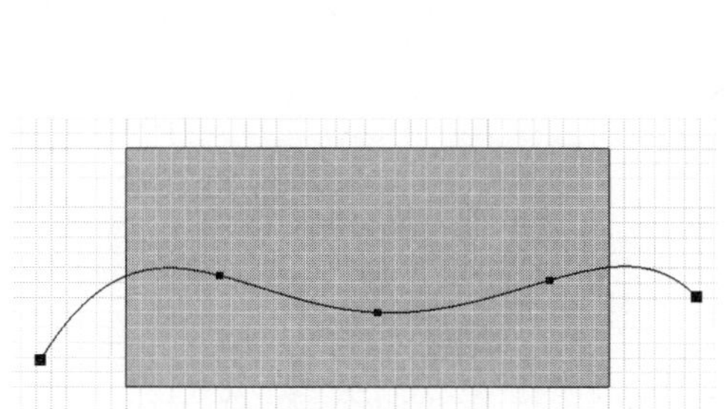

Figure 10.52 - spline sketch on top face

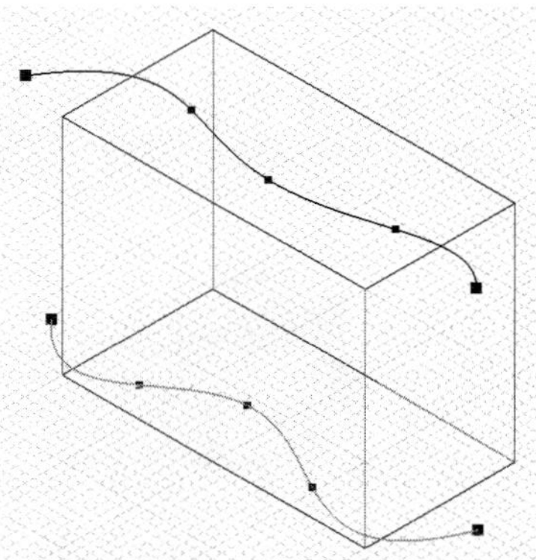

Figure 10.53 - Spline sketch on bottom face

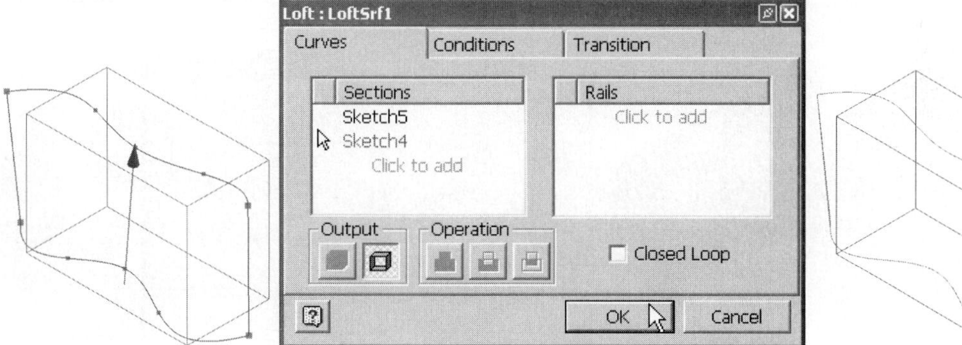

Figure 10.54 - Lofted surface feature operation

Figure 10.55 - Replace face operation

any of the spline control points and drag them to a new location. When satisfied, right-click and select Finish Sketch.

23. Create a new work plane offset a distance of perhaps 2.5 inches from the XY reference work plane. The screen should now resemble Figure 10.56.
24. Create a new sketch on the work plane.
25. Wireframe Display.
26. Look At the sketch.
27. Sketch a circle roughly in the center of the rectangle, similar to Figure 10.57.
28. Finish Sketch.
29. Isometric View.
30. Shaded Display.
31. Extrude the circle so that it joins the camera body, as shown in Figure 10.58.

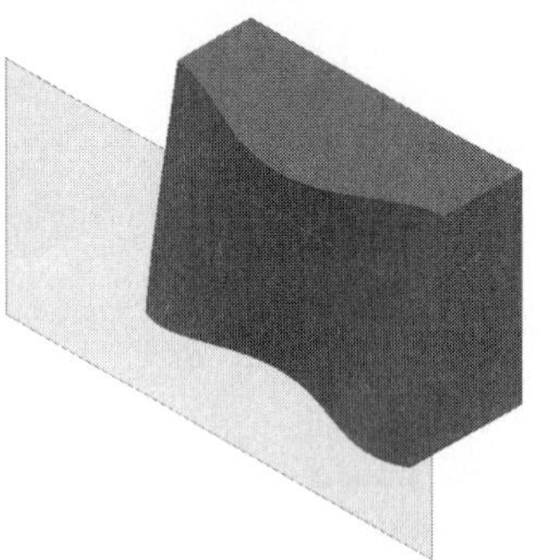

Figure 10.56 - Offset work plane

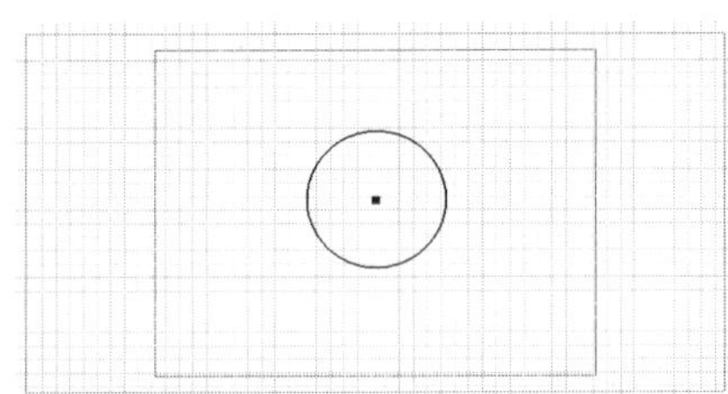

Figure 10.57 - Unconstrained sketch

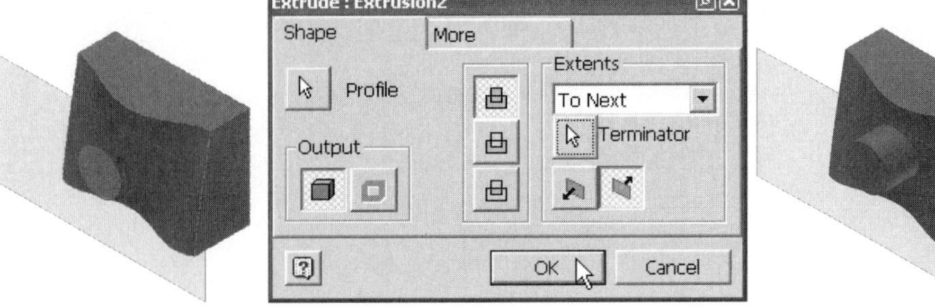

Figure 10.58 - Extrude (join) feature operation

32. Turn off the visibility of the work plane.
33. Use Common View to rotate the view so that the back of the camera body is visible.
34. Use the Delete Face tool to remove the back face of the camera body, as shown in Figure 10.59. Using the Delete Face tool converts the solid part to a surface. We will later use the Thicken tool to change the part back into a solid.
35. Isometric View.
36. Use the Fillet tool to add a 1/4" fillet between the cylinder and the main body of the camera, as shown in Figure 10.60.
37. Add additional 1/8" fillets along the edges of the camera body, as shown in Figure 10.61.
38. Use the Common View tool to change the view so that the back of the camera is visible.

Figure 10.59 - Delete face operation

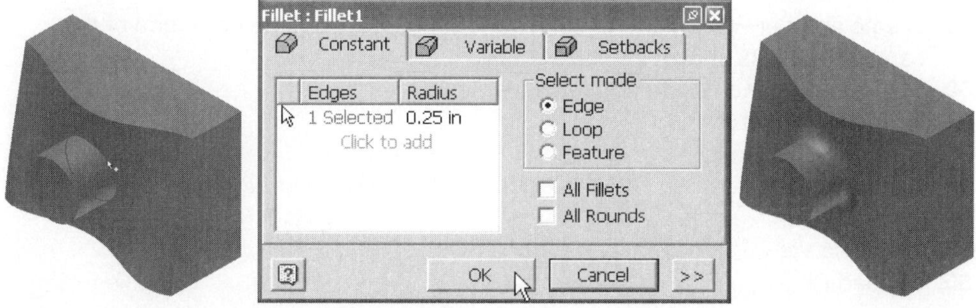

Figure 10.60 - Fillet operation

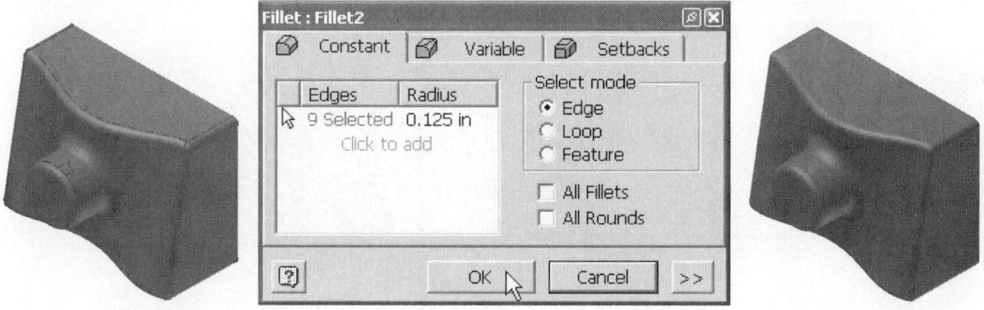

Figure 10.61 - Fillet feature operation

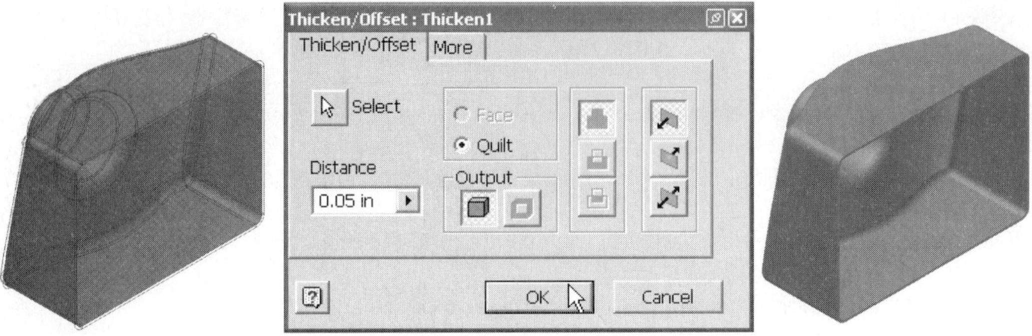

Figure 10.62 - Thicken feature operation

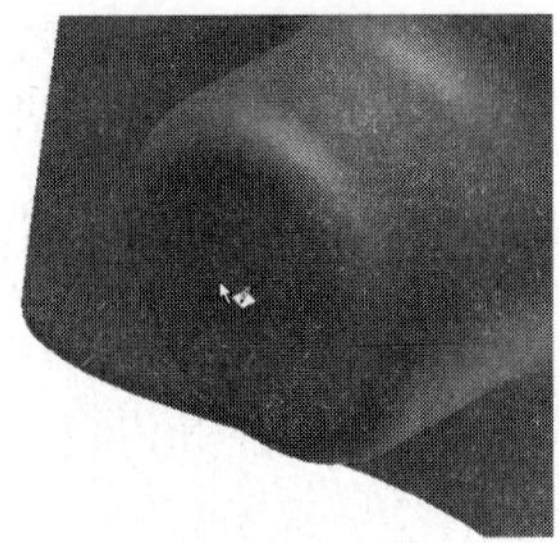

Figure 10.63 - New sketch (on face)

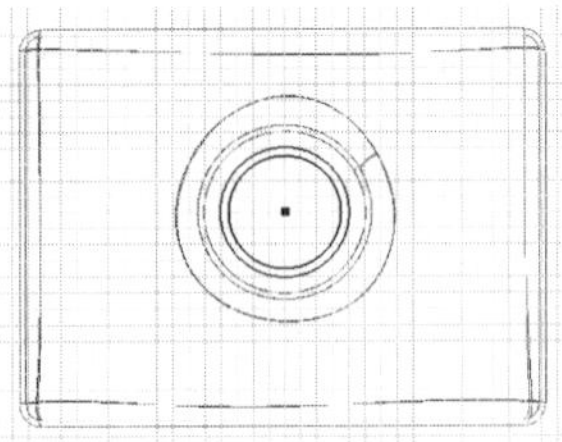

Figure 10.64 - Circle sketch

39. Use the Thicken tool to add thickness to the part, as shown in Figure 10.62.
40. Isometric View.
41. New Sketch - select the face shown in Figure 10.63.
42. Look At the sketch.
43. Wireframe Display.
44. Sketch a circle concentric to, and with a smaller diameter than the circle already visible on the sketch plane (Figure 10.64).
45. Finish Sketch.
46. Isometric View.
47. Shaded Display.
48. Extrude the sketch so that the circular area cuts a hole through the camera body (Figure 10.65).
49. Create another work plane offset by about an inch from the right face of the camera body, as shown in Figure 10.66.
50. Start a new sketch on the offset work plane.
51. Look At the sketch.
52. Wireframe Display.
53. Use the Create Text tool to add the text shown in Figure 10.67.
54. Finish Sketch.
55. Isometric View.
56. Shaped Display.

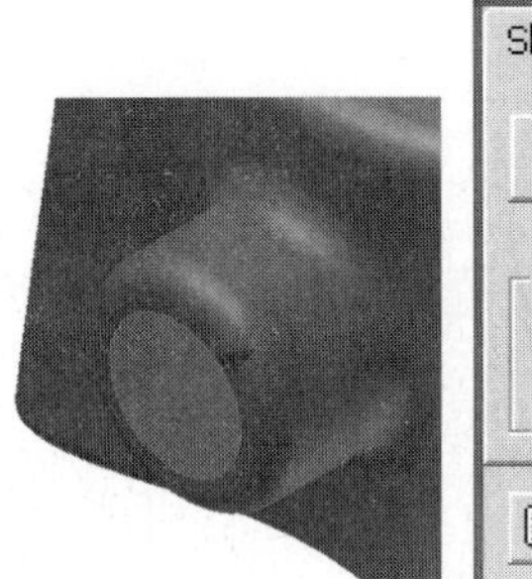

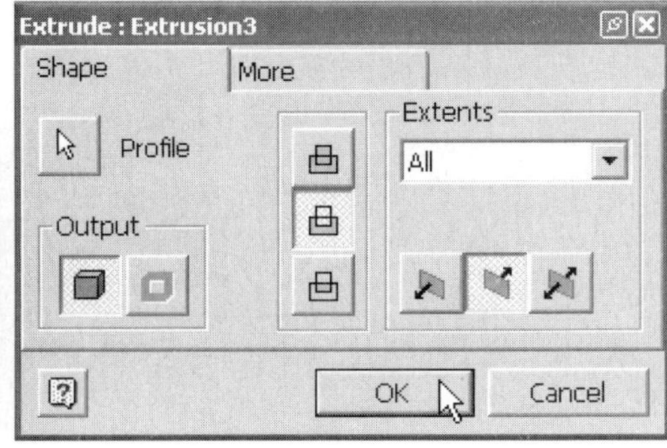

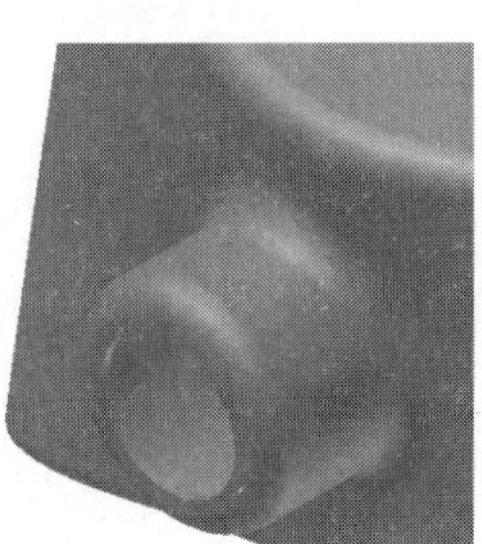

Figure 10.65 - Extruded (cut) feature operation

Figure 10.66 - Offset work plane

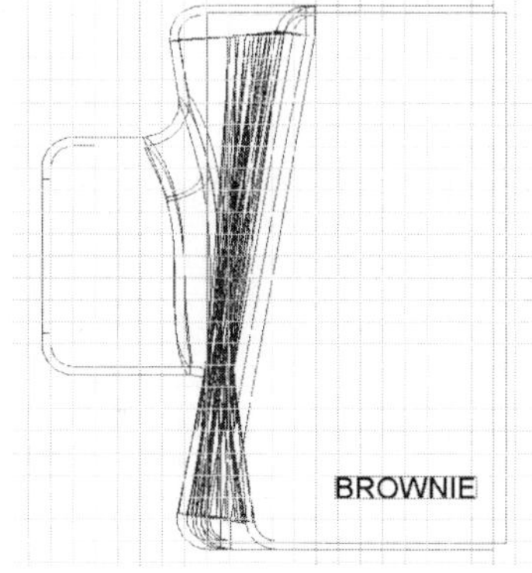

Figure 10.67 - Create text

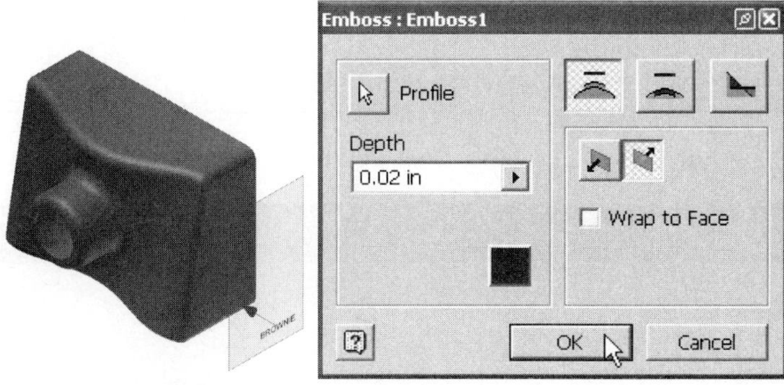

Figure 10.68 - Emboss feature operation

Figure 10.69 - Offset work plane

57. Use the Emboss tool as shown in Figure 10.68. Parameters include the Profile (select the "Brownie" text), a depth of 0.02 inches, a color of brown (flat), and an Emboss from Face type.
58. Turn off the visibility of the recently created offset work plane.
59. Create another offset work plane; this one offset about 3 inches from the XY Plane, as shown in Figure 10.69.
60. Start a new sketch on the offset work plane.
61. Look At this sketch.
62. Select the Insert Image tool on the 2D Sketch Panel. Although any available bitmap (.bmp) file can be used, a file named logo.bmp, available on the CD, will be used here. After selecting logo.bmp, click the open button. Left-click anywhere on the sketch plane to insert the bitmap, and then right-click and select Done. Resize the bitmap image by selecting and dragging an edge. Note that the *aspect ratio* of the bitmap is fixed. After resizing the image, position it over the camera body, as shown in Figure 10.70.
63. Finish Sketch.
64. Isometric View.

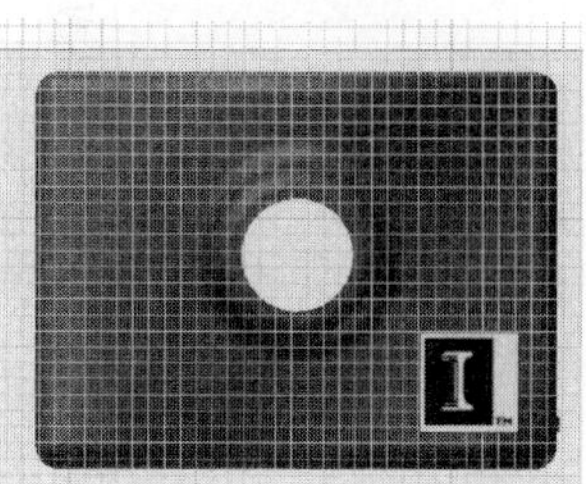

Figure 10.70 - Insert image

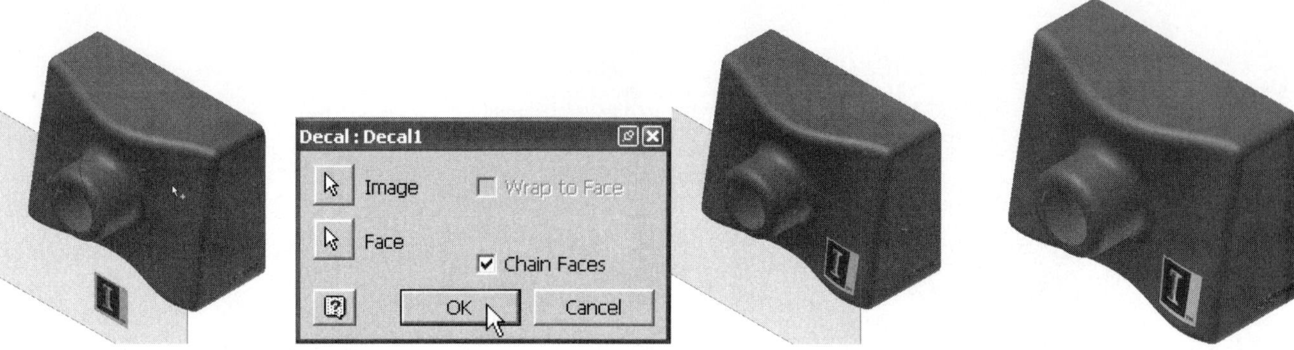

Figure 10.71 - Decal feature operation

Figure 10.72 - Completed part

65. After selecting the Decal tool, first select the bitmap, then select the front face of the camera body, then select OK, as shown in Figure 10.71.
66. Turn off the visibility the offset work plane. The completed part appears in Figure 10.72.
67. Save the file. This completes the camera body tutorial.

TUTORIAL 30 Creamer

Build Strategy

1. Work plane sketches (Figure 10.73).
2. Ellipse work plane sketch (Figure 10.74).
3. Extrude surface (Figure 10.75).
4. Another sketch (Figure 10.76).
5. Extrude surface (Figure 10.77).
6. 3D Intersection (Figure 10.78).
7. Cross section sketches (Figure 10.79).
8. Rail curves (Figure 10.80).

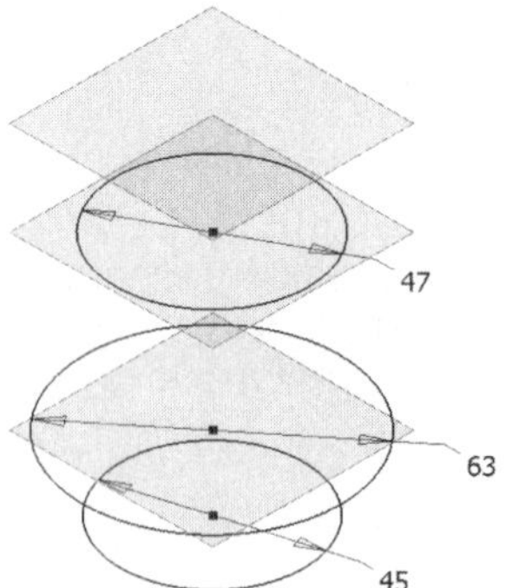

Figure 10.73 - Offset work plane sketches

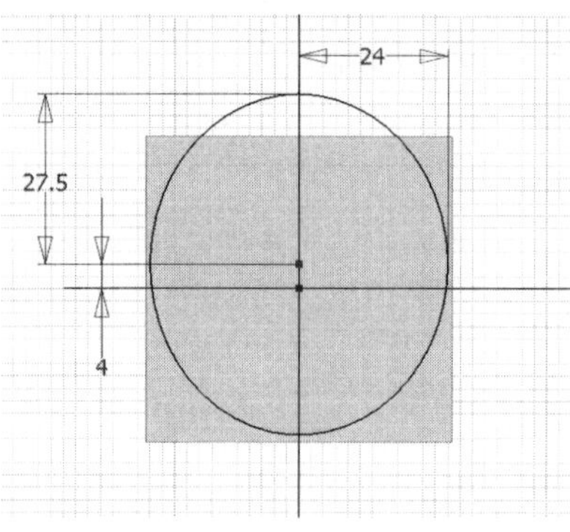

Figure 10.74 - Ellipse work plane sketch

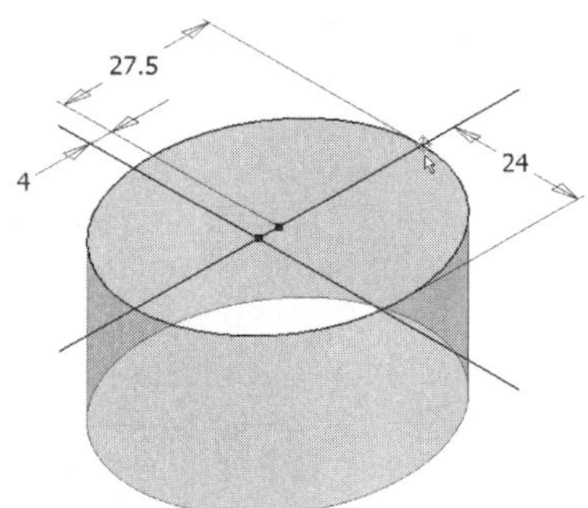

Figure 10.75 - Extruded ellipse surface

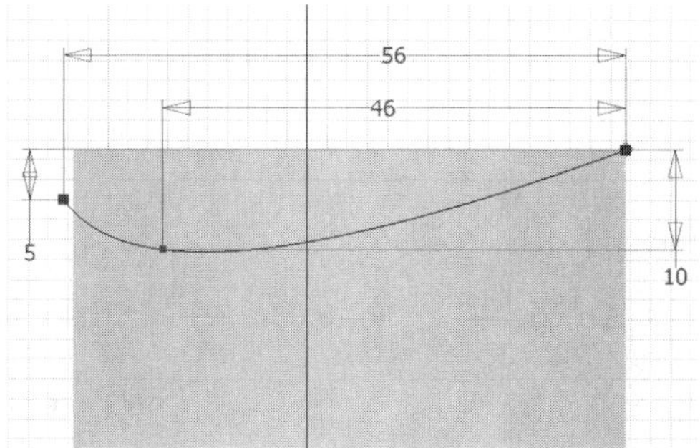

Figure 10.76 - Profile sketch

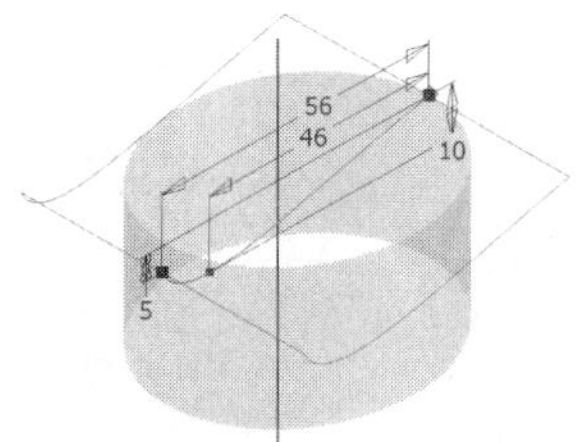

Figure 10.77 - Extrude profile sketch

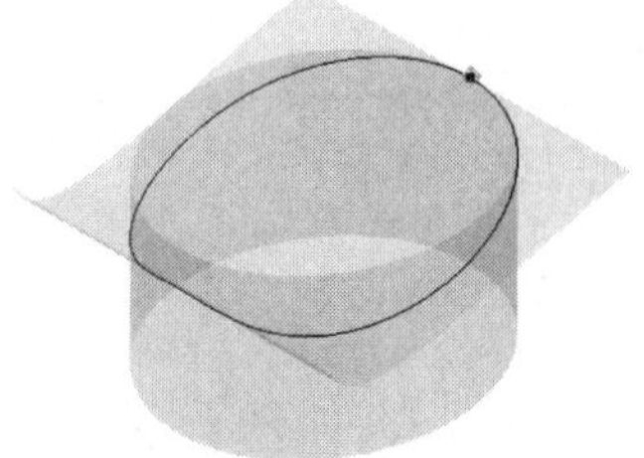

Figure 10.78 - Intersect surfaces; find intersection curve

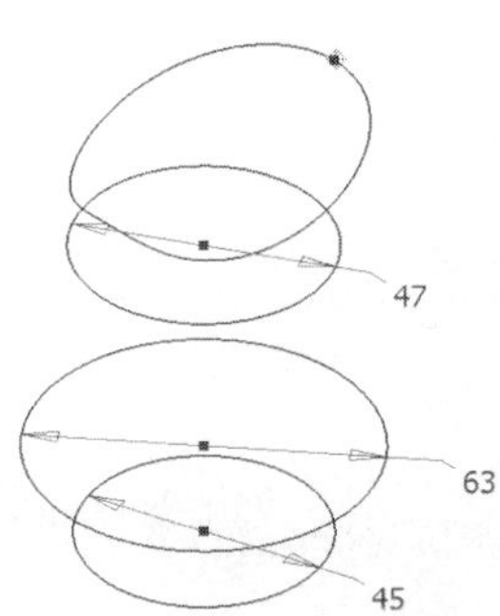

Figure 10.79 - Cross-sectional sketches

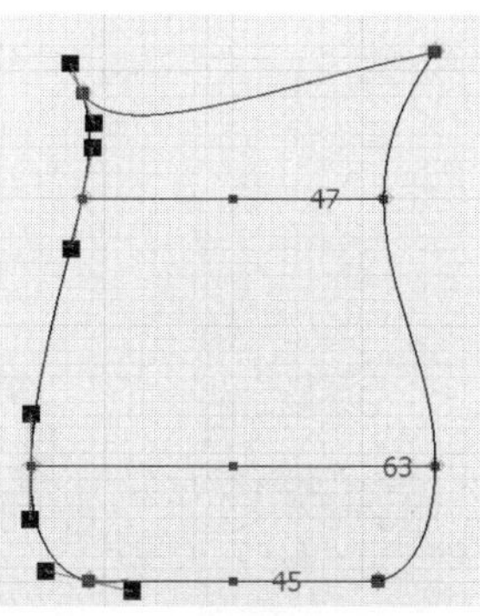

Figure 10.80 - Rail curve sketches (two)

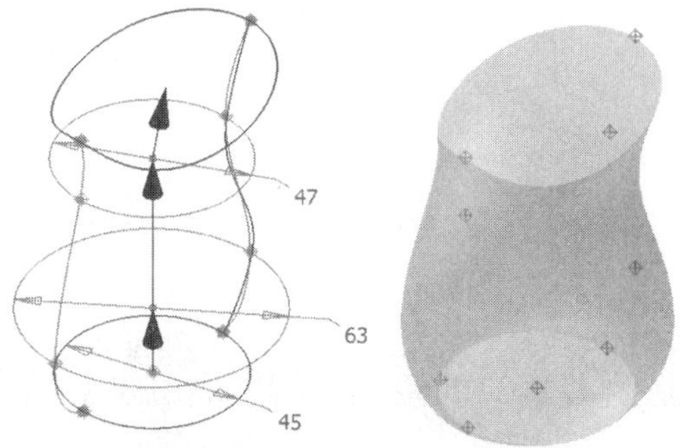

Figure 10.81 - Lofted surface (4 cross-section, 2 rail curves)

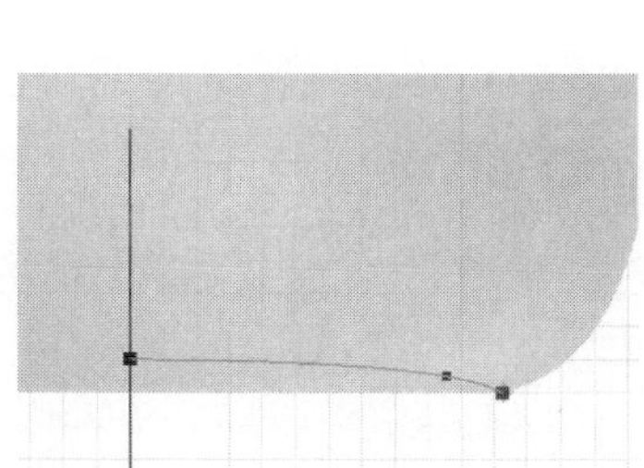

Figure 10.82 - Bottom spline curve

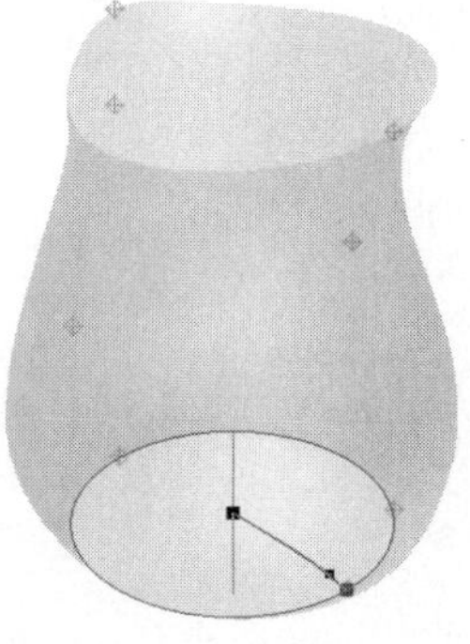

Figure 10.83 - Revolved bottom surface

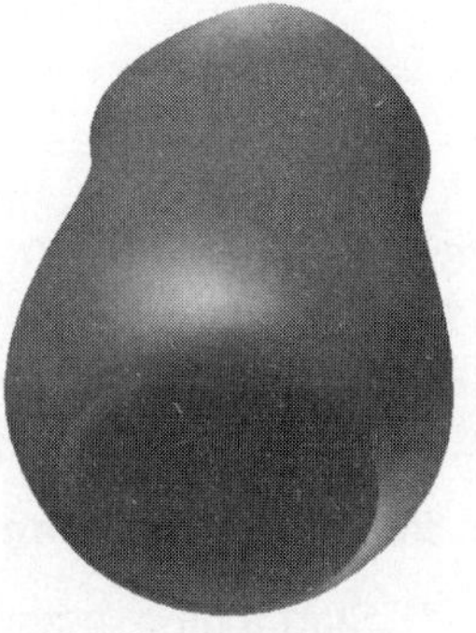

Figure 10.84 - Stitched surfaces (loft + revolve)

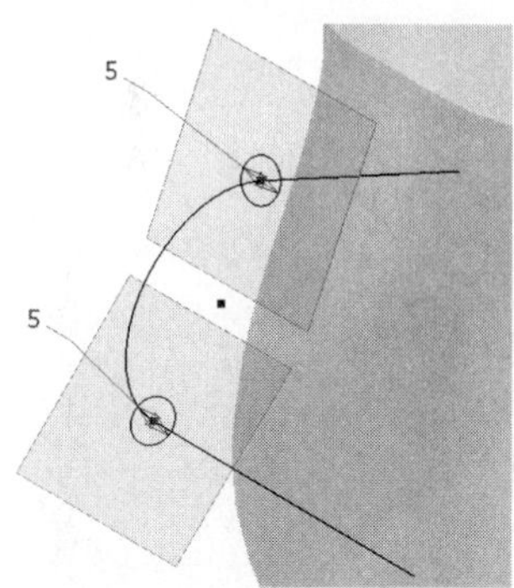

Figure 10.85 - Preliminary handle sketch, work planes, cross-sections

9. Loft (Figure 10.81).
10. Bottom spline curve (Figure 10.82).
11. Revolve surface (Figure 10.83).
12. Stitch surfaces (Figure 10.84).
13. Preliminary handle sketch, work planes, cross-section sketches (Figure 10.85).

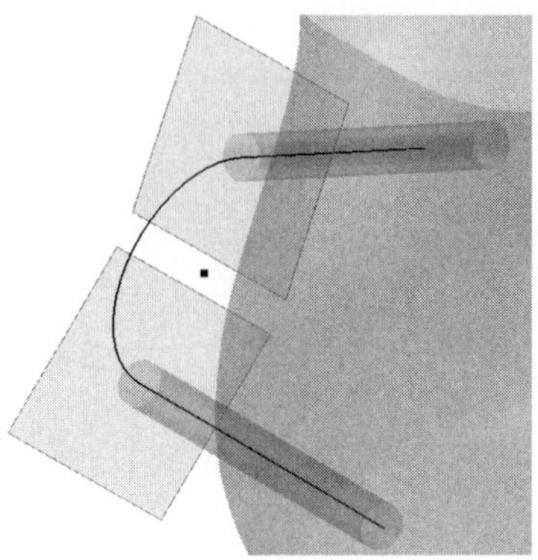

Figure 10.86 - Extrude cross-section curves

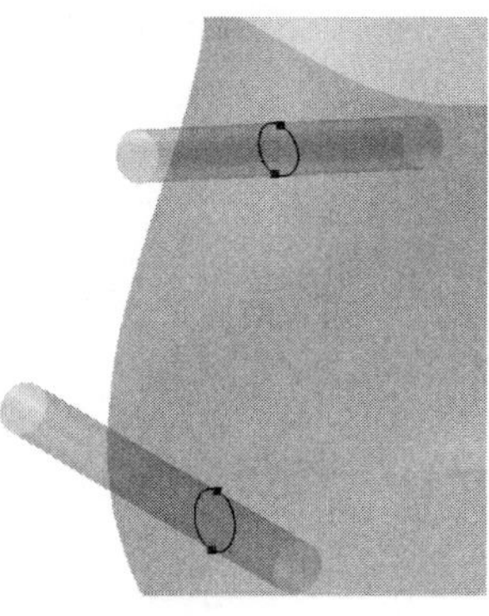

Figure 10.87 - Intersection curves

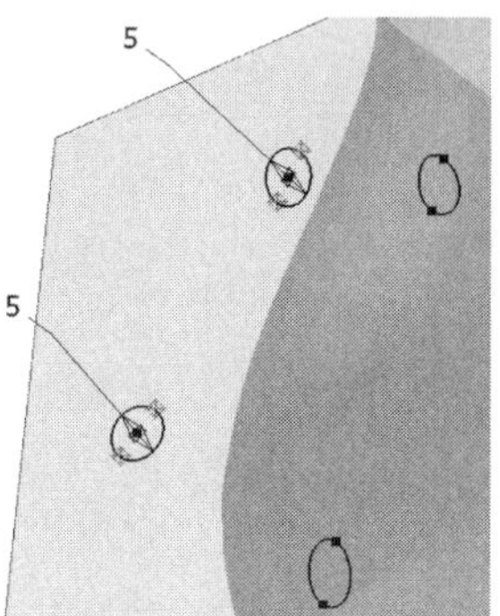

Figure 10.88 - Handle cross-section curves

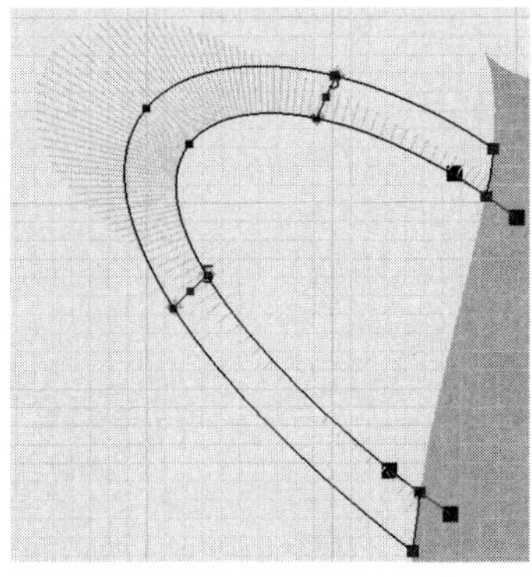

Figure 10.89 - Handle rail curves

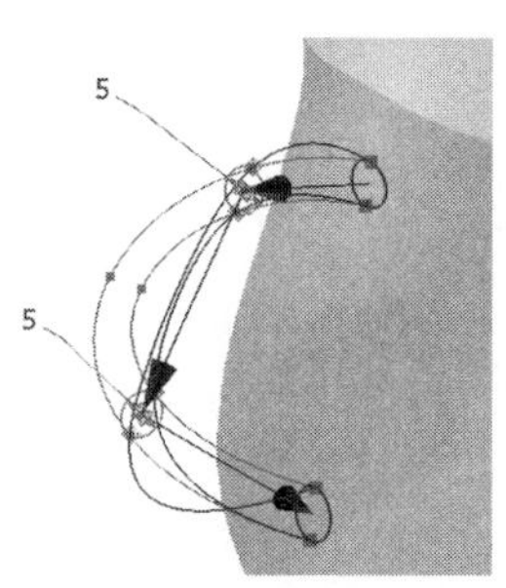

Figure 10.90 - Lofted handled surface (4 cross-sections, 2 rails)

Figure 10.91 - Completed creamer

14. Extrude surfaces (Figure 10.86).
15. 3D intersection curves (Figure 10.87).
16. Cross-sections (Figure 10.88).
17. Rail curves (Figure 10.89).
18. Loft (2 rails) (Figure 10.90).
19. Completed part (Figure 10.91).

Detailed Modeling Steps

1. Start a new metric part file.
2. Project the Center Point onto the sketch.
3. Use the Center point circle tool to sketch a circle with its center at the projected center point.
4. Dimension the circle (Diameter = 45).
5. Finish Sketch.
6. Isometric View.
7. Rename the first sketch; call it Base.
8. Create three work planes, all of them offset from the XY Plane by the following distances; 18, 60, 83.

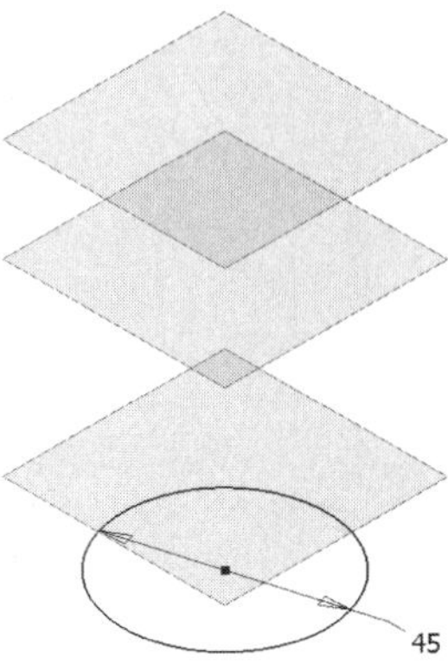

Figure 10.92 - Offset work planes

9. Zoom all. Your screen should now resemble Figure 10.92.
10. Rename the work planes as follows:
    - Waist_wp – 18 mm above XY Plane
    - Neck_wp – 60 mm above XY Plane
    - Top_wp – 83 mm above XY Plane
11. Save the file as **Creamer.** Place the file in a folder of your choice. It is recommended that the file be regularly saved every five to ten minutes.
12. Start a new sketch on Waist_wp.
13. Project the Center Point onto the sketch.
14. Sketch a circle with a diameter of 63 and its center coincident with the projected center point.
15. Finish Sketch.
16. Rename this sketch Waist.
17. Start a new sketch on Neck_wp.
18. Project the Center Point onto the sketch.
19. Sketch a circle with a diameter of 47 and its center coincident with the projected center point.
20. Finish Sketch.
21. Rename this sketch Neck. The screen and browser should now be similar to Figure 10.93.
22. Turn off the visibility of the three sketches created thus far; Base, Waist, and Neck.
23. Start a new sketch on Top_wp.
24. Look At the sketch.
25. Project the X and Y Axes and the Center Point onto the sketch plane.
26. Use the Ellipse tool (click the arrow next to the Center point circle tool) to create an ellipse similar to the one shown in Figure 10.94. Note that the center of the ellipse lies above the origin, on the vertical Y Axis.
27. Add the dimensions shown in Figure 10.95. To dimension an ellipse, first select the General Dimension tool, then select the ellipse. By moving the cursor it is possible to dimension both the major and minor axes.
28. Finish Sketch.
29. Isometric View.
30. Rename this sketch "Top".

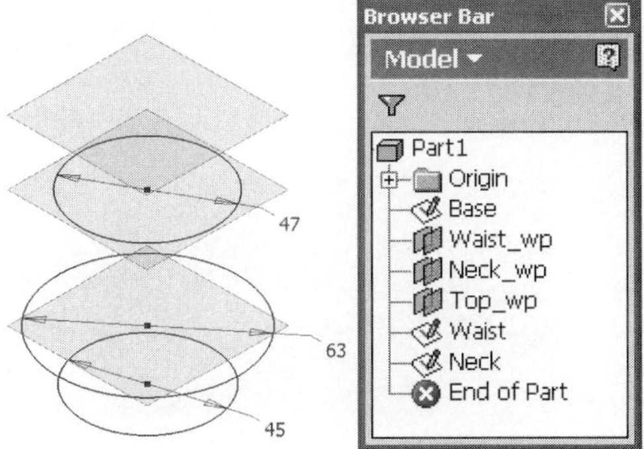

Figure 10.93 - Screen and browser

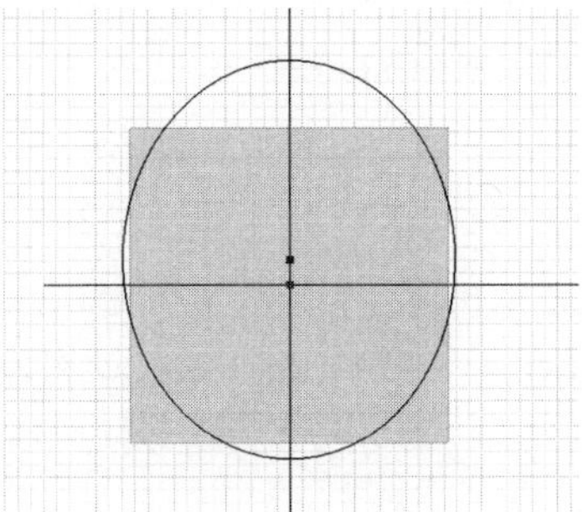

Figure 10.94 - Ellipse sketch

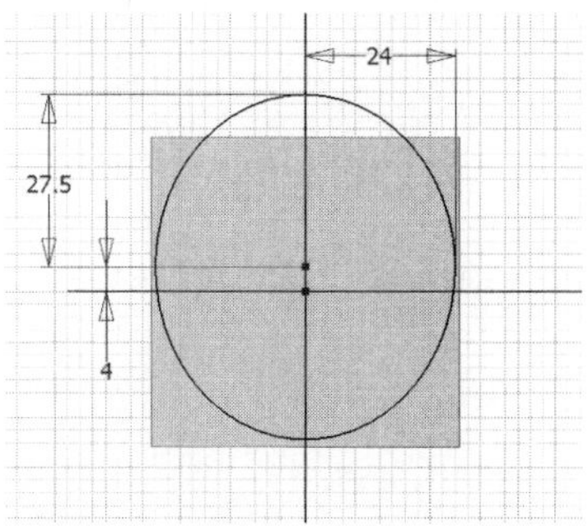

Figure 10.95 - Fully constrained sketch

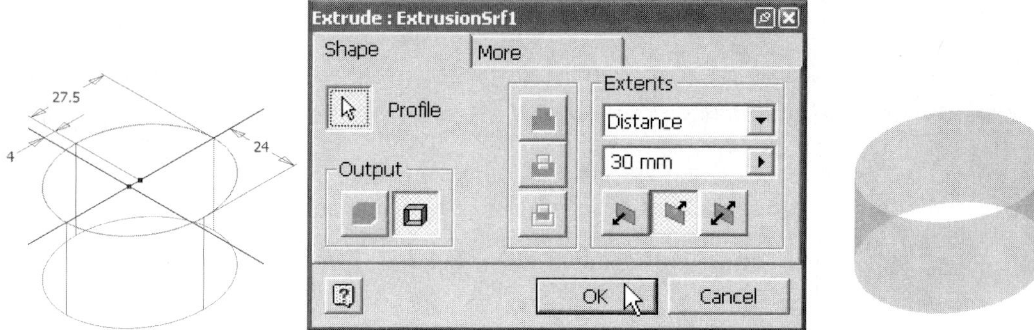

Figure 10.96 - Extrude surface operation

31. Turn off the visibility of the three work planes.
32. Extrude the Top sketch, using the parameters shown in Figure 10.96. Note in particular that the Output is set to Surface. The next step is to sketch a spline on the YZ Plane. Before doing this, however, it will be useful to place a work point at the right vertex of the previously drawn ellipse. This work point will then serve as a starting point for a spline on the next sketch.
33. Turn on the visibility of the Top (ellipse) sketch (you must first expand the extruded surface in the browser).
34. Select the Work Point tool from the panel bar. First select the major axis of the ellipse in the upper right part of your screen, and then select the ellipse. A work point will be placed at the intersection of the two, as shown in Figure 10.97.
35. Turn off the visibility of the Top sketch.
36. Start a new sketch on the YZ Plane.
37. Look At the sketch.
38. Project the Z Axis *and the work point* onto the sketch, as shown in Figure 10.98.
39. Use the Spline tool (click the arrow next to the Line tool) to sketch a spline similar to the one shown in Figure 10.99. Note that the spline endpoint on the right is coincident with the projected work point.

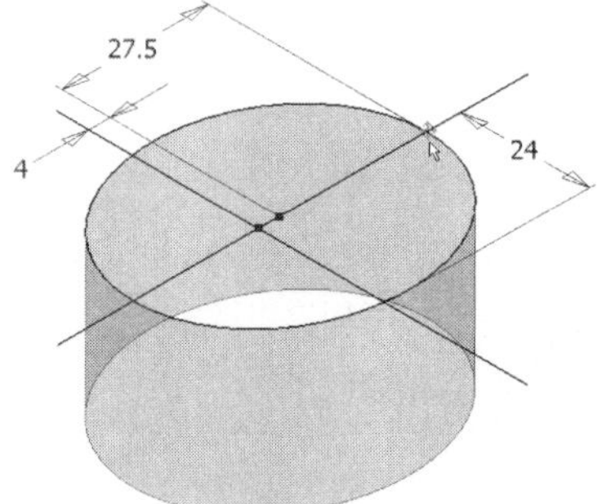

Figure 10.97 - Work point added

Figure 10.98 - Projected axis and work point

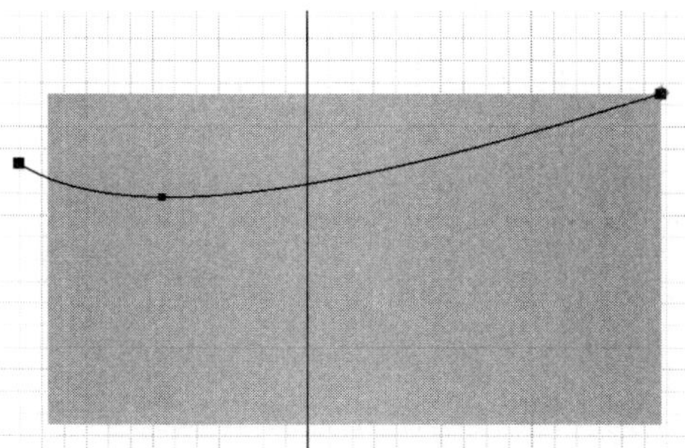

Figure 10.99 - Spline sketch

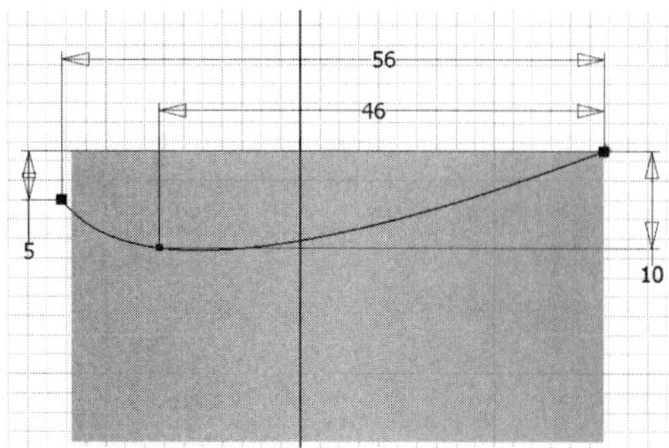

Figure 10.100 - Dimensioned spline sketch

40. Add the dimensions shown in Figure 10.100. We would now like to improve the shape of the spline curve so that the middle spline point is the lowest point on the curve. To accomplish this, we will use one of the bowtie options.
41. Zoom in on the middle and leftmost spline points.
42. After right-clicking on the middle spline point, select Bowtie, then Handle, as shown in Figure 10.101. A bowtie handle should now be visible at the middle spline point.
43. The bowtie handle represents the tangency of the spline curve at a spline point. To ensure that the spline point is the low point on the curve, the tangent line (i.e., the handle) should be horizontal at the spline point. Drag either one of the handle grips and move them up (or down, as the case may be) until the handle is approximately horizontal. Note that it is also possible to drag the handles so that the handle lengthens or contracts. When the handle is lengthened, for instance, the spline curve will maintain the same slope over a greater portion of the curve. When you are satisfied with the shape of the curve, click anywhere in the graphics area. Your screen should now resemble Figure 10.102.

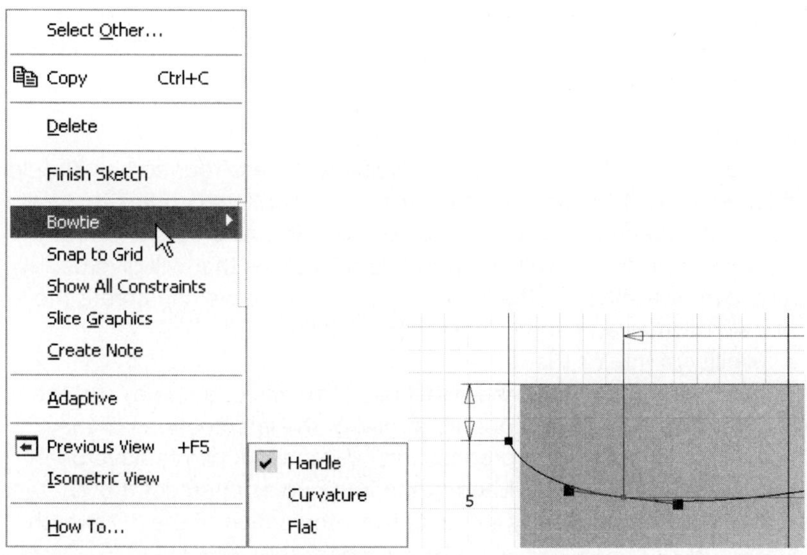

Figure 10.101 - Bowtie handle access and display

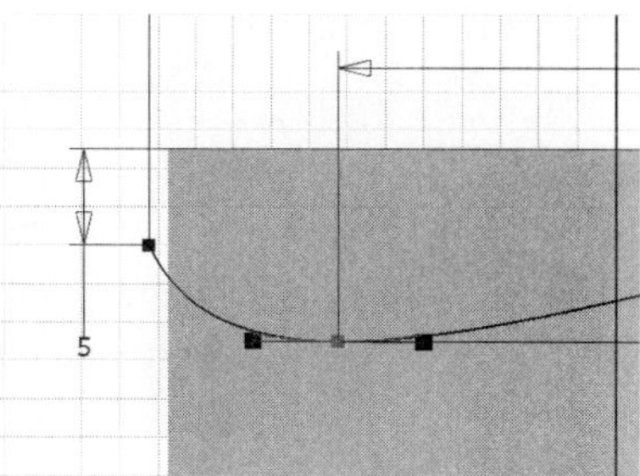

Figure 10.102 - Modified spline

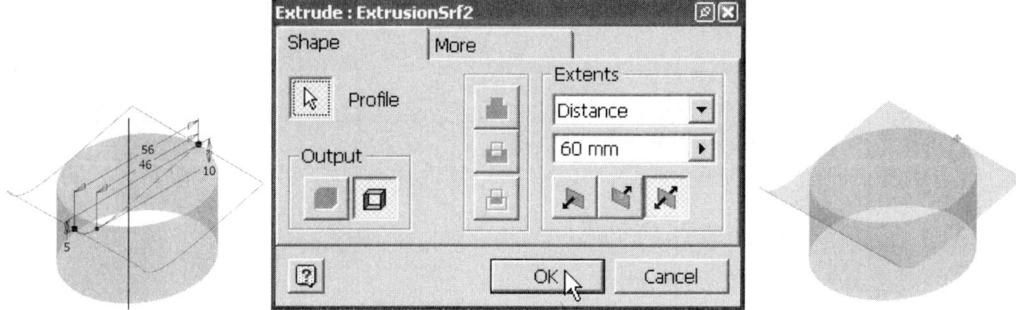

Figure 10.103 - Extruded surface operation

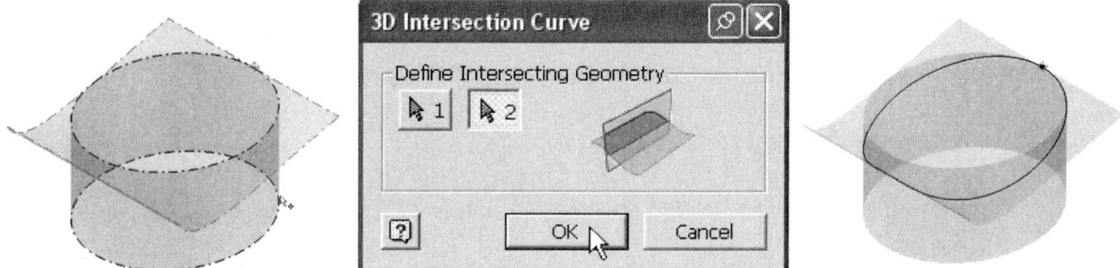

Figure 10.104 - 3D intersection operation

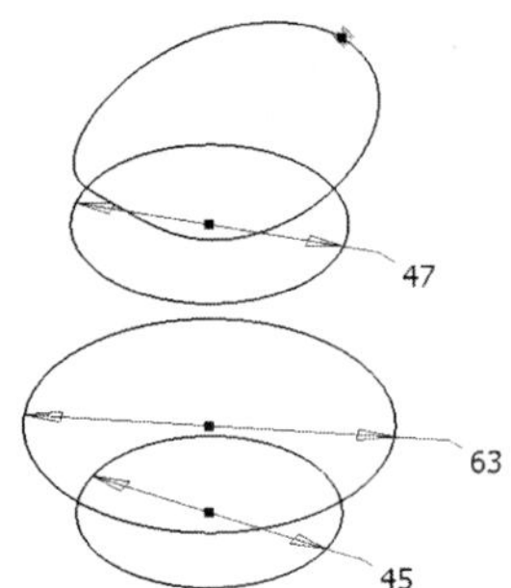

Figure 10.105 - Cross-section sketches

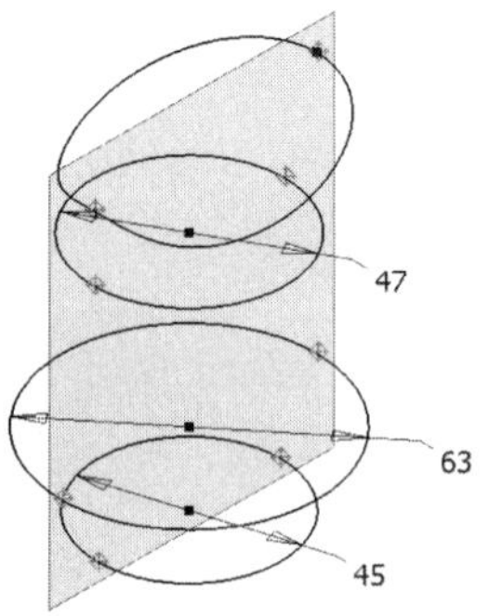

Figure 10.106 - Added work points

44. Finish Sketch.
45. Isometric View.
46. Extrude the open profile spline sketch from the mid-plane, as shown in Figure 10.103. We will now find the intersection of the two surfaces that have been created thus far. To do this, we must first create a 3D sketch.
47. Right-click while in the graphics area and select New 3D Sketch. The panel bar changes to display the 3D sketch menu.
48. From the panel bar, select 3D Intersection . After selecting the two surfaces, click OK, as shown in Figure 10.104.
49. Finish 3D Sketch.
50. Turn off the visibility of the two extruded surfaces.
51. Turn on the visibility of the Base, Waist, and Neck sketches. Your screen should now resemble Figure 10.105. We now have the cross-sectional curves that will form the body of the creamer. Although it is now possible to create a lofted surface using these curves, we will first add two rail curves that will give us greater control over the shape of the lofted surface. Before we can create the rails however, we need to add some more work points.
52. Turn on the visibility of the YZ Plane.
53. Select the Work Point tool from the panel bar. Now select a curve, and then select the YZ Plane. A work point appears at one of the intersections of the curve and the plane. Continue this process until seven work points have been added, as shown in Figure 10.106. Each of the four curves intersect the YZ Plane twice; one work point already exists at the intersection of the 3D sketch and the plane.
54. Start a new sketch on the YZ Plane.

55. Project the four work points on the right side of the screen onto the sketch plane.
56. Use the Spline tool to pass a spline through the four projected work points.
57. Look At the sketch. Your screen should be similar to Figure 10.107.
58. Use the bowtie handles to modify the forward rail curve, as shown in Figure 10.108.
59. Finish Sketch.
60. Start a new sketch, again using the YZ Plane to sketch on.
61. Project the remaining four work points on the back (left) side of the creamer body.
62. Pass a spline through the four work points.
63. Modify the back rail curve using the bowtie handles as shown in Figure 10.109.
64. Finish Sketch.
65. Turn off the visibility of the YZ Plane.
66. Isometric View.
67. Select the Loft tool from the panel bar. Select Surface Output. In the Sections area, click where it says Click to Add. Now select the four cross section curves in order from bottom to top (Base, Waist, Neck, 3D Intersection). Now click in the Rails area where it says Click to Add. Select the two rail curves, and then click OK. See Figure 10.110. We will now create the bottom of the creamer body.
68. Create a new sketch on the YZ Plane.
69. Look At the sketch.
70. Project the lower forward work point and the Z Axis onto the sketch plane, as shown in Figure 10.111.
71. Zoom in on the bottom of the creamer.
72. Use the Spline tool to sketch a spline similar to that shown in Figure 10.112. Note that the spline has three points, and that the spline endpoints are coincident with the projected work point and the Z Axis.
73. Use the bowtie handles to modify the shape of the spline, as shown in Figure 10.113. Also note that it is also possible to move the middle spline point by dragging it.
74. Finish Sketch.
75. Isometric View.

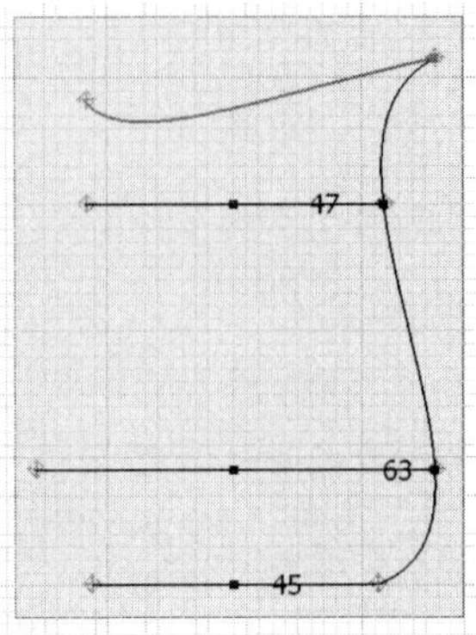

Figure 10.107 - Forward rail curve

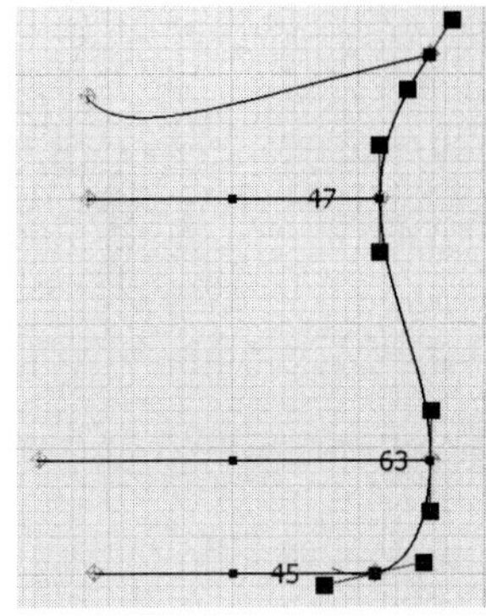

Figure 10.108 - Modified forward rail curve

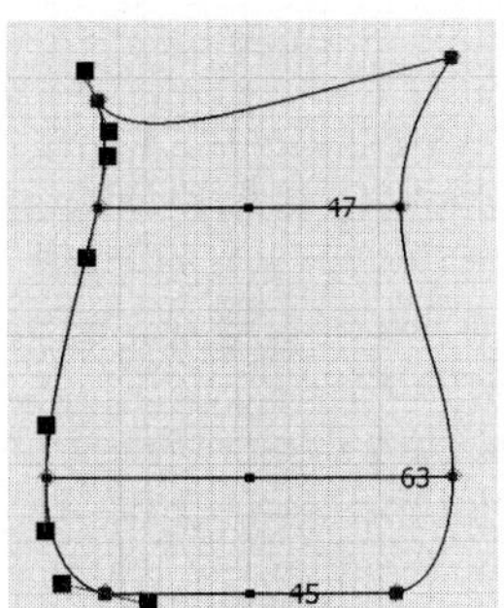

Figure 10.109 - Modified back rail curve

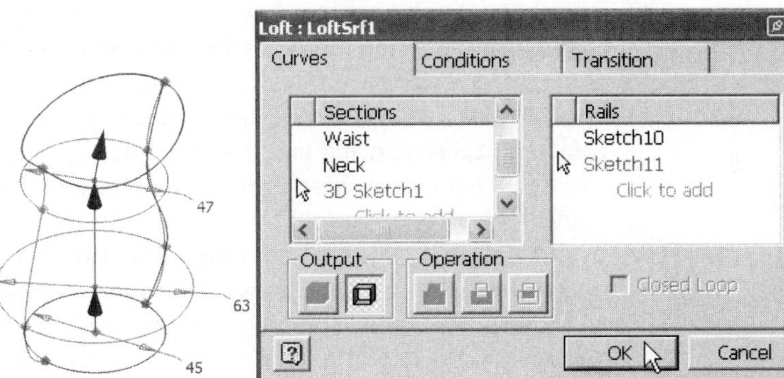

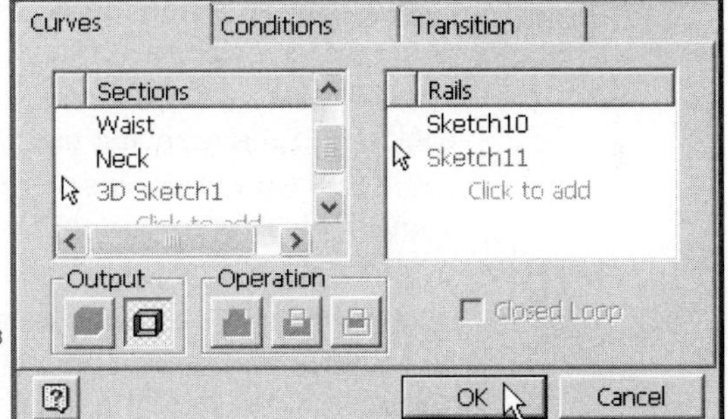

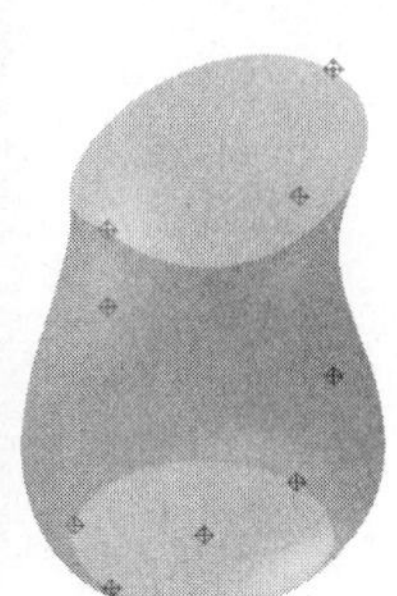

Figure 10.110 - Lofted surface operation (2 rails)

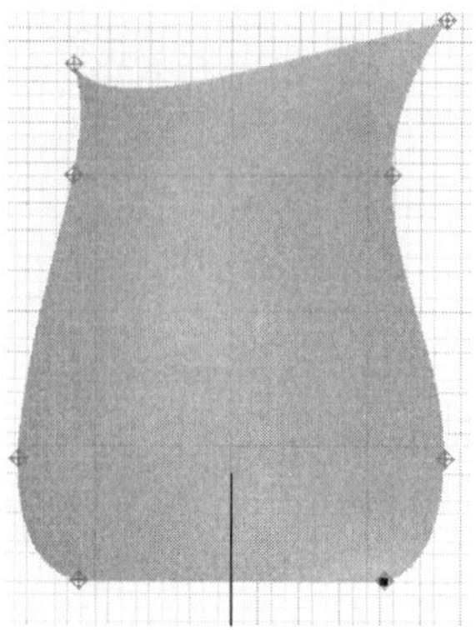

Figure 10.111 - Projected axis and work point

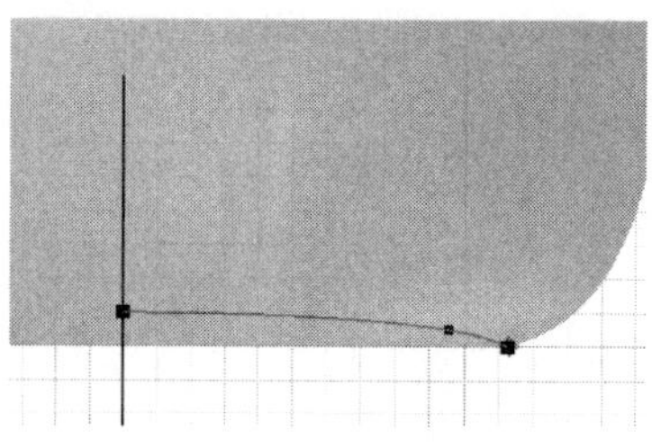

Figure 10.112 - Spline sketch

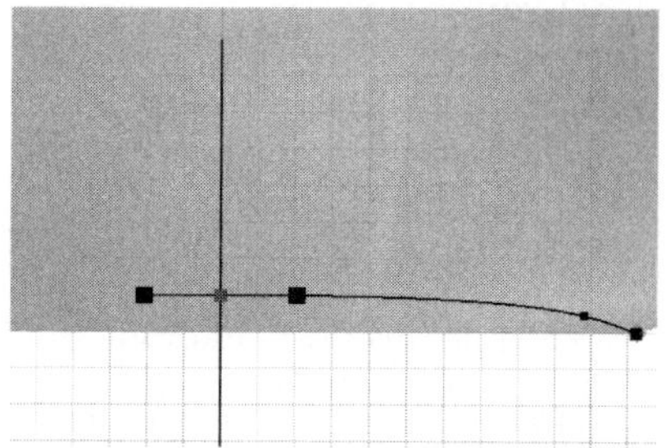

Figure 10.113 - Modified spline sketch

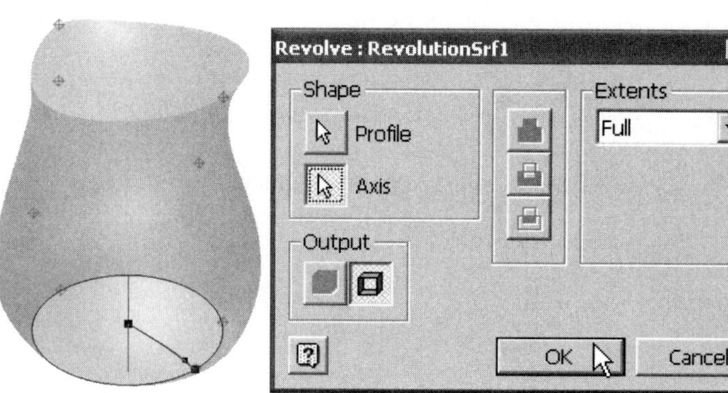

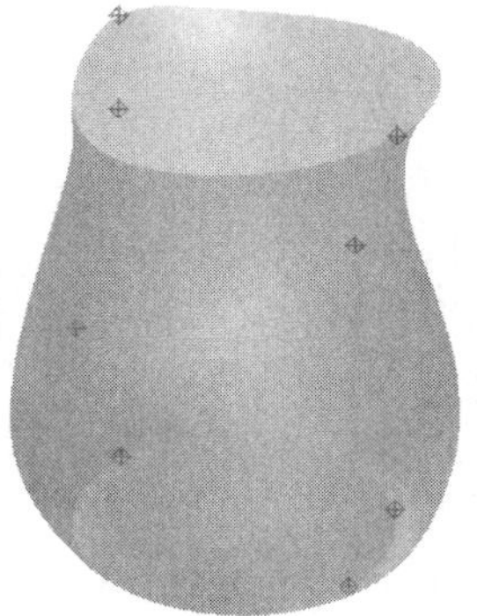

Figure 10.114 - Revolve operation

76. Use the Common View tool to look at an isometric view of the creamer from below.
77. Select the Revolve tool. As the Profile, select the spline; the Z Axis is the Axis. Click OK when ready, as shown in Figure 10.114.
78. Turn off the visibility of the work points.
79. Turn off the default translucency of the two surfaces (loft, revolve) by right-clicking on the revolved surface in the browser, and de-selecting Translucent.

    If you now look closely at the intersection of the two surfaces, you will see some "holes" or imperfections between the surfaces. We will now address this by stitching the two surfaces together to create a single surface.
80. Select the Stitch Surface tool from the panel bar. Select the two surfaces, and then select OK, as shown in Figure 10.115. At this point the "pin pricks" or imperfections are still visible.
81. Turn off the visibility of the two separate surfaces (lofted, revolved).
82. Turn off the translucency of the stitched surface. Your screen should now resemble Figure 10.116. Note that the imperfections are gone.

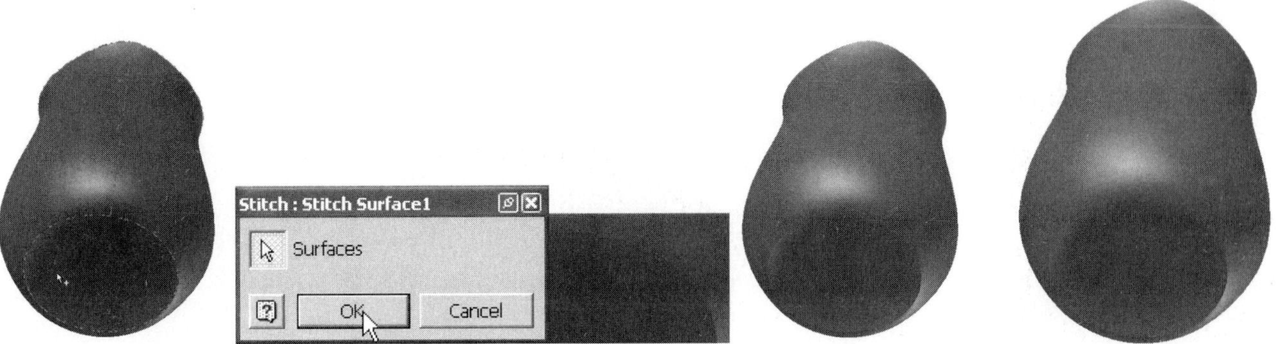

Figure 10.115 - Stitch surface operation

Figure 10.116 - Stitched surface

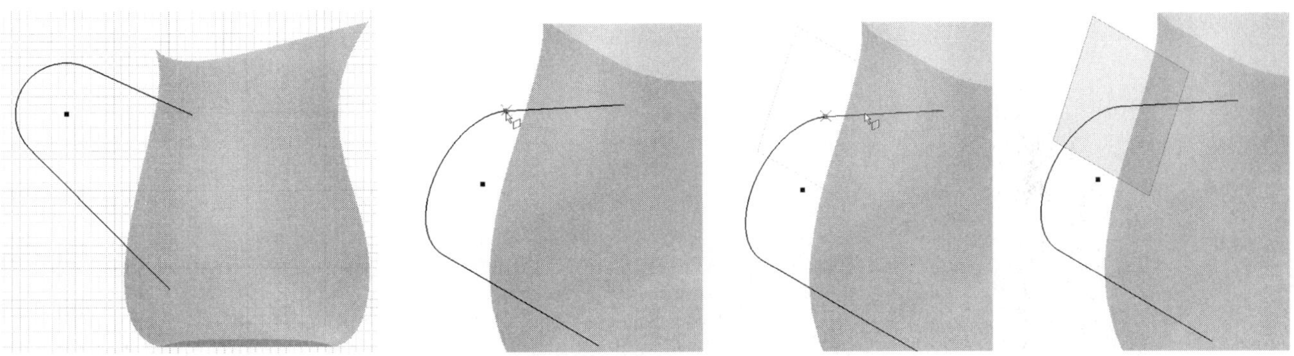

Figure 10.117 - Sketch geometry

Figure 10.118 - Work plane (point on plane, line normal to plane)

83. Isometric View. We are now ready to start modeling the creamer handle.
84. Change the stitched surface display back to translucent.
85. Start a new sketch on the YZ Plane.
86. Look At the sketch.
87. Use the Line tool to sketch a polyline (line, arc, line) similar to that shown in Figure 10.117.
88. Finish Sketch.
89. Isometric View.
90. Use the Work Plane tool to create a new work plane. First select the line endpoint as shown in Figure 10.118 (on the left), and then select the line segment (middle). The result is the work plane shown on the right.
91. Repeat the process described in the previous step, this time using the lower line segment. The result is shown in Figure 10.119.
92. Start a new sketch on the upper work plane.
93. Project the line (normal to the plane) onto the sketch.
94. Sketch a circle with its center coincident with the projected point.
95. Dimension the sketch as shown in Figure 10.120.
96. Finish Sketch.

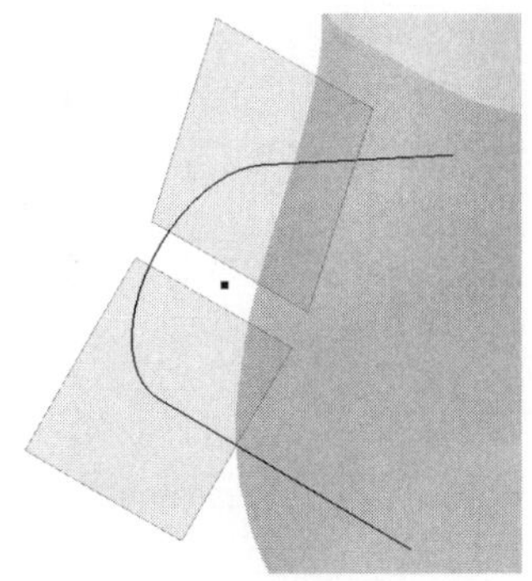

Figure 10.119 - Another work plane (point on plane, line normal to plane)

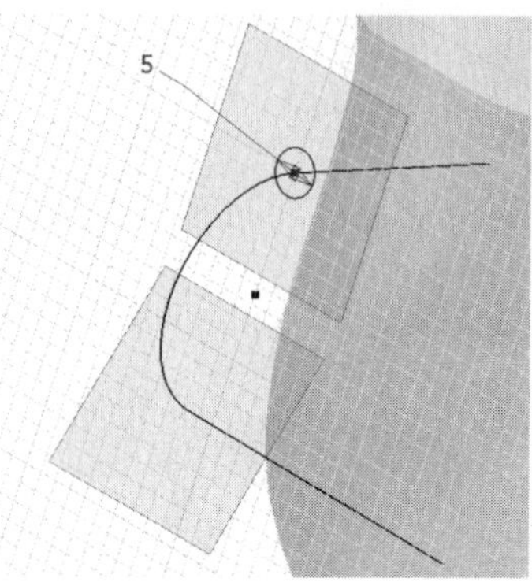

Figure 10.120 - Dimensioned sketch

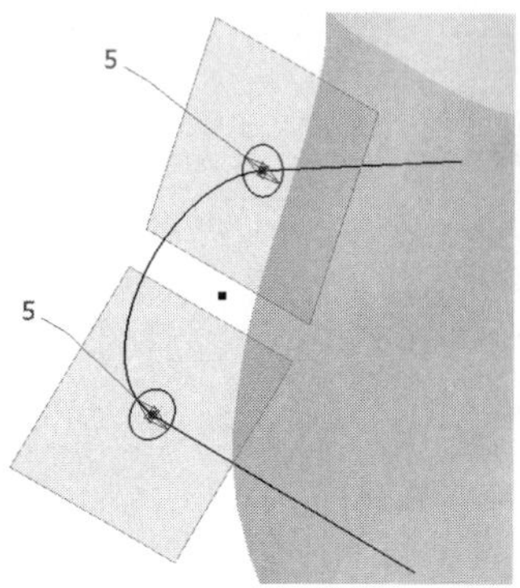

Figure 10.121 - Another dimensioned sketch

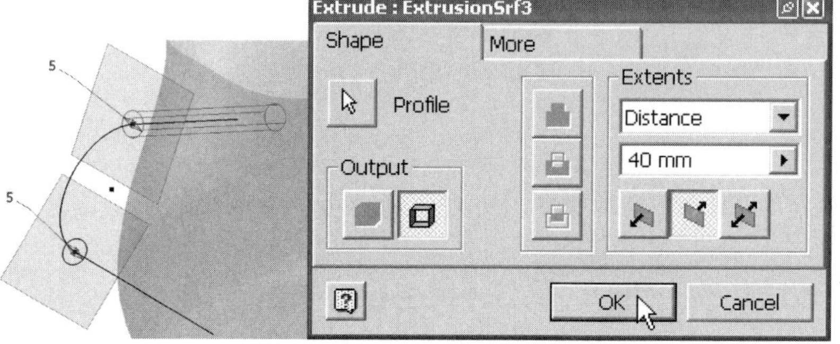

Figure 10.122 - Extrude (surface) operation

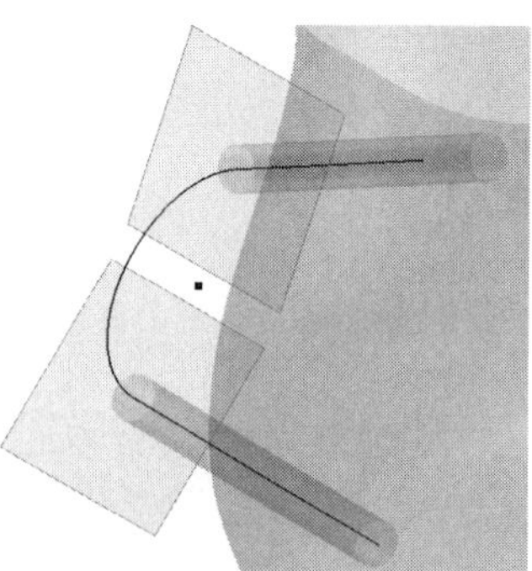

Figure 10.123 - Another extrude surface operation

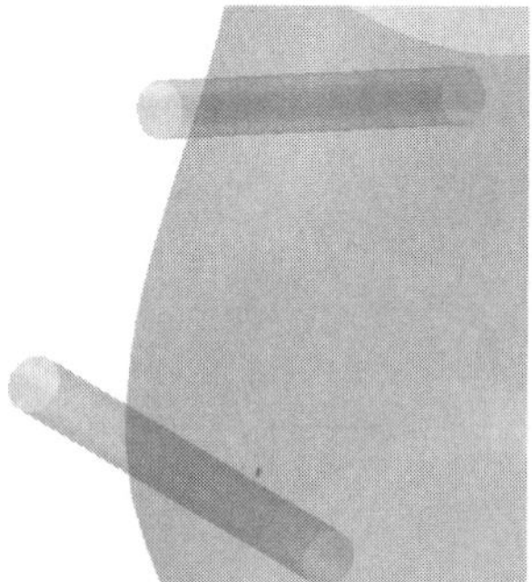

Figure 10.124 - Hide work planes, sketch

97. Repeat steps 92 to 96 on the lower work plane. The screen should now resemble Figure 10.121.
98. Extrude the upper circle as a surface so that it intersects the creamer body, as shown in Figure 10.122. It may be necessary to rotate the view to ensure that the extrusion intersects with the creamer body.
99. Repeat the previous step for the lower circle. The results are shown in Figure 10.123.
100. Turn off the visibility of the work planes and the polyline sketch (line, arc, line), as shown in Figure 10.124.

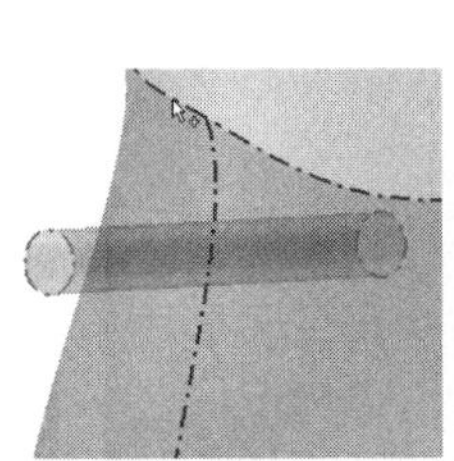
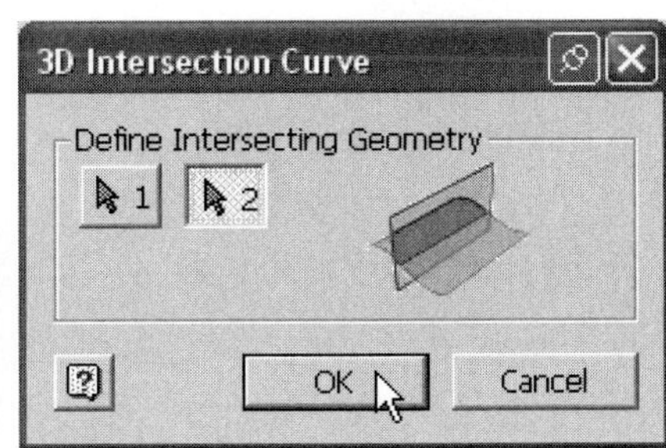

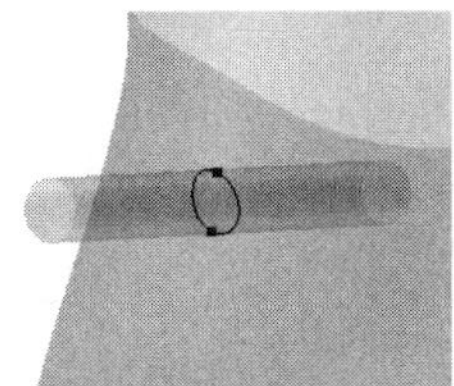

Figure 10.125 - 3D intersection curve operation

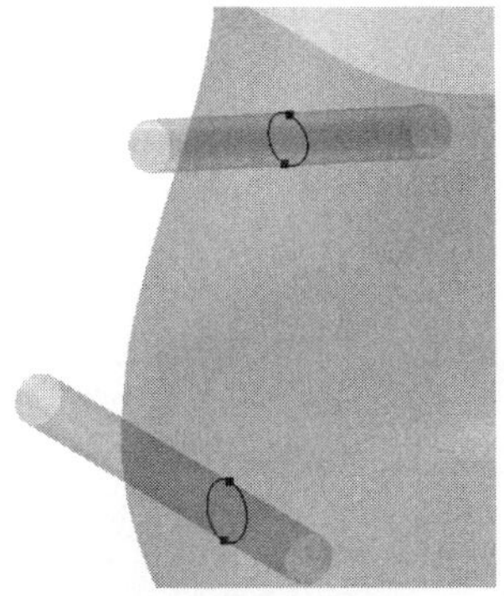

Figure 10.126 - Another 3D intersection curve operation

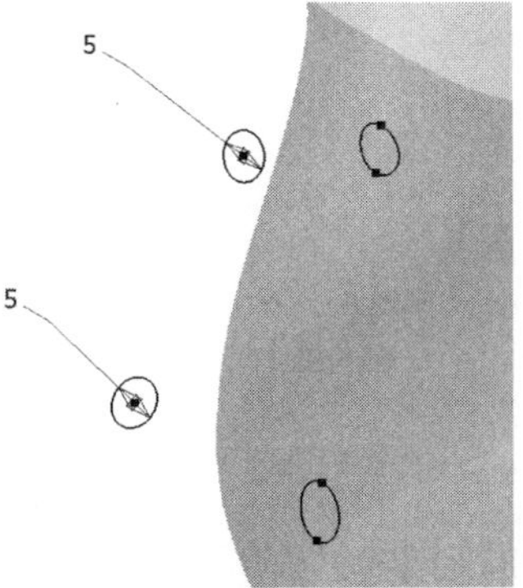

Figure 10.127 - Modified feature and sketch visibility

101. Start a new 3D Sketch.
102. From the 3D Sketch menu on the panel bar, use the 3D Intersection tool to find the intersection of the upper cylindrical surface and the creamer body. See Figure 10.125.
103. Finish 3D sketch.
104. Repeat steps 101–103, this time finding the intersection curve of the lower cylinder and the body. See Figure 10.126 for the result.
105. Turn off the visibility of the two cylindrical surfaces.
106. Turn on the visibility of the two circle sketches, as shown in Figure 10.127. At this point we could loft a surface through the circles and 3D intersections, but the results would be less than satisfying. As with the loft of the creamer body, we will create two rail curves to give us greater control of the shape of the handle.
107. Turn on the visibility of the YZ Plane.
108. Use the Work Point tool to create work points at the two intersection points of each circle with the YZ Plane, as shown in Figure 10.128. Note that it will be necessary to rotate the view in order to select the second intersection point on each circle.
109. Start a new sketch on the YZ Plane.
110. Look At the sketch.
111. Project the outer quadrant points of the 3D intersection curves and the outer work points on the circles onto the sketch plane (Figure 10.129).

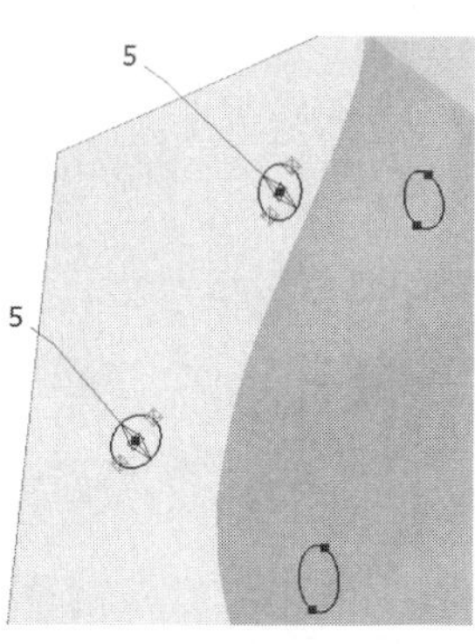

Figure 10.128 - Work points added at circle quadrants

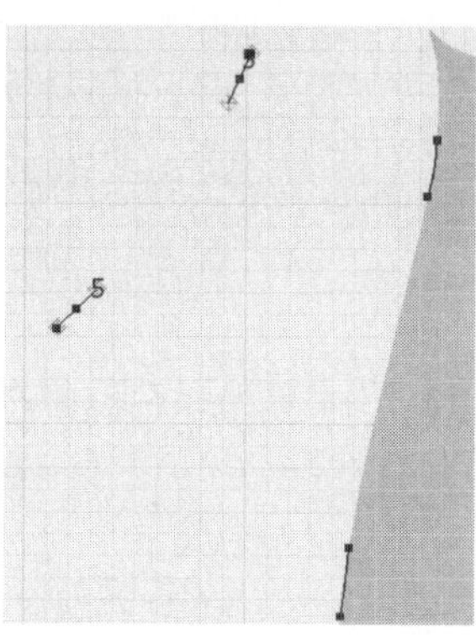

Figure 10.129 - Projected points

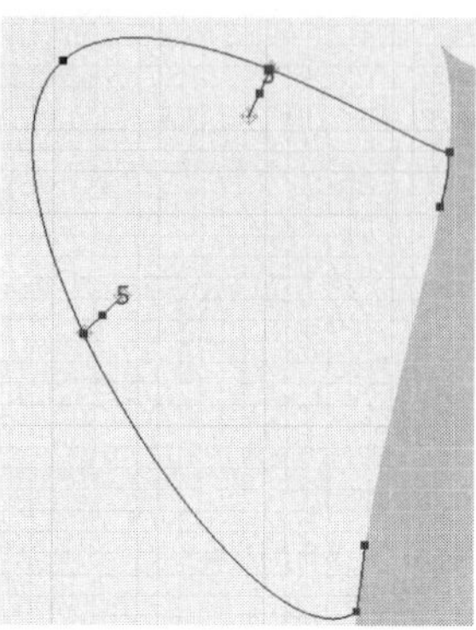

Figure 10.130 - Preliminary spline (outer rail) sketch

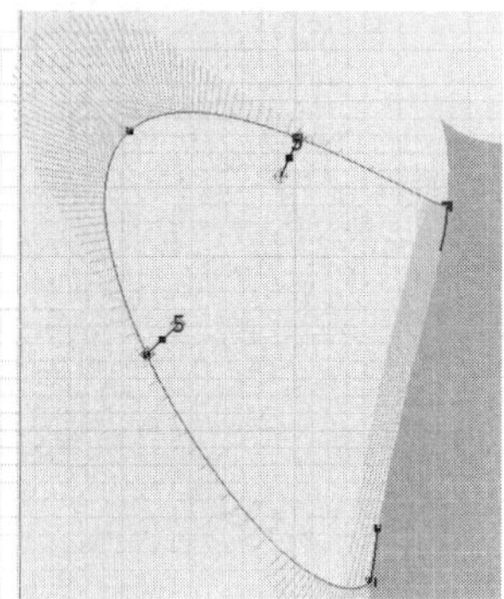

Figure 10.131 - Spline curvature comb displayed

112. Use the Spline tool to pass a spline through the projected points. Note that there is an additional spline point between the work point projections, as seen in Figure 10.130. At this point the outer rail curve is not very pleasing to the eye. We will use several Inventor spline tools to improve the shape of this curve.

113. Right-click on the spline, then select Display Curvature. A *curvature comb* representing the spline's curvature is now displayed, as shown in Figure 10.131.

114. Next use the bowtie handles to adjust the shape of the curve (Right-click on the spline endpoints, then select Bowtie, then Handle). Use the handle grips to modify the spline tangency at these endpoints.

115. Modify the location of the unconstrained middle spline point by dragging. After these modifications, the curve appears as in Figure 10.132.

116. Turn off the spline curvature comb (right-click on spline, de-select Display Curvature).

117. Finish sketch.

118. Start a new sketch on the YZ Plane.

119. Project the inner quadrant points of the 3D intersection curves and the inner work points on the circles onto the sketch plane.

120. Use the Spline tool to pass a spline through the projected points. Note that there is an additional spline point between the work point projections, as seen in Figure 10.133.

121. Use the techniques employed in steps 113–115 to improve the shape of the inner rail curve, and also to match it by eye with the outer rail curve. The resulting curve is shown in Figure 10.134.

122. Turn off the curvature comb (right-click on spline, de-select Display Curvature).

123. Finish Sketch.

124. Isometric View.

125. Turn off the visibility of the YZ Plane.

126. Select the Loft tool from the panel bar. Select Surface Output. In the Sections area, click where it says Click to Add. Now select the four cross section curves in order starting with the upper 3D intersection curve. Now click in the Rails area where it says Click to Add. Select the two rail curves, and then click OK. See Figure 10.135.

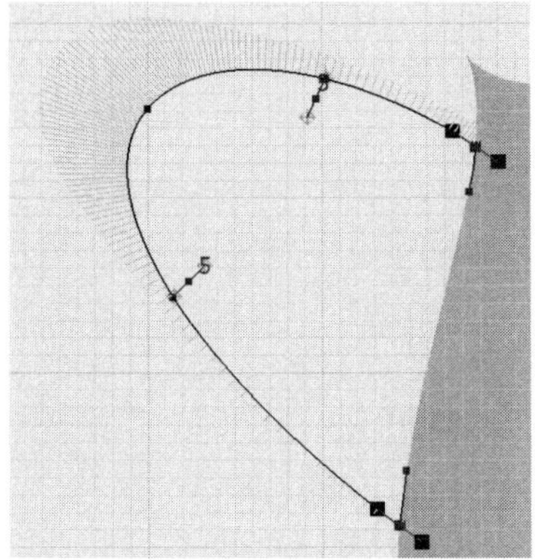

Figure 10.132 - Modified spline (outer rail) curve

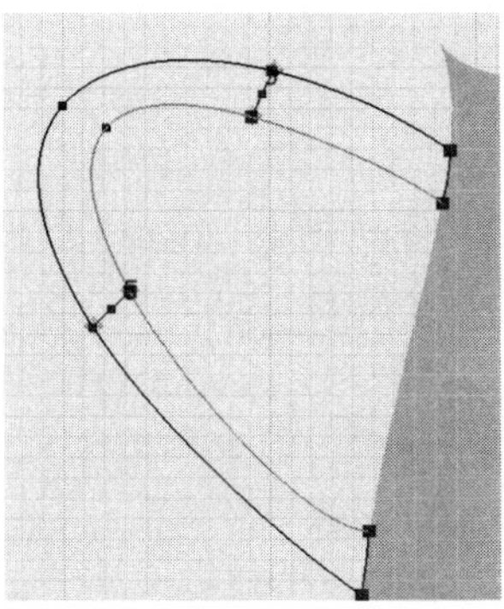

Figure 10.133 - Preliminary spline (inner rail) curve

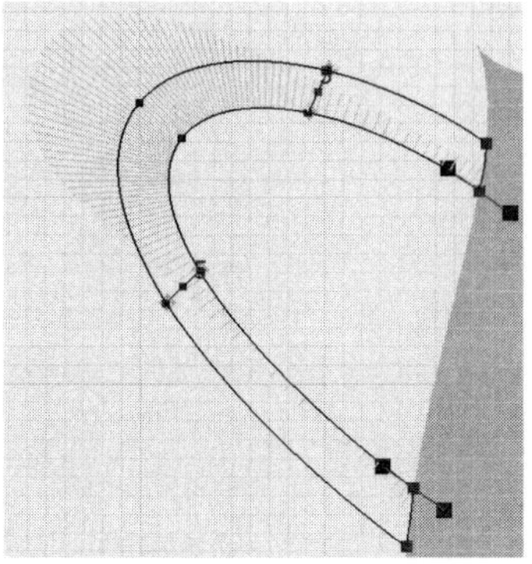

Figure 10.134 - Modified inner rail curve

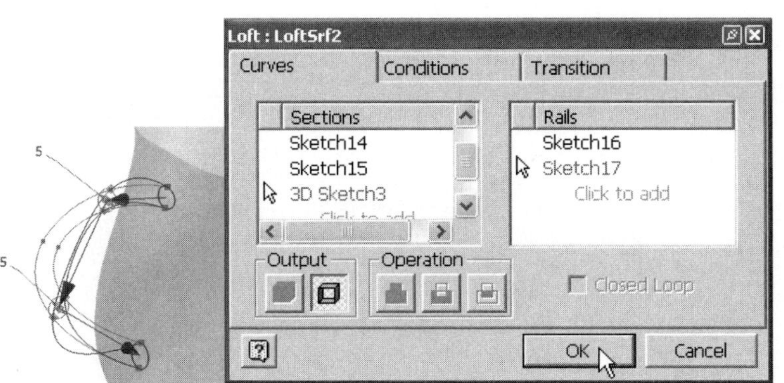

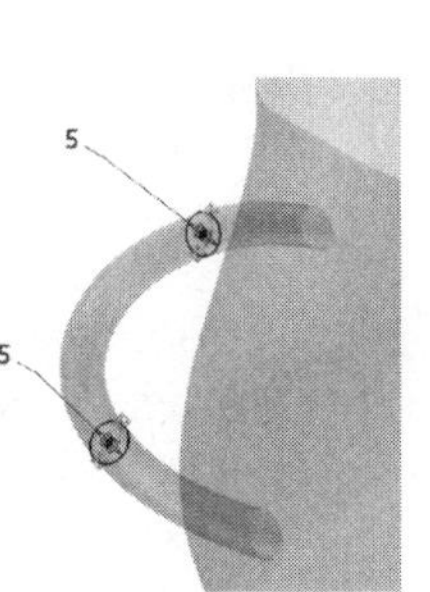

Figure 10.135 - Lofted surface operation (2 rails)

Figure 10.136 - Completed creamer

127. Turn off the visibility of the two circle sketches, as well as the work points.
128. Turn off the Translucency of both the stitched (body) surface and the lofted with two rails (handle) surface.
129. Isometric View. The completed creamer should be similar to Figure 10.136.
130. Save the file.

QUESTIONS

1. Which of the following is not a tool used to control the shape of a spline:
   a. Close Spline
   b. Insert Point
   c. Stitch Method
   d. Spline Tension
   e. Fit Method
2. T F The Bowtie Method for spline control allows the user to visualize the smoothness and/or discontinuities in a curve.

3. Which of the following sketched features cannot be used to create a surface feature?
   a. Revolve
   b. Coil
   c. Sweep
   d. Loft
4. T F The Thicken/Offset tool can be used to both add and remove thickness from an existing solid.
5. T F The Delete Face tool is used to delete the face of an existing surface; in the process, the surface is converted to a solid.
6. T F The Replace Face tool is used to replace the face of an existing part with another face.
7. T F The Stitch Surface tool is used to combine faces on two existing parts.
8. T F Rail Curves are used to define the boundaries of a lofted feature.
9. T F The Decal tool allows for the creation of a raised or recessed feature on an existing face.
10. What tool can be used to import a bitmap file (picture) onto a sketch plane?
11. T F Using the 3D intersection tool it is possible to trim surfaces in Inventor.

PROBLEMS

1. Using the file Hybrid.ipt (contents shown below on left), create the solid part shown on the right. HINT: Use the following commands - Extrude, Replace Face, Delete Face, Thicken/Offset.

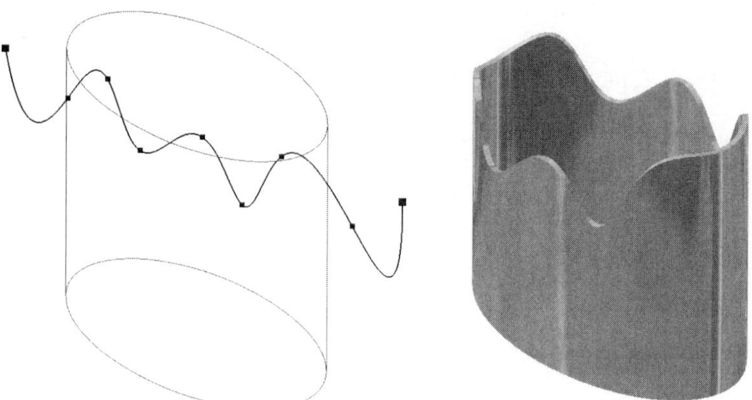

2. Use the file Sweep1.ipt to sweep the open profile arc along the spline path. After the sweep is created, use the Thicken tool to convert the sweep surface to a solid. The file contents are shown in the figure on the left.

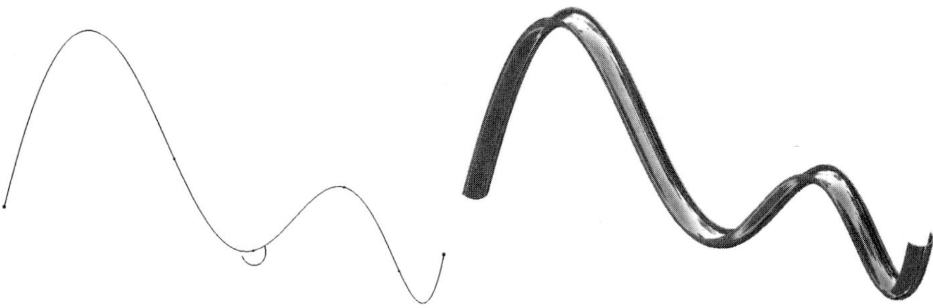

3. Use (eight) grounded work points to help create the path for the toast rack's base sweep feature. The height of the vertical legs is 15 mm. The sweep profile is a circle with a diameter of 3 mm.

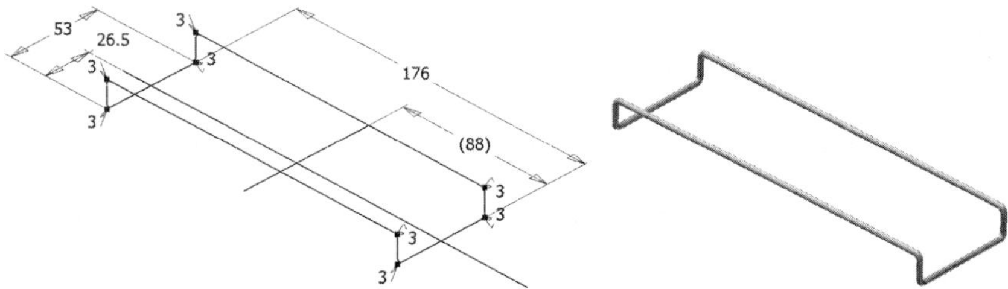

4. Find the intersection curve between the two surfaces contained in the file 3dintersection.ipt. The surfaces are shown in the figure below.

5. The file Loft1.ipt contains three sketches, as shown in the figure. Use this file to create a:
   - Lofted solid.
   - Lofted surface. Now thicken the surface, converting it to a solid.
   - Lofted solid with a rail curve. HINT: start by adding work points at the intersections of the closed curves and either the YZ or the XZ Plane.

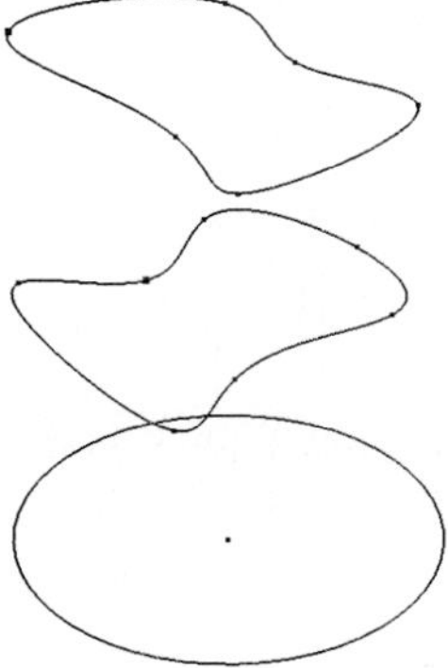

INDEX

D

E

F

T

U

V

W

X

Z